Systems Practices

As Common Sense

by
Walter Sobkiw

Published by

C~B
CassBeth, Cherry Hill, NJ

Published by Cassbeth, Cherry Hill, New Jersey, USA

For information contact www.cassbeth.com

First Edition

Library of Congress Control Number: 2011900101

ISBN: 978-0-9832530-8-2 (hardcover)

This book is dedicated to my wife Claudia.

It is for my children, their families, and their world.

Preface

On July 16, 1945 as part of the Trinity nuclear test the first nuclear bomb was detonated at the White Sands Proving Grounds in New Mexico. This test was born from the horror of World War II and it changed the world forever. Dr. J. Robert Oppenheimer, scientific director of the Manhattan Project recalled the Bhagavad Gita: "If the radiance of a thousand suns were to burst at once into the sky, that would be like the splendor of the mighty one." and "**Now I am become Death, the destroyer of worlds.**" The Trinity detonation was estimated to be the equivalent of 20 kilotons of TNT.

This is a picture[1] of the Trinity explosion about 0.016 seconds after detonation. The fireball is approximately 200 meters (600 ft) wide. Trees can be seen as black objects.

Imagine the feelings rushing through those witnessing this test. On the one hand was the possibility of ending World War II and on the other was this terrifying explosion. For the first time the release of power came from the splitting of the atom. The analysis showed that splitting a single atom released millions of times more energy than the breaking of any chemical bond. Now here was proof of what the equations showed.

By November 1, 1952 the United States in Operation Ivy had tested the first fusion bomb. It was on Elugelab Island in the Enewetak Atoll of the Marshall Islands. The explosion yielded the equivalent of 10.4 megatons of TNT. This time instead of splitting a hand full of atoms they were fused. The energy released was beyond anyone's imagination; it was completely outside the experience of humanity. The genie was out of the bottle; technology was heading down its normal path of maturity. It could not be stopped. Imagine at this point what people may have been thinking. Humanity could now actually destroy the world.

After the atomic bomb events, the end of World War II, and the birth of the hydrogen bomb, two massive developments surfaced. The first was Interstate Freeways and the second was the Semi Automatic Ground Environment (SAGE)[2].

The Manhattan project created the atomic bomb, employed approximately 130,000 people and cost approximately $1.89 billion in 1944 dollars or $20 billion in 1996 dollars. The SAGE project was created to defend against the nuclear threat

---

[1] Photo: Trinity Site 16 July 1945 United States Federal Government.

[2] From Whirlwind to MITRE: The R&D Story of The SAGE Air Defense Computer (History of Computing), Kent C. Redmond, Kent C. Redmond, Thomas M. Smith, The MIT Press, 2000, ISBN 0262182017.

cost approximately $55 billion in 2000 dollars.

SAGE was the first computer[3] based system for tracking and intercepting enemy bomber aircraft. The aircraft that could deliver atomic and then hydrogen bombs. By 1963 the system consisted of 22 Sector Direction Centers and three similar Combat Centers linked to North American Aerospace Defense Command (NORAD) and another site at Canadian Forces Base (CFB) North Bay in Canada.

Each computer, the AN/FSQ-7, used 55,000 vacuum tubes, took about half an acre (2,000 m²) of floor space, weighed 275 tons, and consumed three megawatts of power. Performance was about 75,000 instructions per second (75 KIPS). The software consisted of approximately 500,000 lines of assembly code.

Photo: AN/FSQ-7 computer United States Air Force

The SAGE engineering effort was huge and it gave birth to our modern computers. Prior to this time computers were viewed only as calculators. SAGE introduced the concept of situational awareness and presented alphanumeric data and graphics to operators sitting at display consoles. The consoles had keyboards

---

[3] Broadband analog RADAR data was digitized and transmitted via modem to the SAGE computers, which would process the data and display it as symbols on a display console. The situation display console (not shown) used a light gun to select items on the display and access information as needed

for entering data and pointing devices to select objects on a screen. The command centers even included large screen displays. This technology eventually transferred to air traffic control. The last SAGE site was decommissioned in 1983.

As the technology transitioned from vacuum tubes, transistors, integrated circuits, to microprocessors, the air defense and air traffic control systems across the planet continued to evolve. Each time computers, displays, communications, and software technology pushed to the next levels.

SAGE is one of the most advanced and successful large computer systems[4] ever developed. From it flowed our modern air traffic control and air defense systems. It also resulted in the largest machine ever built by humans, the Internet. Something as simple and yet profoundly powerful as packet switching was born from the need to deal with low reliability components. Think of it, computers, displays, keyboards, pointing devices, software, inter-computer connections, and packet switching; everything that you now use in your daily activities.

Besides SAGE and Interstate freeways, something else happened. This thing called Systems Engineering emerged. It was matured and honed to such a level that many in the field started to view it as common sense. Unfortunately what might be considered common sense or just training from one perspective might be viewed as significant knowledge and education from another perspective.

The reality is after World War II the people were terrified of the power of a nuclear world and they created systems, processes, and methods, which led to spectacular system engineering efforts and machines. However, this body of work was done and captured in the complex collection of for profit companies, non-profit companies, think tanks, and national labs. Some of this work never made it back to the universities. Now we are faced with the dilemma that many of the institutions that created this body of knowledge have disappeared and many of the people have died. This book is an attempt to capture some of this knowledge and pass it to the next generation.

Why is this knowledge so important?

---

[4] Photo: SAGE control room United States Air Force. The large screen display shows the east coat of North America from Philadelphia to Nova Scotia. Two targets are being tracked off the coast near Cape Cod.

It is important because something special happened after World War II. For the first time in thousands of years humanity was uplifting itself at such a pace that many dared to dream of actually populating the heavens. Today we have satellites that provide us worldwide communications. They look at our land, help us find needed minerals, and study our geological past. They look out into the universe and let us know there are many things we may never imagine or fully understand. They even help us find our way as we travel to see our parents, children, and friends.

However people have started to take it all for granted. The technologies and systems born from the terror of nuclear war are starting to show their limitations. These two conditions are not good for our future. So in my humble opinion it is important for us to take a look at what happened after World War II especially our Systems Engineering work.

I claim that the only way we can survive and prosper in the future is through the application of Systems Engineering. We stumbled into it while desperately trying to survive after the splitting of the atom. There are no governments, institutions, companies, markets, management techniques, political dogma, ideologies or other engineering scientific or technical approaches that will get us through the next 100 years as we learn to replace our old systems and technologies.

Air defense was developed out of a serious need but have we developed new systems that are needed today? Air Defense was inserted with computer technology and SAGE transitioned from tubes to chips, but have we taken other critical systems and inserted technology to make them better?

Systems practices were used extensively after WWII and yielded our modern world of water, sewage, electricity, telephones, radios, televisions, airplanes, cars, dishwashers, highways. It was common sense born from the need to survive as a world was on the brink of destruction. Somehow, so far, we survived but it took all our systems practices to get us to this point in history.

Figures

Tables

Examples

1 Systems Practices

What kind of a title is Systems Practices? There are practitioners who refer to "systems" as systems engineering, systems thinking, and systems science. There are applications like air traffic control systems, stereo systems, healthcare systems, energy systems, etc. If you get a group of self-professed systems people together they will eventually start to try to discern the differences between each form of systems. However there are some fundamental elements of Systems Practices that are shared between all these systems people.

Imagine a place where you can create things and make decisions where there are no hidden agendas and all stakeholders are treated equally. How would potential approaches surface, how would they be narrowed and selected, how would decisions be made? What tools and techniques would be used if they were not the greatest moneyed interests, the most politically powerful, or the most dangerous?

How about using logical methods based on reasonable techniques understood by reasonable people in a process that is fully transparent and visible to everyone. Everyone has a view of all the alternatives. Everyone has a view of all the decision paths. Everyone has an opportunity to impact the alternatives and decision paths.

Do not fall for the rhetoric that this is mob rule or design by committee. These are reasonable people using thousands of years of tools, techniques, processes, and methods to make informed decisions. There are no hidden agendas with vested interests or people who just give up and go silent or worse compromise. Everyone is comfortable with the decision because it is intuitively obvious to all. Everyone obviously has responsibility in such an endeavor. No one can ignore that responsibility. That in a nutshell is Systems Practice.

The reality is many of us engage in systems practices in our everyday lives when we reject fate and try to control our destinies. We use whatever tools we have at our disposal to make these decisions everyday of our lives. That is why this book is for everyone. Everyone engaged in our modern world should have some say in its evolution and that say should be from the system perspective of his or her view. So we can start at this point by saying that you as a systems practitioner do the following:

1. Always broaden your horizons and perspective. Just when you think you have it, move to a broader view.
2. Reject the status quo. If you find yourself in that mix, step out of it and look for truth. It takes time for a valid solution to solidify.
3. Reject all vested interests including hidden interests.

4. Try to see the forest from the trees but always zoom in onto the leaf of a particular tree and find the cute bug sitting on the edge.
5. Avoid the trap of being set only at one abstraction level, the high level. You must be able to scale all abstraction levels simultaneously. This is hard.

If you had the opportunity to write a book on systems practices what would you focus on? Would you identify the most important and trendy processes, methods, techniques and describe them or would you try to teach how to develop processes, methods, and techniques? My goal is to broaden the discussion and hopefully show how to develop processes, methods, and techniques that are needed for the next generation. That means they should have a view of what happened in the past, appreciate the struggle, and be somewhat proficient in many different processes, methods, and techniques, which I arbitrarily grouped into practices.

This book uses the phrase systems engineering. However the practices are not limited to systems engineers. They are appropriate for all that are engaged in systems, especially large systems. We may be systems users, managers, administrators, policy makers, architects, designers, builders, testers, teachers, doctors, etc in fields as diverse as healthcare to spaceships. But these practices can apply and should be considered when trying to make existing systems better or create new systems. They should become our common sense.

This book was born from my experience of going back to the universities as an advisor on senior class systems engineering projects. After the first year I suspected a text was needed. By the second year I was convinced that a text was critical for the next generation.

This book is an attempt to introduce systems engineering, practices, methods, and experience in a semi academic text. In many ways it is based on a format of an advisor sitting at the table with students. It includes the mechanics of executing a practice, offers examples, but then departs from the traditional textbook format and offers opinions. I have tried to delineate where my opinions start but the line is not always clearly drawn. I also included exercises at the end of each chapter to stimulate thinking beyond the content of the chapter. They represent what might be found in a high quality work setting or student advisor session.

Fundamentally the purpose of this book is to introduce my children and future generations to this work. I wrote a less formal book on systems engineering and suggest that you consider both these texts as you enter the systems space and make your contributions[5]. I invite others to improve upon this work and offer their own reference texts.

I hope you enjoy the rest of the book.

---

[5] Sustainable Development Possible with Creative System Engineering, Walter Sobkiw, 2008, ISBN 0615216307

2 Introduction

So is this book about systems engineering or systems or systems science or systems whatever? I decided to name the book systems practices to clearly state that this book applies to all that think they are engaged in systems of any kind. At times systems, systems engineering, and systems practices are used throughout the book and should be considered interchangeable. The goal is to encourage people involved in all types of systems to examine these practices, methods, techniques, thinking for application to their domain. Whether it is a new product, product line, the revamping of a countries health care system, building a house, building a community, cancer research, or upgrading a computer based automation system. So let's begin.

Systems are multi dimension. To understand a system, multiple views must be considered. This means there are different kinds of pictures, analysis, and words all grouped into some cohesive set so that all can understand the system. For example most individuals will attempt to view a system from a financial point of view. But there are also functional and performance points of views in which a system can be optimized.

When designing an amplifier do the designers first focus on the financial aspects of an amplifier or do they focus on the power output, frequency response, and amplifier noise? So why do people today immediately drift towards tracking the money when attempting to design or optimize systems like healthcare, transportation, electric power, fuels, villages or cities? Could it be because they lack basic knowledge in representing systems from a functional and performance point of view? Did people in the past focus on function and performance first when designing cars, houses, villages, radios, televisions, stereos, toasters, etc?

The reality is a system with a given set of functional and performance characteristics will be at a given cost. The designers can choose to reduce performance or functionality to reduce costs, but then they run the risk of creating a useless system from the user point of view, those that must rely on the system. In some cases this reduced functionality and or performance can lead to loss of life. Most focus on examples of airplanes falling from the sky, bridges falling down, or cars killing needlessly. However, we should never forget that people have died from bad water, bad food, bad sanitation, bad drugs, bad health care, and even bad cities that result in crime and destruction of human spirit.

Those that engaged in systems practices in the last century realized that our technologies have become so powerful that we could literally destroy our planet. The start of a nuclear world was the first time this realization surfaced and was unfortunately true. By the 1970's people started to realize the ramifications of our actions on our fragile global environment and future generations. The realization that there is no place to run, there is no hidden valley protected from our actions coupled with our massive technology. Shangri-La does not exist.

So people started to practice this thing called systems engineering. From this flowed various systems practices with vast application potential. Did they invent systems engineering or did it just naturally emerge in a complex space where nothing else would work? Does it matter? The point is systems engineering was studied, practiced, and honed to such a high level of maturity that people could clearly point to systems engineering as the only way to get a job done and deliver working solutions.

Not everyone was or is engaged in systems engineering activities. Those that were touched by the power of our post World War II technologies engaged consistently in systems engineering; for example, those that had seen the nuclear tests and the promise of powerful technologies but at huge risks and costs. For them, systems engineering or systems practices had become common sense. Some but not all went on to build houses, appliances, roads, satellites, and many of the systems we all enjoy. The dilemma is that not everyone was or is now involved in systems and some outside the peer group erroneously feel it is foreign and complex.

There are many traps that people can fall into when they first enter the systems engineering or practices space. For example most believe systems engineering will result in the perfect optimized solution, architecture, or design – the system. The reality is that multiple architecture approaches will work, but these architectures must be born from the fire of systems engineering. Each architecture or system will shine in a particular area and be weak in another area. Money can be applied to the weak architecture areas to bring it into the desired performance level. Also any of these architectures will cost about the same in the end. This realization was born from empirical data at Hughes Aircraft[6], Fullerton, California circa 1982 and communicated to me by a very senior Hughes person. He had worked with Howard Hughes in the early years. His name was also Howard and yes he was brilliant.

Unfortunately little work transitioned back to the universities from institutions like Hughes Aircraft. These institutions have disappeared and the senior people who gave birth to modern systems engineering or practices have started to pass on. I am younger and thus just an echo of these people and their work. Everywhere I have gone I have tried to pass on this body of knowledge.

Let us begin.

---

[6] Hughes Aircraft is now owned by Raytheon, See the University of Las Vegas Howard R. Hughes College of Engineering archives.

The Early Bird was world's first communication satellite. NASA launched the satellite built by Hughes Aircraft Corporation[7] on April 6, 1955 at 6:48pm EST from Complex 17a at Cape Kennedy, Florida. Early Bird was built for the Communications Satellite Corporation and weighed about 85 pounds after being placed in a synchronous orbit of 22,300 miles above the Earth. It was positioned over the Atlantic to provide 240 two-way telephone channels or 2-way television between Europe and North America. The outer surface of Early Bird was covered with 6,000 silicon-coated solar cells, which absorbed the sun's rays to provide power to the satellite for its intricate transmitting and receiving equipment.

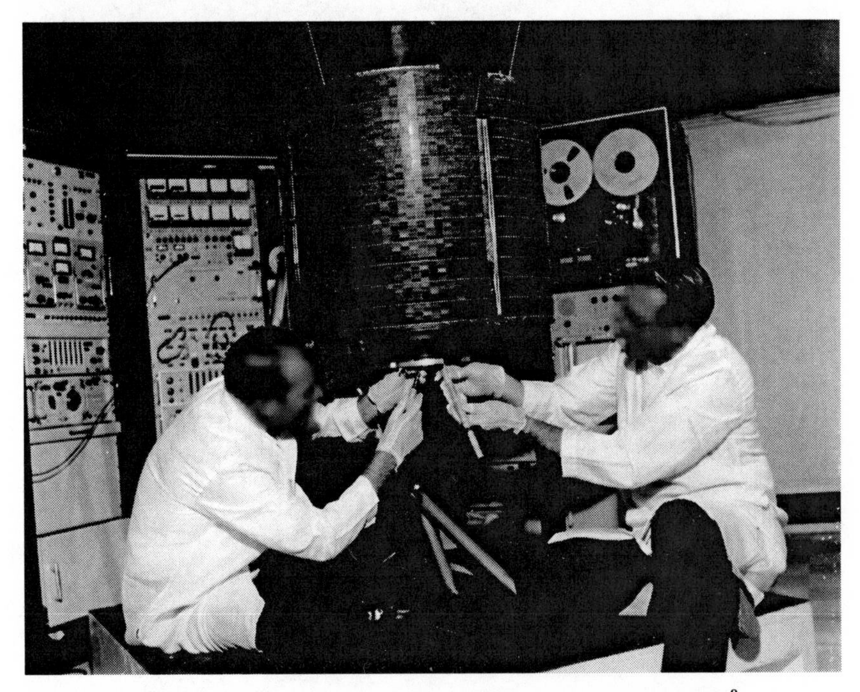

Figure 1 Early Bird - First Communications Satellite[8]

[7] Hughes Aircraft Company was a major American defense contractor founded in 1932 by Howard Hughes in Culver City, California. In 1953 it was donated to the non-profit Howard Hughes Medical Institute.

[8] Photo: NASA. Engineers checkout Early Bird-Communication Satellite.

Intelsat IV in an anechoic (sound absorbing) chamber. Built by Hughes Aircraft it stood over 17 feet tall with a capacity of about 6,000 voice grade circuits or 13 television channels. Intelsat was an international organization of 65 nations that was established August 20, 1964 out of the growing demand for channels of communication and greatly expanded the commercial communications network. The Intelsat IV was placed in a synchronous orbit over the Atlantic.

Figure 2 Intelsat IV Communication satellite[9]

Two beautiful girls standing next to this massive machine. A machine that would bring their worlds together and elevate their understanding and consciousness of the world to such levels that some would have considered it impossible just a few short years ago.

[9] Photo: NASA. Intelsat IV in an Anechoic chamber

3 Roots

Engineering disciplines tend to have a base that is used to capture and understand the design. Electrical engineers have their circuit diagrams and equations based on physics to predict their circuit behavior. Mechanical engineers have their mechanical drawings and equations based on physics. Chemical engineers use chemical equations and process diagrams. Civil engineers were the first and they too are based on physics and mathematics. At some point all these engineering disciplines must interact with more than their own field of study. For example an electrical circuit used to control acceleration in an automobile must be subjected to failure analysis, maintenance analysis, human factors analysis, etc. It is intuitively obvious that we are referring to a system engineering view of the simple accelerator pedal. So what do systems engineers or practitioners use to understand their designs?

In recent years many analysis techniques have surfaced. However engineering and systems engineering is based on science and mathematics. Science and mathematics are based on logic and set theory. Much of the early system engineering methods and techniques were traceable to these roots. For example logic is used when creating operational sequence diagrams. Set theory is used in grouping functions and creating subsystems. Mathematics is used in all modeling activities when searching for system instabilities and understanding performance.

This approach to solving problems, using science to postulate then prove or disprove, using logic to walk through a problem, using set theory to group things, using mathematics to predict behavior is key to systems practices. It is proper to create[10] a new method or process, especially for new systems, but those techniques must be based on fundamentals. These ideas are not just limited to systems engineering. For example, as early as elementary school we learned about set theory and logic, which then helped us to develop an outline for our very first composition paper.

There is an old saying that you should consider in all your efforts: sometimes less is more. The computer has allowed the generation of vast amounts of information, but many times the information it generates is not necessarily what is needed, especially in new and innovative areas. For example a simple free hand drawing or a concept diagram not available in any image library can succinctly state a case and show how a significant problem is solved. Hundreds of computer

---

[10] Many prefer to use design or develop and suggest create is inappropriate. I use create to emphasize the creativity and innovation that must be part of each design or development.

generated artifacts cannot replace an image drawn on a napkin born from an instant of genius.

So as you consider the methods, techniques, and processes in this book, realize that you will need to modify them and perhaps create new methods, techniques, and processes. But they must be based on science, math, logic, and set theory. They should not be created to separate you from other stakeholders, especially because you do not agree with their analysis. They should be created when they enhance communications between all the stakeholders and add new system insights. A new process, method, technique is validated and proven valuable when it surfaces a new insight that radically changes the system vision such that it is now obvious to all the stakeholders.

If everyone spends more time on the process, method, or technique rather than on the system solution, something is wrong. The miracle is not in the mechanical pressing of the buttons and turning the process crank. The miracle is from the human who comes up with this thing called an answer, solution, architecture, design, implementation, approach, etc.

3.1 Perspectives, Views, Dimensions, Domains

Circuits in electrical engineering can be viewed from the time domain or frequency domain. We use mathematics to switch between the time and frequency domain. These are different dimensions of the problem, or perspectives. What exists and is readily seen in the time domain does not exist or is seen in the frequency domain. Yet to really understand what is happening in the circuit both perspectives, views, domains, dimensions need to be seen and understood. This is readily summarized in the liberal arts with the simple statement of "walk a mile in my shoes", the implication being that you need another perspective.

Systems Engineering is especially sensitive to these ideas of multiple views. Systems Engineering is not complete unless multiple perspectives are offered. These perspectives come from various words and pictures captured in analysis documents, presentations, white papers and even prototypes. These views, perspectives, dimensions, domains can be surfaced using the practices offered in the following chapters.

For example a functional view of a system is incomplete. It usually must be complemented with an operational view of the system. As the importance and complexity of a system increases so must the number of views increase as embodied in different analysis and their inherent practices. These views include maintenance, training, support, safety, security, logistics, growth, technology insertion, human factors, reliability, peak load scenarios, etc.

3.2 Systems Analysis Levels

The most challenging systems should probably use all the practices outlined in this book plus more, some of which may need to be invented for the new system.

However some systems only need a subset of the practices in this book to gain a reasonable level of success. The following is a road map that can be used to determine the appropriate level of practices to be applied to 3 different system complexity levels.

Table 1 System Analysis Levels

Small	Medium	Large
Context Diagram Functional Block Diagram Concept Diagram Architecture Block Diagram A-Spec System Test Manual Website	From small to everything in the book Use your judgement and common sense to decide	Everything in book plus 5000+ years of other practices plus new practices for new system

3.3 Technology Versus Goal Driven Systems

There are three approaches to create a system. These approaches are based on either being technology driven, mission goal driven, or both. No approach is right or wrong and they need different systems practices to make them effective.

The first approach is to identify the goals / mission / needs requirements and then attempt to create a system with current, modified, or new technologies, machines, and processes. Examples of this might be the goal of sending a man to the moon before the end of decade.

The second approach is to let the technology evolve on its own taking its own path of least resistance and when someone decides, throw it into a real world operational setting, letting the human participants morph it into their work. An example of this might be the Internet[11]. All of the sudden the Internet technology appeared and everyone embraced or rejected it. Those that embraced the Internet technologies have new systems that have transformed their worlds.

The third approach is a combination of being both technology driven and goal driven. This is for extremely complex systems where it is important to not rule out any possible avenues. An example of this might be a new air traffic control system where there is a need to increase capacity, that is a reasonable goal. At the same time there are new technologies that can be studied and tested to determine if they result in increased system capacity. So coming from the requirement driven approach and technology driven approach the most effective system might emerge.

In the end there will be a system whether we like it or not. The system can be great, and everyone will just smile and think, – wow what a great product, system,

---

[11] High Performance Computing and Communications Act of 1991 (HPCA) Public Law 102-194, the Gore Bill. It helped fund the University of Illinois National Center for Supercomputing Applications where programmers developed the Mosaic Web browser.

or solution – or the system will be bad and everyone will be sad if forced to live with the bad system.

3.4 Infrastructure Systems

There are many different systems. Many of us are stakeholders in many systems, so we can immediately start to talk about these systems form our perspectives. The following is a list of such systems:

Transportation System	Healthcare System	Sewage System
Air Traffic Control System	Telephone System	Education System
Interstate Highway System	Cell Phone System	Radio Broadcasting System
Political System	E-Commerce System	TV Broadcasting System
Financial system	Power System	Home Cable System
Banking System	Water System	Home Satellite System

Infrastructure systems are always operational. Simulations exist in offline lab settings. Failure is immediately detected. Military systems are always in simulation mode. Operational settings exist in unique isolated situations. Failure is not always apparent in the confusion of war. The challenge is how do you create, modify, and or expand an infrastructure system that is always online?

3.5 Systems Practitioners Role

The role of the system practitioners is the most important role on any project. They need to ensure that the ideal system is provided. That is not an easy task and it requires significant technical and management skills. Obviously there are different levels of maturity of systems practitioners and as time progresses they move from apprentice, journey-persons[12], to masters. They are always learning and applying their new knowledge towards solving a problem. To do this they have certain characteristics and address certain roles:

- Creator, keeper, and champion of the system architecture
- Search for the forest and leave the trees to others
- Look for key requirements and issues that are system architecture drivers
- Use scientific method to solve technical problems that optimize architecture
- Digest the findings of specialists who provide data critical to the architecture and system vision
- Communicate vision to all of every skill level and background

---

[12] Journey-person or journeymen is able to address their own normal challenges with reasonable success while a master mentors journey-persons and is able to address unusual and difficult challenges.

- Constantly ask "what if" and look for any break through that moves the system closer to the ideal or to a new ideal
- Study the past and present of the system or similar systems and postulate a future state of being for the new system
- Immerse in technologies and postulate what could be

There is always an ideal system. The difficulty is to find the ideal system and then bring it into reality. Sometimes the ideal system surfaces quickly and everyone on the team agrees that it is the ideal system, but they do not know how to get to the ideal system. There could be many limitations that need to be addressed. These limitations include technology, politics, money, time, motivation, education, people, etc. I have no tolerance for politics or money. In both cases these are external forces that can destroy any system.

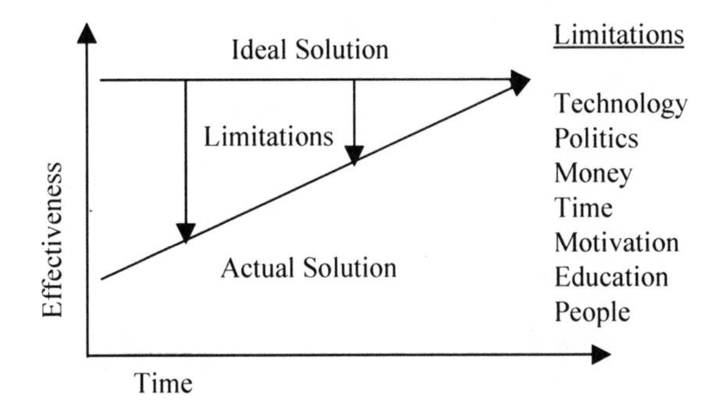

Figure 3 Ideal System Convergence

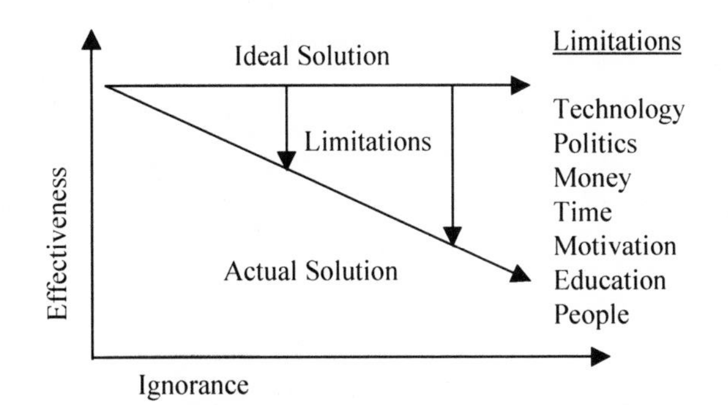

Figure 4 Ideal System Loss

Sometimes the ideal system is in clear sight and the approach for developing the ideal system is clearly visible. The system even may be in place but not necessarily available for this set of stakeholders. However the ideal system and its availability fade as ignorance grips the stakeholders.

The possible scenarios associated with the ideal system and the actual solution can be identified. Realizing these possible scenarios exist will help system practitioners understand their roles in the organization. It will also act as a gage to measure the organizations' ability to actually engage in systems engineering, systems thinking, and systems practices.

The reality is organizations that get lost are compromised by politics and then excuses such as money and schedule are offered. Alternatively they are compromised by politics and then excuses of technology, people, education, and motivation are offered. There is not much system practitioners can do once an organization is compromised other than attempt to point out the condition and the ramifications of the condition or leave the organization.

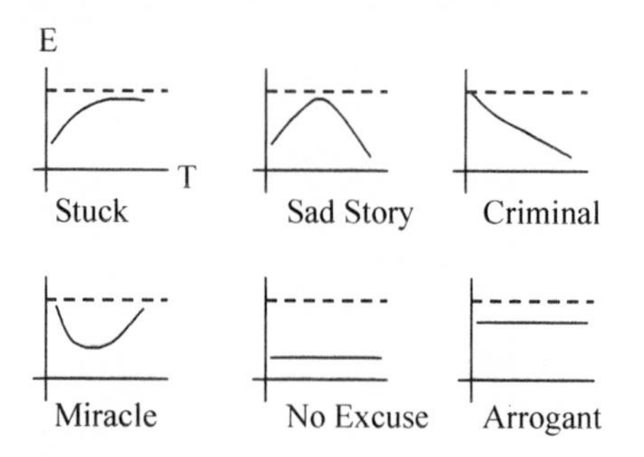

Figure 5 Organizational Failure Scenarios - Effectiveness over Time

3.6 Exercises

1. What are your thoughts on system views and perspectives?
2. What are your thoughts on system analysis levels?
3. Can you provide examples of technology versus goal driven systems?
4. Do you think you are systems driven and if so why?
5. Do you think the Internet would have arrived without the High Performance Computing and Communications Act of 1991 (HPCA) Public Law 102-194, the Gore Bill?
6. Provide examples of good and bad systems? What is the difference between the two?

3.7 Additional Reading

1. Sustainable Development Possible with Creative System Engineering, Walter Sobkiw, 2008, ISBN 0615216307.

4 Fundamentals

Before we move into systems practices we need to establish some common ground with fundamentals. This common ground is observations associated with human behavior and certain observations about nature typically addressed in general education and engineering schools.

4.1 Creativity and Innovation

There is a simple formula that can be applied by anyone to spur creativity and innovation. The formula for personal creativity and innovation is Saturate, Incubate, Synthesize, Optimize, and Select.

1. **Saturate**: Exactly as the word implies. Immerse yourself in the material, no matter how mundane or irrelevant it may seem in the beginning. Talk to everyone about the material and dialog if possible.
2. **Incubate**: Exactly as the word implies. Sleep on it and let the miracle of the human mind go to work on the problem.
3. **Synthesis**: Happens after incubation. You can't stop it; it is human. Eventually approaches will start to surface. Mature these approaches as you continue to saturate and incubate.
4. **Optimize**: As the word implies, take each viable approach and move it to its most elegant limit.
5. **Select**: You guessed it, pick your best approach.

This process of personal creativity dovetails in and out of the systems process that unfolds in this book. It is hard to say if they are separate or one and the same, but I like to think I practice both at the same time. So to expand upon the above simple personal creativity and innovation model to systems:

1. **Saturate**: Immerse the team in the material, no matter how mundane or irrelevant it may seem in the beginning. The entire team is engaged and meets regularly to exchange findings. Humans are left and right brain creatures. To fully understand a situation, words and pictures are needed. The words backed up with mathematics, logic, set theory and a picture born of true inspiration are the heart of communicating our visions. Essentially these all represent different perspectives of a problem and are associated with saturating our minds with the problem. In many ways the practices outlined in this book are a template to help us saturate our minds with the

problem space.

2. **Incubate**: Exactly as the word implies where the team sleeps on it and we let the miracle of the human minds go to work on the problem. A team of people needs time to incubate a problem. It is at this point that a gestalt surfaces and the team starts to become more than the sum of its individual minds. You will not know when incubation will start but you will detect it as the team members start to offer parts of a solution. Gradually more solution parts emerge then someone or some members start to put the whole picture together. These people are typically the systems architects. They emerge with the system solution; you cannot appoint the system architects.
3. **Synthesis**: Just like for the personal case, this happens after team incubation. You can't stop it; it is human. Eventually various approaches start to surface from within the team. Mature these approaches as you continue to saturate and incubate. In many ways systems engineering is purely about synthesis rather than reduction. Throw all kinds of things on the table and start to put them together. Putting together the non-obvious pieces can lead to major breakthroughs. It is at this point that the natural systems architects that emerged start to inspire the team. The rest of the team starts to find their comfort zones in various specialization areas.
4. **Optimize**: Just like for the personal case and is as the word implies where the team takes each viable approach and moves it to its most elegant limit. The specialists really start to take on the load at this time as the system architects maintain the vision and adjust the vision based on the specialists' findings.
5. **Select**: Just like for the personal case the team picks the best approach. The selection is born from intense tradeoff analysis using the Measure of Effectiveness (MOE) equation[13]. The selection is fully transparent and understood by all the stakeholders. It is born of reason and common sense.

4.2 Program Project or Project Program

So which is it? Projects make up programs, period. So programs are bigger than projects. Programs are successful by definition because it takes more than one project to make a program. Programs are like product lines but programs may be made of multiple products and product lines.

It is critical that a valid charter be established for the project or program. The charter needs to be a succinct statement of why the project or program exists and what it is to accomplish. Everyone should be aware of the charter. Day one on the job, as part of briefing the program or project, the charter must be clearly stated to the new members.

Never start a project unless the financial box is big enough to support the charter

[13] Discussed in detail in this text as part of architecture tradeoff and decision making.

and initial goals. If no one is willing to commit reasonable funds to tackle the problem, don't waste the money. Apply the money to an existing project or wait until proper funding with a reasonable schedule can be established. If funds are never made available for reasonable projects then let the organization self-destruct in its own greed. Move to another organization and make sure you tell everyone your story.

Do not confuse these words with individual, grass roots, or mom and pop business operations. These worlds use completely different criteria when deciding to start a project. Usually there is no funding and all work is based on sweat equity.

The problem is when large institutions, governments, and corporations ignore economies of scale and run parallel operations modeled after individual, grass roots, or mom and pop organizations. These Darwinian entities waste enormous resources even though they may show short-term gains to investors. In a twist of language these organizations advertise themselves as ecological organizations not realizing it is about symbiosis.

4.3 Information Chunking and Modularization

Chunking is a principle that applies to the effective communication of information. It is very important in written communication. In 1950 George A. Miller published a landmark journal article entitled "The Magical Number Seven, Plus or Minus Two". He studied short-term memory trying to determine how many numbers people could reliably remember a few minutes after having been told these numbers only once. The answer was: "The Magical Number 7, Plus or Minus 2".

Chunking involves grouping things into 7 groups plus or minus 2, then picking one group and chunking it into 7 groups plus or minus 2. Then the next group is picked and chunked, or decomposed. So there are 7 groups across the top and 7 layers down. This obviously represents a very large number of items that can be easily accessed by a human.

This concept applies in all forms of communications. It even applies to engineering activities such as software, where chunking reduces complexity by increasing understandability, which increases maintainability. This was recognized as Modularization.

Modularization was a big trend in the 1970s and it transformed many things from television sets to major infrastructure systems. Yourdons' famous structured system analysis from the 1970s was based on this finding[14].

The old term was called a rat's nest. This was typically a thing that did not represent or embrace the concept of chunking. There are still examples where people are not familiar with this work even though they have been exposed to outlining techniques in writing and structured system analysis techniques. Usually

---

[14] Tom De Marco applied decomposition to software in 1978 in his book Structured Analysis and System Specification, ISBN 0917072073

this is a sign of a problem in the organization where the solution was never really fully understood and yet a milestone needed to be met as mandated by management. So there is more to chunking than meets the eye.

4.4 Momentum

Momentum is a term found in physics[15] where an object is understood from a speed, direction, and mass point of view. Typically one would calculate the amount of momentum in an object to determine what it would take to change its speed or direction if an external force were applied. For example if you had the unfortunate experience of pushing a car you know that it takes a great deal of effort to get the car moving. Once it is moving, keeping the car in motion requires much less effort. Conversely if you ever needed to stop a car you are pushing, you know that it takes a great deal of effort to stop the moving mass.

This concept of momentum can be quantified in physics once applied to objects. However this same concept applies to organizations and architectures in the abstract. It may not be quantifiable but it most certainly exists and needs to be realized, respected, and qualitatively understood.

4.5 Hysteresis

Hysteresis[16] refers to a condition where application of input is not immediately addressed. There is a time delay. Once the action starts, removing the input does not immediately stop the action.

An example is a heater thermostat. When the temperature is set to 70 the heater will not activate until the temperature falls to 68. Further the heater will run until the temperature reaches 72. This prevents the heater from constantly turning on and off. The delta between the on and off points is set by the heater technician.

In the real world of systems engineering people attempt to track progress using metrics without realizing that different metrics have different Hysteresis characteristics. For example attempting to track projects solely using Earned Value Management System (EVMS)[17] techniques is extremely dangerous. By the time EVMS shows a project is in trouble, severe damage has occurred and recovery will be significantly more challenging. Meaningful non-financial metrics always have less Hysteresis than EVMS but they require a domain expert to understand and

---

[15] Fundamentals of Physics, David Halliday and Robert Resnick, John Wiley & Sons Inc; Revised edition (January 1, 1974), ISBN: 0471344311

[16] Physical and Quantum Electronics Series, Demetrius T. Paris, F. Kenneth Hard, McGraw-Hill Book Company, 1969, Library of Congress CCN 68-8775, ISBN 070484708.

[17] American National Standards Institute/Electronic Industries Alliance ANSI/EIA-748, Establishes 32 minimum management control guidelines for an EVMS to ensure validity of information used by management. The Unites States Federal government adopted the guidelines in ANSI/EIA-748 for use on government programs and contracts.

appreciate. In the ideal situation both EVMS and design metrics are tracked.

The same can be said of operating a live system, tracking its money flow has very large Hysteresis characteristics once compared with the systems non financial metrics. The trick is to find the meaningful non financial metrics, the essence of the system performance.

4.6 Event versus Process and Control Theory

The difference between an event and a process is very important and fundamental. A process always has a feedback loop. The feedback loop is critical and causes the output to modify. In an event there is no desire to modify the output. What surfaces either is useful or not useful. These ideas have their roots in control theory[18].

In control theory the feedback loop is used to stabilize a system. In the case of an amplifier a negative feedback loop will reduce noise. It comes at a cost and that cost is amplifier gain. A positive feedback loop can lead to system instability. A good example of this is a public address system where someone places a microphone to close to a speaker and the audience hears loud painful feedback.

A development process is used to develop a system. The development process must always have feedback. If it is executed as an event, the emergence found during system analysis cannot be factored back into the problem. The development process will fatally break and the system will be fatally flawed.

Many times a system itself must also have feedback mechanisms. These feedback mechanisms cause the system to adjust to the conditions as they unfold. A simple example is the number of cashiers in a store. With a sudden influx of customers the system should adjust by adding more cashiers until the load is worked off and typical system operations resume with a typical system load. This simple idea is also the heart of multiprocessing computer systems and electric power systems.

4.7 Peak Load and Burno ut

Everyone should do the simple experiment of taking an electric light bulb and burning it out by slowly increasing the supplied voltage. This is an incredible lesson. The same should be done with a mechanism of gears such as a mechanical clock where one of the gears is continuously increased in speed until the gears start to strip their teeth and the mechanism stops.

All systems must be designed to withstand the ravages of a worst case peak load. However it needs to be understood that no system can survive continuous operation (100% duty cycle) at its worst-case peak load. Many think that this is inefficient. They do not understand. In complex systems such as a restaurant, the

---

[18] Modern Control Systems, Richard C. Dorf, Addison-Wesley Publishing Company, 1967, 1974, Library of Congress CCN 67-15660, ISBN 0201016060.

peak load happens at breakfast, lunch, and dinner. In between those times there is preparation, maintenance, and the storage of energy. The energy is stored in the humans who must serve the peak load when it happens.

There are statistical methods that can be used to understand the peak loads. Agner Krarup Erlang[19] studied telephone systems and the result is the Erlang distribution, which is tied to a peak time of day, usually before lunch. The Internet exhibits two peak loads during a normal business day. Other modeling and distributions are Poisson process, Bernoulli process, and Markov processes.

All systems exhibit a peak load. There is a desire, when resources are low or policy makers are ignorant to create systems, which assume an average spread of the load. These systems fail or stop providing critical services when the natural peak arrives. There are various approaches to modeling the peak load. The ideal approach is to derive empirical data and then test the system in live situations prior to taking the system live.

4.8 Timing and Sizing

Many people equate timing and sizing with computer systems from the last century. But timing and sizing applies everywhere. A bridge needs to be appropriately sized to handle rush hour traffic in a reasonable way. The same is true for a road. A store needs to have some number of cashiers to process the customers in a reasonable time. The same is true for a restaurant and its count of waiters, tables, and cooks.

This all requires some quantitative understanding of the operation. It can be modeled using a static model implemented in a spreadsheet, but the model must have reasonable worst case scenarios in addition to the typical and lightly loaded scenario cases. These three scenarios should be understood and appreciated as the system is sized.

4.9 Capabilities Functions Features

Capabilities, functions, and features are constantly mentioned in the context of systems. Some go to great lengths to distinguish between them and offer the advantages of each as though there is an inherent approach built into the words. However capabilities, functions, and features are nothing more than different frames of reference or perspectives from different stakeholders.

If you are buying a product you are interested in its features. The features separate one offering from another offering. If the industry is healthy with competition and innovation, the key features are always listed as the vendors strive to find key discriminators to separate themselves from the competition. So features are the finished results of a product that a buying stakeholder can review and make

---

[19] Mathematician statistician engineer, Agner Krarup Erlang (January 1, 1878 to February 3, 1929) invented traffic engineering and queuing theory.

an informed buying decision.

If you are interested in having something built and not sure what you are looking for, you speak in terms of capabilities. The capabilities might be feasible or not and you may be unsure how to implement the capabilities. For example you might want to build a house that will survive a hurricane but you are not sure how that might be implemented. As the analysis unfolds and you are given the alternatives you may decide only a portion of the house should survive a level 5 hurricane. So, capabilities are the early goals of the stakeholders buying the system.

If you have to design and implement something you need to work from blue prints. The blue prints are your requirements. Your requirements may be captured in a set of blue prints, drawings, text-based specifications, or other information products. If you are buying the system you need to understand the link between your original capabilities / goals and the resulting requirements. This understanding is captured in analysis and studies that form the supporting information products for the specifications. So the requirements are given to the stakeholders tasked with designing and implementing the system while the stakeholders buying the system understand the requirements.

4.10 Static versus Dynamic Systems

Many practitioners have attempted to define static and dynamic systems. Traditionally dynamic systems have feedback and are able to change their behavior as a result of that feedback. Traditionally static systems have no feedback. Examples are then offered for static systems such as roads, bridges, and buildings. Then Air Traffic Control is offered as a dynamic system example.

However, there is no such thing as a static system! A system is defined by its boundary. If someone feels they have discovered a static system, what they have really found is the wrong system boundary.

A bridge can be safe or unsafe. This characteristic is critical to a stakeholder attempting to cross the bridge. The stakeholder will visually inspect the bridge. If it looks safe then the stakeholder will slowly enter onto the bridge and listen very carefully for crackling noises. In a healthy society the bridge system will include a maintenance philosophy implemented by a maintenance crew and certified by independent inspectors. All of these are feedback. The feedback can result in decisions to stop crossing over the bridge, fix its aging parts, or to enhance its current structure for a new function such as lightweight rail.

A bridge also has traffic capacity. The traffic capacity can support an average traffic rate, a peak traffic rate, average weight load, and peak weight load. The traffic can be throttled. For example an accident on the bridge can result in one or more lanes on the bridge being closed. Depending on the bridge resilience in the face of an accident, the bridge may need to be closed because all lanes are blocked.

A road provides feedback to its users and maintainers everyday. The way traffic merges, diverges, as the rains fall, and the sun causes glare. The feedback loop might be measured in hours, days, weeks, months, and years rather than seconds or microseconds. The feedback loop impacts the maintainers, users, and even the road itself as potholes and cracks develop and are repaired.

So a road or bridge system is not just the physical road or bridge, just like an Air Traffic Control system is not just a collection of computers, displays, and two-way radio boxes. It is a system in the context of its' use and interaction with the stakeholders. It includes maintenance, growth, technology insertion, people, average and peak performance, etc. During its day to day operation its performance will vary as feedback is given to the users of the road or bridge and the road or bridge maintainers.

The same is true of a house, an office building, a shopping center, a park, a city etc. People interact with these structures everyday and that interaction impacts the people and their use of these structures. The structures change as people adapt them based on their interaction. All systems are dynamic once the boundary is properly considered.

4.11 Cause and Effect

Causality or cause and effect is the relationship of an event and a second event which is the result of the first event. Cause and effect in many cases is a chain of events where a causing event is a resulting event of a previous cause.

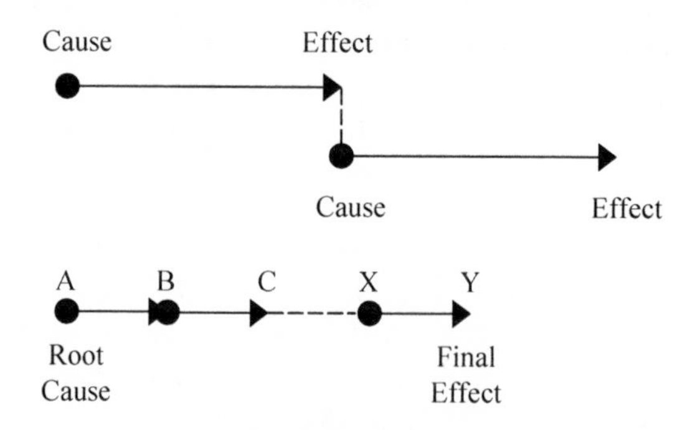

Figure 6 Cause and Effect

One of the biggest issues associated with cause and effect is to find the full chain of events and not be misled to a premature conclusion. There is always the temptation to be superficial and not view a situation from different perspectives. Cause and effect is a fundamental principal found in almost all system practices. Many will point to:

- State transition and mode analysis
- Root cause analysis
- Failure analysis
- Diagnosis troubleshooting debugging
- Sequence Analysis
- Timing sizing and load analysis

Cause and effect also can be used to understand the relationship between different system variables. There can be either a positive or negative effect between these variables, usually referred to as a causal loop. In a positive causal loop or link two nodes or variables increase together or decrease together in a linear or non-linear fashion. In a negative causal link the two nodes change in opposite directions. If one node increases the other node decreases.

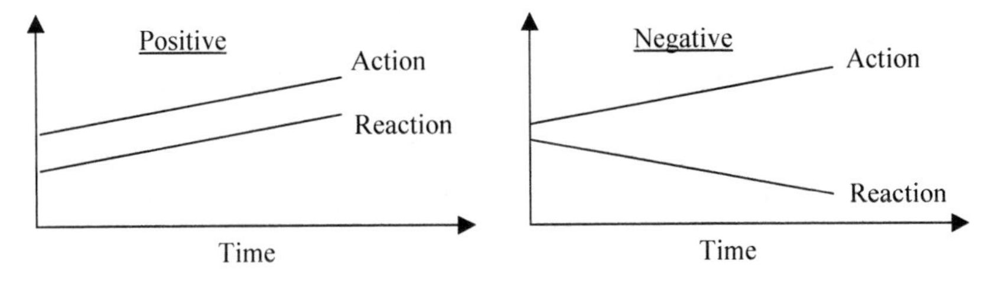

Figure 7 Causal Relationships

A causal loop can change from positive to negative or negative to positive. This is usually an unexpected state change.

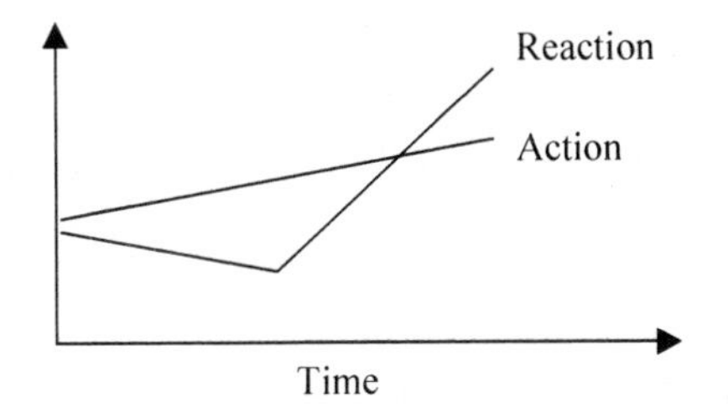

Figure 8 Causal Relationship State Change

It is best illustrated with a social system example. Assume a ruthless dictator applies draconian rules on the population in an attempt to instill fear. Most of the population complies out of fear, however there is a very small segment that always rebels. As more draconian tactics are used then eventually the majority of the population rebels. This is an example of a flip from a negative causal loop where increased draconian tactics reduces the populations desire to rebel, until a state change happens, and then increased draconian pressure leads to increased rebellion.

4.12 Brainstorming and Meetings

People have attempted to define and capture the characteristics of naturally flowing activities. For example there are various words written on how to run an effective meeting. When we are given group projects we meet and try to listen to each other hoping to maximize our own benefits. Ideas are not suppressed unless the group members are hampered with external baggage. There are different meeting types that have been named and codified, such as brainstorming.

Brainstorming is a group activity to generate a large number of ideas for solving a problem[20]. One of the fundamental concepts of brainstorming is to break down fear of expression. Expression by all the members is key because no one individual mind is capable of solving the problem. The problem requires a collection of minds working together and freely in a "multiprocessing mode".

At the time when Osborn wrote his book few were familiar with the term multiprocessing. SAGE was not common knowledge. In fact multiprocessing has only recently entered the mass consciousness with personal computers that use more than one microprocessor. Labor was viewed as something that could be distributed across multiple bodies, but distributing problem solving was and is still new to most people. Academics naturally understand the concept of problems being greater than a single mind and have embraced technical publications as the means to address this issue.

Just as in the case of multiprocessing computing, the processing can be tightly coupled or loosely couples and everything in between. In the traditional academic model of publishing the processing is very loosely coupled. In the case of a brainstorming meeting the processing is very tightly coupled. In the case of an Internet message board the processing is less tightly coupled but can include thousands of minds working 24/7 for weeks or months on a problem beyond the grasp of a single mind. "At what point do the addition of more humans result in no new identification of alternatives?"[21]. There are 4 basic rules in brainstorming.

1. **Keep the ideas coming**: The greater the number of ideas generated, the greater the chance of producing a radical and effective solution.
2. **Withhold criticism**: It is critical that no one is intimidated. Everyone must speak freely. On the surface the most bizarre idea might actually be the solution or lead to the solution when examined in detail.
3. **Welcome unusual ideas**: Breaking down the status quo is hard. If the status quo could solve the problem then there would be no need for the brainstorming session or sessions.
4. **Combine and improve ideas**: Combine ideas. These combinations may lead to new perspectives, which lead to new ideas.

4.13 Doodling Note Taking and Concept Maps

As we go through our formal education we tend to develop certain techniques for remembering things, capturing concepts, and helping us to solve problems. For example, when trying to solve a new problem, we take a piece of paper and just start

---

[20] Osborn, A.F. Applied imagination: Principles and procedures of creative problem solving (Third Revised Edition). 1963. New York, NY: Charles Scribner's Son.

[21] Sustainable Development Possible with Creative System Engineering, Walter Sobkiw, 2008, ISBN 0615216307.

to write various elements about the problem on the paper. When in class while listening to a lecture we typically take notes based on words or phrases. Occasionally we create pictures that capture the essence of a new concept.

Many times we create concept maps. A concept map is a graphic, which shows concepts as nodes and relationships between the concepts via lines. The concept nodes may be represented with words or a small hand drawn picture. Usually there is a starting concept and from it flows other concepts connected by lines which then also connect to other concepts via lines. So distance and concept node hops from a concept tends to be represented. The concept map has notes and boundary lines that are usually added towards the later stages of the concept map development[22].

4.14 Storyboards

Storyboards in this context are associated with the Sequential Thematic organization of Publications (STOP)[23]. The idea[24] is to organize a technical document by two page topics where the left page contains the thesis sentence followed by supporting text, approximately 500 words. On the right page is a supporting graphic. The text and graphic complement and reinforce each other. This obviously leads to other ideas such as how to start the writing process and manage the whole activity. In the middle of this is the STOP storyboard.

Figure 9 Wall Review[25]

[22] A website with a main page and several links to other internal pages is an example of a concept map and its implementation.

[23] Sequential Thematic Organization of Publications (STOP): How to Achieve Coherence in Proposals and Reports, Hughes Aircraft Company Ground Systems Group, Fullerton, Calif., J. R. Tracey, D. E. Rugh, W. S. Starkey, Information Media Dept., ID 65-10-10 52092, January 1965.

[24] Created at Hughes Aircraft Fullerton California in 1963.

[25] A graphic from STOP Manual - notice free hand drawing, its projected imagery and the number of messages being sent.

The STOP storyboard is similar to other storyboards. They both use a graphic. However, STOP storyboards also contain the topic name, thesis sentence, and text phrases representing main points supporting the thesis sentence. In the STOP format there is less emphasis in a high quality graphic. In some instances it just can be a concept of a graphic showing some but not all the elements. Typically a practitioner should spend between 5 and 15 minutes on a preliminary storyboard.

Book: Sustainable Development Possible with Creative System Engineering **Topic:** Introduction **Thesis:** The only way we can survive in the next 100 years is through the application of creative systems engineering.	
Main Points: 1. It's about sustainable development 2. Sustainable development it not new has always existed 3. The only way to do this is through creative system engineering 4. Creative system engineering is not new 5. What is creative system engineering 6. Creative system engineering transformed the world	Graphic Table Image use proposed book cover image # 33 **Caption:** Title **Name:** Walt Sobkiw **Date:** 01/31/2008

Figure 10 Storyboard - Book

The storyboards are posted on a wall for the team to casually review. At some point a wall review is scheduled for all the storyboards. These reviews continue and may split the team into separate reviews until everyone is comfortable with the storyboards. At some point writing can begin.

So the STOP storyboard with its strong thesis sentence, text phrases, and supporting graphic can be used anytime to capture any element of the systems engineering effort. It can show:

- Functions and Performance
- Architecture Alternatives
- Selected Architectures
- Study Findings
- Tradeoff Results
- Computer Screen Layouts
- Implementation Details
- Etc

Today many of us can relate to this form of communication when we build or attend power point presentations. However, whereas a power point slide can be deliberately vague, a storyboard should be very precise and accurate.

Book: Technical Volume on System Tools
Topic: Automated Specification Reviews
Thesis: The Specification Analysis Tool (SAT) uses a computer to quickly analyze specification text that humans find tedious and allow them to focus on what they do best which is creativity and innovation.

Main Points:

1. The current approach is based on manual mentors and checklists

2. There has been significant automation of requirement management, modeling / simulation, but no requirement text authoring

3. Everything is eventually reduced to specification text

4. Machines are great at searching, counting, filtering, categorizing, profiling, and visualizing

5. Humans are great at creativity, critical thinking, inspiration, intuition

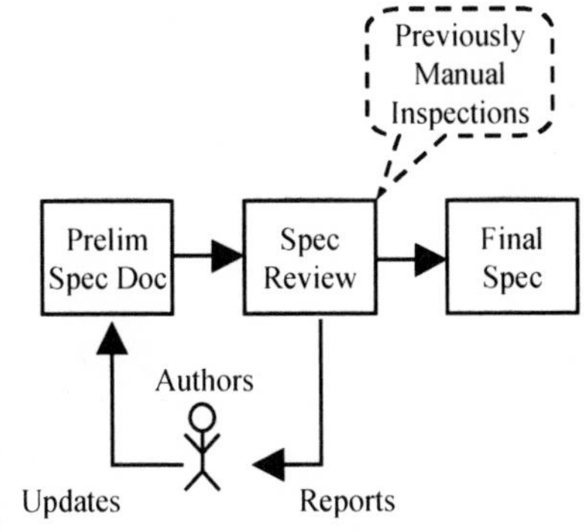

Previously Manual Inspections
Prelim Spec Doc
Spec Review
Final Spec
Authors
Updates
Reports

Caption: The Idea - its time has arrived

Name: Walt Sobkiw **Date:** 03/02/2009

Figure 11 Storyboard - Proposal

The elusive thesis sentence is a challenge[26]. It is even more difficult to support the thesis sentence when the analysis has actually been performed and understood. When supporting the thesis sentence consider:

- Have a point get to it
- Treat it completely
- Keep out extraneous matter
- Relate graphic to the text

Even though STOP storyboards were created to support the development of technical publications after the analysis is complete and clearly understood, they can be used to convey information as the system analysis unfolds. It's just that the storyboard may be incomplete until the end of the analysis. Also, as in the case of STOP, storyboards may be abandoned as dead ends. The thesis may not really have been appropriate and the system went in a totally different direction. This is where storyboards become very valuable. It is much easier to abandon a 1-page storyboard

[26] Is it any wonder that new people at Hughes were told they would be getting the equivalent of a few Ph.D.'s per year?

than a 2-page topic or a 25-page white paper. The lesson is never write text until you can draw a picture, establish a point or thesis, and have 5 to 8 supporting word phrases.

<table>
<tr><td colspan="2">Book: Systems Practices as Common Sense
Topic: Preface
Thesis: Systems engineering at one time was viewed as common sense but now few are familiar with this body of work and less are even aware of some of the common practices.</td></tr>
<tr><td rowspan="2">Main Points:
1. Nuclear bomb detonated in 1945 ended world war II

2. SAGE was created to defend against the abuse of this new technology

3. Systems engineering emerged and honed to such a level some dared call it common sense

4. Problem is all knowledge captured in companies, non profits, think tanks, government labs is being lost

5. This book attempt to capture this knowledge and pass to next generation</td><td>Graphic
Table
Image

use proposed book cover image # 3, 50, 55, or universal man # 2</td></tr>
<tr><td>Caption: Front Cover

Name: Walt Sobkiw Date: 02/10/2010</td></tr>
</table>

Figure 12 Storyboard - This Book

4.15 Previous Work and History

When developing a system, previous work and history must be investigated. Recreating the wheel or falling into traps that others have documented and are easily avoided is inexcusable. Just like all other system practices the previous work and history is approached methodically. It starts simple and grows in complexity as the search for information becomes more intense and thorough.

- **Literature Search:** An analysis of previous work and history always starts with a literature search. In the past a literature search was limited to print and microfiche. Today computer-based searches yield electronic media. The literature search includes popular media, trade journals, conference proceedings, books, magazines, and websites with their own unique content. Although scholarly works need to reference high quality works, other information sources should not be discounted. For a large program usually someone collects relevant articles and makes them available to the team. Do not discount this activity especially when the team has concluded they completed the data gathering activity. There is always new data that surfaces which needs to be considered.

- **Industry and Technology Surveys:** At some point the broad literature search starts to focus on the related industries and technologies. Various trends and relationships are captured and presented to the team. Vendors may be contacted but this is still a library research activity.
- **Vendor Search:** Eventually key venders surface. The venders should be analyzed and presented to the team. In some cases, selected venders are invited to present their products and technologies.
- **Samples and Demonstration Requests:** After selected venders present their information, some can be invited to offer samples and demonstrations. These venders may become future subcontractors offering key elements in the system.
- **Customer and Sponsor Visits:** Visiting the customer allows the systems staff to see perspectives not possible in other setting. There tend to be experts and other valuable stakeholders that will suddenly appear at meetings only at the customer site. There are also resources such as access to documents, facilities, equipment, related work that a site tour usually surfaces and aids in understanding the system needs and customer perspectives. The sponsor may or may not be collocated with the customer. Visiting the sponsor in their environment also provides insights not possible in other settings. Although not necessary, it is efficient to simultaneously visit the customer and sponsor on the same trip. This builds confidence in both stakeholders and allows the systems team to expose the new visit findings to the customer and sponsor, sponsor hears about the customer visit or customer hears about the sponsor visit.
- **Site Surveys:** A site survey is used to gather physical and other information about and from a location. It can be a survey of vacant land to determine its ability to accept new, modified, or expanded structures. It can be a survey of an existing facility and its ability to accept a new, modified, or expanded system. The site survey results feed back into the system development in the same way as any other stakeholder input. In many ways the results of the site survey may be the most important stakeholder input as reality is used to temper the idealistic system visions. A site survey includes a visiting and host team. The visiting team gathers physical information such as space, heating, cooling, power, ingress, egress, storage, humidity, vibration, noise levels, lighting, and other environmental and physical characteristics. The host team gathers information ahead of time to minimize time spent in physical surveys and maximize time spent interviewing the host team. Talking to the host team about the unique needs and previous experience including maintenance and support issues will almost always surface key

system requirements. At the conclusion of a site survey a report is prepared that contains the data gathered and references to information products provided. The site survey report identifies the needs, goals, issues, and requirements surfaced. The new information is clearly noted. The site survey report is shared with the host team so that all information is accurately represented.

4.16 Black Box White Box Analysis

A black box is used to represent a system in terms of its inputs and outputs. The input is subjected to an internal operation that is called a transfer function. At its simplest level it can be represented by the equation $y = F(x)$, which states that the output y, when subjected to an input x, is processed by a function F. For example, in electricity a capacitor subjected to an impulse function has a response over time that can be measured in a black box environment. To understand the input output relationship requires physics and mathematics at the white box level. From a complex system point of view there are multiple inputs, functions, and outputs. Also there might be relationships between the functions.

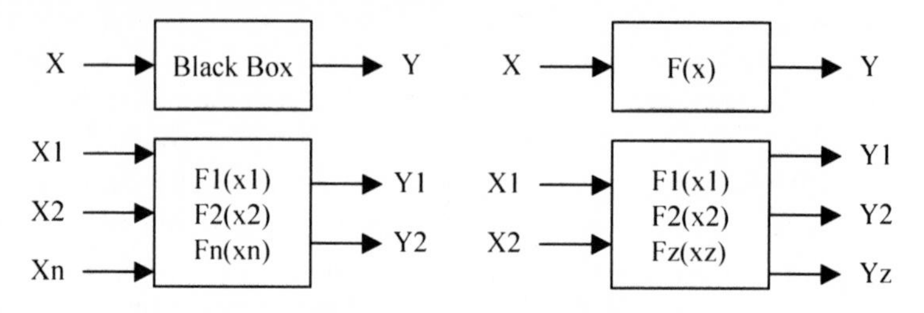

Figure 13 Black Box White Box Analysis

Example 1 Block Box - Radio

In a black box system representation of a radio the input is the power and radio signal while the output is sound. The output sound can selectively originate from many radio stations making different radio signals. The function is to capture all the radio signals, tune to a particular station, amplify the signal, and play the music. This black box radio system also has performance. The amplifier function performance might be a sound quality frequency response of 30 Hz to 15 kHz. A sensitivity of 5 microvolts and a selectivity of 80 dB might be the performance elements of the tuner function.

Black box analysis is used to identify the functions and performance of a system without really knowing how the functions work or even what the functions are that makes up the full system. Various black boxes can be compared. Some are obsolete while others are state-of-the-art from a function and performance point of view. The key functions and performance of black boxes for the same application or need are what separate them from each other when users are trying to decide on which black box solution to accept or purchase. As part of the function and performance decision is the issue of aesthetics or form. In many instances aesthetics plays a major factor in the decision to pick a particular system such as in the case of clothing, jewelry, cars, houses, etc. Sometimes this is referred to form fit function[27] where form represents aesthetics and fit represents performance.

White box analysis looks inside the black box to identify how the functions work. Usually when a black box is opened, it consists of multiple internal black boxes. There is some insight into how the highest level functions work, but the picture is not complete. So the next step is to open these boxes. Eventually the lowest level black boxes in the decomposition show how the most fundamental elements of the functions work. So a white box view is relative. If you can see inside you have a white box view, but it may be just a view of other black boxes until you get to the lowest levels of the system.

Black box analysis is about drawing a black box, identifying the inputs, outputs, surfacing the functions, and identifying the performance characteristics of the functions. It then continues by decomposing the black boxes, which then offer new understandings and views of the system. Eventually there are no more internal black boxes to decompose and the full functionality and performance of the system is understood.

4.17 Exercises

1. Can you provide an example of the technique of creativity and innovation activity and its results?
2. Is this class a project or a program?
3. Give an example of information chunking. Represent the example as a decomposition tree or an outline.
4. Provide an example of momentum and hysteresis. What were the impacts?
5. Provide an example of an event and a process. Are there differences between the event and the process?
6. Have you experienced burnout and if so what did you do to recover? Why were you able to recover? Are other systems able to recover after burnout?
7. Is it possible for a system to operate at 100% duty cycle peak load without premature failure? Can you provide an example?
8. If a system operates at peak load 100% of the time is it possible that there is a higher system peak load not known by the system designers?
9. At what duty cycle, 100%, 50% or 10%, can system peak load be

---

[27] Traditionally form fit function is a manufacturing perspective for an item. Form: Shape, size, dimensions, mass, weight, balance or center of mass, and unique visual parameters. Fit: Ability to physically interface, interconnect or become an integral part; includes tolerances. Function: Actions performed.

reasonably handled? Is it dependent on the level of peak above normal load?

10. Relate timing and sizing and peak load?
11. List related capabilities, functions and features.
12. Can you provide an example of a static system? Expand the system boundary, does it still remain a static system?
13. Provide an example of cause and effect. Are there intermediate effects or follow on effects in your example?
14. Meet with several classmates and brainstorm some of answers to the questions in this section. What are your thoughts?
15. Share your recent doodling and notes with your classmates. Are there any similarities? Are there any differences? Are there any concept-maps?
16. Create a concept map and share it with your classmates.
17. Create a storyboard of this section. Spend only 5 minutes on the storyboard. What are your thoughts on the results?
18. Draw a black box view of your computer. Go inside the computer black box and draw those internal black boxes. Do this on one sheet of paper. Do you have a new understanding of your computer?
19. What are the most important or key functional and performance requirements of your ideal music player, car, computer, and house?

4.18 Additional Reading

1. Applied Imagination: Principles and Procedures of Creative Problem Solving, A.F. Osborn, New York, NY: Charles Scribner's Son, Third Revised Edition 1963.
2. Fundamentals of Physics, David Halliday and Robert Resnick, John Wiley & Sons Inc; Revised edition (January 1, 1974), ISBN 0471344311.
3. Modern Control Systems, Richard C. Dorf, Addison-Wesley Publishing Company, 1967, 1974, Library of Congress CCN 67-15660, ISBN 0201016060.
4. Physical and Quantum Electronics Series, Demetrius T. Paris, F. Kenneth Hard, McGraw-Hill Book Company, 1969, Library of Congress CCN 68-8775, ISBN 070484708.
5. Sequential Thematic Organization of Publications (STOP): How to Achieve Coherence in Proposals and Reports, Hughes Aircraft Company Ground Systems Group, Fullerton, Calif., J. R. Tracey, D. E. Rugh, W. S. Starkey, Information Media Dept., ID 65-10-10 52092, January 1965.
6. Structured Analysis and System Specification, Tom DeMarco, Englewood Cliffs, NJ: Yourdon Press, 1978, ISBN 0917072073.
7. Sustainable Development Possible with Creative System Engineering, Walter Sobkiw, 2008, ISBN 0615216307.

5 The Process, Method, Methodology

DeMarco[28] provides a definition of *methodology*: "A general systems theory of how a whole class of thought-intensive work ought to be conducted." A *technique,* on the other hand, may be regarded as less encompassing or comprehensive than a methodology. A methodology is comprised of one or more techniques, together with concept or theory that makes it cohesive[29].

Process, method, and methodology terminology is used frequently. Many use process and method interchangeably and state that they are a series of steps to accomplish a task. Many use the term methodology to house the description of the approach to be used in an analysis. Methodology is usually viewed as something that is static and does not change.

For this text a process consists of multiple methods. Some will use methodology to represent the collection of methods to accomplish a task. I prefer to use process in place of methodology because a process suggests feedback loops, automation using tools where practical, and change. In systems practices it is critical that feedback loops and iteration exist in the thing (the process) that describes how to perform systems practices for a selected system.

Process is not static. It is like a living thing changing and evolving as other things in the process setting change. Organizations change because they also display the attributes of a living thing, continually changing and evolving. Organization and process affect each other, they cannot be created in isolation. For example the process might state that the test team must be independent. The implications of that statement are that the test team reports to an external authority[30].

When offering a system solution the team needs to identify how they will create[31] that system. They need to identify the information products that are produced, how they tie together, what tools and techniques will be used or created, and what the issues are in building the system. It does not matter if it is a re-spin of something that has existed for 100 years or building something new like going to

---

[28] Peopleware, DeMarco, Tom, Yourdon Press, 1987. Second edition, Dorset House Publishing Company, Inc, ISBN: 0932633439

[29] Sandia Software Guidelines Volume 5; Tools, Techniques, and Methodologies; Sandia Report, SAND85–2348 I UC–32, Reprinted September 1992.

[30] The question is how independent: company president, division president, physical location general manager, engineering head, business area manager, or project manager.

[31] Some suggest that create is not appropriate, instead design should be used. Create is used here to represent design and art in the broadest sense possible, not just some limited phase of engineering or artistic pursuit.

the moon for the first time.

Day one identify how the job will tackled, write it down, start doing it, if it does not work or has problems, modify it, then change the written description of the process. The trick is to not do this in isolation. Every stakeholder needs to own the process. They have to agree to the process, love the process, and make it work. It is not the job of management to force a process on those that will implement the process. However management is an equal stakeholder in the process. The best way to make management a stakeholder in the process is to give them technical tasks that feed the process.

5.1 Idealized Process Description

The process discussion could start with a general process that some refer to as the waterfall model. It has its roots in environments where change is prohibitively costly or impossible. Examples are manufacturing and construction. The implication is that there is no feedback mechanism for change beyond perhaps just the previous level.

Although this may be a simplified idealization of a process, the reality is that these steps are iterated with different feedback loops to any other level. For example before there is a final production facility, there are multiple pre-production prototypes and test production runs. Before there is a building or bridge there are several mockups and scale models before construction begins. Many early software practitioners coming from computer science backgrounds were not exposed to these engineering principals and have misunderstood this simplified ideal process[32].

Engineering disciplines use ideal representations to support engineering analysis that is based on logical and mathematical representations. This same concept applies to the waterfall process representation. It is an idealized representation of a more complex world.

The key issue associated with this model is the number of iterations. In the case of a physical building or manufacturing line there is no degradation with iteration. The iterations are performed on elements not part of the final solution; they are prototypes, mockups, and scale models. With software, the iterations usually happen on the final product unless the team is willing to go through all the code and create a final clean version[33]. The problem is this code will eventually become less maintainable, stable, and flexible with successive iteration. This was known by the late 1970's, with software based systems that evolved through the years and that were then believed to be unstable, inflexible, and difficult to maintain. So in the

[32] Winston W. Royce recognized the issues and proposed a modified waterfall model for software. Managing The Development Of Large Software Systems, Winston W. Royce, Technical Papers of Western Electronic Show and Convention (WesCon), Los Angeles, USA, 1970.

[33] The spiral model suggests multiple iteration based on prototypes.

case of software intensive systems there is a process dilemma that is more significant than other systems[34].

Recently software practitioners have adopted other models, which basically represent passing through the waterfall model several times. With each pass through there is the risk that the code can become less stable as additions are attached to earlier versions of the design rather than considered holistically at the early stages of the effort. This is a challenge because using the waterfall model in its idealized state adds other complications that some suggest have led to significant software project failures.

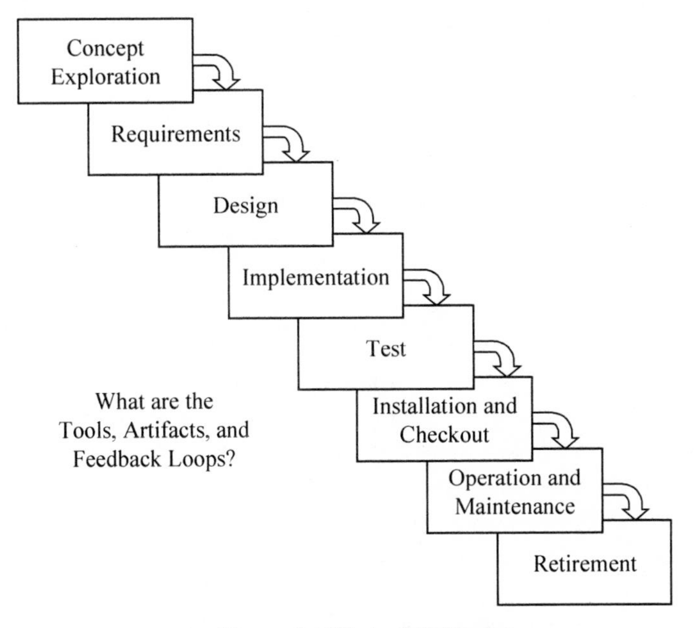

Figure 14 Waterfall Model

So, the team must be able to create process, methods, methodology and it must match the proposed system solution. Not only are the sequence of activities important, but also the tools and artifacts are key to the process. The tools are unique to each domain and change with time. The process is not complete until all the tools are identified and related to the process elements. Also the process is not complete until all the artifacts are identified and related to the process elements. Finally the feedback loops must be defined.

5.2 Small Project

The following is offered as a quick start guide for a small system. It is based on

[34] This was a major reason the Federal Aviation Administration (FAA) used in the 1980's to justify the Advanced Automation System (AAS) Program.

small projects. It is an iterative process. Each process is unique and bound to a solution and organization. These are broad suggestions for consideration.

- Define the problem
- Identify the goals and objectives
- Collect data and information
- Define the process or methodology
- Perform system analysis and modeling
- Identify evaluation criteria
- Develop alternative solutions
- Evaluate alternatives against evaluation criteria
- Finalize selected alternative
- Implement and test
    - If needed change evaluation criteria
    - If needed reevaluate alternative as needed
- Final solution – a prototype

5.3 Large Scale Systems

The following is offered as quick start guide for very large-scale systems. This is based on a process used at Hughes Aircraft Fullerton[35]. The applications were air defense and air traffic control. These systems were delivered around the world in various physical and cultural environments. They were and are some of the most complex systems known to humanity. Each process is unique and bound to a solution and organization. These are broad suggestions for consideration.

- Identify system users, maintainers, and managers
- Identify system boundary and why it is the boundary
- Identify key requirements and why they are selected (top 10)
- Identify key issues and why they are selected (top 10)
- Identify alternatives and why they are selected (4 to 9)
- Draw picture of each alternative (1 page)
- Succinctly state essence of each alternative (1 page)
- Tradeoff each alternative (include support for 100 years, etc)
- Apply science and engineering to each spin of tradeoff (don't cook the books or lie to yourselves)
- Identify how each approach can be built (process, methods, tools, etc)

---

[35] This process was suggested by Walt Sobkiw for sustainable development: Sustainable Development Possible Sustainable Development Possible with Creative System Engineering, Walter Sobkiw, 2008, ISBN 0615216307. I had the privilege of being introduced to systems engineering at Hughes Aircraft circa 1980's.

- Build proof of concept prototype (from paper to computer simulations to physical models)
- Mature prototype at a small operational setting (try it before you make a mistake)
- Roll out solution to the infrastructure (slowly and learn)
- There is beauty in diversity (nothing wrong with multiple approaches and companies)
- Do this like your life depends on it

5.4 Sustainable Development

The following process is offered as a quick start guide for sustainable development programs and projects. This process is focused on the front end[36] to emphasize the importance of applying systems practices to sustainable development rather than magic products. Each process is unique and bound to a solution and organization. These are broad suggestions for consideration.

Plan Project

- Define scope, objectives and constraints
- Establish project structure, schedules, roles responsibilities
- Prepare a Systems Engineering Plan

Define Stakeholder Requirements

- Identify stakeholders
- Elicit gather infer requirements and constraints
- Establish measures of effectiveness
- Generate and model alternative scenarios operational concepts

Analyze Requirements

- Identify system boundary and architectural constraints
- Specify system functional and performance requirements
- Define verification criteria
- Maintain traceability to stakeholder requirements

Design Architecture

- Define alternative logical architectures
- Partition system functions and allocate or derive requirements
- Evaluate alternative designs
- Maintain traceability to system requirements

---

[36] Offered by Dr. Peter Scott for sustainable system senior design projects at University of Pennsylvania. Projects proposed by Walter Sobkiw and Dr. Scott (2008 to 2011).

- Define verification strategy

<u>**Assess Project**</u>

- Review progress and status
- Update and verify the Systems Engineering Plan

5.5 Life Cycle Based

Many times the process is described in terms of the system life cycle where the system starts its life with the articulation of the system goals and missions. These representations are usually tightly bound to the team and what they think is important for the process.

- Operational Concept
- Requirements Analysis
- Functional Analysis
- Performance Analysis
- Architecture Synthesis
- Design
- Implementation
- Test
- Validation
- Maintenance
- Decommissioning
- Disposal

In this case the emphasis is on requirements, functional, and performance analysis. However these are just placeholders for the analysis that will eventually unfold as described further in this text.

5.6 Summary

The following is offered as a summary of many front-end system processes. Transparency is key because the problem is bigger than a single Human Mind!

- Identify Stakeholders
- Surface Requirements
- Identify Alternatives
- Perform Modeling
- Identify Tradeoff Criteria
- Develop Life Cycle Costing
- Perform Tradeoff with MOE
- Select Best Approach
- Optimize Selected Approach

5.7 Systems Fundamentals

There are many elements to systems. The following is offered as fundamentals to systems processes.

- Identify Stakeholders
- Gather Requirements
- Identify Trade Off Studies
- Be Transparent
- Be Holistic
- Based on Science, Engineering,

- Identify Architecture Alternatives
- Select Optimum Architecture
- Design, Implement, Verify, Validate
- Maintain, Upgrade, Transition
- Decommission, Dispose

And Art

- Develop Full Life Cycle Costs

5.8 Where is the Heart?

There is a theme starting to surface relative to systems and system process. However, where is the heart of the systems process? The heart is a function of the project and organization. It does change from system to system.

- Stake Holders
- Requirements
- Trade Off Studies
- Architecture Alternatives
- Architecture Select
- Design, Implement, Verify, Validate
- Maintain, Upgrade, Transition
- Decommission, Dispose
- Key Issues
- Key Requirements
- Modeling
- Simulation
- Configuration Management

5.9 Value Systems

Eventually decisions must be made. These decisions are based on values. Most approach this part of the process from a financial point of view. However, value is more than just financial. Some value a day at the beach more than a day at the mountains. This can be captured with the Measure of Effectiveness[37] (MOE) where: MOE = sum of tradeoff criteria / total cost.

Table 2 Financial and Non-Financial Value Systems

Lowest Cost	EVA - Economic Value Added
Highest Profit	ROV - Real Option Value
ROI - Greatest Return on Investment	ROA - Return on Assets
TCO - Total Cost of Ownership	ROIE - Return on Infrastructure Employed
IRR - Internal Rate of Return	MOE - Measure of Effectiveness

5.10 Cost Analysis

The cost analysis needs to include all the costs. Especially indirect costs shifted to other systems and stakeholders. The Life Cycle Costs (LCC) include: LCC =

---

[37] MOE was used extensively at Hughes Aircraft to select architectures. The MOE is discussed extensively further in this text.

R+D+P+O+M+W+S+T, Where:

R = Research
D = Development
P = Production
O = Operation
M = Maintenance
W = Waste
S = Shut Down and Decommissioning
T = Disposal

An alternative view is: LCC = R+D+P+O+M+S+I+E or PROMISED Where:

R = Research
D = Development
P = Production
O = Operation
M = Maintenance
S = Shut Down Decommissioning & Disposal
I = Infrastructure
E = Environment & Waste

5.11 Measure of Effectiveness

Measure of effectiveness is a critical measure used in the last century to evaluate systems and make major decisions associated with extremely large projects. It is a measure of goodness for each dollar spent. Measure of Effectiveness (MOE) = sum of tradeoff criteria / total cost.

5.12 Systems Practices and Development Phases

All efforts can be divided into two broad phases. There is a front-end phase where the initial system is being conceived. An architecture concept is being developed during the front end. The other is back end where the architecture is converted to a working reality. Within the back end systems engineering phase, there is a design and implementation phase where the system vision or architecture is converted to a physical implementation. Then there is the test phase where the system is verified against the original system vision and validated in a real operational setting.

5.13 Front End Systems Practices

During front-end systems engineering the emphasis is on developing the system architecture. There are several artifacts that may be appropriate to the project or program and the organization. The more important information products are:

- Operational Concepts
- Studies, Prototypes, Models
- High Level Architecture Drawing and Description
- Key Requirements

Some suggest that systems engineering stops after front-end activities, however that is not the case. Systems practices are used throughout the life cycle.

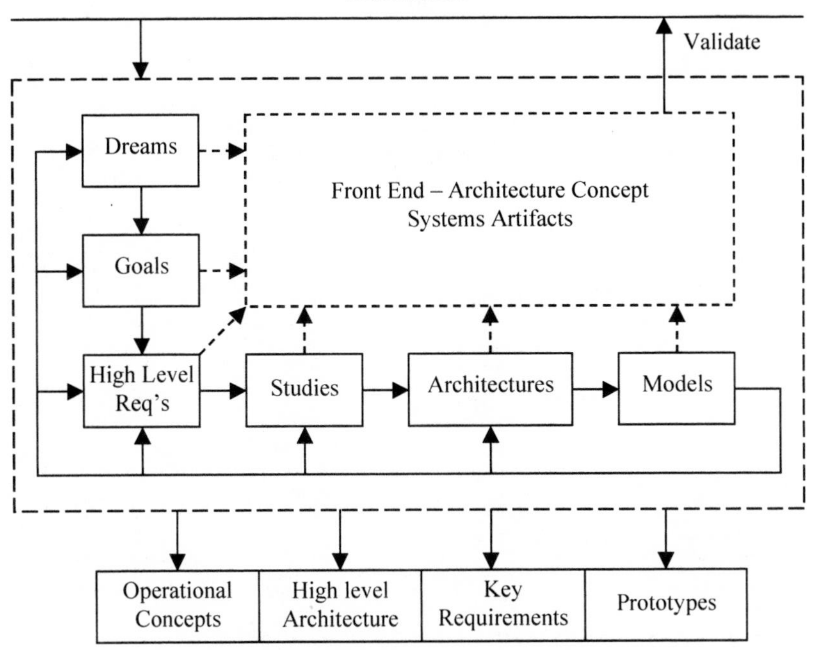

Figure 15 Front End Systems Practices

5.14 Middle Systems Practices

During middle systems engineering the emphasis is on designing and implementing the system architecture. The studies and plans have significantly slowed down or have been completed. There are several artifacts that may be appropriate to the project or program and the organization. The more important information products are:

- Detailed Architecture Description
- Specifications
- Design Documents
- Test, Verification and Validation Procedures
- Training Manuals
- Maintenance Manuals

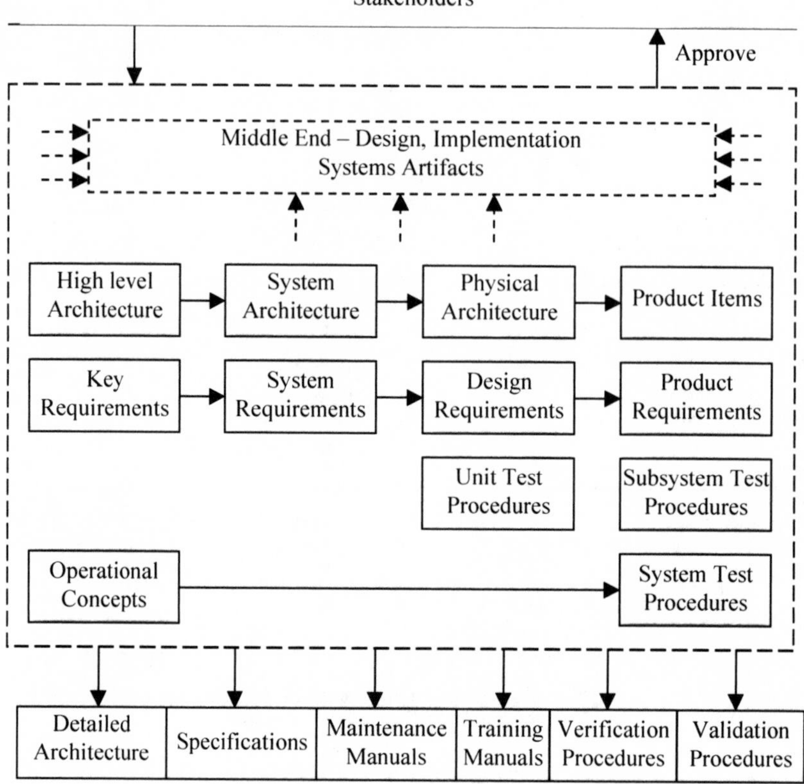

Figure 16 Middle Systems Practices

5.15 Back End Systems Practices

During back end systems Engineering the emphasis is on testing, verifying, validating, installing, and taking the system to a full up operational condition. The more important information products are:

- Test, Verification, and Validation Results
- Training Artifacts
- Installation and Maintenance Manuals
- Offline and Shared Operations
- Certification That The System Can Go Live
- Operational Switchover

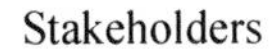

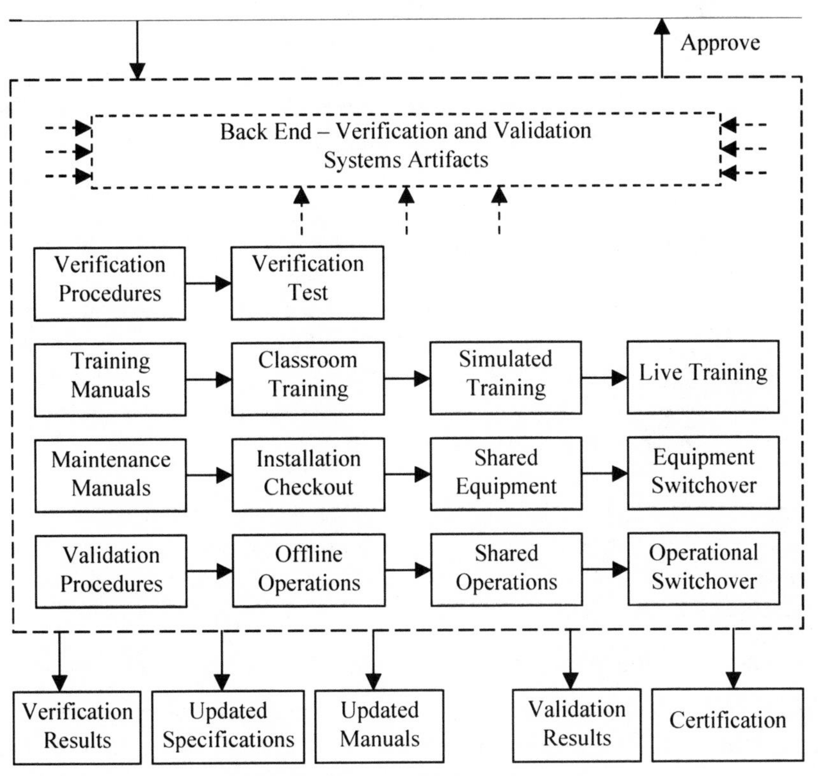

Figure 17 Back end Systems Practices

5.16 Capability Maturity Models

Some point to a failure of process when studying failed projects or programs. In the early 1980's the concept of capability maturity models[38] surfaced[39] and some think that an organization must be at a certain capability maturity level for certain efforts.

There are fundamentals when assessing an organization. Different assessment organizations may have different names for each maturity level, but they tend to fall into the following levels of consciousness:

1. **Chaotic**: Everything is random, you don't know what anyone will do next,

---

[38] Founded 1984 Carnegie Mellon Software Engineering Institute (SEI) federally funded research and development center.

[39] Managing the Software Process, Watts S. Humphrey, Addison-Wesley Professional, 1989, ISBN 0201180952.

things are very political, lots of liars and bullshit artists[40], and honest folks are afraid, management prays on the innocent.

2. **Repeatable**: Things tend to get done in a repeatable way, everyone knows what everyone else is doing and how their part fits into the collective.
3. **Documented**: You have documented processes that people follow and the documents are maintained, in other words they always represent more or less what is happening no matter the time of the year or what year it happens to be.
4. **Measurable**: You actually start to measure stuff that matters, everyone knows those measures and are not afraid of the measures because everyone is enlightened enough to not hurt anyone in the group, so it's not about shallow productivity asset stripping schemes by poor management.
5. **Predictable**: Based on your metrics you evaluate new techniques, technologies, tools and you are able to make qualitative and quantitative predictions about their introduction into the process.

These basic principles apply to all organizations and all time frames. They are not hundreds of pages, they do not take special training, and they do not use armies of strangers looking at a proprietary process, the billion-dollar process that separates the organization from everyone else in the world.

Example 2 Solution and Process are Inseparable

Assume you are in pre proposal phase or you are conceptualizing a new product. When your technical staff prepares their technical presentation, have them identify how they think the job will be tackled. If they can't identify how the job will be done to build this new thing they are proposing then maybe the solution is not viable. So the two are tied together - the technical solution and can it be built using either current approaches or a new equally break through approach. Perhaps new tools need to be developed. Perhaps new technology needs to be matured and converted to a tool or set of tools. Perhaps there are feedback loops that most consider to costly.

The reality is breakthrough solutions also require break through processes: methods, tools, techniques, etc. It's not about Gantt charts from project management worlds.

Example 3 Process and Practitioners are Inseparable

If you document a process that works and is proven over years of operation you cannot move it wholesale to a new organization. That process is bound to that organization. The reality is if an organization has anything of value its people transition from apprentice to journeymen to master over the span of years. It takes

---

[40] On Bullshit, from Princeton professor Harry G. Frankfurt, Winner of the 2005 Bestseller Awards, Philosophy Category.

anywhere from 2-5 years to transition from apprentice to journeymen. So no matter how detailed the process description, everyone new to the process is an apprentice regardless of experience in other organizations. This is an organization that has real value to offer - for example top three in an industry where there are many players. That is why when organizations document their processes then attempt to replace staff they self-destruct. So why should you take the time to document your processes? It is about bringing in new apprentice staff and getting them on board as fast as possible and it is about visualizing the operation so that continual improvement can follow new technologies, products, and techniques.

5.17 Organizational Capabilities

An organization's process capability can be examined and assessed. However that is not a complete view of the organization. This becomes very apparent when an organization is subjected to a proposal or vendor selection process. Typically the selection process includes the technical approach, company assessment, and cost. The company assessment includes items such as:

- Years in business
- Size of company and project size relative to company
- Company financial stability
- Company employee turn over rate and layoff history

When a company is viewed from its technical ability to take on a particular job with a particular set of needs, its capabilities are reviewed. This obviously includes listing the unique products and technologies the company offers and its match with the need. There is however a generic approach to view a companies capabilities. The following are ordered lists of company capabilities from the most capable to the least capable:

- Develops and owns successful significant technologies
- Develops and owns successful product lines
- Develops and owns successful products
- Delivers successful projects
- Has a collection of existing people willing to work
- Has a vessel that can employ a collection of people
- Owns its land buildings and equipment
- Leases its buildings owns its equipment
- Leases its buildings leases its equipment

5.18 Vetting and Pedigree

Many systems need to show a pedigree to support activities such as certification by external entities. This is normally accomplished with information products or documents. The documents include plans, studies, white papers, analysis, published papers, specifications, test reports etc. This is typically referred to as documented evidence.

The information products need to be properly vetted for accuracy, completeness, and consistency. All claims need to be supported with tangible evidence. There should be no over reaching blind faith conclusions that are not really supported.

To support these needs, an internal process is used to fully vet the information products. This is typically called a review. This is accomplished by using senior individuals with guidelines that help identify potential problems for classes of information products.

For example if a paper is submitted to a conference, the paper needs to be submitted to internal impartial third party reviewers within the organization. Their task is to ensure that proprietary data does not leak from the organization and that the paper is a valid work with no unsubstantiated claims that can damage the reputation of the organization and the individuals.

If an information product is destined for a customer the same level of review is needed with the same goals of ensuring the body of work is correct. In this case the reputation of the organization and individuals is extremely important because it directly protects the business and helps in winning future proposals.

If there is a failure and an information product leaves the organization that is seriously flawed, it is a serious process failure and not the sole fault of the original authors and contributors. The organization seriously lacks something important in the review and vetting process.

5.19 Enterprise Process

The enterprise process is the universal process that exists for an entire organization. Both the process and the organization support all the programs, projects, product lines, products, and technologies. It is generic in nature and allows unique process adjustments for each solution offered by the organization.

Process should never be developed or forced from the top down. A project or program team needs to study the available set of processes[41] and select what best fits their unique needs. That may include developing new process elements.

Once a process is used, tuned, and hardened on a successful project, then it needs to be submitted to the enterprise for possible inclusion in the enterprise process. If this is the first time an enterprise is doing something, then there is no enterprise process until the first successful project or product effort is complete.

At no time should enterprise processes be accepted from a source and offered from the enterprise level down to the projects or program until they have been

[41] New personnel may arrogantly conclude the enterprise process does not exist, is inadequate, flawed, or too costly. This can lead to disaster.

proven on a selected project in the organization[42].

5.20 Exercises

1. Do you think a single universal process can be codified that can be used to develop any system?
2. Pick a project and define the process at the highest level to execute the project.
3. Do you think you need to be a process expert to engage in serious systems engineer? If so why? If not why not?
4. What is your new business development process? How does it compare with other students? How does it compare with systems engineering or development processes? Keep your answers for future comparison you will be asked these questions again.
5. What is your research and development process? How does it compare with other students? How does it compare with systems engineering or development processes? Keep your answers for future comparison you will be asked these questions again.
6. Is it possible to create a process for a high technology organization so that people who have no education, experience, or background can successfully execute the process? What about a low-tech organization?
7. How can compensation incentives be changed to force people to execute an accepted process? What can you do if the compensation incentives encourage departure or bypassing of standard process elements?

5.21 Additional Reading

1. Managing the Software Process, Watts S. Humphrey, Addison-Wesley Professional, 1989, ISBN 0201180952.
2. On Bullshit, Harry G. Frankfurt, Princeton University Press, January 2005, ISBN 9780691122946.
3. Peopleware, Tom DeMarco, Yourdon Press, 1987. Second edition, Dorset House Publishing Company, Inc, ISBN 0932633439.
4. Sandia Software Guidelines Volume 5; Tools, Techniques, and Methodologies; Sandia Report, SAND85–2348 I UC–32, Reprinted September 1992.
5. Sustainable Development Possible Sustainable Development Possible with Creative System Engineering, Walter Sobkiw, 2008, ISBN 0615216307.
6. Technical Papers of Western Electronic Show and Convention (WesCon), Winston W. Royce, Los Angeles, USA, 1970.

[42] Developing process in an ivory tower will lead to disaster.

6 Stakeholders and System Participants

The term stakeholder is relatively new and comes from the project management world[43]. The idea is that project success is based on the acceptability of all that have a vested interest in the project. The traditional systems engineering viewpoint was to consider all those who will interact with the system. This is a subtle but important distinction. For example what is the role of a company president in the operation of an infrastructure system like Air Traffic Control. Yet the roles of air traffic controllers, pilots, maintainers, and system managers immediately start to become apparent. The old terms were system users, maintainers, and operational managers.

So the dilemma is to make sure that at the start of a project the stakeholders in the system are identified and separated. There are stakeholders that may have the ability to kill a project but they will never interact with the system. The system engineering team should not get lost in the noise of project management issues and instead must focus on identifying the true stakeholders in the system. One way to do this is to identify all the system external interfaces. Most individuals normally identify people interacting with a system as an external interface. As the system evolves, it becomes apparent that many of these people interacting with the system are a part of the system and they get moved back into the system boundary to become internal interfaces interacting with a machine subsystem or other people.

6.1 Possible Stakeholders

The following is a generic list of stakeholders. It can be used to stimulate discussion and identify stakeholders for the system being developed. Each system is unique and it is critical that the system stakeholders be identified.

- Customers
- Competitors
- Markets
- Sponsors
- Visionaries
- Prototypes
- Previous Systems
- Interfacing Systems
- Policy Makers
- Operators
- Users
- Maintainers
- Administrators
- Regulators
- Producers
- Providers

---

43 Strategic Management: A stakeholder approach, Freeman, R. Edward, Pitman Publishing, 1984, ISBN 0273019139.

- Investors
- System Opponents

Once the stakeholders are identified, the needs for each stakeholder are understood and clearly stated. The needs are then converted to key requirements. This is the start of the top-level requirements for the system.

Example 4 Stakeholder Representations

One way to begin the process of identifying the stakeholders is to list the high-level system requirements and associate them with a stakeholder. As the stakeholder list starts to unfold, other stakeholders start to surface. The requirements of the new stakeholders are then identified. This is a chicken and egg sequence and is not complete until the analysts exhaust their list of stakeholders and their requirements.

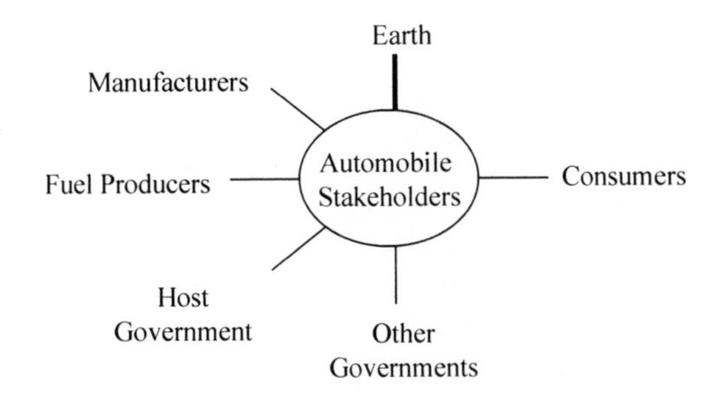

Figure 18 Less Obvious Stakeholders

A stakeholder diagram looks similar to a system context diagram but it shows stakeholders as external interfaces to system. This diagram can exist in parallel with the system context diagram, or if the stakeholders are external interfaces, they can be shown grouped together on the context diagram.

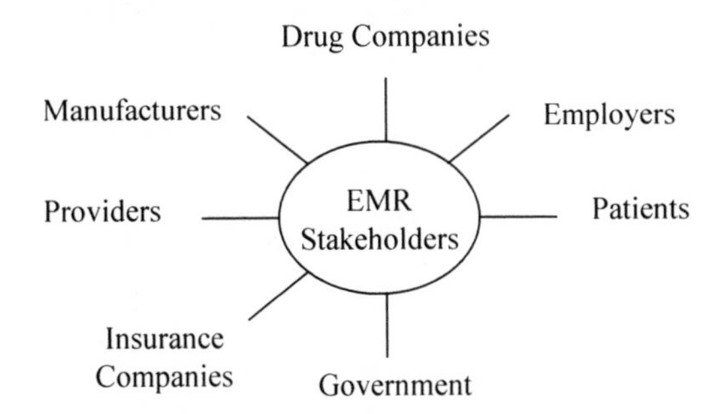

Figure 19 Precisely Defined Stakeholders

It is easy to consider people as stakeholders in a system. But are there other stakeholders and do they matter? In early 1970's people became aware of their impact on the planet[44].

When identifying the stakeholders care should be taken to use precise words to represent a group of stakeholders. This can be a topic of significant debate.

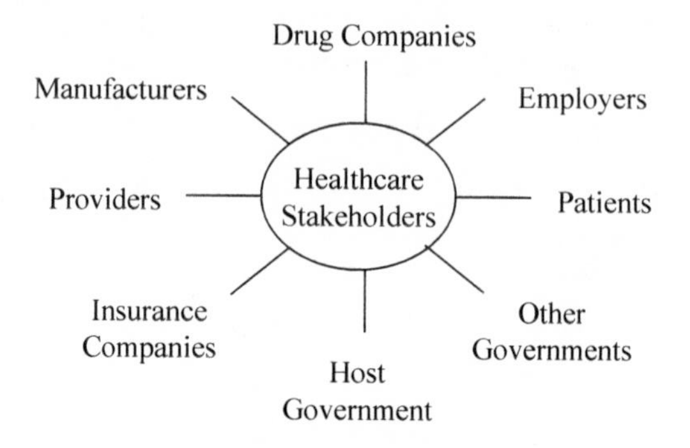

Figure 20 Groups and Benefiting Stakeholders

Sometimes it is easy to group the stakeholders because they share similar interests. Sometimes some stakeholders do something to benefit other stakeholders. Placing these stakeholders strategically on a diagram may help to aid the discussion. Initially those relationships may be unclear. However, as the stakeholders are identified and understood there may be benefit to show a left right relationship or other special representations of the stakeholder groups in the diagram.

Many times some stakeholders have divergent requirements. For example health insurance companies have the primary requirements of increasing earnings and profits while the policyholders have the primary requirements for good health and long life.

Figure 21 Competing Stakeholders

These are divergent requirements. Some will attempt to argue that maximizing earnings and profits in a health insurance company will result in maximum health and long life to the customers, but there is no analysis that can show this linkage.

[44] United States Senator Gaylord Nelson (Democrat) started Earth Day on April 22, 1970 to raise awareness and appreciation for the Earth's natural environment. President Richard Nixon (Republican) proposed the EPA, which began operation on December 2, 1970.

6.2 Addressing Divergent Stakeholder Needs

Sometimes the needs and requirements match between stakeholders. At other times the needs and requirements are complementary. Yet at other times the needs and requirements are conflicting. With conflicting needs and requirements, choices must be made, however if the conflicting needs and requirements are reasonable they may be carried forward as the system is studied and an architecture concept is produced. During this process it may be possible to find a solution that addresses the divergent stakeholders. However, the system designers must not be delusional. If a stakeholder must be removed from the system or modified in such a way that the divergence is not present, then that must be the solution that is offered otherwise the system will be broken.

6.3 Technical Director

When stakeholders are excluded from the systems effort the entire effort is catastrophically compromised. It is no longer a systems driven activity, it becomes a management driven activity. What is buried in this concept is the idea of who drives the systems activity. Should the technical director be one of the stakeholders or someone outside the stakeholder set? Should the technical director be a subject matter expert or a technologist?

Systems efforts are always compromised if the system driver is a subject matter expert (SME) who has risen through the ranks or a technologist who has risen through the ranks. That does not mean that SMEs and technologists cannot make good technical directors. It just means that special care is needed when selecting someone to represent and maintain the systems perspective. The reality is that if the technologists[45] and SMEs[46] ignore each other then the system approach is compromised[47].

Author Comment: One of the most important stakeholders are the innovators and creative people who can punch through to new levels of thinking. If you are using systems practices and there are no innovative creative voices then you will fail or at most develop a marginally successful system. Many managers with no background in systems or experience only in low technology established settings view multiple stakeholders and innovators as a distraction. They will try to remove dissenting opinion and believe they have successfully managed a team to closure and removed risk.

---

[45] Circa 1985 many Hughes technologists were upset that air traffic controllers (SME) were calling all the shots.

[46] Circa 1981 many air traffic controllers (SME) were upset that technologists were calling all the shots.

[47] The Fires: How a Computer Formula, Big Ideas, and the Best of Intentions Burned Down New York City-and Determined the Future of Cities, Joe Flood, May 27 2010, ISBN 1594488983

6.4 Exercises

1. Can the Earth be a stakeholder or is a group of people who speak for Earth the stakeholders?
2. Can animals, plants, and bacteria be stakeholders?
3. Identify the stakeholders in a village, town, city, or nation state. Draw a stakeholder diagram and make a stakeholder list that categorizes the stakeholders.
4. Identify the stakeholders for a system you are considering. Draw a stakeholder diagram and make a stakeholder list that categorizes the stakeholders.
5. Identify the needs of each stakeholder in your system.
6. Identify the key requirements of each stakeholder in your system.
7. Are there any divergent needs and key requirements in your system? If no are you sure you are capturing all the needs of your stakeholders? Have you missed any stakeholders?
8. What can you do if your stakeholders are in constant disagreement and unable to converge?
9. What can you do if your stakeholders are preventing the small group or individual innovator / genius from rising and punching your system through current levels to a whole new level that will make it all great?
10. What can you do if management locks out stakeholders?
11. What can you do if management locks out innovators and creative stakeholders?

6.5 Additional Reading

1. Systems Engineering Handbook, A What To Guide For All SE Practitioners, International Council on Systems Engineering (INCOSE) INCOSE-TP-2003-016-02, Version 2a, 1 June 2004.

7 System Boundary

The start of all systems engineering efforts is to identify the system boundary. The system boundary can be shown with a context diagram. Without a context diagram you will never know what you are building. More importantly, no one else will know what you are trying to build. There is no common frame of reference from which to start communicating with other system designers. So, it is critical to establish the system context. This is the boundary that defines the system.

7.1 Context Diagram

Draw a square or a circle on a piece of paper. Put the name of the system in the circle. Label the inputs and outputs of your system. Explain why the system boundary is the system boundary. If you are given a system boundary, a context diagram, convince yourselves the system boundary is valid. If it is not the correct system boundary, then fix it. Always include the system boundary and its rationale with all discussions and presentations of the system architecture. It is the highest level view of a system.

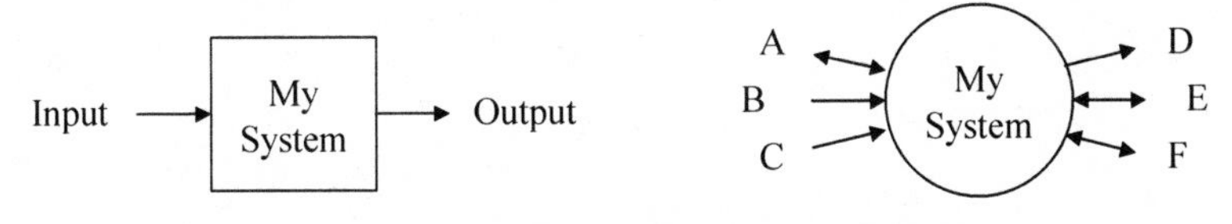

Figure 22 Generic Context Diagram

Your system is shown surrounded by external systems that interface to your system. These systems are not part of your system, but they interact with your system via its external interfaces. The external systems can impact your system, and your system does impact the external systems. They play a major role in establishing the requirements for your system. Systems further removed are those in your system's context that can impact your system but cannot be impacted by your system. These systems in your system's context are responsible for some of your system's requirements.

Your context diagram changes significantly during the early stages of a project as participants attempt to bound the system. At some point stability is achieved. That stability is based on various arguments. Those arguments must be captured and presented each time the context diagram is shown and the system boundary presented. At some point near the implementation phase someone eventually starts to think about a marketing flyer for the system. The context diagram always appears in some form in a good marketing flyer. Just as project participants needed to understand the boundary of the system, so do the clients who will buy, use, and maintain the system. So the context diagram never goes away.

Example 5 Context Diagram - Vacation House

Suppose you are building a vacation house. What would the system boundary of the vacation house show? Would the house use city sewer, cesspool, or an outhouse? Who would be interested in the decision? What about heating, will it be oil based? Will food be stored in the house or will everyday be an adventure to the local restaurants? Is the land an external interface? Will the house be built on flat dry land, a hillside, swampland, or on water?

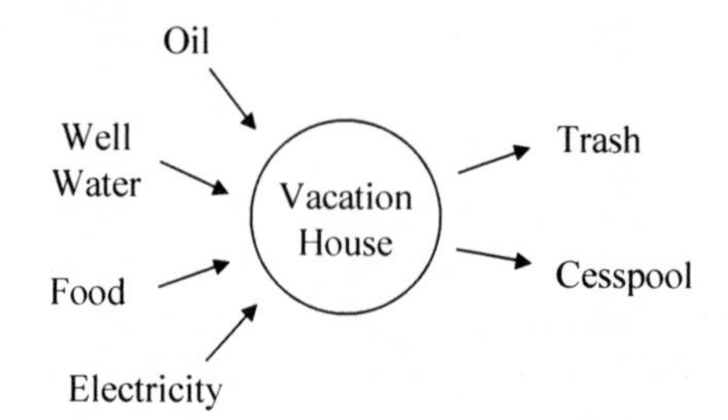

Figure 23 Context Diagram - Vacation House

Example 6 Context Diagram - Primary House

Suppose you are building your everyday house. How long will it take you to get to its final form? Is transition strategy important?

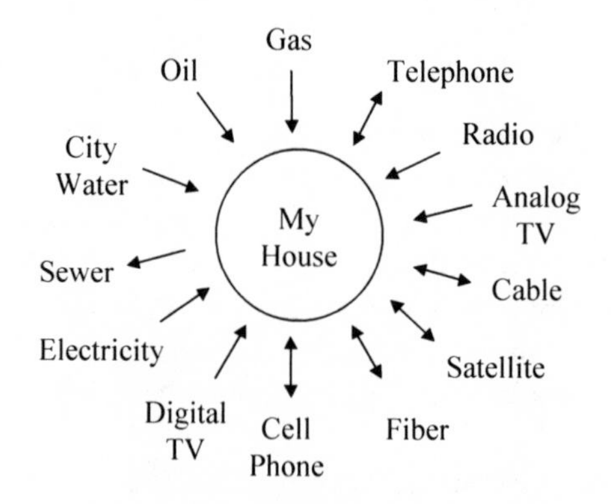

Figure 24 Context Diagram - My House

Do you need backup mechanisms for some services? For example is cell phone communications your only source of traditional voice communications or do you also need a telephone landline? Should the traditional telephone be self powered via the telephone line or rely on external power? That question is interesting because the argument may be associated with the internal functionality of the telephone system rather than the context diagram. However, if the telephone within the house needs power from the landline then the telephone external interface needs to provide that power source.

Is this all common sense and so a context diagram is not needed? Is an artist rendering of the house and land more appropriate with a mini spec sheet listing the rooms and their sizes better? Are both needed when trying to understand the house system boundary?

Example 7 Context Diagram - Other Houses

What if the house is on a cruise ship, on a space station, on Mars, or in a poor country? What if the house is an office building, industrial park, shopping mall, condominium complex, apartment buildings, or a Planned Urban Development (PUD) community?

The previous house context diagrams are potentially at a view that is lower than we may need at this time. For example, what is the house context in reference to the community? Is the house in a city, suburb, or rural area? What are the community elements that you in the house will interface with and interact? These elements might be schools, shopping, health care providers, hospitals, employment, houses of worship, restaurants, malls, industrial parks, air quality, water quality, crime, community involvement, etc. What are the walking distances, driving times, and driving distances? Is the community flat, hilly, wooded, or sandy? Where is the nearest mountain, lake, or ocean? What are the weather and the seasonal variations?

What is the style of house and is that really part of the house context? For example is the house a colonial, split level, bi-level, rancher, or other style? Is the house attached and is the attachment just another house in a twin arrangement or is it a row of townhouses? If the house is attached is there a firewall and is that part of the external interface or do you own the firewall? Is the house a collection of rooms in a shared building and is it on the 50th floor of a building?

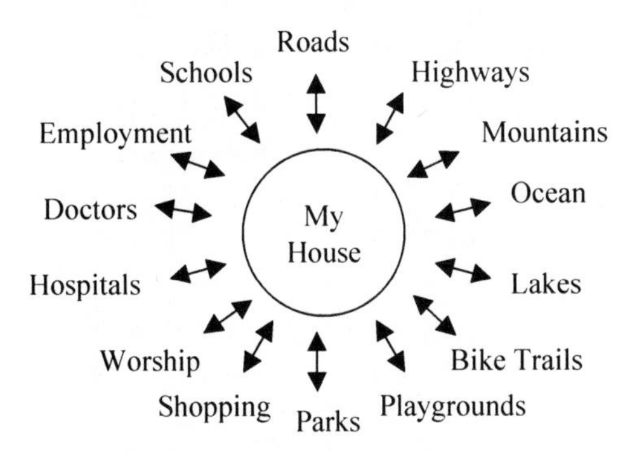

Figure 25 Context Diagram - My House in the Community

In many ways these elements are the connections that the house inhabitants share with the community. Does the community really facilitate these connections or make them difficult? So the house is affected by the community interface. Do you need to live near an airport? If so, add it to your context diagram.

Does it make sense to build houses where there are no schools, shopping, or industry? What is the impact on surrounding communities? Does it make sense for a community to not have a master plan or architecture vision if houses are being built with no supporting community elements? Is it ethical to operate in that fashion and do ethics and morals enter into systems solutions?

7.2 Concept Diagram

The system boundary also can be shown with a concept diagram. The system concept diagram contains pictorials, or icons, of the system and its interfaces. A simple picture readily recognizable by the team can represent each interface.

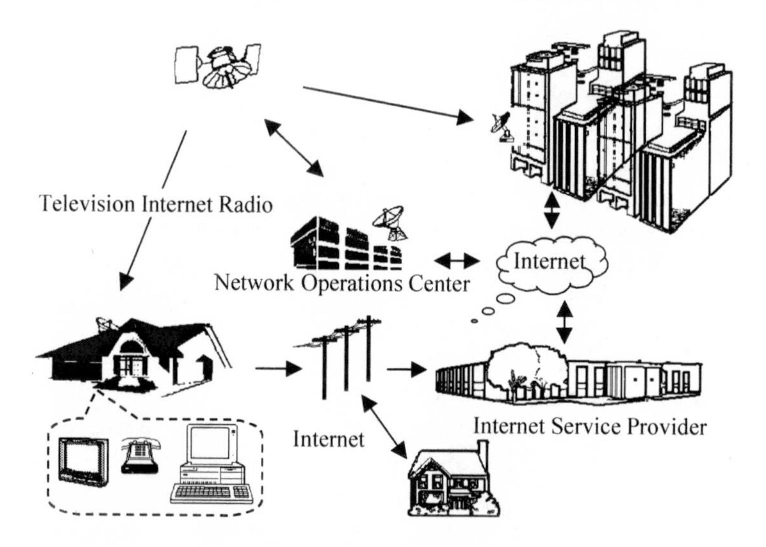

Figure 26 Satellite System Concept Diagram

One suggestion is to draw a system concept diagram by hand and scan the image. As the concept matures and is accepted someone within the organization will eventually clean up the picture. Another suggestion is to find a similar image in a library and modify it to support your needs. However, you may find yourself compromising on your vision, as your try to modify the picture and the modification is difficult and not exactly in line with your original vision. Do not compromise. Your vision is key and must be captured for all to see and understand.

In many ways this simple issue of being able to create a free form concept diagram, an artists rendering, is at the heart of many engineering issues today. Since so much has been created in the past 50 years, reuse has become common practice. However, as these systems have aged and become stressed new visions and concepts are needed to fill the void. This means someone somewhere must create the first picture. Are you working with artists that are trying to understand the vision?

Example 8 Interface Diagram - Stereo Receiver

Concepts can be conveyed with an Interface diagram. Suppose you are creating a new audio product, an amplifier and receiver or suppose you are shopping for an audio receiver for a new house that you are building for a client. What interfaces will the product support? Is satellite radio important? What about Internet radio? Notice they are both missing from this diagram. Is it important to show the individual channel speakers? Does the system support 5 channels or more? How many Reel-to-Reel Tape Decks will the system support and should they all appear in the diagram? Is a 4 channel Tape Deck different from a 2 channel tape deck?

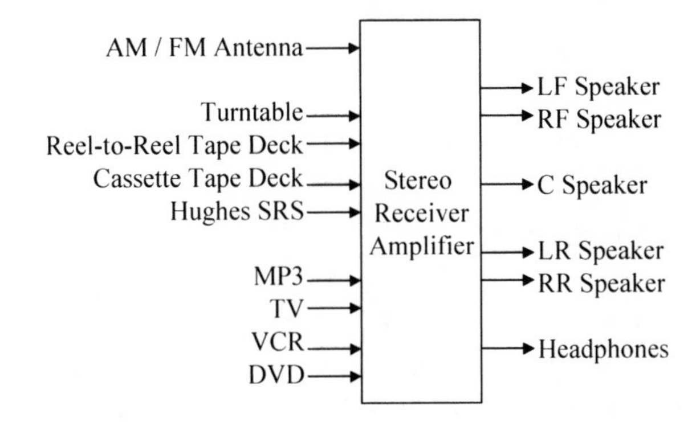

Figure 27 Interface Diagram - Stereo Receiver Amplifier

7.3 Level of Understanding

When attempting to define the system boundary consider what is inside the system boundary. This is the next level of analysis as the system onion is peeled. However, just because there is a jump into the next level of detail, it does not mean that level of detail is represented in the current view. This is an important concept. To have a reasonable representation of the current level, work must happen at the current level plus some aspects of the next level down in the system view. This means that without continuing to read the rest of this book there may be difficulty creating context diagrams and system concept diagrams.

For the next level down from the context diagram, identify the key system goals, functions, and requirements of the system. This is a list of 7 to 10 items that capture the essence of the system. When drawing the next level down diagram, list the key goals, functions, and requirements. Group the goals, functions, and requirements together so that stakeholders and system designers are not confused.

Example 9 Concept Diagram - Health Care

Some systems are physical like cities. It is difficult to make changes to a city if it is broken. It takes generations to build roads, utilities, and buildings. Other systems are more abstract and changes to optimize the system are not physical. Health care may fall into this category.

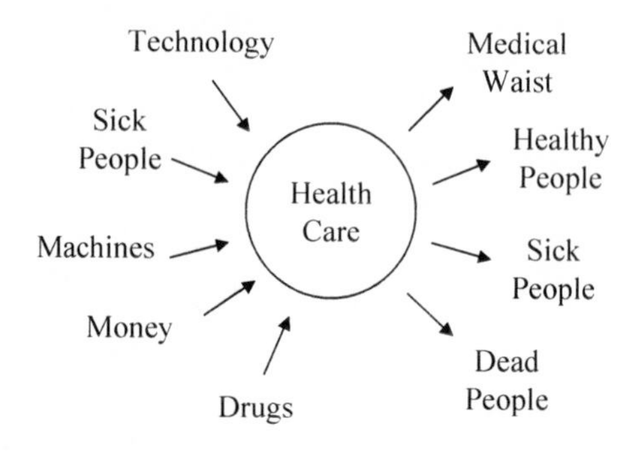

Figure 28 Context Diagram - Health Care

7.4 Human Elements

When identifying the system boundary the human elements need to be considered. Are the humans internal or external to the system? Many practitioners show humans as external system interfaces to emphasize the system significantly interacts with people and that the system will fail if this interface is not properly addressed. For example the Air Traffic Control system is highly dependent on the air traffic controller.

The term stakeholder is relatively new and comes from the project management world. The idea is that project success is based on the acceptability of all that have a vested interest in the project. However, a great system stands alone and must be great via its own virtues. What does this mean? It means that the system interaction with the humans must be considered from a functional point of view. For example, systems have users, maintainers, and managers. So from a systems engineering point of view the team should consider all those who interact with the system. This is a subtle but important distinction. For example what is the role of a company president in the operation of an infrastructure system like Air Traffic Control. Yet the roles of air traffic controllers, pilots, maintainers, and system managers immediately start to become apparent. Thus the humans to consider are system users, maintainers, and operational managers. The operational managers track the system metrics and adjust the system as needed.

So the dilemma is to make sure that at the start of a project the stakeholders in the system are identified and separated. There are stakeholders that may have the ability to kill a project but they will never interact with the system. The system engineering team should not get lost in the noise of project management issues and instead must focus on identifying the true stakeholders in the system.

One way to do this is to identify all the system external interfaces. Again, most individuals identify people interacting with a system as an external interface. As the system evolves, it becomes apparent that many of the people interacting with the system are a part of the system and they get moved back into the system boundary to become internal interfaces interacting with a machine subsystem or other people. It is irrelevant if a project stakeholder with extreme power "likes" the system but will never interact with the system. The system must function and perform properly or the consequences could be devastating.

Example 10 Context Diagram - Sustainable City

Do systems emerge or are they created? If a system has been observed to emerge can it be recreated? If recreated will it be better or worse than the previous system? Can a system like Air Traffic Control that took decades to evolve be recreated using new technology in 5 years? These are some of the questions that surface when we think about cities, villages, nation states, and the planet.

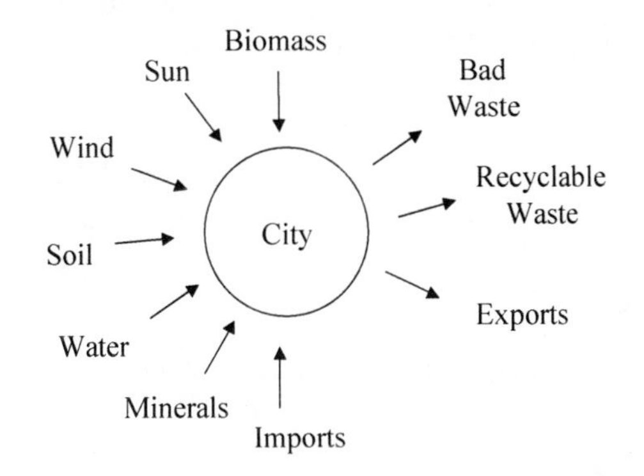

Figure 29 Context Diagram - City, Village, Nation State

If you were asked to create a new sustainable city how would you approach the problem? What elements would you consider? After World War II people went about to build new cities. Some of these new cities work and include diverse housing for all, schools, municipal facilities, parks, libraries, shopping, industrial parks, varied forms of transportation, houses of worship and many other things most people take for granted. Other new cities were poorly planned, especially in the last 2 decades of the last century. There is no shopping, industry, or even effective transportation. These cities had no plans, no system visions, and no common sense.

Example 11 Context Diagram - Sustainable Power

Today we take electric power for granted in many countries. However it was less that 100 years ago that most people lived in the dark. Rural electrification[48] and the construction of hydroelectric dams[49] transformed the USA. No one should take this massive system for granted.

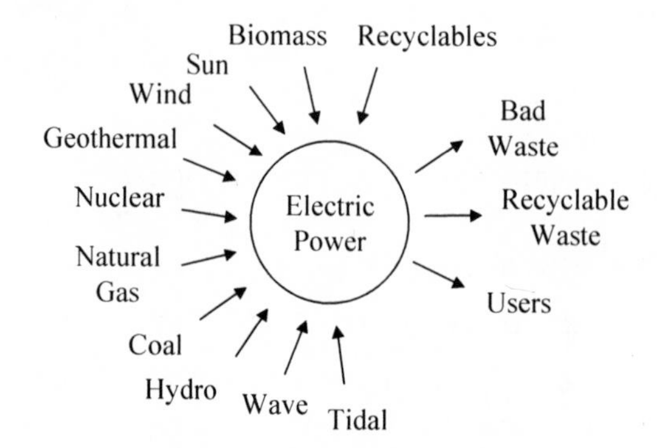

Figure 30 Context Diagram - Electric Power

So what is the system boundary of the electric power generation system? What are the outputs of the system? Is one of the outputs users or megawatts per hour? What are the issues associated with the users? Do they need power when they are awake or sleeping during the night? Designing a system for peak load applies everywhere. What is the peak hour for a telephone system? Are the peak load characteristics for a telephone system the same as for electric power? Should you design for a peak or use regulations or incentives to restrict use and level the peak load? Is peak load inherent in the universe and does it make sense to redistribute your load?

Example 12 Context Diagram - Personal Transportation

Personal transportation transformed our world. There is no question that the automobile was key to allowing our modern world to emerge. Without the automobile less of us would be alive today. So given that the case for the automobile can be made, what is the context diagram of the automobile? Is the propulsion subsystem of the automobile included inside the automobile boundary or outside of the automobile context diagram? Does the context diagram represent a single automobile or hundreds of millions of automobiles? If it includes hundreds of millions of automobiles is it based on different geographic regions, or nation states, or both nations states and their geographic regions?

---

[48] President Franklin Delano Roosevelt issued Executive Order 7037 May 11, 1935 establishing the Rural Electrification Administration. Congress passed the Rural Electrification Act in 1936. The Rural Electrification Act offered low cost loans for electric power in rural areas. It was expanded in 1949 to include telephone services.

[49] On May 18, 1933, Congress passed the TVA Act. The Tennessee Valley Authority (TVA) a US federally owned corporation was created.

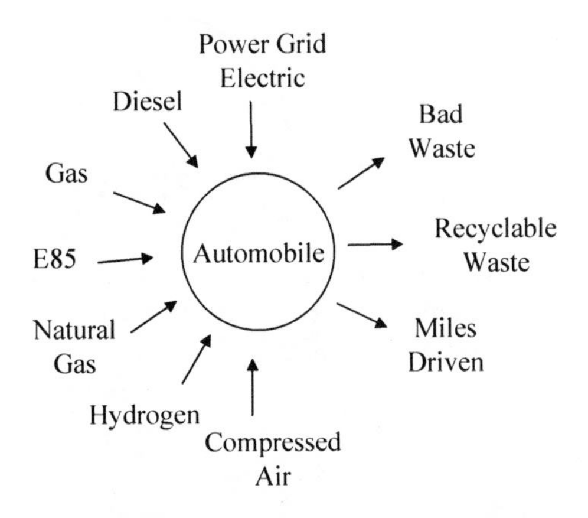

Figure 31 Context Diagram - Automobile

7.5 Exercises

1. Identify a system and create a context diagram for the system. Identify the system boundary and why it is the boundary by stating it on 1 presentation slide. Present it to your classmates and defend you stance or be prepared to modify your system boundary.
2. Identify the key goals, functions, and or requirements in your system.
3. Identify the stakeholder in your system. Separate the stakeholders that do not interact with the system but may influence the project in a negative way.
4. Review a classmates system boundary and context diagram prior to any formal class presentations. Be prepared to defend your critique with sound logic and reason.

7.6 Additional Reading

1. Integration Definition for Function Modeling (IDEF0), Federal Information Processing Standards Publication FIPS 183, 21 December 1993.
2. NASA Systems Engineering Handbook; NASA/SP-2007-6105 Rev1; December 2007.
3. Systems Engineering Handbook, A What To Guide For All SE Practitioners, International Council on Systems Engineering (INCOSE) INCOSE-TP-2003-016-02, Version 2a, 1 June 2004.

8 Functions Performance Subsystems

What does a reductionism[50] thing like functions have to do with systems? Interesting question and its answer is in the way humans solve problems. Before any problem solving can begin saturation needs to happen. Saturation is a process where information is accessed and digested by people. The information can be directly or indirectly related to the problem. Functions are a way for humans to saturate themselves with the system under study. It is a hub from which many paths can be followed and considered.

Once a preliminary system boundary is drawn, the system functions and performance need to be identified. A function is not attached to any physical element. It is abstract and tends to be defined with a noun and an action verb. Many times a single word can represent a noun and action verb phrase. System functions can be listed or they can be represented in block diagrams. Performance is stated in quantitative terms and can be measured.

For an existing system the functions and performance are characteristics of the system. The performance is also equated with the metrics from the system. For new non-existing systems, function and performance become requirements that the team tries to implement. There are also characteristics for new non-existing systems that are requirements, such as identifying a color that must be used. So context is important when referring to characteristics.

For example, a health care system exists in many countries. It has functions and performance, which can be identified, understood, and potentially adjusted to make it more effective. Operational managers gather metrics, which are the performance measurements from the system, and they use those metrics to track conscious changes in the system. If the operational managers are not tracking the functional and performance characteristics of their system, then the system is very low on its capability maturity curve.[51]

Some feel that only the money flow through a system needs to be tracked to optimize a system, however that does not yield an engineering or scientific understanding of the system. Further, such a limited view may actually break the system and make it dangerous or of no value to the stakeholders.

---

[50] Reductionism is a method or theory of reducing complex things to elements that are less complex, however systems are more than the sum of their parts.

[51] Capability Maturity is rated in terms of level 1 to 5. CMMI for Development, August 2005; Version 1.2, CMU/SEI-2006-TR-008, ESC-TR-2006-008; Capability Maturity Model(s) form Carnegie Mellon University (software engineering institute) is an example.

The key issue is that a system must have a certain level of function and performance and the level of function and performance will be at a cost. The cost may be excessive because the system may be inefficient. But implementation efficiency should not be tied to function level and performance. Poor efficiency can be the result of excessive system overhead or poor technology, which a good system functional and performance analysis should surface. For complex systems with limited or no competition there is no relationship between functional performance and maximizing financial results.

8.1 Top Level Functional and Performance Analysis

Begin the functional analysis by looking at the system context diagram and listing the system functions. As part of the process it is natural to surface functions at different levels. This is not unlike developing an outline for a composition paper in grade school. Use a single sheet of paper to list the functions. The ideas will flow, be nonlinear, and be on different levels. If a system concept diagram exists, look at it and continue to identify functions on the same sheet of paper.

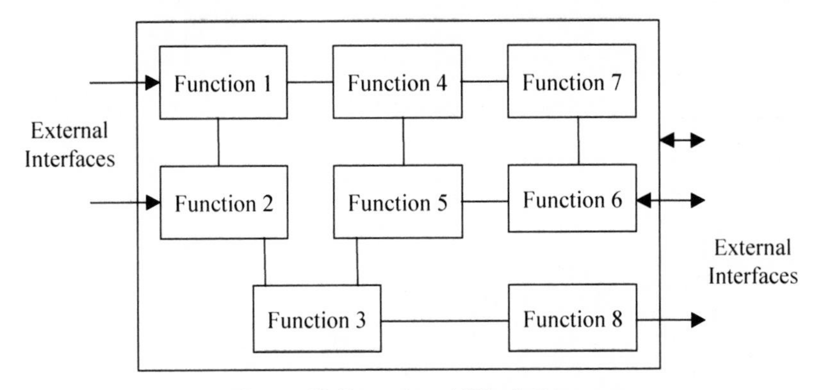

Figure 32 Functional Block Diagram

Take the paper and start grouping the functions. At some point you start to slow down in the function identification rate. Stop after approximately one hour. Group the lists into 7 plus or minus 2 functions and draw your Level 1 functional block diagram. Try to identify connections between your functions. Label connections where you can at this point in the analysis and leave the others unlabeled.

Depending on the depth of the analysis needed, you can stop at this point. The analysis will have errors, which will surface, but this is all you may need at this time. To continue to refine the analysis you need to move to a more formal method called Functional Decomposition. This method has roots in set theory and logic.

Once you have a set of functions, start to identify key functional and performance requirements. These functional and performance requirements should be list oriented and clearly grouped by the functions. From this list, abstract and list key system functional and performance requirements. These requirements are associated with your context diagram and system concept diagram.

Example 13 Functional Block Diagram - Stereo Receiver

Suppose you are designing a new amplifier receiver. What is the functional block diagram for the new system? Do you immediately equate the functions to the subsystems in new product? A subsystem is not a function. A subsystem is a collection of functions starting to be grouped into a physical entity. For example the AM receiver subsystem performs the functions of RF reception, tuning, and demodulation. The function could be implemented as a subsystem using tubes, transistors, chips, or a single chip which performs both the AM and FM receiver function.

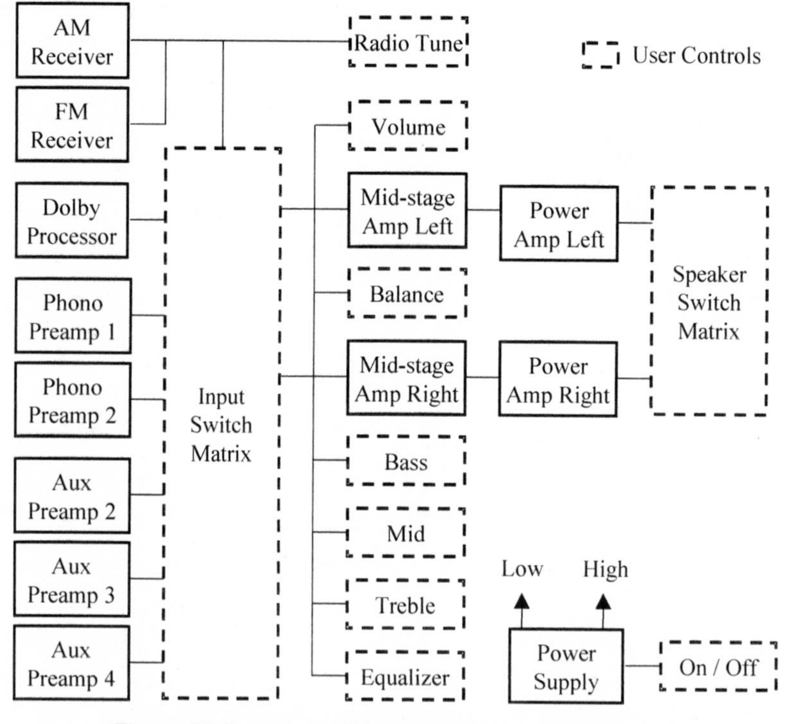

Figure 33 Functional Block Diagram - Stereo Receiver

Each of the subsystems has performance characteristics. For example the power amplifier has a frequency response of 20-30Khz with a power output of 100 Watts RMS. Conversely you could specify that the power amplifier have a frequency response of 20-30Khz with a power output of 100 Watts RMS at .01% THD. The .01% THD might exceed all other product offerings; thus the power amplifier subsystem will require a new and innovate design to meet the requirement. You may have actually developed that new and innovative design. You now have a performance discriminator that separates you from the competition.

So capturing functionality and performance in a system is key to understanding a system. The system requirements associated with the context diagram and or concept diagram for this amplifier receiver might be:

- AM / FM quartz tuner with 25 channel presets
- 120 / 220 VAC, 60 / 50 Hz
- 3 auxiliary inputs and outputs RCA plugs
- 2 Phonograph Inputs
- 6 external speakers with ABC switch selection
- 1 head phone jack

A good product brochure will state the key functional and performance characteristics of this amplifier. It will clearly show the discriminators that separate it from the competition. The brochure front might have a picture of the product, which is a finished concept diagram of the packaged amplifier receiver. It might have a few words about its' superior functionality and appeal more to the aesthetic side of the potential customer. The back of the brochure might show the context diagram, concept diagram, or functional block diagram. It might then list the performance requirements and clearly show the performance discriminators. If a new function is offered, it might be highlighted. For example, perhaps this is the first product to offer an equalizer in addition to bass, treble, and midrange controls.

Example 14 Functional Block Diagram - City

Is systems engineering taught or does it emerge when challenges are great and people rise to meet the challenges? After World War II thousands of people went to work building cities. Some went into undeveloped land and created things like Planned Urban Developments (PUDS). Some went into their existing communities and created master plans.

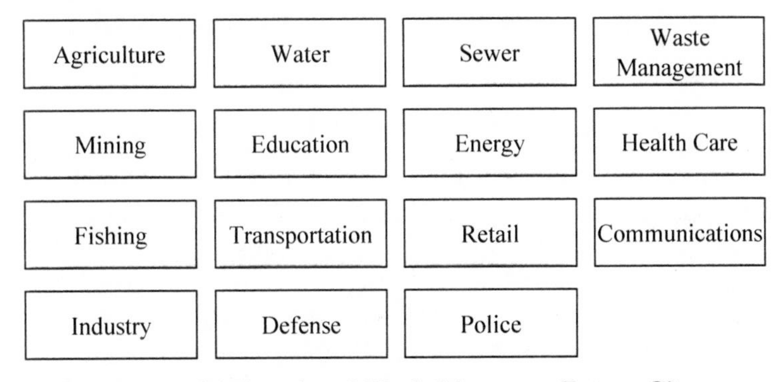

Figure 34 Functional Block Diagram - Future City

What would the functional block diagram look like for your future village, town, city, state, or country look like? What are your key city functions and their relationships?

8.2 Functional Decomposition

After you have performed a top level functional and performance analysis you might find there are too many questions. At this point what follows is referred to as functional decomposition and leveling.

Draw your functional block diagram and label each function with a number. Then proceed to decompose each function at each succeeding level. At some point a natural leaf is reached in a great tree where no more decomposition is possible.[52] As you traverse the levels, functions may move between levels and across function boundaries. For the first time practitioner this unconscious act has roots in set theory and a concept referred to as coupling and cohesion. The child diagram is considered leveled when its inputs and outputs match the parent view. For example Function 1 decomposes into Functions 1.1, 1.2, and 1.3. Further, the interfaces to Function 1 are shown in the child functions.

The simplest approach to do the full decomposition is with a stack of paper, pencil, and an eraser. There are tools available to perform this activity, however there is nothing like taking 100 pictures and posting them on the walls in a room. The first sheet contains the Context diagram or the level 0. The next sheet contains the Level 1 diagram. The next series of sheets contain the Level 2. The next series of sheets contain the Level 3. The decomposition continues until all the leaves are reached. As the process unfolds functions are moved across boundaries and across levels as previously mentioned.

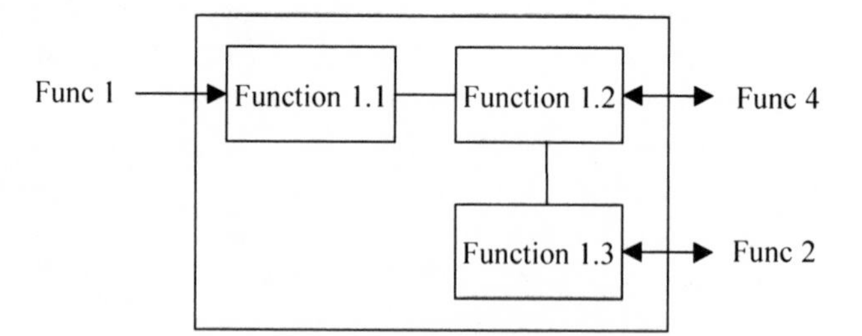

Figure 35 Functional Decomposition Level 1

The functions are identified, named, and numbered. The numbers form the paragraph numbers of your requirement specification. Each section of the specification is represented by decomposition at a certain level. The indentured list and numbering conventions allow the designers to trace the specification to the functional block diagram elements. In many cases only the Concept Diagram, Context Diagram and the Function Level 1 diagrams are shown in the specification.

---

[52] The concept of a tree applies everywhere. For example, there are fault trees, troubleshooting tress, organizational trees, specification tress, etc.

This is especially true for a high-level system specification.

This technique is rigorous and methodical. As you go through successive levels, previous levels will be modified as you learn more about the system. Eventually the context diagrams will be modified to reflect the new insights about the system. You should not spend more than 1 hour per level. Also go with your initial views, they will be the most reasonable and require the least modification as you decompose to the leaf levels.

1.0
1.1 Resulting Specification
1.1.1 Table of Contents
1.1.2
1.1.3
1.2

Figure 36 Functional Block Diagram and Specification

How log will this take? Lets work backwards from a software system that has 1 million lines of code. Given that for maintainability: 1 function should have 100 lines of code and each level of decomposition is on average 10 functions. Then there will be 100,000 functions and 1000 sheets of paper representing the functional decomposition. At 1 hour per page plus some level of rework as the analysis unfolds new insight, the activity is large but well within the capability of human organization and capacity.

This idea of decomposition is not new it has deep roots.[53] We catalog everything: library books, biological species, diseases, device parts, etc. It is found in the concepts of writing where note cards are used to track the personalities of characters in a book. In film, storyboards are used to capture the major themes and threads of a movie. Even in proposals from the last century used a technique called Sequential Thematic Organization of Proposals (STOP) to create important proposals that gave us our modern world.[54]

8.3 Functional Flow Block Diagram

The Functional Flow Block Diagram (FFBD)[55] is also called a Functional Flow Diagram. It adds more formality to the functional block diagram with the introduction of "and / or" logic blocks. It is time sequenced and based on step by step flow through the system functions. The system functions are at the same level and multiple paths are simultaneously represented. It is based on the concepts of

[53] Tom De Marco applied decomposition to software in 1978 in his book Structured Analysis and System Specification, ISBN 0917072073

[54] STOP was created in 1965 at Hughes Aircraft

[55] Systems Engineering Fundamentals; Defense Acquisition University Press; January 2001

decomposition, leveling, and traceability described in the previous section on Functional Decomposition. So FFBDs can be developed for each level of functional decomposition.

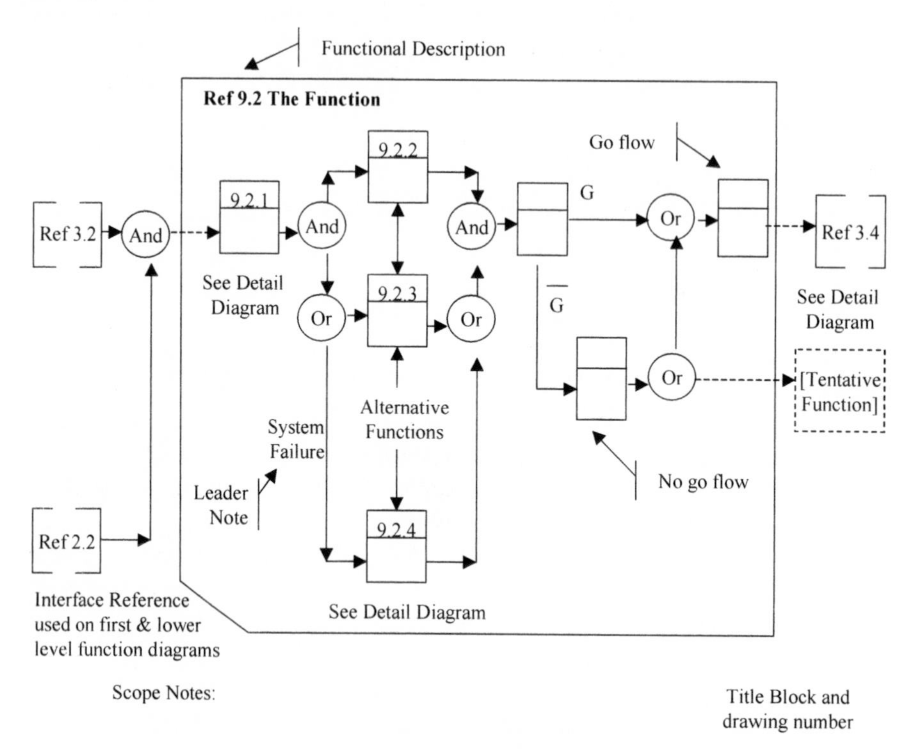

Figure 37 Functional Flow Block Diagram (FFBD)

The goal of the functional flow block diagram (FFBD) activity is to help describe the system requirements in functional terms. This can be a large task depending on the level of completeness desired. The activity should address the life cycle functions, identify and define the system elements (e.g. prime equipment, training, spare parts, data, software, etc.), address system support functions, and show proper sequencing of activities and design relationships including interfaces.

The FFBD is a functional representation of the system not a physical, implementation, or a solution view of the system. The focus is to identify the functions and figure out how to build or design and implement them later. The process of defining lower-level functions is functional decomposition and sequencing the relationships is part of the FFBD. The FFBD shows traceability between functions vertically through the levels. The numbers also should cleanly trace to the specification requirements. There are standard symbols and rules used in the construction of FFBDs. As with all formal and semi-formal methods the designers may expand these rules and symbols to meet their unique need. The basic symbols and rules are:

1. **Function Block:** Each function on an FFBD should be separate and be represented by single box (solid line). Each function needs to stand for a definite, finite, discrete action to be accomplished by system elements.
2. **Function Numbering:** Each level should have a consistent numbering scheme and provide information concerning the function origin. For example the top level is 1.0, 2.0, 3.0, n.0; the first indenture (level 2) is 1.1, 1.2, 1.3, 1.x; the second indenture (level 3) is 1.1.1, 1.1.2, 1.1.3, 1.1.y. These numbers establish identification and relationships that carry through all Functional Analysis and Allocation activities. They support traceability from lower to top levels.
3. **Functional Reference:** Each diagram should contain a reference to other functional diagrams by using a functional reference (box in brackets).
4. **Flow Connection:** Lines connecting functions should only indicate function flow and not a lapse in time or intermediate activity.
5. **Flow Direction:** Diagrams should be laid out so that the flow direction is generally from left to right. Arrows are often used to indicate functional flows.
6. **Summing Gates:** A circle is used to denote a summing gate and is used when AND/OR is present. AND is used to indicate parallel functions and all conditions must be satisfied to proceed. OR is used to indicate that alternative paths can be satisfied to proceed.
7. **GO and NO-GO paths:** "G" and "bar G" are used to denote "go" and "no-go" conditions. These symbols are placed adjacent to lines leaving a particular function to indicate alternative paths.

8.4 Functional Sequence Diagram

A Functional Sequence Diagram (FSD) takes a functional block diagram, drops a single input into the diagram and shows the path followed by the input through the functional view. All of the functions are at the same level and the diagram shows the context of the other functions not impacted by the input. This is a way to test the functional block diagram and gain further insight into the system.

The analysis shows the picture and includes words describing what is happening at each stage of traversing through the system via the functions. Many times when modeling results are described, FSDs are shown to help understand the model results.

Performance can be tracked through the system using this approach. For example time budgets, a performance measure, can be assigned to each function and an overall time to traverse through the system can be identified. The concept of performance budget allocations is critical to understanding the performance of the system. Other performance characteristics are reliability, noise, death rate, cure rate, energy loss, lines of code, processing load, and yes cost of operations.

The team eventually starts to ask questions about the system operation. Usually this happens when questions are asked about the humans in the system. Some of these questions lead to simple answers and others are more complex. Eventually a set of operational sequences start to surface that many on the team agree need formal capture via FSDs.

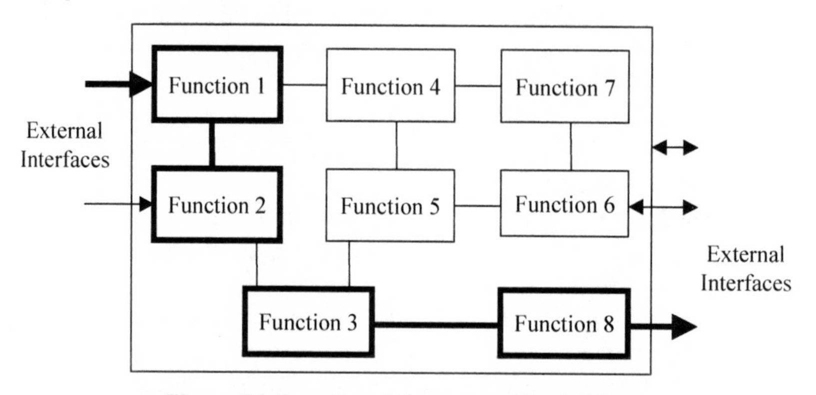

Figure 38 Functional Sequence Block Diagram

The FSD has roots in Operational Sequence Diagrams (OSD). The OSDs show what operators do when interacting with a system. So from a purest point of view OSDs are used to show human functions and interactions in a system. This may include how the humans interact with non-human elements of a system.

Some people will use the term Operational Sequence Diagrams, when referring to Functional Sequence Diagrams. Operational Sequence Diagrams are described in the Human Factors Section.

8.5 Functional Thread Diagram

A Functional Thread Diagram shows the movement of input through the system functions, which are at different levels. The levels and view are based on questions or issues the team is trying to address. For example, response time may be an issue and the time allocation to each function needs to be clearly seen and understood. These allocations form the sum of the performance of the system.

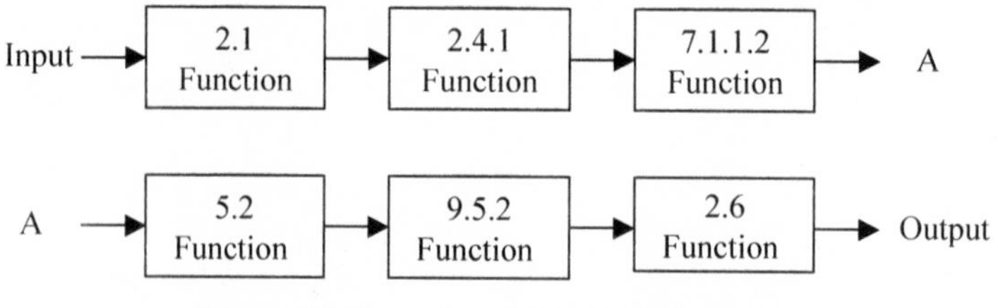

Figure 39 Functional Thread Diagram

Functional thread diagrams also help to identify functional allocation issues and gain system insights. For example if we are tracking input associated with the power train of an automobile and we all of the sudden find ourselves in the middle of the stereo radio function, we have a problem. In fact in the new century two US car companies have done this, such that the car cannot be used if the stereo radio is removed or malfunctions.[56]

In another automobile example, the crossing of traditional functional boundaries yields a better system. By mixing the power train function with the braking function the energy is restored to the system when the driver applies the brakes. The implementation is an electric car with pancake motors used to both move and stop the car. While stopping, the electromagnetic energy is captured by pancake motors and stored back into the system.

8.6 Formal Analysis Dangers

Many people can eventually relate to a single functional block diagram. However once the number of functional block diagrams start to grow people become confused and lost. It becomes worse when AND / OR logic and flow is introduced. However, before you decide not to use a functional view for your system, realize that as soon as you introduce a technique that is not based on words and a simple picture, you start to lose 98% of your stakeholders. It becomes worse if the technique is not based on logic or set theory. The only people who can relate to the formal methods are those trained in the formal method and who use the formal methods on a daily basis. The advantage of functional analysis is that you can control the level of formality and rigor.

It does not mean you can't use other formal methods to try to understand the problem from a different perspective. It just means that it becomes a detailed study that needs to be summarized with words and a picture that everyone can understand. That typically will translate to a high-level functional block diagram.

Running through formal functional analysis surfaces a more correct and complete picture of the high level system that everyone needs to see and understand. Functional decomposition also helps to surface requirements and show traceability of the system functions to the requirements.

8.7 Coupling and Cohesion

A strategy is needed when decomposing and grouping functions. Typically the strategy is based on the concept of coupling and cohesion. Coupling and cohesion are conceptual measures of how tightly connected two things are in a system.

Depending on what you are trying to accomplish you might want tight coupling

---

[56] Such a decision is not based on sound engineering principals. This approach violates basic maintainability characteristics of the system. Regardless of the motivations, it is an example of gross misuse of technology and breaks the product.

or lose coupling. Coupling implies that you have control over the connectivity. In other words you have control on the level of coupling between functions.

Cohesion implies that you have no control of the degree of connectivity between functions. High cohesion suggests that the functions naturally need to be grouped together and if you separate them, the system performance will degrade.

Functions with high cohesion are associated with robustness, reliability, reusability, and understandability while low cohesion is associated with undesirable traits such as being difficult to maintain, difficult to test, difficult to reuse, and difficult to understand. However this is from the point of view of the functions that are grouped. From the functions that are not part of the same group, you are looking for low coupling or cohesion.

For example from a failure point of view low coupling is the desired attribute. The goal is to slowdown or stop fault propagation in the system. This also applies when considering security and how security breaches may propagate through the system. The functions may naturally display high cohesion but you must break that cohesion with low coupling to stop the propagation of an undesirable event or element in the system.

From a performance point of view, such as response time, high coupling is the desired attribute. However the functions in this group may actually exhibit low cohesion, but you need to somehow tightly couple them to maximize performance. Also, there may be a tradeoff that needs to happen between response time and failure propagation or security.

1. **Functional Cohesion**: Functions are grouped because they all contribute to a single well-defined activity. For example the RF front end, mixer / oscillator, and demodulator form the FM receiver function.
2. **Temporal Cohesion**: Functions are grouped because they conceptually or actually perform at similar times.
3. **Logical Cohesion**: Functions are grouped because they logically do the same thing but they are different by nature. For example grouping all radio functions together in an amplifier receiver even though they are AM or FM or Short Wave, etc.
4. **Procedural Cohesion**: Functions are grouped because they always follow a certain sequence of operation. For example a login function takes the user name and compares it with the password.
5. **Sequential Cohesion**: Functions are grouped because the output from one function is the input to another function. For example the FM receiver always has an RF front end to amplify the incoming signal, passes the signal to the mixer / oscillator to extract the appropriate channel, then passes the signal to the demodulator to extract the audio.

8.8 Performance

Performance can be identified for one or more functions, however the performance may exceed the state-of-the-art for the subsystem or subsystems that will support the functions. Yet the desired performance may be critical to the system. For example a serious automobile accident on a highway may block traffic for miles. Yet people are critically injured. The introduction of a helicopter into the health care system seems like common sense today but it was a major breakthrough for the previous generation. Only 50 years ago there was no solution. It was beyond the state-of-the-art.

New systems push the state-of-the-art in one or more areas. The challenge is to identify these areas and apply resources so that the system can become a reality.

Finding the quantitative measures of a system is not a simple task. Pioneers in previous centuries showed us how to extract and analyze performance numbers for electrical circuits and mechanical structures. Today we have a plethora of legacy systems and knowing the key performance numbers is critical so that various alternatives can be understood, compared, and form the foundation for new systems.

Example 15 Performance and Quality

Performance is a requirement for a system being developed. It is a metric that should be measured and tracked for an existing system. For new or modified systems, existing systems should be studied and their performance should be extracted and applied to the new system vision. The following are some examples of system performance measures:

1. **Amplifier**: Power output in watts, distortion in %, noise in dB, frequency response in Hz, total harmonic distortion in %, inter modulation distortion in %, and dynamic range in dB are the performance measures for an amplifier. As the amplifier evolved over time new performance measures were added. For example, power output was a primary performance measure. Then people realized that power output was useless if the distortion was high. And so the performance metrics continued to evolve until an amplifier could be fully characterized and compared with different implementations.
2. **Radio**: The radio has an amplifier subsystem and a receiver subsystem. Sensitivity in microvolts, selectivity in dB, noise in dB, and stereo (channel) separation in dB quantify the receiver performance. The functionality of a radio went from AM to AM FM to AM FM stereo. Each of those functional elements has performance characteristics that matter.
3. **Tape Recorder**: The tape recorder has an amplifier subsystem and a mechanical subsystem. The mechanical subsystem performance is quantified with the steadiness of the speed as music is recorded and played back. This is referred to as flutter and wow both identified in terms of

percent. High quality studio recorders would measure speed on the fly and adjust as needed to minimize the flutter and wow. That is the level of detail applied to something simple like the flutter and wow performance number.

4. **Color Televisions**: Screen size, brightness, contrast, detail, color purity, color depth, stability, linearity, resolution, apparent resolution, and convergence stability are some of the performance numbers for a television video subsystem. Linearity is interesting and is based on drawing a circle and moving it across the screen. All TVs that try to fit the image in the useable screen area have terrible linearity if the original content does not match the aspect ratio of the TV screen. Apparent resolution is even more interesting. If projection TV with continuous phosphor Red Green Blue CRTs is compared to non-continuous displays such as LCD, Plasma, or 3 gun CRT, the resolution will appear to be greater on the projection TV. This appearance can even be measure as the continuous phosphor CRT paints a line or dot on the screen. Then there is the perception of resolution. The human eye sees and the mind perceives things at 120 (peripheral vision), 60 (typical), and 1 to 6 (when focused) degrees. If the screen changes format from the 3 X 4 X 5 triangle and the image is distorted or shrinks, the ability to discern a human facial elements, such as the eye, falls dramatically.
5. **Car**: Acceleration usually 0-60 in x seconds, breaking distance in feet, efficiency in miles per gallon, wheel base inches, trunk space in cubic feet, passenger compartment in number of passengers head room leg room, engine power in horsepower, turning radius in feet.
6. **MP3 Player**: Number of songs in terms of memory in gigabytes, quantization distortion[57] in dB (no one currently offers this performance), bit sample rate are the performance characteristics associated with MP3 players. Notice the traditional audio performance characteristics are missing. The primary reason is based on the quantization distortion, which is based on the number of bits used to capture a single value in the digital to analog converter. The quantization distortion is so high that capturing the other metrics would be a severe embarrassment. The lay person has detected something is not right and there have been movements back into analog audio. However this is probably more of an anomaly. Music has actually changed to hide the issues associated with digitization and the audience has come to accept the "computer metallic sound" from these devices.
7. **Flat Screen Televisions**: Screen size, resolution, screen update, and power consumption and the performance characteristics associate with new flat screen TVs. Sometimes consumers are given performance numbers that do not capture the system performance properly. For example the human eye can discern movement at a 30 Hz rate yet modern TV manufacturers

[57] Discrete Time Systems, James A. Cadzow, Prentice Hall, 1973, ISBN 0132159961.

advertise screen refresh rates well beyond that number. What is really hidden is the performance number for being able to update the entire screen with new content. There is the issue, the entire screen cannot be updated at a 30 Hz rate with completely new content, so false boundary conditions are created and a performance number is inflated that makes no sense.

So searching for performance in a system is not a marketing activity. It is an engineering activity that must be never compromised. Without meaningful performance measures, systems will never evolve and get better. In some cases the system may not even work or cause damage.

What are the performance characteristics associated with your house, community, and heath care system? If you are a homeowner you probably are able to quickly surface the performance characteristics of your house and community that most will understand. If you built your house and you were involved in the details you will surface performance characteristics of your house and community that most people will find foreign. It will not be common sense until you describe them in some setting.

All systems start relatively small and evolve. The focus changes and things that were once impossible are added to the system as the decades unfold. In the beginning the story of the automobile was about speed, then durability, then suspension, aesthetics in form versus function arguments, safety like lights and brakes, comfort like windows, then wipers, heat, air conditioning, radio, FM radio, stereo, power windows, power seats, cruise control, and now GPS. In many ways this is like a pyramid model. You start with a base and build upon each layer until a pinnacle of perfection is reached both in terms of function and performance.

It is always sad when progress is lost and things go back in terms of function or performance. In many ways the digital revolution is an example of this scenario as high quality analog solutions were abandoned for the flexibility or quantity of digital. In some areas like healthcare these resets may be intolerable.

Buried within the concept of performance is quality. A high quality solution is a solution that does everything exceptionally well.

8.9 Subsystems

When performing a functional analysis it is sometimes difficult to separate the subsystem view from the functional analysis. Don't try to separate the functions from the subsystems. You are human and a non-linear thinker. Don't fight your natural tendencies. Non linear thinking is good but you should keep track of your work. That means you need to separate the functional and subsystem information products into two groups when you are ready to present the functional findings.

Example 4: Healthcare systems exist around the world. They have functions, which are grouped into subsystems and the performance can be measured. There is performance data in the form of metrics gathered for various health care systems from around the world. But how do you start to represent these health care systems from a functional point of view? Is there a generic functional block diagram that can represent all of the healthcare systems? Is the performance driven by the implementation of these generic healthcare functions? Are there systems that have more functionality than other systems? Is more functionality associated with a more capable healthcare system? Are the connections between functions different for different healthcare systems?

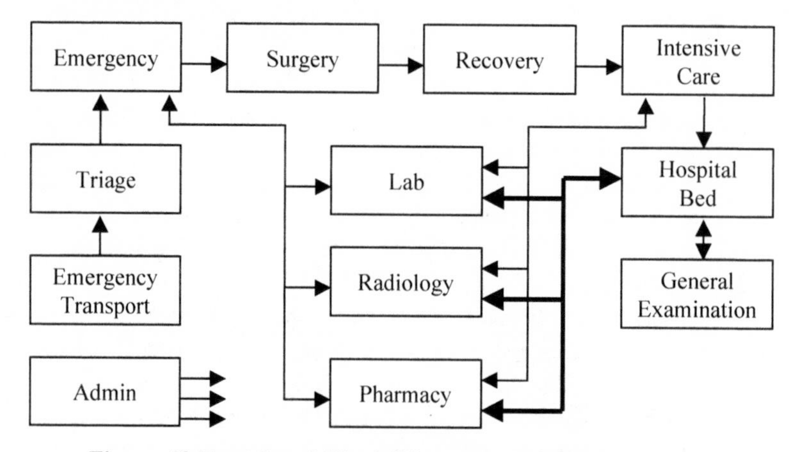

Figure 40 Functional Block Diagram - Health Care View 1

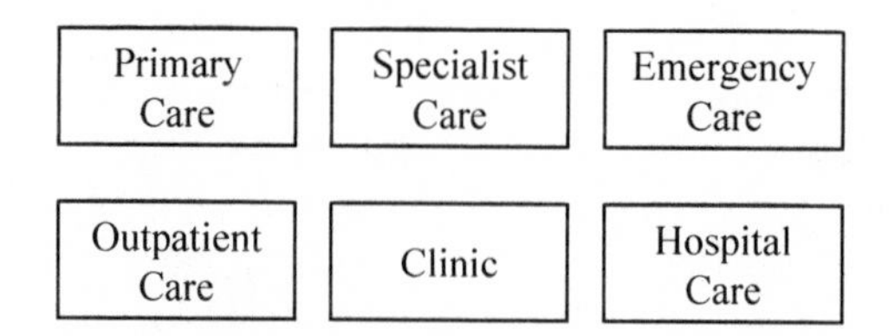

Figure 41 Functional Block Diagram - Health Care View 2

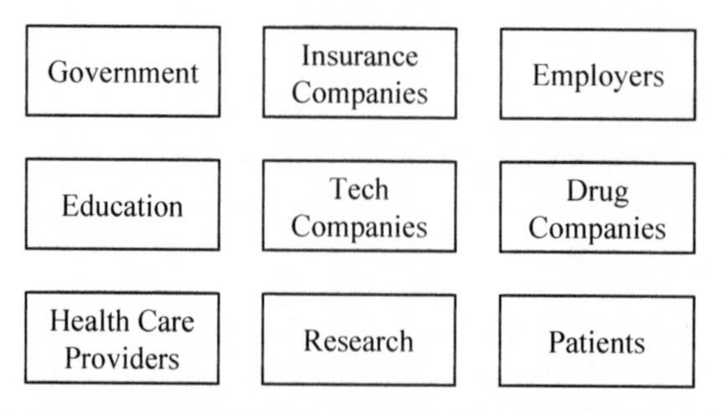

Figure 42 Functional Block Diagram - Health Care View 3

As a first step in identifying the functions it is easy to move to the subsystem level. It is also easy to start identifying the stakeholders. The same word or word phrase can easily be interpreted as a stakeholder, function or subsystem. That is why clear supporting text is important when presenting any analysis.

There is always a danger of using vague words or phrases. This is usually because there is an inability to commit. Commitment is very important. The team cannot delay commitment and move to the next level. It is like building a house on sand. A poor foundation will lead to collapse.

Example 16 Abstraction Level Needed for Current View

Sometimes when we start to identify the system functions we plunge a level or two down, but then we have enough information to abstract the level 1 functions. However when we look at the functions they are extremely generic and might apply to other systems. For example if we look at electric power generation, a view develops that includes extraction, distribution, refining, use, and waste. The focus shifts to the fuel that powers the system. The power generation and its applications fall into a "use" block.

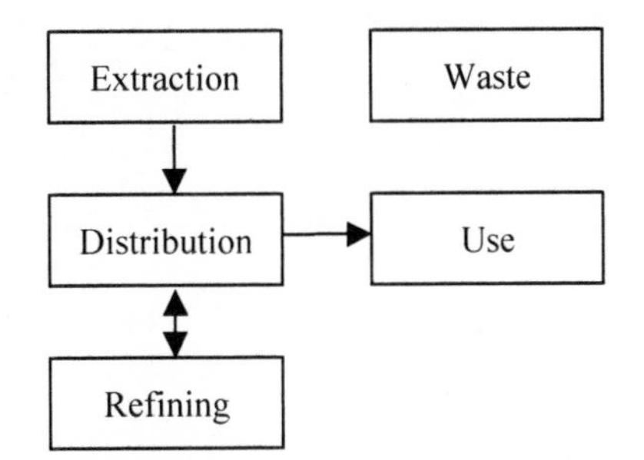

Figure 43 Electric Power Generation / Automobile View 1

This same view can apply to a system of personal transportation, the automobile. The automobile system can be reduced to its source of fuel. Not only can the functional block diagram be similar between power generation and automobile transportation, but also the high-level sequence diagram can be similar.

The differences between the power generation and automobile transportation systems start to surface when lower level functions are identified.

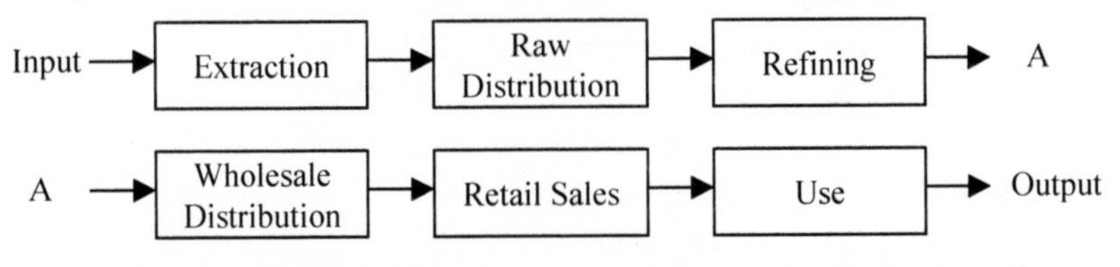

Figure 44 Electric Power Generation / Automobile View 2

Subsystems are the next level of understanding in a system. Functions are allocated to subsystems. Performance is analyzed within the context of a subsystem technology and its selected size. This is the start of architecture analysis. The details of the architecture and its subsystems are described in other sections.

The intent in this section was to show that the line between function and subsystem can be hard to control and that we should not fixate on that line or else we may get stuck in our analysis. The goal is to always push forward and realize that you will approach the problem from different abstraction levels because the problem solving sequence is not linear. You are forced into non-linear thinking, accept it, and manage it by keeping all the data.

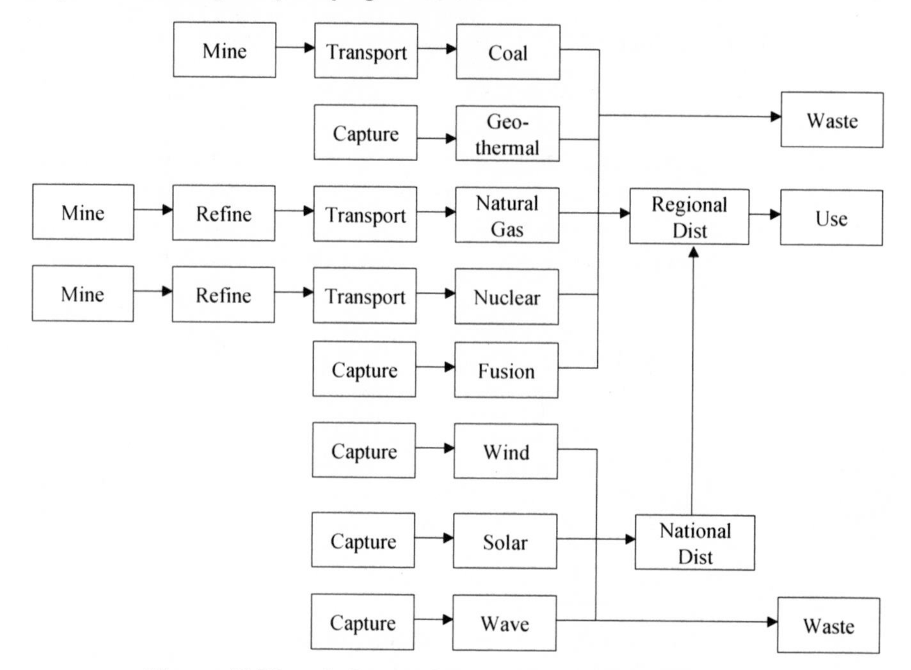

Figure 45 Electric Power Generation / Automobile View 3

8.10 Exercises

1. Draw a detailed functional block diagram of a national level transportation system.
2. Draw a functional sequence block diagram of a health care system as you, friend, or a family member has experienced it.
3. Identify possible performance requirements for a healthcare system and discuss how those performance requirements might be achieved by maximizing profits. If there is no relationship between achieving the performance requirements and maximizing profits, state why.
4. Does it make sense to draw a functional block diagram of a house or should a traditional artist rendering and floor plan be used in all settings? Discuss your choice and be prepared to defend your positions with the class.

8.11 Additional Reading

1. Discrete Time Systems, James A. Cadzow, Prentice Hall, 1973, ISBN 0132159961.
2. Sequential Thematic Organization of Proposals (STOP), Hughes Aircraft Fullerton, 1963.
3. Special Issue on STOP Methodology, edited by R. J. Waite, Journal of Computer Documentation, Volume 23/3, August 1999.
4. Structured Analysis and System Specification, Tom De Marco, 1978, ISBN 0917072073.
5. The Beginnings of STOP Storyboarding and the Modular Proposal, Walter Starkey, APMP Fall 2000.
6. Two Approaches to Modularity: Comparing the STOP Approach with Structured Writing; Robert E. Horn; Visiting Scholar; Stanford University; This appeared in Journal of Computer Documentation, 1999.

9 Operations and Threads

A system can be viewed and analyzed from an operational point of view. The approach is to describe the operation of the system at a high level, then identify a number of operational threads or uses of the system. Once the threads are clearly identified then they are described. This is not unlike functional analysis and decomposition where the system is viewed at different abstraction levels. This is called thread-based development.

Systems are rarely developed with just a thread-based view. Typically the threads augment other analysis techniques, especially context, function, and performance analysis. However, you can begin with thread based development then augment the analysis with other practices.

One of the advantages of thread based development is that subject matter experts easily relate to the activity. In many ways it is a product of the subject matter experts and other non-technical stakeholders in the project.

There are many ways to capture the operations and threads in a system. The key is to realize that this information will surface in the course of the system development. The issue is how the information is captured and treated. If this information falls on the floor then the project will significantly suffer. If this information is partially captured it will be reflected in some portion of the test program, especially during validation. An approach is to task the team with reflecting operations and threads in all their information products. However that does not necessarily mean that subject matter experts and other stakeholders are able to reflect their views of the system.

9.1 Concept of Operations

The Concept of Operations (ConOps) is a standalone document and it is the highest level view of the operation of the system. It can be reflected in a picture or in a collection of pictures. However, there are usually words that the subject matter experts and other stakeholder need to offer to fully express their views. The information products are a ConOps document and the presentations that reflect the body of work.

The ConOps document describes the system from the viewpoint of the users who interact with the system, operators in their daily work activities, maintainers, and the system operations managers. One of the most important parts of the ConOps is the list of operational threads or operational concepts. The list results in standalone Operational Concepts (OpsCons) that are captured in other stand-alone presentations and document information products. The operational threads described in the child OpsCons information products are also the basis for test threads used in the test program.

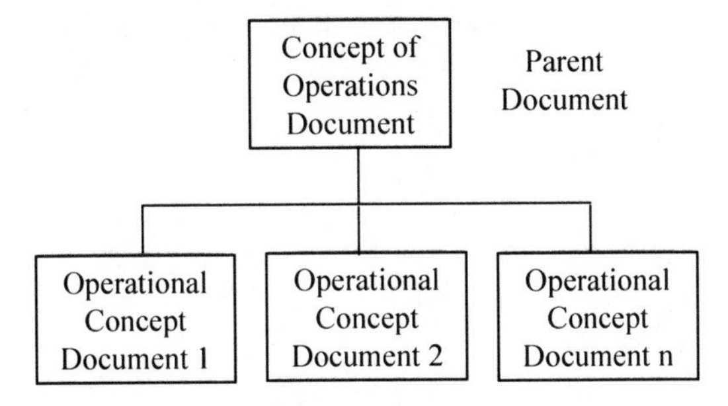

Figure 46 Concept of Operations Document Tree

The ConOps bridges the gap between the users operational needs and visions and the technical specifications, without becoming stuck or lost in detailed technical issues. It also describes the system and the operational needs so the system can be understood without needing any special knowledge except knowledge of what is required to perform normal job functions.

The ConOps documents the stakeholder desires, visions, and expectations without needing complete, consistent, quantified, testable specifications. For example, the users could express their need for a highly reliable system, and their reasons for that need, without having to produce a testable reliability requirement. A statement such as, the brakes or steering in an automobile should never fail, might capture their views.

The ConOps also allows all stakeholders to express their views on possible solution strategies, offer design constraints, the rationale for those constraints, and indicate the set of acceptable solution strategies. For example the normal hydraulic brakes in an automobile should have a backup hydraulic mechanism and a third manual cable parking brake as a last resort to stop the car.

The ConOps is one the earliest information products on a system project. It can be treated as a seed that kicks off all other analysis or it can be treated as a living document capturing the system views for all stakeholders as the system development evolves. This introduces overhead, because the document needs to capture information that is in other analysis information products. It also is a source of error as multiple "books" of the system are maintained. The error can stem from time delay or the ConOps document owner(s) not realizing their role shift to system historian(s) rather than system drivers.

Example 17 Concept of Operations and Operational Concept

As always there is the big picture and the details. The big picture is the ConOps feeds the OpsCons and they all feed the other information products. The details include relationships between all these products and the feedback loops. The details are that any information product can impact any other information product. In fact this feedback might result in the consolidation of operational scenarios and the resulting OpsCons.

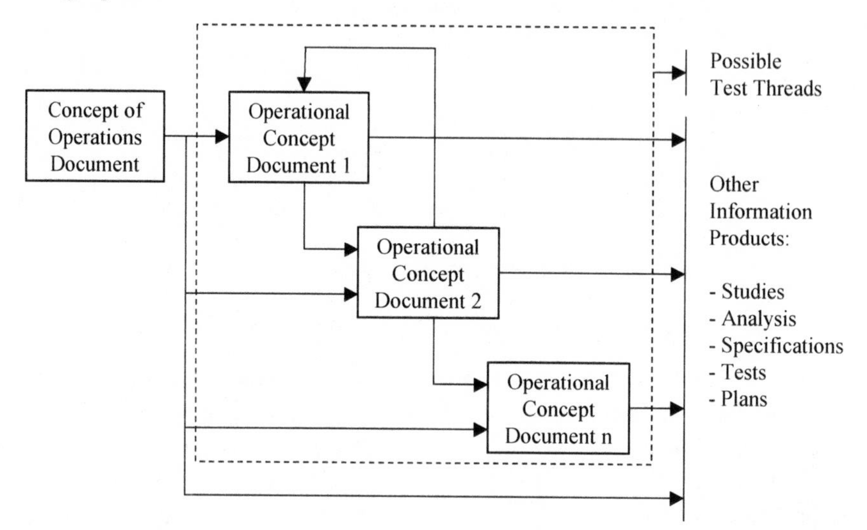

Figure 47 Operational Document Interactions

For example, OpsCon 1 might be the logon / logoff description for a web portal. OpsCon 2 might be the user password recovery description. The iteration shown in the figure might be so significant, the coupling so tight, that the OpsCons should be merged. That means the ConOps document will need to be updated to reflect this new understanding that was not seen when the ConOps was being developed. This simple process scenario needs to be understood and respected. As soon as eternal forces are applied to shut down this emergence the system will suffer and may actually lead to catastrophic results.

If the ConOps is a seed why is it offered in this section of the book? The issue is that someone needs to identify the system boundary and the high-level system functions. Without this view most stakeholders get lost and are unable to even surface the system boundary. That does not mean that all stakeholders are not part of the early system boundary and functional analysis? To the contrary everyone is key to the effort and their inputs must be solicited throughout the process. In many ways this is a chicken versus egg question. What comes first: OpsCons, Boundary, or Functions and Performance? In this case they happen simultaneously.

The following is a possible table of contents for the ConOps[58]:

---

[58] Outline based on United State Government data item description, Operational Concept Description (OCD) DI-IPSC-81430.

Table 3 Concept of Operations Outline

Concept of Operations Outline
1. Introduction
2. Referenced Documents
3. Current System
3.1 Background, Objectives, and Scope
3.2 Operational Policies and Constraints
3.3 System Description
3.4 Modes of Operation
3.5 Users and Stakeholders
3.6 Support Environment
4. Justification and Description of Changes
4.1 Justification of Changes
4.2 Description of Desired Changes
4.3 Change Priorities
4.4 Changes Not Included
4.5 Assumptions and Constrains
5. New System
5.1 Background, Objectives, and Scope
5.2 Operational Policies and Constraints
5.3 System Description
5.4 Modes of Operation
5.5 Users and Stakeholders
5.6 Support Environment
6. Operational Scenarios
7. Summary of Impacts
7.1 Operational Impacts
7.2 Organizational Impacts
7.3 Development Impacts
8. System Analysis Record
8.1 Summary of Improvements
8.2 Disadvantages and Limitations
8.3 Alternatives and Tradeoffs Considered
9. Notes
Appendices
Glossary

9.2 Operational Concepts

Operational Concepts (OpsCons) are standalone documents that describe how a system is used in various operational scenarios. OpsCons are created to help solve problems in current operations, to take advantage of new knowledge or technology that enables improvements in current operations or to understand operations in a new system.

They are developed with the subject matter expert stakeholders and derived by first analyzing the user services. The user services discuss how to improve operations, interpret stakeholder goals needs and requirements into guiding principles, and apply knowledge about the state of the existing and emerging technologies. The combination of the desired operations improvements, guiding principles about making those improvements and the reality of technological advances are reflected in the operational concepts.

Many times a stand-alone architecture document uses the OpsCons for the next level architecture analysis. The following are examples of OpsCons outlines.

Table 4 Operational Concept Outlines

1. System Service Overview
2. Reference Documents
3. Input Output Messages
4. Initial Conditions
5. Detailed Service Description
6. Error Handling
7. Diagram Showing Behavior

<u>**Or**</u>

1. Introduction
2. Referenced Documents
3. System Description
    - 3.1 Scope and Context
    - 3.2 Operational Policies and Constraints
    - 3.3 Users and Stakeholders
    - 3.4 Modes of Operation
    - 3.5 Goals Needs Requirements
    - 3.6 Support Environment
4. Configuration
    - 4.1 Segments
    - 4.2 Subsystems
    - 4.3 Interfaces
    - 4.4 Concept of Operations
5. Roles and Responsibilities
    - 5.1 Operators
    - 5.2 Administrators
    - 5.3 Support
    - 5.4 Policy Makers
6. Scenario and Thread
    - 6.1 Description
    - 6.2 Graphic
    - 6.3 Resulting Goals Needs Requirements
7. Technology Assessment
    - 7.1 Impacts on Operations
    - 7.2 Reasonableness

Table 4 Operational Concept Outlines

9. Notes Appendices Glossary

9.3 Operational Scenarios

An operational scenario is a description of how a portion of the system will accomplish a function while interacting with a user. It can be described as a list of sequential steps or shown as functional blocks using interconnecting lines flowing in one direction. It should trace to or use the functional block diagrams developed for the system. Ideally a pictorial using the functional block diagrams described by text and summarized by sequential steps.

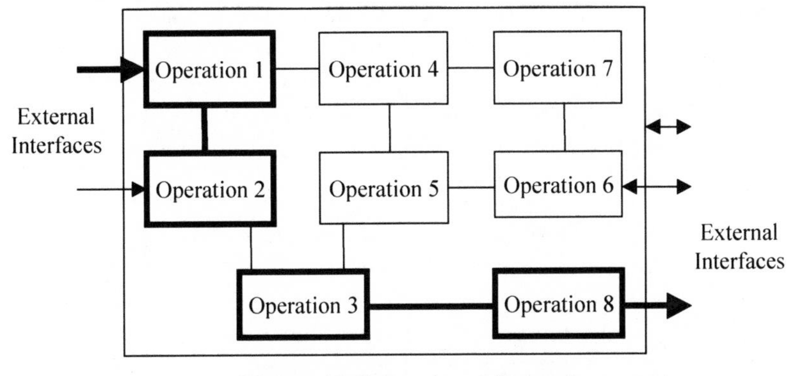

Figure 48 Operational Scenario

It some instances it may make sense to create operational sequence diagrams. These diagrams are similar to functional block diagrams but capture the system from an operational point of view. In this case the operational sequence diagrams augment the functional block diagrams and both should be consistent with each other. Initial inconsistencies are a good sign suggesting the analysis is worth the effort. The inconstancies must however be addressed.

Various performance parameters can be allocated to each element in the operational scenario elements just like in functional block diagrams. These performance numbers are typically timing, but they can be failure rates, reliability, error introductions, noise, frequency response loss, power loss, energy loss, death rate, cure rate, lines of code, processing load, cost of operations or any performance number needing to be allocated for some important insight into the system.

9.4 Operational Threads

The Operational Threads conceptually are like the operational sequence diagrams except they do not show the context of other functions, which are not stimulated. Threads tend to be associated with a human initiating an activity with a system. Identifying the system threads from a user point of view essentially groups many functions together to accomplish a desired result.

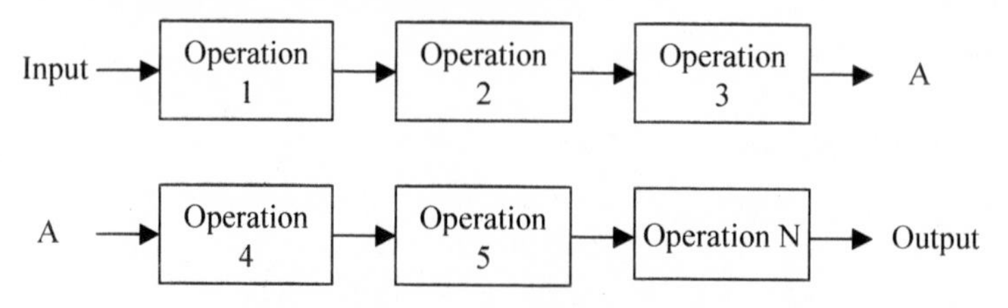

Figure 49 Operational Thread Diagram

Example 18 Operational Thread and Budgets

Suppose you would like to understand the Operational Thread of "Check Email" and its impact. You would start with a description of why the "Check Email" thread exists. Perhaps it is used for support. It might include how many emails typically arrive and other relevant information. Analysis of the thread might yield an Operational Thread Diagram with the following sequence and timing:

Table 5 Operational Support Thread

Operation	Minutes
Arrive from day job	Start
Get a drink of soda	0.5
Attend to biological needs	5
Power on computer	3
Log on to computer	1
Connect to WIFI network	1
Connect to email POP account	1
Sort messages	1
Delete SPAM based on subject and sender	5
Open first email	0
Compose solution	5
Open last email of the session	0
Compose solution	5
Disconnect from email POP	0
Power off request	0
Disconnect from WIFI	1
Log off	1
Power off	1

Who would have thought that something as simple as providing email support is so complex and time consuming? We see that it takes 8 minutes to log on and off to the computer. We also see that it takes 10 minutes to attend to 2 email requests. So we see the level of overhead is high in this case. Perhaps we should check the email support every 2 days or couple the email support activity with other activities and get economies of scale.

9.5 Task Analysis or Operational Sequence Diagrams

Operational Sequence Diagrams (OSD) are referenced in many analysis techniques. For example tracing an operational thread through a functional block diagram is often referred to as an OSD. However Kurke developed Operational Sequence Diagrams (OSD)[59] in 1961. So OSD is an overloaded term.

The Kurke OSD is typically used to perform a task analysis. The OSD graphically represents information, decision sequences and complex multi person tasks. The output of the OSD shows the tasks performed and the interaction of the operators. The level of detail of the OSD is based on the goal. A simple OSD can just show the order of the tasks. A complex OSD might show a complete team and their interactions. The OSD can be used in complex command and control analysis including air traffic control and air defense and simpler settings such as fire service, energy distribution, etc. The method was originally used in the nuclear power and chemical process industries[60].

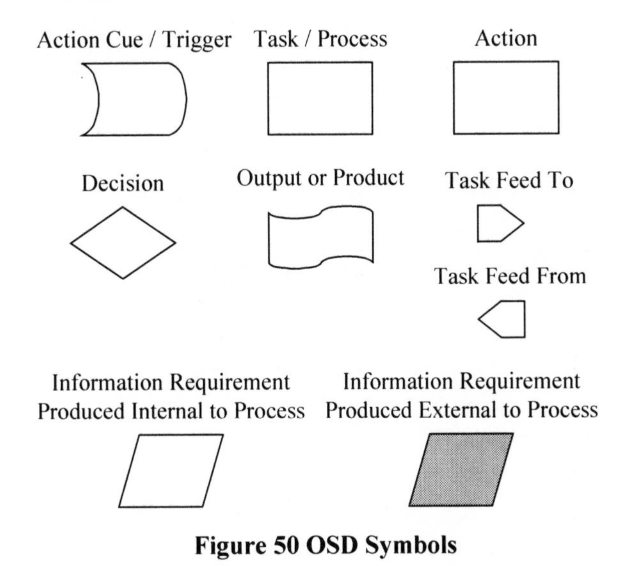

Figure 50 OSD Symbols

A task unit is typically represented in one diagram. The order of the symbols in a task unit is a grouping of the OSD elements; it does not represent flow or order.

---

[59] Operational Sequence Diagrams in System Design, Kurke, M. I., 1961.

[60] Human Factors Methods: A Practical Guide for Engineering And Design by Neville A. Stanton, Paul M. Salmon, Guy H. Walker, and Chris Baber, 2005.

Typically the first element in a task unit is the title or description of the task and it is represented by a rectangle. The second element is a group of separate pieces of information or other items needed to accomplish the task or make a decision and it is represented by one or more parallelograms. The third element is a decision and is represented by a diamond. The final element is the output or product of the task or decision and it is represented by a wavy rectangle. This is only a typical representation, and task units will vary with each task analysis using the OSD.

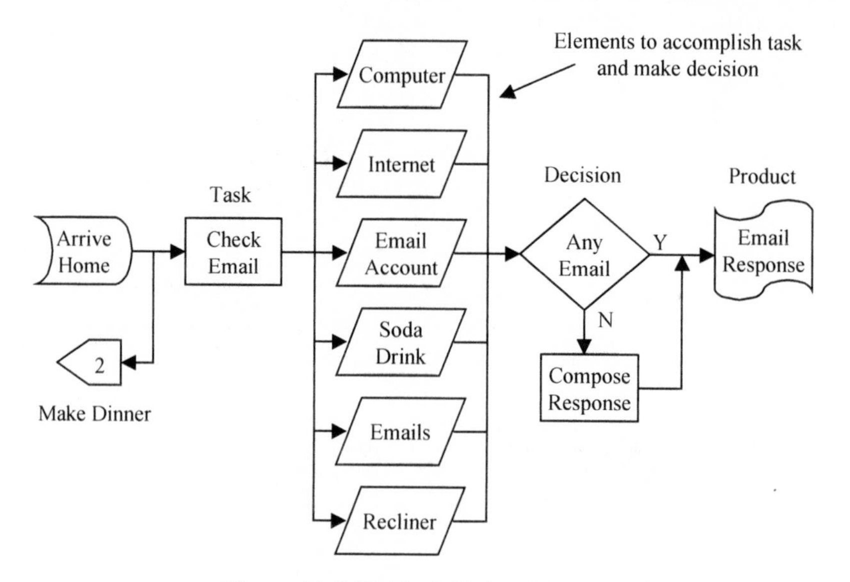

Figure 51 OSD Task Unit - Check Email

The OSD analysis can be simple and represented on a single sheet of paper or complex and represented in a book with many pages. The OSD diagrams reference other OSD diagrams using the task feed symbols. Numbering schemes and naming conventions are used for easy reference and maintaining consistency. Notes can be added to each task unit. The notes can point to other task units. Notes also can be added to the task unit elements. As part of reviews, questions can be written directly on the diagrams with pointers to offending elements. A wall review with all the diagrams displayed in a room can be used to find missing items, inconsistencies, and errors. It also can be used to encourage discussion and stimulate movement to the next levels of system understanding.

Task analysis is most effective when working with physical items such as assembling or disassembling equipment. As soon as task analysis is applied to abstract cognitive activities such as writing a book, painting a picture, or engineering a system, etc the most important element of the task, the creativity and innovation element cannot be defined and the subject is best left to philosophy. However process, including engineering process, can be defined using the OSD approach just as easily as with any other approach.

9.6 Use Cases

A use case is an isolated operational view of the system. It can be represented with a use case diagram and supporting text. It can be incorporated into the ConOps, OpsCon, operational thread and operational scenario information products. The use case diagram is similar to the use case diagram associated with SysML and UML[61].

The diagram shows the system functions and operators with links showing a relationship between the operators and system functions. It is a visual slice used to describe an important story about the system. UML 2 does not allow associations between Actors. However, the relationship between actors is useful in understanding the behaviors between actors.

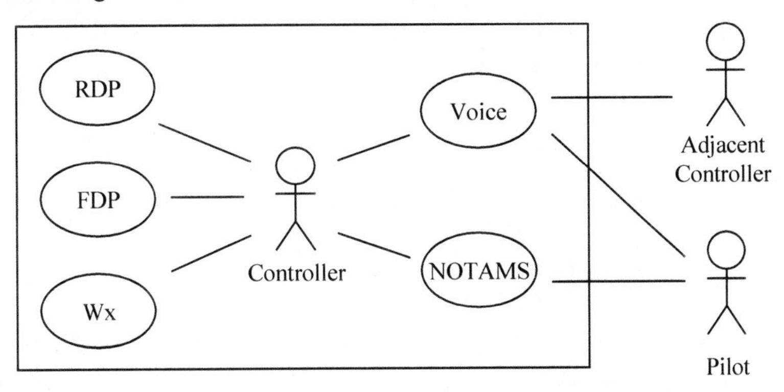

Figure 52 Generic Use Case - Air Traffic Controller

9.7 Transactions

A transaction is an exchange between two entities such that the originator can verify that the recipient received and processed the request. This is typically done with a receipt that summarizes the key elements of the request. Transactions exist everywhere. When you go to a restaurant and order a meal, when done eating the meal, you pay for the meal, and the restaurant gives you a receipt.

The difference between a transaction and a communication with an ACK[62] and NACK sequence, is that the transaction includes the processing, not just the communication. The processing can be withdraw or deposit of money in a bank or the request and acceptance of a flight plan change in an air traffic control computer system. In both cases the request is successfully completed and a receipt is offered to the originator.

---

[61] OMG an international, open membership, non-profit computer industry consortium since 1989. Modeling standards include Unified Modeling Language™ (UML®) and SysML.

[62] In communications the ACK or acknowledgement is sent when a it is received without error. The NACK is sent if something damaged the communication. Parity, Cyclical Redundancy Check (CRC), byte count, etc are used to check the communications integrity.

The key element of a transaction is that both parties are able to verify the exchange, not just the recipient. For example, with the birth of the Internet a new form of business relationship was pioneered called an Associate Program[63] or Affiliate Program. A website provides specially encoded web links that take a user to another website offering a product for sale. If the user buys the product, then the originating website, by prior agreement receives a percentage of the sale. However, there is no way for the originating website to verify if a sale actually occurred. It must rely on the honesty of the website making the sale. This is NOT a transaction.

Micro transactions involve very small sums of money. They were viewed as a method to sell content on the Internet instead of using advertising or associate programs. In the late 1990s, the World Wide Web Consortium (W3C) worked on making micro payments part of HTML but the W3C activity was eventually closed. So there is an overhead associated with transactions. The overhead is processing, communications bandwidth, administrative, monitory, etc.

Today many transactions have a computer based automation system in the middle of the transaction activity. To deal with the possibility of computer system failures, typically data is entered, then once complete the transaction request is submitted for processing. An internal database commitment sequence is initiated. This may take several seconds and the operator is displayed a status indicating the transaction is being processed. If the internal database commitment is successful the transaction status indicates success and a receipt is printed. If not successful, the internal database commitment may be followed by an internal database rollback to restore the account to its previous state. If two transactions attempt to modify the same account at the same time, this is called a deadlock. The internal processing detects the deadlock condition and either one or both transactions are rejected, the database record is rolled back, and the transaction sequences reinitiated with different timing sequences.

9.8 Ladder Diagrams

Ladder diagrams were originally used to show sequences, such as found in transaction processing and communication protocol exchanges. With the rise of object oriented software techniques, such as UML, sequence diagrams have been suggested for software analysis and design. However, it is important to note that sequence diagrams (ladder diagrams in this case) where borrowed by software and they have use in understanding various aspects of a system.

A ladder diagram uses vertical lines to represent interacting elements and horizontal connecting lines to represent activities. It is read from left to right then top to bottom.

---

[63] Patent 6,029,141 Internet-based customer referral system. Internet-based referral system that enables individuals and other business entities ("associates") to market products, in return for a commission, that are sold from a merchant's Website. 2000-02-22 1997-06-27

In some instances a table format is used where the columns represent interacting elements and horizontal lines crossing columns representing the activities or items of interest. Table versions of the ladder diagram were used extensively in capturing keyboard input and display output sequences in display intensive systems of the 1970's. In some cases sample display symbols and text was placed in the columns as needed. The table column format also allowed the addition of free flowing text and notes.

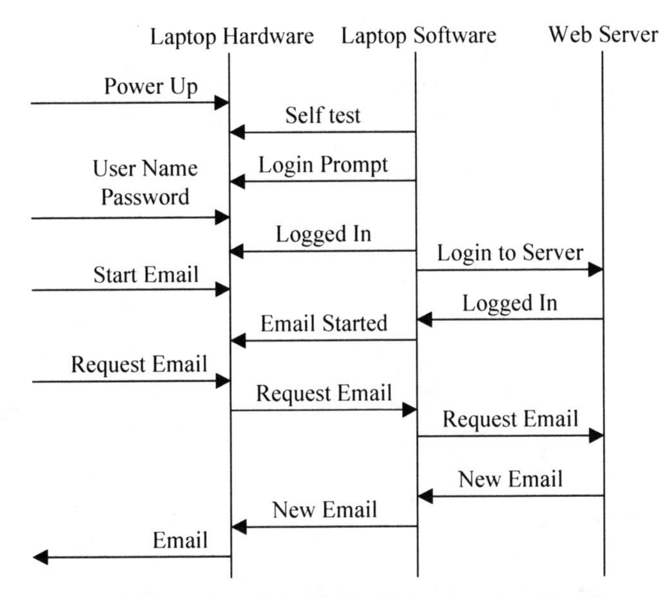

Figure 53 Ladder Diagram - Checking Email

Ladder diagrams were used in air defense and air traffic control systems as early as the 1960's. They were also used extensively to describe the signal line sequence of mainframe computer interfaces in the early 1970's.

Example 19 Thread Based Development

Ladder diagrams can be used to capture the operation of a system in development and the system used to develop the system. The system used to develop the system includes practices that group into methods, and tools that implement the practices. A ladder diagram can quickly and easily capture this information.

Thread based development focuses the team on how the system is used in various settings. The use is captured in OpsCons and then tested on the back end as part of system tests. In the middle an engineering activity, such as software engineering may interpret the OpsCons anew using a formal method such UML. In this example Use Cases are created to fill in the holes from the parent OpsCons. The Use Cases then feed various Object Interaction Diagrams (OID), which form the basis of communication between the software staff for the software design.

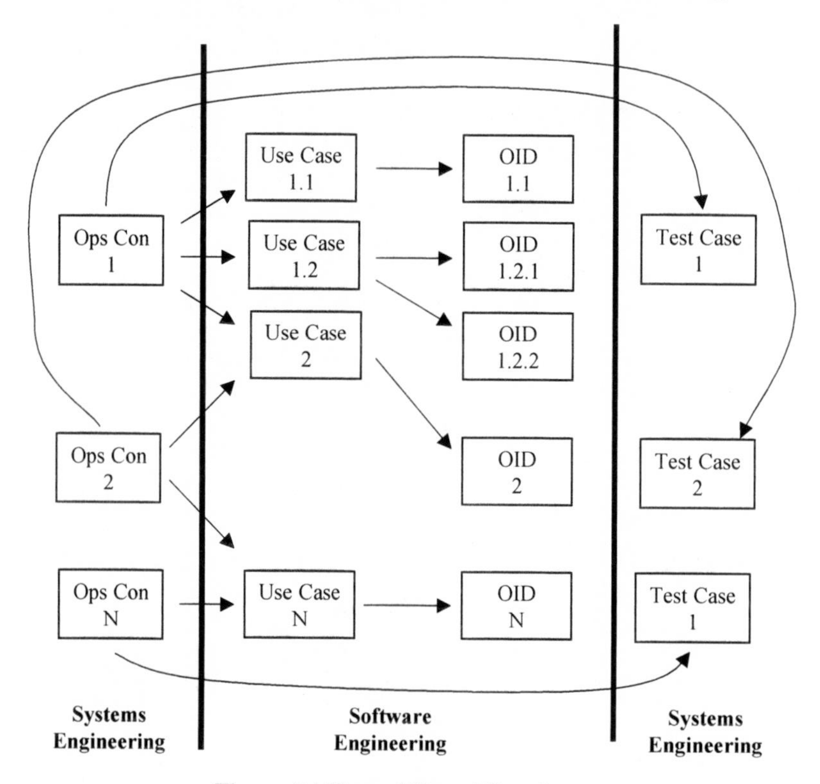

Figure 54 Thread Based Development

9.9 Timing Diagrams

Timing diagrams are used to represent sequence and relative timing. Unlike sequence diagrams, which are text based with no emphasis on timing, timing diagrams emphasize critical timing needs. They were primarily used to show signal timing needs for logic components and computer interfaces. As such they tend to mimic what is shown on an oscilloscope. A timing diagram usually contains many rows one of them being a clock.

For example when an interrupt is enabled, the receiver needs to respond with an interrupt accept within T1 time. If the interrupt is not accepted within T1 time then the sender knows there is a problem. It can then disable the interrupt and then try again some random time later with another interrupt. Alternatively after several attempts to get a successful interrupt accept, the sender can generate an error status. If the interrupt is acknowledged then the sender starts sending the data T2+T3 time later. When the interrupt line drops the data should stop. The clock shows that the edges of the signals coincide with the clock rising and / or falling edges. The enabling signals should always fall within the rising or falling edges of the clock plus some safety time S1.

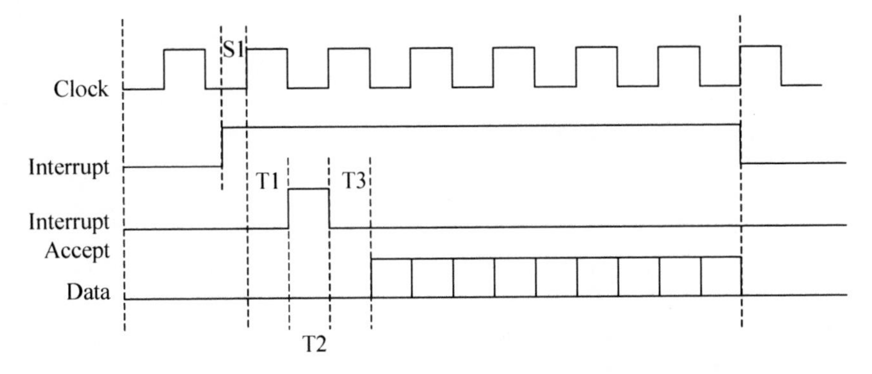

Figure 55 Timing Diagram - Simple Interface Protocol

There are several questions raised by this timing diagram. When does the interrupt signal go down and the data transfer stop so the receiver can be released? Are there other devices on this backplane that need to communicate with the receiver? Are there other devices on the backplane that need to communicate with other devices and if so how are those devices selected? Are there different interrupt priority levels and what happens if a higher priority interrupt is raised?

A backplane is a series of connectors with multiple pins that are connected together so that each pin of each connector is linked to the same pin of all the other connectors. They can be used to form a computer bus and the computer using the bus usually drives the pin signals. There are custom computer buses associated with various computer manufacturers and several popular open computer buses such as S100[64], VME[65], PCI[66], etc. The concept of cards and connector slots was eventually adopted for many non-computer devices such as consumer televisions and audio electronics[67].

---

[64] The S-100 bus was an early computer bus (1974) for the Altair 8800. It was designed for hobbyists so it can be considered the first personal computer bus. It was the first industry standard bus for microcomputers. The S-100 bus consisted of the Intel 8080 pins connected to the backplane.

[65] VME bus is a computer bus standard, originally developed for the Motorola 68000 line of CPUs. It was widely used for many industrial and military applications and standardized ANSI/IEEE 1014-1987.

[66] PCI bus (Peripheral Component Interconnect) was introduced with personal computers. Circuits can be mounted on motherboard, called planar devices in PCI specification, or cards can fit into a connector slot. PCI-X and PCI Express replaced PCI.

[67] The Akai 1730D SS (1971) uses an audio bus to accept multiple common amplifiers implemented as standalone cards.

9.10 Exercises

1. Is a concept of operations similar to a user manual? How are they different?
2. Are operational concepts similar to user guides or frequently asked questions? How are they different?
3. Develop an operational scenario and operational thread diagram for making a dinner. How are they different? Allocate time budgets to the thread.
4. Perform a task analysis by developing an operational sequence diagram for preparing dinner. Does the introduction of formal symbols add to the understanding?
5. Develop a use case diagram for preparing dinner. How is it different from operational, thread, and OSDs?
6. Did you include kitchen clean up after preparing dinner? If not go back and modify the diagrams? If not why do you think you neglected that operation?
7. Are there any transactions in the dinner preparation scenario? Did you include the purchase of the ingredients?
8. Develop a ladder diagram for preparing dinner. Does it make sense to develop a ladder diagram for preparing dinner? How about if the dinner scenario is on a ship, camping, on a space ship?
9. Is it easy to see the link between operational concepts, use cases, object interaction diagrams and test cases? If so why or if not why not?
10. Develop a timing diagram for preparing dinner. Are drawing issues surfacing associated with different time lengths? Can time slices be reasonably represented by breaking then restarting lines?
11. Is the financial system a subsystem, a bus, or both a subsystem and a bus? Is a bus a subsystem?

9.11 Additional Reading

1. CMS Requirements Writer's Guide Version 4.11, Department of Health and Human Services, Centers for Medicare & Medicaid Services, August 31, 2009.
2. Human Factors Methods: A Practical Guide for Engineering And Design by Neville A. Stanton, Paul M. Salmon, Guy H. Walker, and Chris Baber, 2005.
3. NASA Systems Engineering Handbook, NASA/SP-2007-6105 Rev1, December 2007.
4. Operational Concept Description (OCD), DOD Data Item Description, DI-IPSC-81430 1994, DI-IPSC-81430A 2000.
5. Operational Sequence Diagrams in System Design, Kurke, M. I., 1961.
6. Systems Engineering for Intelligent Transportation Systems, Department of Transportation, Federal Highway Administration, Federal Transit Administration, January 2007.

10 Human Factors

When we think of human factors we focus our attention on the modern machine age and how humans interact with our modern day machines. However, human factors are as old as our civilization. You know when you have properly fitting garments. You know when you sit in a comfortable chair. You know when you sit at a nice table of food. How was the decision made for the size of a plate, fork, and knife? How was the decision made for the size of a door, a window, a room size? We can usually detect really bad human factors design decision. However, sometimes we cannot detect subtle human factors design errors. This can result in loss of health such as damaged eyes, ears, vocal cords, muscles, tendons, bones, nerves, etc.

After World War II there was great interest in understanding how we humans interact with our environment. Labs were established and extensive studies were performed in these great labs. Sometimes the studies were trivial and common sense and at other times the studies were significant. In all cases they needed to be reflected in our systems. In 1958 Public Law 85-726 known as the Federal Aviation Act of 1958 established the Federal Aviation Administration (FAA). The following is an excerpt from the Federal Aviation Act of 1958:

"The Administrator shall develop, modify, test, and evaluate systems, procedures, facilities, and devices, as well as define the performance characteristics thereof, to meet the needs for safe and efficient navigation and traffic control of all civil and military aviation except for those needs of military agencies which are peculiar to air warfare and primarily of military concern, and select such systems, procedures, facilities, and devices as will best serve such need and will promote maximum coordination of air traffic control and air defense systems. The Administrator shall undertake or supervise research to develop a better understanding of the relationship between human factors and aviation accidents and between human factors and air safety, to enhance air traffic controller and mechanic and flight crew performance, to develop a human-factor analysis of the hazards associated with new technologies to be used by air traffic controllers, mechanics, and flight crews, and to identify innovative and effective corrective measures for human errors which adversely affect air safety. The Administrator shall undertake or supervise a research program to develop dynamic simulation models of the air traffic control system and airport design and operating procedures which will provide analytical technology for predicting airport and air traffic control safety and capacity problems, for evaluating planned research projects, and for testing proposed revisions in airport and air traffic control operations

programs. The Administrator shall undertake or supervise research programs concerning airspace and airport planning and design, airport capacity enhancement techniques, human performance in the air transportation environment, aviation safety and security, the supply of trained air transportation personnel including pilots and mechanics, and other aviation issues pertinent to developing and maintaining a safe and efficient air transportation system."

We all know that we have our five senses of sight, sound, hearing, taste, and feeling[68]. We also know that our physical bodies can only carry so much, reach so far, withstand certain temperatures, take a certain amount of G forces, pressure, and loss of atmosphere. We also know our bodies need a certain amount of water and food and must excrete a certain amount of waste. This is all common sense for the modern industrial human. However, when we think of our systems, this common sense can easily be lost. Further, the details may be unknown to those outside the specialty-engineering role of human factors.

10.1 Human Factors Basics

When the system boundary is drawn broad enough there is always human interaction. Human factors address physical and cognitive needs of humans. Social behavior and psychological needs are also elements that many tend to ignore. This human interaction or human factors is an extremely important aspect of all systems. Although MIL-STD-1472[69] is broad and deep, it is an excellent starting point for all teams to review. Some of the more important human factor topics include:

- Vision and Displays
- Hearing and Audio
- Voice and Phraseology
- Touch and Tactile Feedback
- Ergonomics
- Taste and Smell
- Sensory Overload
- Feed Back
- Response Time
- Vision Versus Sound
- Visual and Sound Noise
- Circadian Sleep Cycles
- Stress Sources
- Ethics

10.2 Vision and Displays

We can use our basic understanding of system decomposition to identify our bodies' functions and their subsystems. The eye is an example of a subsystem. Understanding this subsystem is key to integrating it into our systems. When we

[68] Human Information Processing, Peter H. Lindsay, Donald A. Norman, Academic Press, 1977.

[69] Human Engineering Design Criteria, DOD, MIL-STD-1472C May 1981, MIL-STD-1472D March 1989, MIL-STD-1472E October 1996, MIL-STD-1472F August 1999.

address vision there are key topics that should be considered:

- Night Vision
- Display Flicker
- Field of View
- Color Use
- Contrast
- Display Distance
- Display Surface Area
- Text Size

An image is projected on our retina, which is lined with rod and cone cells. The eye uses rods and cones to detect light. Rod density is greater in the peripheral retina than in the central retina. The rods are more sensitive than the cones but are only able to detect what we perceive as monochrome or black and white[70] images. The cones require more light but are able to detect color and fine detail.

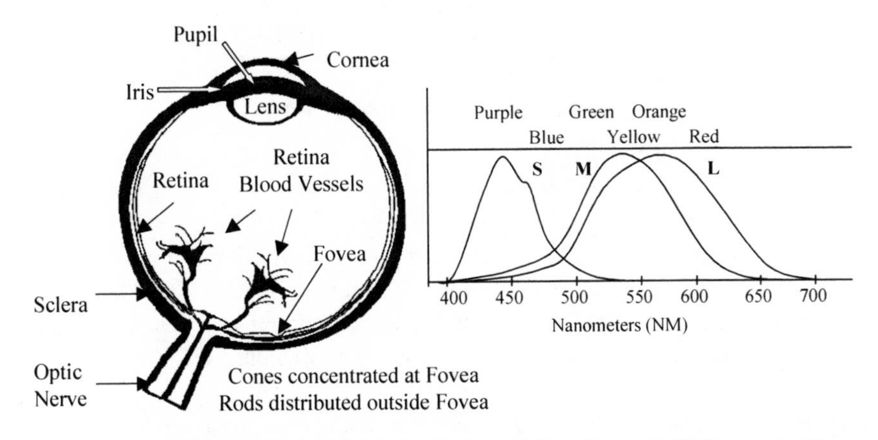

Figure 56 Human Eye as a Subsystem and Eye Cone Cell Response Curves

The cones are densely concentrated near the fovea. There are few cones present at the sides of the retina. We are able to detect more details in the center or the fovea region. Images are also most sharply in focus at the fovea, when directly looking at an object. Not far from the fovea, near the center of the retina there are nerve connections, so there is a blind spot in this region. The eye compensates for this blind spot and different detection characteristics by involuntarily movement. If the eye were to become perfectly stationary the rods and cones would stop firing and the image would fade.

There are three types of cone cells that respond to different frequencies of light. There is overlap between the response curves of the cells. The greatest overlap is with the cells that detect red and green. Cone cells that respond mostly to long light wavelengths are designated L for long and peak in the red region. Cone cells that

[70] Monochrome has shades of gray while black and white (B&W) technology is just black or white. Many use B&W when referring to monochrome images (e.g. B&W television).

respond mostly to medium light wavelength are designated M for medium and peak at green. Cone cells that respond mostly to short wavelength light are designated S for short and peak in the blue region. The signals received from the three cone cell types let the brain perceive all possible colors. Rod cells have peak sensitivity at 498 nanometers (nm) this is approximately the Blue-Green region.

Table 6 Eye Cone Cells and Color Ranges

Cone Cells	Color (nm)	Range (nm)	Peak wavelength (nm)
S	470 Blue	400–500	430
S + M	493 Blue-Green	-	-
M	500 Green	450–630	540
M + L	535 Yellow-Green	-	-
M + L	570 Yellow	-	-
M + L	600 Orange	-	-
L	700 Red	500–700	570

10.2.1 Night Vision and Eye Rods and Cones

As the brightness increases our rods saturate and stop detecting the image while the cones start to activate and begin to detect the image. If the light is suddenly reduced, the rods will not detect any images until they have time to recover. This is referred to as night vision recovery and it can take several minutes for the rods to recover and allow someone see in the low light condition. Light in the red band does not impact rod night vision. So instrument gages in aircraft use red lighting. In facilities where there is the possibility of the loss of all lighting, emergency lighting is red. Maintaining night vision in such settings may save lives.

Automobiles use red lights on the back to indicate their presence and to show when a driver is stopping. If the lights were "white" then the night vision of the person following the vehicle would be lost. The white lights are needed on the front of a vehicle so that the driver can see the road, but now there is the possibility of damaging night vision of drivers from ongoing vehicles. So there is a delicate balance where the lights are directed down towards the road. Some light is permitted to go above the road so signs with light reflecting paint can be seen. The light reflecting paint in the signs allows for less light to intrude into the visual space of the ongoing vehicle. People who do not maintain their headlight alignments or use illegal lights where the brightness is beyond the limits of maintaining night vision capability of on coming vehicles is against the law. It is amazing how something as simple as vehicle lighting results in very complex system interactions that even interface to our system of law.

10.2.2 Display Flicker and Eye Persistence

Our eyes have image persistence. So if an image is flashed very quickly, the image slowly fades. It is this persistence that allows us to see when the eye is moving, viewing film in a theater (that is projecting images at 30 times a second), or viewing Cathode Ray Tube (CRT) images (at some refresh rate). The image persistence for most people is about 1/30 of a second and decays at some rate.

There are two artificial light technologies that are used extensively in our workspaces. The first is incandescent and the second is fluorescent. Incandescent light is a constant light. The brightness does not change or go on or off. Fluorescent light actually flashes and the flash rate is based on the power source. Our power source in the USA is alternating current with a 60 Hz frequency. As a result our older fluorescent lights flash at a 120 Hz rate, twice the rate of the power source. We do not see the flash because our eye has image persistence of 1/30 of a second or 30 Hz. If our electric AC frequency were to slow down, then at some point near 30 Hz many people would start to detect flashing lights from fluorescent fixtures.

In our systems when we couple old florescent lights with CRT displays we will detect display flicker as the two light sources synchronize. Changing the CRT refresh rate reduces the flicker as synchronization is minimized, but not eliminated. Newer fluorescent lamps use high-frequency electronic ballast's that power the light tube at 5 kHz. This is faster than the natural decay rate of the gas in the tube so the light production becomes continuous.

Display flicker is annoying. Further in multiple human-factors lab studies it has been shown that perceived and unperceived display flicker leads to significant human fatigue. This fatigue in an air traffic controller can lead to an erroneous clearance, which can lead to loss of life. So command and control centers are dark with no florescent lights.

Example 20 Office Computer Display Flicker Abuse

Computer CRTs paint an image across a phosphor screen with some persistence level. The slower the phosphor (more persistence) the more smeared a fast moving image appears. The faster the phosphor (less persistence) the more "black time" between refreshes and thus greater the chance of perceiving flashing or display flicker. The phosphor and display refresh rates are tuned to match the eye persistence and minimize flashing. The problem is what happens if a light source that also flashes is introduced into the system (fluorescent lighting). This was a massive problem when computers were introduced that used display monitors based on flashing images (CRT technology) into offices with flashing fluorescent lights.

Studies in Air Traffic Control and other systems in the 1970's showed that display flicker introduced massive user fatigue. It increases stress levels and reduces cognitive ability. If the beating frequency between the two light sources approaches certain rates it could induce epileptic seizures[71]. Although various communities knew this in the 1970's it was ignored as personal computers were rolled out in the

---

[71] Photosensitive Epilepsy, suggestions are to avoid screen flicker rates of 2-55Hz.

1980s. This condition lasted up until approximately 2008[72].

Author Comment: It is unclear what the results were in office settings where people were subjected to fluorescent lights and computer displays. It is also unclear how many users did not know enough to change their display refresh rates to minimize display flicker. Further it is unclear why many computer support departments in these environments did not attempt to deal with this issue that at one time was considered common sense. Simple procedures passed to employees or performed by support staff could have mitigated the display flicker by just changing the CRT display refresh rate. That did not happen and flickering displays were the norm in most settings. It's unclear what the health care and system error costs were as a result of this destructive arrogance[73]. Today Liquid Crystal Displays (LCD) are used in offices. The LCD technology uses static images; there is no refresh rate or flashing images painting at 30 times a second. However as technology shifts display flicker may resurface.

10.2.3 Field of View and Eye Rod Cone Distribution

The visual field of view and what we actually see is determined by the eye rod cone distribution. There are approximately 125 million rods and 6 million cones. The fovea has the highest density of cones and there are no rods at the fovea.

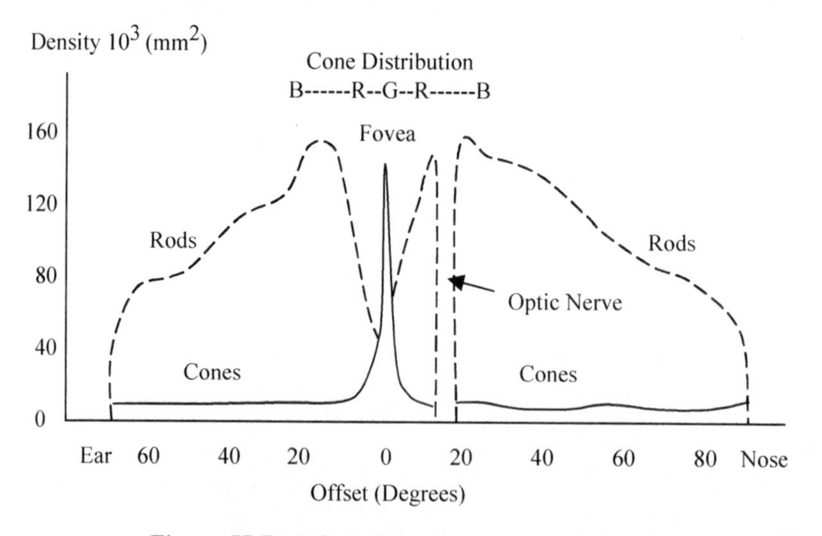

Figure 57 Rod Cone Distribution and Field of View

[72] Many CRT monitors were being replaced by LCD displays by 2008.

[73] It is one thing if a society does not know something, such as display flicker pre 1970's. It is another if it has the data and decides to ignore it and ignore specialists. This is the worst form of ignorance.

The fovea has the highest cell packing density allowing the highest visual acuity or detail detection. The cone cells in the fovea region have smaller receptive fields, than the periphery cone cells. The smaller receptor fields require more light. So there is a tradeoff between light sensitivity and detail detection. Fovea vision is optimized for fine details and peripheral vision is optimized for coarser information. Fovea vision is used to see highly detailed objects while peripheral vision is used for organizing a large scene and for seeing large objects.

The percentage of cones is blue (4%), green (32%), and red (64%). The cones also have a retinal distribution resulting in asymmetrical color detection. The center is primarily green cones (most densely packed for detail), surrounded red cones, with the blue cones on the periphery. The center of the retina has no blue cones. The rod cone distribution[74] and field of view can be graphically represented.

We have two eyes and our maximum field of view is 180 degrees. Our field of view is a function of our eyes, eye movement, and head movement. However the vision is significantly different as the field of view changes. Our eyes have overlapping visual coverage that is approximately 120 degrees. As the field of view is extended beyond 120 degrees the binocular vision is lost. So the ability to judge distance only exits within the 120 degree range. Since the cones are concentrated at the fovea, fine detail vision or focused vision is only within 1 degree[75]. Since the rods are distributed to the periphery of the cornea, as the field of view is extended the image is perceived less in color and more in monochrome or black and white. There are no cones on the periphery. The rods are less densely packed on the periphery so only movement and not shape can be detected at the extremes of the field of view.

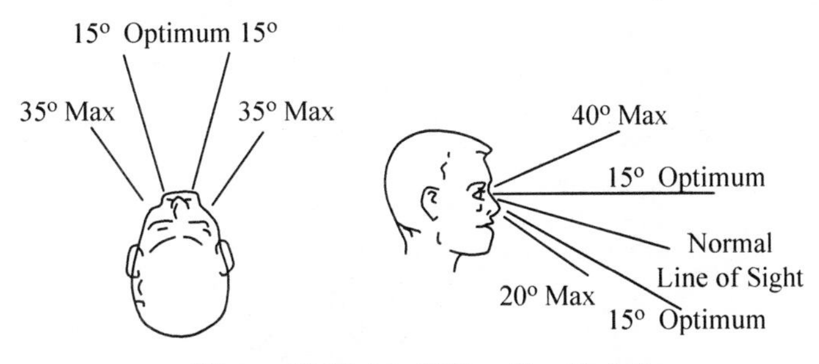

Figure 58 Field of View Eye Rotation

Even though our field of view approaches 180 degrees, what we perceive as

[74] Optics, Eugene Hecht, 2nd Ed, Addison Wesley, 1987, ISBN 020111609X.

[75] From Medical Facts for Pilots, Federal Aviation Administration, Publication AM-400-98/2, August 2002: "To fully appreciate how small a one-degree field is, and to demonstrate foveal field, take a quarter from your pocket and tape it to a flat piece of glass, such as a window. Now back off 4.5 feet from the mounted quarter and close one eye. The area of your field of view covered by the quarter is a one-degree field, similar to your foveal vision."

images and detailed color images is much smaller. This has enormous implications in photography, film, display layout, and anything where people visually interact with a system.

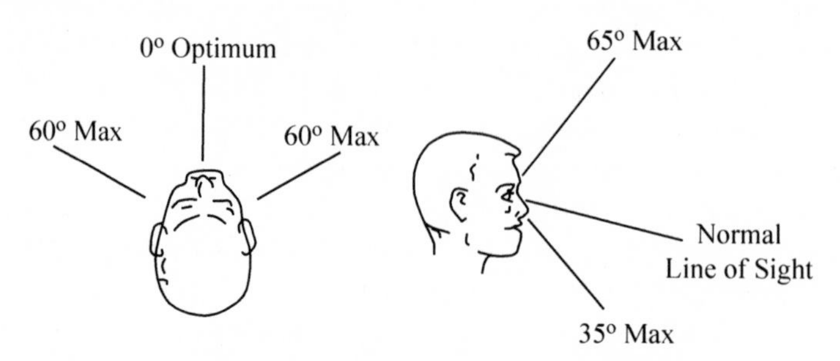

Figure 59 Field of View Head Rotation

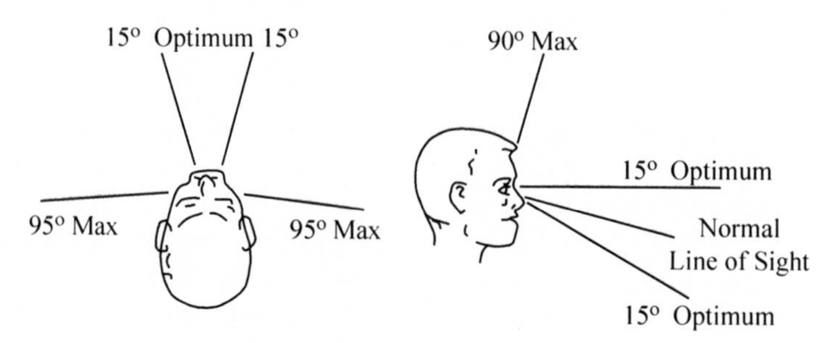

Figure 60 Field of View Eye and Head Rotation

10.2.4 Color Selection Eye Focal Points

The eye lens focuses incoming light on the retina. Different wavelengths of light have different focal lengths. The lens must change its shape so that the light is focused correctly.

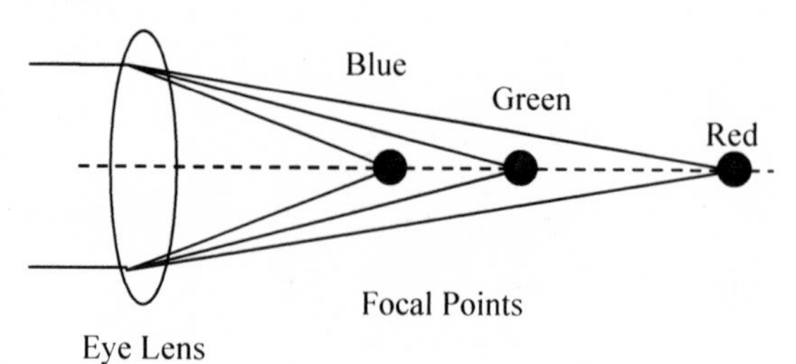

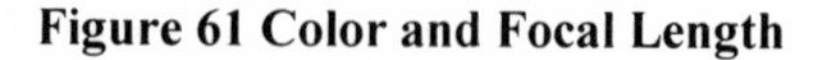

Figure 61 Color and Focal Length

Longer wavelengths have a longer focal length. Red is the longest focal length and blue is the shortest. For an image to focus on the retina, the lens curvature must change with wavelength. Red light needs the greatest curvature and blue light the least curvature. So when pure blue and reds are intermixed, the lens is constantly changing shape and the eye becomes fatigued.

The eye lens is like a prism. When the light enters it is bent. If you look at the prism you will see that blue is bent the most. That means when we look at something with blue, the eye will focus on the majority of the light on the retina but the blue items will be in front of the retina. If there is sufficient blue then the eye will constantly shift focus to place the blue on the retina (everything else is out of focus) then place the other items on the retina (blue is then out of focus).

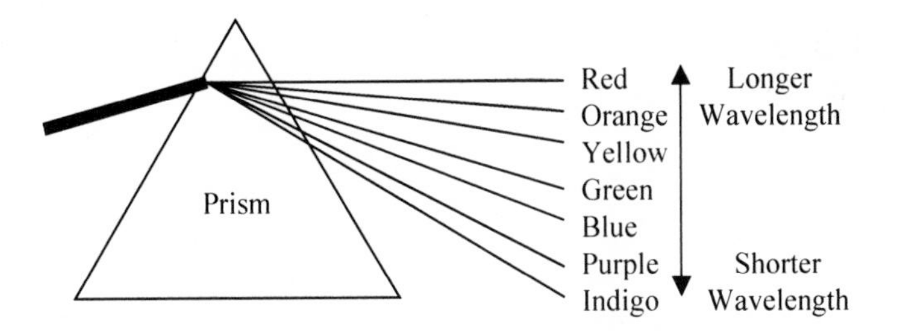

Figure 62 Prism and Light Bending

Notice green is in the center on the prism. Green boards replaced black boards in the 1960's. Studies showed that the green boards led to less eye fatigue, which makes sense based on where the focal length falls relative to white chalk. Also green activates the color detecting cones and rods rather than just the monochrome rods. As we know from eye function the cones are also able to process more detail than the rods.

The lens absorbs about twice as much in the blue region as in the red region. As we age the lens yellows, which means it absorbs more of the shorter wavelengths. So older people are more sensitive to longer wavelengths (yellows and oranges) than they are to the shorter wavelengths (cyan to blue). The eye fluid also absorbs light and it absorbs more light as we age especially blue. So older people are less sensitive to light especially blue. Their apparent brightness level decreases especially blue. This translates to important guidelines:

- Avoid the simultaneous display of highly saturated, spectrally extreme colors. This causes the lens to rapidly change shape and tires the eyes. Use colors that are close together in the spectrum. De-saturate colors wide in spectrum, but pay attention to contrast. Reducing contrast will reduce perception and also cause serious stress. Transparent displays should never be used where operators must safely interact with a system.
- Since there are no blue cones in the center of the retina, Pure blue should not be used for text, thin lines, and small shapes, they are difficult to see. Using blue background color may blur the raster lines on a poor display, but the cost is enormous as the eye constantly tries to focus. This is a serious

stress issue and may lead to eye problems.

- Avoid colors adjacent to blue. Blue does not contribute to brightness and creates fuzzy edges.
- As eyes age higher brightness levels are needed to distinguish colors.
- Colors change in appearance as the ambient light level changes.
- The magnitude of a detectable change in color varies across the spectrum.
- Since there are few cones on the periphery of the retina, avoid red and green in the periphery of large displays.

10.2.5 Contrast

As contrast drops operator fatigue increases. This can lead to system errors. So it needs to be addressed in the system displays.

Contrast is affected by ambient light conditions. The brighter the ambient light the less contrast in a display, it gets washed out. As display brightness increases in an ambient light setting, its contrast improves. However there comes a point where the display brightness will overload the operator eyes. This suggests that maximum contrast options should be selected when laying out alphanumeric symbol based (text) display information.

When displaying text, maximum contrast always should be used. Black background should use full brightness green and or red text. Full black (not gray or almost black, but black) text should use full brightness white background. A gray background can use black, red, green, and yellow text.

Transparent windows have very poor contrast and include visual noise (unnecessary visual data) that can lead to fatigue or visual sensory overload. The following are some quantitative contrast suggestions:

- Color options should have a color contrast ratio greater than 1:5.
- Contrast between light characters and a dark screen background should be greater than 6:1 and 10:1 is preferred.
- Contrast between dark characters on a light screen background should be greater than 1:6 and 1:10 is preferred.
- In bright ambient light to attract attention and to sharpen edges contrast ratio should be greater than 7:1.
- In dark ambient light with continuous reading, contrast ratio should be greater than 5:1.
- To hide image or display problems and smooth edges contrast ratio should be less than 3:1.

10.2.6 Color Use

Color is a good coding approach, but it needs to be used in a way so that if color is lost the information can still be provided. There are color perception differences between people and some users are unable to reliably differentiate color codes in a display. Using another code, such as text redundantly, along with the color is the best way to ensure that all people can detect the information.

To preserve night vision red background lighting (greater than 620 nm) should be used. Adding blue and white lights, especially strobe lights, not only damages night vision but also causes the eye to severely defocus. The defocusing is most severe in nighttime conditions.

For black background displays green text, symbols, and lines provide the highest visual perception. Offering red and variants of red green mixture such as orange and yellow are the most effective use of color in dark ambient light conditions with black or gray display surfaces.

When offering color pure blue must be avoided because the eye will defocus, visual acuity will be lost. Using blue background to defocus the eye so that a defective background, such as poor display resolution or noisy display circuits, is not only unethical but it also degrades system performance as the eye is constantly refocusing. This significant unnecessary stress results in fatigue, eyestrain, headaches, visual perception errors, and impaired cognitive ability, which can lead to system errors.

Blue detecting cone cells are further away from the fovea suggesting blue visual alerts at the visual periphery extremes can be used to attract the attention of an operator. Blue light emitting diodes (LED) use very little power for the relative amount of light they emit. Ideally white light should be used to attract attention to extreme visual peripheries unless the background is bright and the white alert contrast is poor. However something as important as an alert always should be placed within the most sensitive field of view area. That is 6 degrees, followed by 60 degrees, and at the worst case 120 degrees of the visual field of view.

If blue is used it should be mixed with red and green. This will reduce some of the focus issues, however adding mixtures of the other base colors to a primary color moves the color to more of a pastel shade. This is good for an interior room wall, but is a problem when high contrast ratios are needed such as when displaying text on a background.

Status indicators need to use consistent color coding schemes based on Red, Yellow, and Green. These colors are closest in depth of field and are thus in reasonable focus on the retina.

- **FLASHING RED**: Used for critical alerts and failures. Indicates emergency conditions, which require immediate operator action to avert impending personnel injury, equipment damage, or both. Indicates personnel or equipment disaster.
- **RED**: Used for critical alerts and failures. Master summation if there is a no-go, error, failure, or malfunction. It alerts the operator that the system or any portion of the system is inoperative, or that a successful operation is not

possible until appropriate corrective or override action is taken.

- **YELLOW**: Used for warnings. Indicates marginal condition, caution, or impending danger. Alerts the operator to situations where caution, recheck, or unexpected delay is necessary.
- **GREEN**: Used for system normal. Indicates a function is activated, equipment is in tolerance, ready, or a condition is satisfactory and that it is all right to proceed. The green indicator should be larger, and preferably brighter, than all other indicators.
- **WHITE**: Used to show system settings. Indicates normal system status conditions that do not have right, wrong, or failure implications. These normal status conditions are alternative functions (FM selected) or state conditions (on, off, test in progress, function available).
- **BLUE**: Used for advisory. Blue should be avoided because of the focus issues. If blue must be used it should be mixed with red and green (pastel version of blue) minimizing the focus issues.

10.2.7 Blink or Flash Rates

Blink or flash rate should range from 2-5 Hz. The higher the blink rates the more urgency perceived by the operator. When developing a status indicator blink or flash rate concept, a common blink rate should be used for all the indicators. The number of blink rates should be limited to three blink rates (slow, medium, and fast) although MIL-STD-1472 suggests only two flash rates. The following general practices should be used:

- **Duty cycle**: A measure of the on and off time. A 50% duty cycle is preferred. The percentage of "on" time should be equal to but not less than the percentage of "off" time.
- **Flash rate**: No more than two flash rates should be used according to MIL-STD-1472, however this text suggests three flash rates may be used in some applications. The flash rates should differ by more than 2 Hz. The higher flash rate reflects more critical information and should be less than 5 Hz (6 Hz with three flash rates). The slower flash rate should be greater than 0.8 Hz. Flashing should be synchronized or massive cognitive overload may happen and the operators could freeze.
- **Text**: Characters that must be read should not flash. An adjacent flashing symbol or flashing background can be used on text that must be read.
- **Flash suppression**: Event acknowledgment or flash suppression control needs to be provided.
- **Flashing area**: Only a small area of a display should flash at anytime.
- **Flasher device failure**: If the display is energized and the flasher device fails, the light should illuminate and burn steadily preferably at higher

brightness level.

10.2.8 Warning and Caution Displays

A warning or caution display should be presented in a location so the operator has a greater probability of detecting the triggering condition than normal observation would provide in the absence of the display. Warning displays are red and caution displays are yellow.

- **Warning Displays**: Warning displays are flashing red. The flash frequency is 3-5 Hz with a 50% duty cycle. The flash rates for all warnings need to be synchronized. If used with caution displays, warning displays should be coded to be easily distinguished from caution displays.
- **Caution Displays**: Caution displays are yellow. A minimum of two different characteristics (text and color) is used for rapid identification and interpretation of caution displays. If used with warning displays, caution displays should be half the intensity of the warning signal. If cautions take the form of flashing text, the text should flash at a rate not greater than 2 Hz with a duty cycle (on off interval) of approximately 70% on.
- **Text Height**: Text height for visual warning and caution displays are 8.7 to 17.4 mrad (30 to 60 minutes of subtended arc) as measured from the longest anticipated viewing distance, with the larger size used where conditions may be adverse.
- **Co-location**: Warning displays and the information needed to respond to them should be grouped in a single location. When textual information about warning conditions are listed in a single location, warnings and caution information should be grouped separately and the user should have the option to list warning and caution messages in priority, chronological or last in order.

10.2.9 Display Distance Placement and Text Size

Display distance impacts the field of view, display size, minimum size of text and symbols, and the eye's degree of adjusting to close up vision. As display distance decreases a point is reached where the eyes start to excessively cross to focus on an object; This is fovea based vision. This will cause fatigue, eyestrain, and headaches. These all will reduce system performance and impair cognitive abilities. As display distance decreases, display quality issues surface, such as poor display resolution, repeatability, focus, etc.

- **Minimum Viewing Distance**: This is a subject of much controversy but the bottom line is that the greater the distance from the display surface the less eyestrain for a user[76]. This needs to be balanced with the proper size of display objects and the reach of the controls around the display. The following are various guidelines that exist in the community:

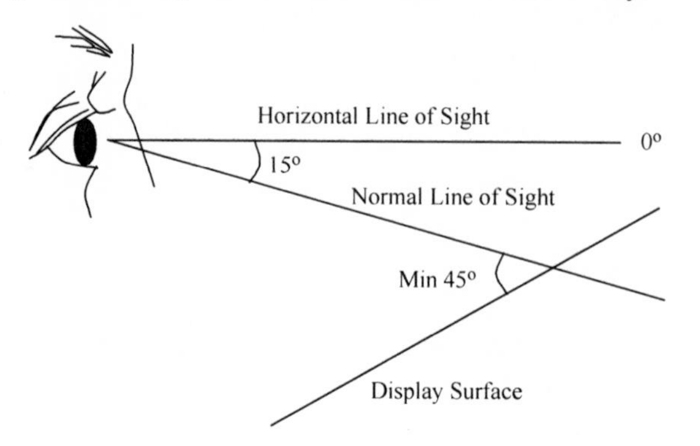

Figure 63 Line of Sight

- ❑ The effective viewing distance to displays, with the exception of cathode ray tube displays and collimated displays, should be more than 330 mm (13 in) and preferably greater than 510 mm (20 in).
- ❑ Some suggest preferred viewing distance is 20 to 40 inches others suggest 18 to 24 inches. Notice the significant difference.
- ❑ Computer users should view their display from a safe distance, which is between 21 and 27 inches. Closer distances risk serious eyestrain, headaches, and other problems that can result from overuse of the eyes.
- ❑ A 50 cm (20 in) viewing distance should be provided. When periods of display observation is short, or when dim signals must be detected, the viewing distance may be reduced to 25 cm (10 in). Design should permit the observer to view the display from as close as desired. Displays which must be placed at viewing distances greater than 50 cm (20 in) due to other considerations should allow the user to modify display size, symbol size, brightness ranges, and resolution.
- ❑ Air defense and air traffic control consoles in fixed facilities used displays that were approximately 32 inches from the operator eyes to the center of the display. This text recommends this distance. It is proven over decades of safe operations with tens of thousands of personnel spending their entire careers at these displays effectively performing their tasks with minimal side effects.

- **Viewing Angle**: The viewing angle is the normal line of sight of an operator. This is 15 to 20 degrees down.
- **Angle of View**: Displays should be perpendicular to the operator's line of

[76] It is well known that people who work close-up on objects eventually develop vision problems.

sight (have a 90° viewing angle at screen center) whenever feasible. No part of any display including secondary displays should present a viewing angle of less than 45° from the operator's normal position.

- **Participants Angle of View**: All critical observers need to view the display as perpendicularly as possible. In no case should observers be forced to view displays at an angle of less than 40° from the plane of the screen.
- **Color Object Size**: For accurate color perception the major dimension of isolated large symbols subtends greater than 8.7 mrad (30 min) of visual angle, preferably 13.1 mrad (45 min). The height of small symbols and characters subtend greater 5.8 mrad (20 min) of visual angle, as measured from the longest anticipated viewing distance, and should have a color contrast ratio not less than 1:5. For fovea vision under relatively high levels of illumination use symbols that subtend greater than 5.8 mrad (20 min) of visual angle.
- **Alphanumeric Size**: Alphanumeric characters should be scaled to subtend at least 15 minutes of arc; other complex shapes should subtend at least 20 minutes of arc.
- **Character Height**: The visual angle subtended by height of black-and-white characters should be greater than 4.6 mrad (16 min) with 5.8 mrad (20 min) preferred. The visual angle subtended by height of color characters should be greater than 6.1 mrad (21 min) with 8.7 mrad (30 min) preferred.
- **Character Stroke Width**: This applies if stroke display technology is used. Stroke line width is greater than 1/12 but less than 1/6 the number of pixels used for character height. This offers a open character while also offering a reasonable line width.
- **Font Characteristics**: Font style should clearly differentiate similar characters (letter L/number 1, letter Z/number 2). A common standard font should be used. Where users must read quickly under adverse conditions (poor lighting), a sans serif style should be used. Text should contain a conventional mix of uppercase and lowercase letters. The use of all capital letters should be limited to abbreviations and acronyms.
- **Grayscale**: There should be a minimum of at least five grayscale levels. For handwriting, resolution of fine detail, or complex image interpretation, a minimum of eight distinguishable grayscale levels should be provided.
- **Geometric Stability (jitter)**: Over a period of one second, the movement of a picture element is less than 0.2 mrad (41 sec) of visual angle.

Character height can be calculated using Tan(theta) = S/D where theta is the visual angle in radians, degrees, minutes and seconds, S is the character height, D is the viewing distance normal to the screen. So S = (Tan (theta))/D. Character height should never be less than 2.5 mm or 0.098 inches[77].

10.2.10 Display Technologies

When we develop systems and they use some form of visual interface a display technology or approach is used. There are two approaches to offering a displays: continuous and dot based. An example of a continuous display is a painting produced by using brush strokes. An example of a dot-based display is a painting, which uses a paintbrush but the paintbrush creates an image by using "dots". As the user steps away from the painting the dots are integrated by the eyes and a continuous image is offered. The display resolution of a continuous display is driven by the size of the paintbrush and the ability to steadily hold the brush with minimum pressure. The display resolution of a dot-based display is driven by the size of the dot. So if a paintbrush and its user has the ability to paint a small dot or line then the continuous display will offer a higher resolution display. The dot-based display resolution is cut off at the level of the size of the dot. If someone can paint smaller than the dot, then information or resolution is lost. That is a very important relationship, and needs to be considered at all times.

Displays can be illuminated from the back and front or they can operate from ambient light. A poster uses ambient light in the daytime. During the night artificial front lighting illuminates the poster. In the last century front illuminated labels evolved to back illuminated indicators, to Cathode Ray Tubes (CRT), various forms of Liquid Crystal Displays (LCD), Plasma Displays, Light Emitting Diode Displays (LED), and Digital Micromirror Device (DMD) displays or Digital Light Processing (DLP)[78], Liquid Paper. The following are various display technologies and their fundamental characteristics:

- **CRT Monochrome**: A monochrome CRT uses a continuous phosphor painted on the screen surface. It is a continuous display surface. Its ability to paint progressively smaller lines and symbols is driven by the bandwidth of the deflection electronics. When this technology was developed and standards established mid last century the deflection electronics were implemented with tube technology. By the turn of the century the tubes were displaced with exotic semiconductors with bandwidth to drive CRT technology to very high levels of resolution.
- **CRT Color**: A color CRT no longer uses a continuous phosphor. Instead phosphor dots are painted on the screen surface that produce red, green, or blue light. Mixing the three colors produce the desired color. This display resolution is limited to the size of the dots. Eventually the dots were

---

[77] Electronically / Optically Generated Airborne Displays, DOD Handbook, MIL-HDBK-87213A February 2005, MIL-HDBK-87213 December 1996.

[78] DLP is a trademark of Texas Instruments.

replaced with small vertical rectangles. This eliminated the dead space between the dots and offered a brighter picture. It also was less prone to color errors with less bleed across the different colors.

- **CRT Color Projection**: Color projection uses 3 contiguous phosphor CRTs (red, green, and blue) to project an image[79]. The electronics evolved to such a high level of precision and accuracy that it was possible to align the CRT images and offer extremely high-resolution contiguous stunning color displays.
- **Liquid Crystal Light Valve**: This large screen display uses 3 contiguous phosphor CRTs (red, green, and blue) to modulate a liquid crystal light valve illuminated by an arc lamp. This display is used for large audience venues and in command and control centers. In recent years the large venue audience displays have shifted to DLP color displays[80].
- **LCD Color**: Color LCD technology moved into television and computers by the early part of the new century. They are a dot-based display. The resolution of these displays is limited to the dots per inch. There are also mismatches between image decay, the human eye, and the standards. As a result fast moving images appear unnatural. This is being addressed in part by increasing the display refresh rates. Their image quality is significantly lower than CRT projection displays but they offer a small package. It is easy to pack three or more LCD units in the same shipping space as a single projection unit
- **Plasma Color**: Plasma display use small gas tubes that emit light when stimulated. They are a dot-based display. The resolution of these displays is limited to the dots per inch. They are being displaced by LCD displays as LCD display characteristics such as contrast ratio are improving.
- **DLP Color**: This is an interesting mix of dot-based display and an apparent continuous image. It uses mirrors that represent dots to reflect light. The reflection can be modulated and there is light bleed. Dots are not visible with these displays. However bright scenes tend to throw large amounts of light and distort the brightness of the scene. DLP color displays have started to move into movie theatres. The resolution is less than film, but the image is steadier than an old worn film projector. Contrast ratio is not as good as film and there are issues with bright scenes in the theatre applications.
- **Electronic Paper**: Electronic paper used ambient light. There is no back lighting. They are ideal for any light conditions including in bright sunlight such as on a beach. They have started to move into electronic book readers. They are monochrome and update too slowly to support moving images.

---

[79] Hughes Aircraft drove this technology, it eventually moved to home television.

[80] Hughes Aircraft was the primary supplier of liquid crystal light valve large screen displays. They were extremely popular in the 1980's.

- **Film Projection**: Film is over 100 years old but its display characteristics exceed all other display approaches. A 35MM image can be projected on a movie theatre screen 100 feet across and there are no lines, dots, or grain. Film can resolve 5000 lines per inch[81].

It is possible to quantify various display characteristics however as technology changes the comparisons become tricky and new measures of performance may be needed for effective comparison. Some of the performance characteristics to consider when selecting a display approach are:

- **Spot Distortion**: Spot diameter should not vary by more than a ratio of 3:2, for any two points on the screen. Distortion should not be sufficient to cause obvious non-linearity anywhere on the screen when viewing alphanumeric formats or picture images. Circles should appear as circles across the entire screen and not as ovals or other variants across the screen. The image should be a faithful presentation, circles are circles and not ovals (wrong aspect ratio used).
- **Geometric Distortion**: The combined effects of all geometric distortion should not displace any point on the display from its correct position by more than 5% of the picture height.
- **Brightness Linearity**: Brightness linearity is a measure of the uniformity of the brightness across a display surface. It is also a measure of the faithful reproduction of the bright ranges once compared to the original source. In a digital system this is driven by the number of bits dedicated to the brightness channel and associated analog compression expansion prior to digitization. In analog systems this is driven by the linearity of the analog circuits and the display.
- **Position Linearity**: This is a measure of the ability for the display to reproduce positions of elements once compared to the original source.
- **Contrast Ratio**: Contrast ratio is an example where care needs to be taken to develop an effective metric and test. One reasonable test approach uses alternating white and black squares to determine contrast ratio. However measurements are affected by ambient light. A display may offer a better contrast ratio under high ambient light conditions than another display with less maximum brightness output but the lower brightness display may offer a superior contrast ratio in low ambient light conditions. The best displays from best to worst are Film Projection, CRT, Plasma, DLP, and LCD.
- **CRT Safety**: When using Cathode Ray Tubes (CRT) a transparent safety

---

[81] FUJICHROME Velvia 50 Professional [RVP50] has a resolution of 160 lines/mm or 6299 lines/inch. The 5000 line per inch number was used for film at Hughes circa 1981 as a goal for one day offering electronic displays with the same resolution and color as film.

screen, which may be part with the CRT faceplate, is used to prevent implosion injury. Protection is also needed against low-intensity X-radiation.

- **Glare**: Glare management is a large issue in low and high light conditions. This is handled by adding coatings to a display surface or Fresnel lens that not only directs the emitted light but also reduces glare.
- **Bezel**: When using multiple displays a matted surface finish should be used on the bezel to avoid reflection. The reflection adds visual noise.

10.3 Hearing and Audio

Just as the eye and its ability to see is related to displays, the ear and its ability to hear is related to audio. The ear is able to hear from 20 to 20,000 Hz. We have two ears with a visible area called the pinna or auricle or auricula. The pinna is one of the mechanisms used to help localize the source of a sound.

A simple experiment anyone can perform is to slide the middle finger across the thumb to make a noise. And sweep it around your head while making the noise.

Starting at the right sweep to the left and notice how the sound location changes. Sound travels at a fixed speed. So there is a small time delay of sound arriving at the farthest ear because it needs to travel further. Most of the location information is associated with the arrival time of the sound at each ear. As the location changes the frequency response also changes ever so slightly. At the extreme sides the signal contains more high frequency sounds. So localizing sound in front uses not only arrival time of sound at each ear but also the subtle frequency shift and the perceptual template residing in the brain from previous experience.

Continue around your head and sweep from left to right behind your head. Notice how you can determine that the sound is coming from the back. That is because the signal has less high frequency content. The pinna filters the higher frequencies more than the lower frequencies, as it needs to travel through the tissue.

In addition to the pinna there are other elements that affect our perception of sound. These elements are part of the environment and the ear operation. Other elements to consider that are fundamental to audio and hearing:

- **Spatial Separation**: The ear cannot determine direction of low frequency content below approximately 100Hz. Spatial separation increases with increasing high frequency content.
- **Air Attenuation**: The air tends to attenuate higher frequencies more than lower frequencies. It is very good at removing distortion such as clicks, pops, hiss, and even amplifier clipping which introduces square waves into the signal.
- **Voice Optimization**: The ear pinna and ear canal are optimized to the frequencies associated with normal speech frequency response.

- **Ear Damage**: The ear can be damaged by loud noise. It also can be damaged by long exposure to noise, not necessarily very loud, of the same frequency content, such as a room loaded with machinery.

10.3.1 High Fidelity Recording Playback

When we go to a concert we hear sound coming from the front of the stage and we hear sound coming from reflections in the auditorium, including the back walls. Music recording engineers understand these issues and try to duplicate the characteristics of world class auditoriums. This is an art of multiple microphone placement and mixing of the various sound sources[82].

If you place two microphones, one at each end of a stage the sound will arrive at different times to each microphone. For example a guitar signal on the left will arrive at the left microphone first then the right microphone. In addition echo from the various auditorium structures will arrive at each microphone.

If you mix a portion of the left microphone signal on the right there is a component that arrives on the left and right channel at the same time. Now here is where it gets interesting. You can leave the mix as is, with the very low-level echo or you can combine the left and right channel, but flip one of the channels by 180 degrees. This can be accomplished with a transformer or amplifier stage. This will cancel the primary signal coming from the stage of the left guitar and what will be left are all the sounds arriving at different times. This is called ambient sound. This ambient sound can then be sent to the left, right, or both channels.

When played back on a stereo the user can modify the recording engineers' mix of ambient sound by flipping the phase of one of the channels and mixing it into the primary channels. In the past this was easily accomplished with a resister between the negative amplifier terminal and both speakers' negative terminals. Today's amplifiers use dual push pull amplifiers with a floating negative and the technique cannot be used. However, these capabilities have been added in the preamplifier stages of many modern amplifiers. In addition, characteristics of the pinna have been simulated and offered in sound processing systems[83]. So the user can turn a knob and change their location in the auditorium from the front to the back as desired.

The details of high quality recording and mixing can only be appreciated when the actual signal is preserved. There are various metrics that evolved with the discipline and they include:

---

[82] Some recording studios in the 1950's and 1960's had basements with cinderblock tunnels to try to simulate the effects of auditoriums.

[83] The Hughes Aircraft Sound Retrieval System (SRS) used in high fidelity simulator sound systems was commercialized and offered in many consumer products starting in the late 1980's.

- **Frequency Response**: This is a measure of a system to faithfully reproduce audio frequency ranges. Even though our ears hear between 20 to 20,000 Hz, some suggest there is interaction between higher frequencies and so they should be captured. Also the frequency response is rarely flat and so a system that has a broader frequency response will have more of a flat response in the hearing range. Typical goal 25 to 33,000 Hz +/-3 dB.
- **Wow and Flutter**: In the past audio was captured on electromechanical devices. They were based on motors and transport mechanisms that had speed variations or vibrations from less than perfect mechanical parts (bumps and not round). Typical goal less than 0.03% WRMS.
- **Signal to Noise Ratio**: Recording media inherently has some level of noise. The noise is increased by electronics. The higher the signal to noise ratio the better. Compressors and expanders can significantly improve signal to noise ratio but they add non-linearity to the audio level. Typical goal uncompressed better than 65 dB, compressed expanded greater than 128 dB
- **Dynamic Range**: This is the range of loudness available in the system. The higher the dynamic-range of the system the better the sound reproduction. Typical goal dynamic is 128 dB with compressor expander with 1 kHz peak recording level.
- **Harmonic Distortion**: When a signal is introduced into an audio system the audio system produces artifacts not present in the original signal. This is determined by submitting a signal such as a 1 kHz Sinusoidal wave and measuring the amplitude of other frequencies that are added by the system. The typical goal is less than 0.4%.
- **THD at RMS Power**: Total Harmonic Distortion (THD) at Root Mean Square (RMS) power is a measure of harmonic distortion of a power amplifier. As the power is increased the THD increases. The RMS power is the maximum power before the amplifier begins to produce square waves. Square waves are a sign of massive THD. They contain many frequency components at high amplitude levels. The typical goal is less than 0.7%.
- **Intermodulation Distortion (IMD)**: IMD is a measure of the introduction of unwanted signals when two signals are input into the audio system. Typical goal should be less than 0.1%.
- **Channel Separation**: All systems have cross talk. The level of cross talk increases with modulation techniques and fancy mixing of channels on physical media. Typical goal greater than 65 dB.
- **Quantization Distortion SQNR**: This is associated with converting an analog signal to digital format. It is a result of the number of bits available with each sample. The total number of bits represents the loudness levels in

> the system. Fewer bits lead to higher quantization distortion[84]. For example 16 or 24 bits is very bad for quiet passages. An approximate formula is SQNR = 6.02 X number of bits. The problem is the quiet passages have huge levels of quantization distortion (24 dB). Some suggest this poor quality has changed the nature of music and moved it to highly compressed signals with lots of boom-boom. All ambience or ambient sound is lost.

Care needs to be taken when using metrics or specifications when comparing different technologies. These specifications or metrics come with various test approaches that may need to be expanded when a different technology is introduced.

For example the frequency response of FM radio and television is 30-15 kHz, CD is 20-20 kHz, professional and audiophile audio 20-22 kHz, and AM radio extending up to 5 kHz. However the CD frequency response is very misleading. The Nyquist[85] sampling rate of 44 kHz drives the CD frequency response. So the maximum frequency response is 22 kHz, beyond that there is severe noise and so an analog extreme cutoff high-pass filter is used to remove frequencies past 20kHz. This is in severe contrast to analog systems, which have a gentle roll off of frequency content. So in a CD based system there is nothing after 22 kHz. In an analog system there is content well beyond its rated response. An analog system rated at 20-22 kHz will have content at 23 kHz, 25 kHz, 30 kHz albeit at ever lower sound levels.

Another example is that a different technology may call for the introduction of a new metric that just does not apply with other technologies. For example a digital system uses some number of bits to capture the audio level at an instant in time (a sample at the 44kHz rate). If there are not an infinite number of bits, then the level will not follow the signal smoothly but will jump based on the next bit level. This is referred to as quantization distortion. For very quiet passages the quantization distortion can approach infinity for small bit systems such as 16 and 24 bits. The approach to dealing with this is to use an analog Compressor and Expander, Compander[86], to keep the analog to digital converter full with bits. The problem is this approach introduces non-linearity into the signal and does not fully address the quantization distortion issue. Analog systems have no quantization distortion.

---

[84] Discrete Time Systems, James A. Cadzow, Prentice Hall, 1973, ISBN 0132159961.

[85] The Nyquist–Shannon sampling theorem establishes that to reproduce a signal the sampling rate needs to be twice the rate of the highest frequency component in the signal.

[86] In a Compander the signal is sent to a non-linear transformation prior to transmission. A reverse of the transformation happens at reception. The transformation boosts the quiet portions and reduces loud portions. Noise is reduced because the quiet signals are louder, compared to the noise in the transmission channel. In this case more bits are used to represent a quiet signal.

Author Comment: Some suggest that moving to digital media has changed the nature of music. To reduce quantization distortion and file storage size the analog signals are severely compressed prior to digitization. This reduces the quantization distortion but introduces analog compression expansion non-linearity, another form of distortion. So everything is at the same loudness levels. Also there is a shift to very low bass content which is has better reproduction because of massive over sampling. Additionally, the severe cutoff of frequencies at the sampling rate to prevent noise from insufficient sampling has shifted the focus away from producing sound with high levels of separation and ambience.

10.3.2 Voice Grade Application

Voice grade applications include telephones, public announcement (PA) systems, and automated announcement systems. There is less of a need for high fidelity because the human voice tends to be range only up to 8 kHz.

- **Telephony**: The US telephone useable frequency range is 300-3400 Hz. Empirical data before the breakup of AT&T suggested 220-3.3 kHz with 3.2 kHz for short and medium connections and 2.7 kHz in long distance connections. The allocated bandwidth for a single voice channel is usually 4 kHz, including guard bands, allowing a sampling rate of 8 kHz for digital transmission.
- **Voice Frequency**: Voice Frequency (VF) or Very Low Frequency (VLF) is the band in the electromagnetic spectrum allocated to 300-3000Hz.
- **Human Voice Range**: The human voice range is approximately 200-7000 Hz. There are differences that occur with age and gender, 120 Hz for men and 210 Hz for women.
- **Normal Speech Rate**: A normal speech rate is 150 to 180 words per minute. Slow speech rate is approximately 100 and maximum intelligible speech rate is approximately 300 words per minute.
- **Dropouts**: Dropouts are a recent phenomenon associated with cellular phones and early IP phones. The longer the dropout and greater the number of dropouts the less intelligible the signal. With a normal speech rate of 150–160 words per minute a 1-second dropout is significant.
- **Intelligibility**: Speech intelligibility decreases with decreasing bandwidth. For single syllables, 3.3 kHz bandwidth yields an accuracy of only 75 percent, as opposed to over 95 percent with 7 kHz bandwidth

10.3.3 Noise Response

When exposed to loud sound such as loud music we tend to get used to the loud sound. When we leave the loud sound setting we notice that normal sounds are very quiet and our ears may ring. Eventually, if there is no damage to the ears our hearing returns to normal. The basic mechanisms inherent in this ability allow us to filter noise.

When we are exposed to sound with pink noise in the background such as static or hiss, the static or hiss tends to eventually fade and we just hear the sound. This characteristic is most pronounced when both ears are subjected to the same sound environment. This phenomenon is at the heart of many systems, which use sound for detection, such as early metal detectors and shortwave telegraph tones. If the signal is a pure tone (sinusoid) or a complex tone (combination of sinusoids), the ear can detect it. The ear also efficiently detects periodic modulation in the very low frequency range and responds to variations in intensity or frequency.

10.3.4 Sidetone

Part of the noise canceling ability of the ears is associated with both ears picking up the same sound signals. Although sidetone was initially introduced to get people to talk louder into telephones, as the systems improved it was found that sidetone had another benefit. It tended to cancel the background noise and allowed the listener to more clearly hear and understand the speaker on the other end of the phone line. The following are sidetone considerations:

- **Speaker Sidetone**: The speaker's verbal input should be in phase with its reproduction as heard on the headset. This sidetone should not be filtered or modified before it is received in the headset.
- **Feedback**: For sound powered telephones[87] in-phase feedback or sidetone level can be used to control talker's vocal effort within limits. Feedback or sidetone is 3 to 6 dB lower than the produced speech. This increases the vocal effort and improves the signal-to-noise ratio.

Author Comment: In elementary school I was introduced to sidetone by using a small phone system in the classroom. We were asked to make conversation with our hands on and off the handset microphone. There was also a switch to enable disable sidetone. When the microphone was covered or the switch used to disable sidetone, the background noise was eliminated and it was difficult to hear the person at the other end of the phone line. Background noise was coming in the open ear and someone talking was coming in the ear covered by the handset.

This was a simple but effective experiment that introduced us to hearing and perception. The sidetone needs to be sent to someone speaking while only the speakers' voice should be sent to the listener at the other end of the telephone link. It is interesting how cell phone systems ignored the whole sidetone feature - can you

---

[87] These are telephone systems used in extreme emergency situations and are found on ships and industrial settings. They do not need power, they work exclusively on the electromagnetic characteristics of the transducers.

hear me now - was a popular cell phone advertising phrase used to target signal dropouts, not sidetone. In a mobile phone setting it is more important to cancel the background noise than offer a side tone. However a sidetone can easily be added with another microphone.

10.3.5 Ambient Noise

Ambient noise can be a nuisance needing techniques such as sidetone or it can be a health hazard. Sound is measured in units called decibels. On the decibel scale, an increase of 10 dB means that a sound is 10 times more intense, or powerful. This is perceived as sounding twice as loud. Refrigerator humming is 45 decibels, normal conversation is approximately 60 decibels, and city traffic can reach 85 decibels.

Long or repeated exposure to sounds at or above 85 decibels can cause hearing loss[88]. The louder the sound the shorter the time period before the hearing loss. Sounds of less than 75 decibels, even after long exposure, are unlikely to cause hearing loss. Noise sources that can cause Noise Induced Hearing Loss (NIHL) include motorcycles, firecrackers, and small firearms, all emitting sounds from 120 to 150 decibels. The following are some system considerations:

- **Noise Danger Signs**: Permanently post noise danger signs in areas with noise levels above 100 dBA or 140 dBP.
- **Noise Hazard Area Signs**: Permanently post on or in equipment noise hazard caution signs if steady-state noise levels are 85 dBA or greater. This posting is regardless of exposure time or duty cycle.
- **Manuals**: Manuals identify and discuss the system noise danger (100 dBA or 140 dBP) and hazards (85 dBA). Manuals include operator, maintenance, observer, field, technical, etc. The discussion topics include hearing protection locations, recommended hearing protection devices, noise levels of equipment and the distance at which the noise is below 85 dBA.

10.3.6 Audio Signals

Audio signals are machine generated audio signals that provide status, warnings, or alarms to operational, maintenance, or bystander personnel in the system. Audio signals are provided under the following conditions:

- Call attention to imminent or potential danger.
- Short, simple, and transitory information needing an immediate or time-based response.

---

[88] Noise-Induced Hearing Loss, U.S. Department of Health and Human Services - National Institutes of Health - National Institute on Deafness and Other Communication Disorders, Publication No. 08-4233, December 2008.

- Visual display is restricted. Visual sensory overload, excessive ambient light, limited mobility, degraded vision, anticipated operator inattention.
- As part of a redundant notification element for a critical event. The audio signal backs up and reinforces a flashing alarm or warning display.
- Warn, alert, or cue operator for subsequent additional response.
- Custom or usage has created anticipation of an audio display.
- Voice communication is necessary or desirable.

The ear is a good detector of periodic signals in noise. Even when signals are considerably weaker than the background noise, if the signal is a pure tone (sinusoid) or a complex tone (combination of sinusoids), the ear can detect it. The ear also efficiently detects periodic amplitude and frequency modulation in the very low frequency range and responds to variations in intensity or frequency. The following guidelines are offered for audio signals:

- **Signal meaning**: Each audio signal should have only one meaning.
- **Discomfort**. Audio warning signals should not cause discomfort or "ringing" in the ears. Levels should not exceed 115 dB at the ear of the listener. This is especially important if personnel must pass close to the sound emitter such as a fire alarm.
- **Audibility**: A signal-to-noise ratio of at least 10 dB should be provided in at least one octave band between 200-5,000 Hz at the operating position of the intended receiver. Signal to noise ratios can be greater as long as the levels do not exceed 115 dB at the ear of the listener.
- **Attention and Startle Reaction**: Startle reaction should be minimized. Signals with high alerting capacity should be provided when the system or equipment requires the operator to concentrate attention. Signals should not be so startling that they hinder appropriate responses or interfere with other functions by holding attention away from other critical signals. The increase in sound level during any 0.5 seconds period should be less than 30 dB to minimize startle reactions. The first 0.2 seconds of a signal should not be presented at maximum intensity, use square topped waveforms, or present abruptly rising waveforms.
- **Differentiation**: Audio alarms intended to bring the operator's attention to a malfunction or failure should be different than routine signals, such as bells, buzzers, and normal operation noises.
- **Prohibited Signals**: Signals that may cause confusion with the operational environment should not be used for alarms or warnings. The following are examples:
  - ❑ Modulated or interrupted tones that resemble navigation signals or coded radio transmissions.
  - ❑ Steady signals that resemble hisses, static or sporadic radio signals.

- Trains of impulses that resemble electrical interference, whether regularly or irregularly spaced in time.
- Simple warbles that may be confused with the type made by two carriers when one is being shifted in frequency (beat-frequency-oscillator effect).
- Scrambled speech effects that may be confused with cross modulation signals from adjacent channels.
- Signals that resemble random noise, periodic pulses, steady or frequency modulated simple tones.
- Signals similar to random noise generated by air conditioning or any other equipment.

- **Apparent Urgency**: The attention gaining characteristics of signals should match the relative priority of the signal. They include rapidity of pulse pattern, frequency, and intensity.
- **Frequency Range**: The frequency range should be 200-5,000 Hz. The range of 500-3,000 Hz is preferred. When signals need to travel over 300 m (985 ft), sounds with frequencies below 1,000 Hz should be used. Frequencies below 500 Hz should be used when signals must bend around obstacles or pass through partitions. The selected frequency band should differ from the most intense background frequencies.

10.4 Video and Audio Compression

Switching from analog to digital has given system managers options in managing the system that severely impacts the quality of audio and video. Digital compression reduces file size, which reduces the size of transmission bandwidth and storage. However compression also reduces the quality of the signal. The problem is there are no accepted metrics to measure the effects of compression. Some of the effects of video and audio compression are:

- Severe pixelation with fast moving images, dark night time scenes, fog, water, cloud filled skies
- Loss of detail especially facial detail where color becomes uniform and there are no bumps, lumps, pores as found in a normal face
- Artifacts around facial features especially around the eyes
- Chopping of video frames to fit time slots
- Loss of stereo separation as audio is converted to mono
- No dynamic range as all frequency content is compressed to one loud level
- Severely reduced frequency response as sampling rate is severely reduced

10.5 Voice and Phraseology

Many systems with significant cognitive load and or communications that may have extreme noise use an established phraseology. The phraseology not only uses words that offer less confusion but also phrases that are succinct and minimize communication time. Air traffic control uses phraseology between the pilot and air traffic controller[89].

The idea of using certain pronunciations of numbers and letters surfaced with early radio communications where excessive noise may garble the messages. As the systems evolved and operator cognitive load increased it became obvious that efficient unambiguous communications via an established phraseology would reduce system errors. The following number character pronunciation is an early example and forms the start of the concept of using a phraseology:

0 Zero ZE–RO	4 Four FOW–ER	8 Eight AIT
1 One WUN	5 Five FIFE	9 Nine NIN–ER
2 Two TOO	6 Six SIX	
3 Three TREE	7 Seven SEV–EN	

A Alfa ALFAH	K Kilo KEYLOH	T Tango TANGGO
B Bravo BRAHVOH	L Lima LEEMAH	U Uniform YOUNEE FORM
C Charlie CHARLEE	M Mike MIKE	V Victor VIKTAH
D Delta DELLTAH	N November NOVEMBER	W Whiskey WISSKEY
E Echo ECKOH	O Oscar OSSCAH	X X–ray ECKSRAY
F Foxtrot FOKSTROT	P Papa PAHPAH	Y Yankee YANGKEY
G Golf GOLF	Q Quebec KEHBECK	Z Zulu ZOOLOO
H Hotel HOHTELL	R Romeo ROWME OH	
I India INDEE AH	S Sierra SEEAIRAH	
J Juliett JEWLEE ETT		

10.6 Touch and Tactile Feedback

Haptics is the sense of touch. Many refer to this as tactile feedback. Tactile feedback is a mechanism offered by a system element to signal a user via the sense of touch. It usually is associated with forces, vibration, and motions offered to a user by a control such as a push button. However touch can be used to detect temperature, texture, force, vibration, motion, and motion distance. The feedback can be extreme and result in pain that triggers an involuntary user response to remove the pain. The fingertips are the most sensitive to touch feedback and are easily used for system interaction. Tactile feedback can take many forms.

10.6.1 Vibration

Vibrating the hand controls of an aircraft prior to its stall conditions where wing

[89] Air Traffic Control, FAA, 7110.65P February 2004, 7110.65T August 2010.

lift is suddenly lost because of high climb rate and low airspeed is a form of tactile feedback. Once a stall happens all airlift is lost and the airplane nosedives towards the ground. Vibrating the controls is a huge tactile feedback event that matches the huge consequences if the airplane is permitted to stall.

Audible alarms are also used to indicate if an airplane is approaching stall speed. However the audible alarm may be washed out by sensory and cognitive processing overload of the pilot. Because of the constant physical link between the pilot and the airplane yoke, it is unlikely that vibrating controls will be masked in this scenario.

10.6.2 Force and Push

Force is used in robotics to signal the user on the level of force being applied by a robotic tool. This first surfaced in the 1950's to signal operators remotely handling radioactive material.

Many systems use switches to interact with the system. In a noisy environment a switch that distinctly needs more pressure and then clicks into position as it is pressed offers an indication that does not need visual or audio feedback that the switch has been pressed. It is a very strong tactile feedback mechanism.

The same concept can be applied to a rotary valve or a lever. The use of increasing the force as in the case of a rotary valve may be unneeded. In fact defective valves which are stuck because of rust or other conditions may have the opposite force relationship. The force is extensive until the valve is freed. This is an important tactile feedback indication, a stuck valve.

10.6.3 Snap Action

Snap action is a variant of push tactile feedback. Rather than offer progressively more force as a switch is pressed, the force is relatively the same all the way through the push distance, but there is a distinctive snap or sudden increase in force then decrease and stop when the switch is activated. The snap is reinforced with a mechanical click sound.

10.6.4 Detents

Detents mechanically resist or arrest the rotation or sliding of a control. The detent positions can easily communicate five positions: 0%, 25%, 50%, 75%, and100% or off, one quarter, half, three quarters, and full. In the case of a push switch there are two positions, depressed (or pressed) and not depressed (or not pressed). Monetary switches are only depressed when the operator is applying pressure. As soon as the pressure is removed the switch returns to the normal position (not pressed). A spring-loaded lever with detents is a variant of a momentary switch but has multiple push states that provide operator feedback via the detents.

10.6.5 Position Size Shape Texture Movement

As an operator interacts with a system the position of a particular interface is imprinted. Maintaining the position via a fixed console position aids in the imprint. This is reinforced with visual feedback.

Size of a control relative to other controls signals the operator on the control type. This is initially reinforced with visual feedback. As the user continues to interact with the system the position and size work together to offer tactile feedback that is sufficient for the operator to take action with the control.

The shape of a control such as edges, curves, lumps, convex, concave, flat, square, round, button, lever, wheel, vertical pedal, horizontal pedal offers operator feedback. Different shapes and bump use can separate a control from a group of controls on a panel. They can signal the relative position such as when trying to locate the center of a logical collection of buttons. It is easy to identify three-dimensional characteristics in controls.

Texture such as roughness, smoothness, hardness can be used to separate different controls. A variant of this also can include temperature.

The attachment of the control provides further information as it changes position based on its attachment. For example a foot-peddle attached at the top, bottom, left, right, or center.

10.6.6 Physically Disparate

Physically disparate tactile feedback is not logically associated with the physical control that is being activated. For example a touch screen offers no tactile feedback as part of its physical characteristics. However the touch screen can be coupled to a vibrating device to offer tactile feedback when an operator makes a touch entry. This type of control is extremely versatile because the touch screen can offer an infinite choice of labels, which are used as part of visual feedback that reinforce the tactile vibrating feedback mechanism[90].

10.6.7 Mission Critical Tactile Feedback

In critical applications all feedback needs a backup mechanism. The backup feedback mechanism for tactile feedback is usually visual, such as enabling a light indicator, offering a response in a display area, or changing some other system visual output such as a dial or an entire display area.

Another backup mechanism is sound. Sound is extremely effective on keyboards, touch telephones, pointing devices such as a mouse. A digital camera uses a prerecorded sound of a mechanical camera shutter to provide feedback that the camera take-picture button has been pressed.

In mission critical systems special care needs to be taken when there is no tactile

[90] Cell phones have recently coupled touch entry with vibrating tactile feedback.

feedback. Many of these systems approach the load limits the human sensory inputs and in a crisis the traditional visual and audio feedback mechanism may be lost as information overload sets into the operator. An approach to consider is to couple a vibrating device either on the chair, headset, or some body location on the operator.

10.7 Ergonomics

Just like our eyes and ears have characteristics that we need to understand before we start to offer system interface approaches, our bodies have characteristics that need to be identified and understood. Ergonomics starts with the physical body, its size, movements and continues with design that addresses the fit between humans and machines in the operating environments. It considers the user's capabilities and limitations to ensure that tasks, equipment, information and the environment are safe and effective. Poor ergonomics design can result in fatigue, which can impair cognitive ability leading to system errors. It also can lead to health issues or even loss of life. This analysis should be used to develop effective design solutions and offer guidelines for system users and maintainers to reduce risk of injury[91].

10.7.1 Anthropometry

Anthropometry is measurement of the human body to understand human physical variation[92]. Anthorprometric data can be grouped and organized in many ways. The following grouping is offered for body physical characteristics and their impact on a system:

- **Foot Size and Ankle Movement**: Affects pedal designs and other foot related controls.
- **Sitting Knee Height and Clearance**: Affects seating and shelf height. Affects objects to the left and right that might bang into knees.
- **Sitting Leg Length**: Affects space offered to stretch legs and feet.
- **Sitting Eye Height**: Affects the placement of displays and controls.
- **Arm Length**: Affects the placement of controls so that they can be reached without stretching.
- **Folded Arm Length**: Affects length of shelves and side arm rests.
- **Arm Length**: Affects reach distance. Stretching and twisting should be avoided.
- **Standing Head Height**: Affects placement of hanging lights, ceiling height, door height, etc.

---

[91] The US Department of Labor Occupational Safety and Health Administration (OSHA) offers ergonomic guidelines for different industrial settings.

[92] Anthropometry of US Military Personnel, DOD-HDBK-743A, February 1991; Body size information in the form of anthropometric data.

- **Standing Eye Height**: Affects the placement of displays, wall lighting fixtures, wall art, light switches, etc.
- **Shoulder Height**: Affects the height placement of controls and physical items. Moving arms above shoulder height stresses the shoulder.
- **Floor to Hip Distance**: We tend to use our hips to move objects such as opening push doors and closing doors.
- **Floor to Hanging Hand Distance**: Affects the placement of standing controls such as door openers and light switches.
- **Handgrip Size**: Affects the size of an object that can be gripped by one hand. It is the ability for the fingers to lock onto an object.
- **Finger Tip Surface Area**: Affects the size of buttons and other finger-manipulated devices. Many small keypads use some form of tactile and visual feedback to make sure a user presses the desired key even though a finger or thumb tip may cover multiple keys. It is a delicate balance of pressure points and feedback.

Anthropometric data changes with time and population[93]. For example the recent trend has been for humans to be taller than in the past. To address these differences especially in a workstation or cockpit setting, the positions of various elements can be adjustable. For example in an automobile the seat position, seat height, seat angle, back angle, lumbar support, and thy supports can be adjustable. The wheel height and length can be adjustable. The foot pedals can adjust to accommodate the leg length of the driver.

10.7.2 Ergonomic Risk Factors

Ergonomic risk factors have some probability of causing muscular or skeletal problems. Some of the risk factors are carpal tunnel syndrome, tendentious, rotator cuff injuries (shoulder problem), epicondylitis (elbow problem), trigger finger (repeated use of a single finger) and, muscle strains and back injuries. They are associated with repetitive, forceful, or prolonged exertions of the hands or feet, frequent or heavy lifting, pushing, pulling, carrying, prolonged awkward postures, vibration and cold.

- **Force**: physical effort needed to perform heavy lifting, pushing, pulling or to maintain control of the equipment or tools.
- **Repetition**: performing the same motion or series of motions frequently for an extended period of time.
- **Awkward Prolonged Static Postures**: assuming positions that place stress on the body. Examples include repeated or prolonged reaching above the

[93] Ergonomic data can be accessed but it is also appropriate to gather measurements on you and the team members as a starting point. You are part of the ergonomic sample set.

shoulder height, using a tool with wrists bent, head turned to one side, leaning over, bending forward or to the side, twisting, twisting torso while lifting, kneeling, or squatting.

- **Contact Stress**: using the hand, foot, hip as a hammer. Pressing the body or part of the body such as the hand, foot, and hip against hard or sharp edges.
- **Vibration**: whole body vibration may damage joints of the skeletal system. Vibrating tools such as sanders, chippers, drills, grinders, or reciprocating saws may result in fatigue, pain, numbness, increased sensitivity to cold, and decreased sensitivity to touch in fingers, hands, and arms.

Before departing proven safe standard norms of ergonomic design, anthroprometric data needs to be accessed and analyzed so that reasonable justification is offered. Moving from accepted practices in even the most benign looking elements can lead to severe consequences including loss of life.

Author Comment: Shirking seat size on an airplane to get a few more passengers on a flight restricts leg and body movement that then restricts blood flow and circulation. Deep Vein Thromboses (DVT) blood clots form deep inside legs and form with immobility such as after surgery, long plane flight, or car ride. If they dislodge and travel to the lungs a pulmonary embolism develops and is potentially fatal. About 20 to 25 percent suffer sudden death. A human that is unable to stretch and move legs while seated is being abused and may die.

10.7.3 Ergonomic Design Starting Points

There are many existing systems which can be used as a starting point for the ergonomic design of a new system. The following are starting guidelines that can be used to stimulate the ergonomic considerations of many systems:

- **Weight**: Repetitive lifting should be less than 10 pounds. Rarely lifted items should be less than 40 to 50 pounds.
- **Reach**: The maximum viewing distance to displays located close to their associated controls should not exceed 635 mm (25 inch).
- **Work Surface Height**: Desk tops and writing tables are 74 - 79 cm (29 to 31 in) above the floor.
- **Writing Surfaces**: If consistent with operator reach requirements, writing surfaces on equipment consoles are not less than 40 cm (16 in) deep and should be 61 cm (24 in) wide.
- **Seat Pan And Vertical Adjustment**: The seat pan has an adjustable height of 38 to 54 cm (15 to 21 in) in increments of no more than 3 cm (1 in) each. If seat height exceeds 53 cm (21 in), a footrest is provided and single pedestal seats have a 5-legged base. The seat pan needs a 0 - 7° adjustable tilt rearward, be between 38 - 46 cm (15 - 18 in) wide, and should be 40 cm

(16 in) deep.

- **Backrest**: A supporting backrest that reclines 100° - 115° should be provided. The backrest engages the lumbar and thoracic regions of the back, and supports the torso in such a position that the operator's eyes can be brought to the eye line with no more than 8 cm (3 in) of forward body movement. The backrest width is 30 - 36 cm (12 - 14 in).
- **Cushioning and Upholstery**: Where applicable, both the backrest and seat are cushioned with at least 25 mm (1 inch) of compressible material. Upholstery is durable, nonslip, and porous.
- **Armrests**: Armrests that are integral with chairs are at least 5 cm (2 in) wide and 20 cm (8 in) long. Modified or retractable armrests are provided when necessary to maintain compatibility with an associated console and are adjustable from 19 to 28 cm (7.5 to 11 in) above the compressed sitting surface. Distances between armrests are not less than 46 cm (18 in).
- **Seat Base**: Chairs have at least four supporting legs. Swivel chairs should have five supporting legs. The diameter of the seat base of swivel type chairs should be 46 cm (18 in).
- **Footrests**: Footrests contain nonskid surfaces and are adjustable from 2.5 to 23 cm (1 to 9 in) above the floor, not less than 30 cm (12 in) deep, and 30 - 40 cm (12 - 16 in) wide. Footrest inclination is 25 - 30°.
- **Knee Room**: Knee and foot room is not less than 64 cm (25 in) high, 51 cm (20 in) wide, and 46 cm (18 in) deep beneath work surfaces; however, if a fixed footrest or a foot-operated control is provided, the height dimension is increased.
- **Display Placement, Normal**: Visual displays mounted on vertical panels and used in normal equipment operation are located 15 - 117 cm (6 - 46 in) above the sitting surface.
- **Display Placement, Special**: Displays that must be read precisely and frequently are located in an area 36 - 89 cm (14 - 35 in) above the sitting surface, and no further than 53 cm (21 in) laterally from the centerline.
- **Warning Displays**: For seated operations, consoles needing horizontal vision over the top, critical visual warning displays are mounted more than 57 cm (22.5 in) above the sitting surface.
- **Control Placement, Normal**. Controls mounted on a vertical surface and used in normal equipment operation are located 20 - 86 cm (8 - 34 in) above the sitting surface.
- **Control Placement, Special**. Controls that need precise or frequent operation are located 20 - 74 cm (8 - 29 in) above the sitting surface.

10.8 Sensory Cognitive Data Information Overload

It is possible to overload a human with too much system information that may or may not need a reasoned and correct response. The problem is when the human is overloaded, this can lead to serious health consequences and erroneous human responses. Erroneous human responses can lead to increased costs, damage to property, physical harm or death.

There is a distinction between data and information. Information is collections of data that have some reason to be grouped together. This group may result in a conclusion. This is typically refereed to as fusion. Many people in the loop systems use humans to perform data fusion. This requires use of the senses and cognitive processing, each of which can be overloaded.

Sensory overload happens when one or more senses reach their load limits. The following are examples of sensory overload. Display too much information in an area that exceeds the normal field of view of the operator, rapidly changing and scrolling displays. Too much audio such that there is no dead time to offer a voice or pointing hand response, the responses just stack up while the audio continues to stream to the recipient. Constantly speaking unable to receive audio input from a half-duplex communications channel. Even if the eyes and ears are not overloaded the limbs may be unable to respond quickly enough with button presses, button pushing, touch entry, mouse moving, joystick movement, trackball spinning, etc.

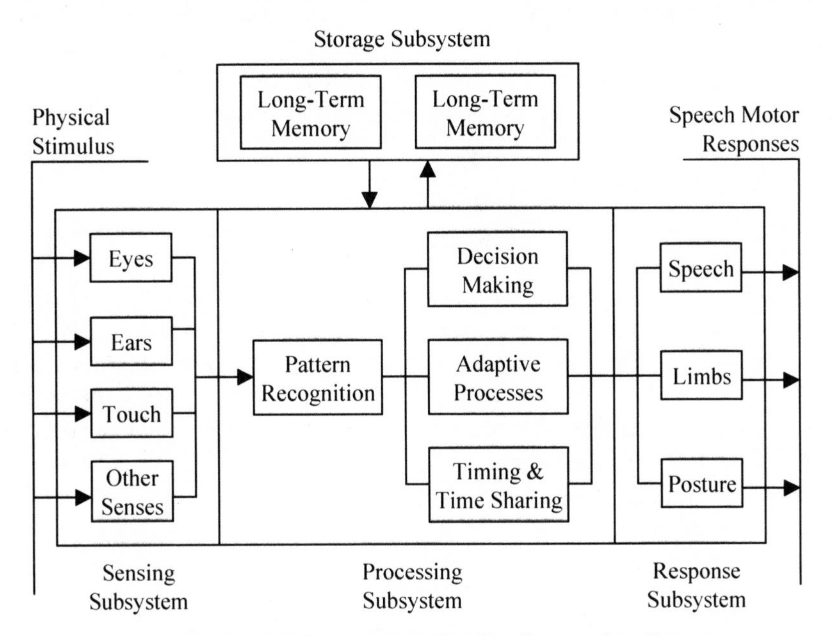

Figure 64 Human Information Processing System

Typically long before the senses are overloaded the user hits their maximum cognitive processing ability. They are unable to process the individual data elements and offer responses based on internal rules, heuristics, algorithms, experience, and or instinctive response.

Data overload is a condition where the user is unable to form meaningful information. There is no fusion or conclusion. There is too much noise or extraneous data preventing the formation of information or data fusion.

Information overload is a condition where the system presents too much noise and the user is unable to filter and get to the most important information that needs immediate action. Even though the user is presented highly processed and fused information, the effect is the same as with too much raw data. The user is unable to detect what needs detection from what becomes a vast noise pool.

Thrashing is a condition where a processing element starts to process an input but is unable to complete the processing because another input arrives that displaces the processing of the first input. Either the first input or another input then displaces the second input processing. In this trashing condition processing never completes on any input. People just like machines can enter into a thrashing condition.

10.9 Feedback

When a human is introduced into the system a feedback loop is established. Negative feedback reduces noise in a system. Positive feedback adds noise to a system. This is a critical issue that needs to be understood in the machine human feedback loop of the target system. If at anytime the system moves from a stable negative feedback condition to a positive feedback condition the human will eventually be swamped with sensory and or cognitive overload.

10.10 Response Time

When computers were first introduced in the 1960's there were studies to determine operator stress levels as a function of computer response time. The time was categorized into preview area, transaction response time, and busy time. The preview area echoed the characters typed by the operator. It was found that the preview area should offer a sub second response time. A transaction is a completed entry that is submitted to the computer. If it took more than 3 seconds for the computer to respond a busy light started to flash and it should respond within 6 seconds. If the response time exceeded 10 seconds the busy light should go to a steady on state and stay illuminated until the computer offered the response[94]. If the response time exceeded 60 seconds a count down display was offered. The following is an updated set of response time suggestions:

- **Key Response**: Key depression until positive response, appears to represent mechanical keyboard tactile feedback, e.g., "click" 0.1 seconds
- **Key Print**: Key depression until appearance of character, appears to be

[94] The FAA, IBM, and others performed these studies with articles published in trade journals during the 1970's.

similar to a preview area, 0.2 seconds

- **Page Turn**: End of request until first few lines are visible, 1 seconds
- **Page Scan**: End of request until text begins to scroll, 0.5 seconds
- **XY Entry**: From selection of field until visual verification, 0.2 seconds
- **Function**: From selection of command until response, appears to be similar to a transaction response time, 2 seconds
- **Pointing**: From input of point to display point, 0.2 seconds
- **Sketching**: From input of point to display of line, 0.2 seconds
- **Local Update**: Change to image using local database, e.g., new menu list from display buffer, 0.5 seconds
- **Host Update**: Change where data is at host in readily accessible form, e.g., a scale change of existing image, appears to be similar to a transaction response time, 2 seconds
- **File Update**: Image update requires an access to a host file, appears to be similar to a busy condition, 10 seconds
- **Inquiry (Simple)**: From command until display of a commonly used message, appears to be similar to a transaction response time, 2 seconds
- **Inquiry (Complex)**: Response message requires seldom used calculations in graphic form, appears to be similar to a busy condition, 10 seconds
- **Error Feedback**: From entry of input until error message appears, appears to be similar to a transaction response time, 2 seconds
- **Slow Response Time**: If computer response time will exceed 15 seconds, the user should be given a message indicating that the system is processing.
- **Standby**: When system functioning requires the user to stand-by, a WORKING, BUSY, or WAIT message or appropriate icon or indicator should be displayed until user interaction is again possible. If delay is likely to exceed 15 seconds, the user should be informed.
- **Long Delays**: For delays exceeding 60 seconds, a countdown display should show delay time remaining. This is especially important in automated telephone based interactions.
- **Long Operations**: For long operations a progress indicator should be provided.

10.11 Vision Versus Sound

There is an old saying that a picture is worth a thousand words[95]. However pictures can be open to interpretation unless precise symbols are used. Today it is easy for many to relate to download time because of the Internet. There is a

[95] At Hughes Aircraft when STOP was first introduced they performed several studies to determine how many words were used to augment a facing graphic in a topic. The number was approximately 500 words.

download time associated with different sensory inputs. For example a movie takes significantly less time to download than a book. Although the experience is different each is able to provide the same messages.

When interacting with a system the vision and sound inputs need to be balanced. More information can be consumed via the visual path than the sound path. If the visual path is reinforced with the sound path the input is even faster and less prone to error. The same is true with tactile feedback. A click and force difference with an indicator change is extremely fast and effective.

10.12 Visual and Audio Noise

When presenting data and information to a user it is important to minimize the visual and audio noise. Noise increases the recognition time and is a source of significant errors.

Display noise can be reduced with user selectable display filters, display offset, and range scaling. Display filters allow the operator to quickly enable disable categories of information. Ideally a single user action such as a button press enables disables a display filter. Display offset allows the information to be moved into the fovea field of view. Range scaling removes extraneous information and allows other relevant information to be provided.

Sound noise can be reduced with sound contouring via an equalizer that changes frequency content, clever noise removal mechanisms during clicks and pops, and squelch circuits that mute audio when then is no sound. In digitized audio systems, dropouts can be minimized with buffering.

Even though display and audio noise is a serious operator load issue, the only alternative may be to subject the operator to the noise environment. Humans are one of the most effective approaches for detecting patterns in a visual smudge or noisy sound. Part of the human cognitive processing mechanism tends to eventually "tune out" the noise so that when a different pattern arrives, it is detected. This may be associated with the fundamentals of a neural network that always seeks the lowest energy-state. The neurons fire when initially subjected to a constant signal, but then slowly stop firing as the lowest energy-state is achieved. So pink noise in the form of sound static or visual noise in the form of snow and double images tends to fade and only the sound and pictures are detected. Broadband sensor RADAR (Radio Detection and Ranging) and Sonar (Sound Navigation and Ranging), operators use this characteristic.

10.13 Circadian Sleep Rest Cycles

Disruption of the circadian rhythm results in fatigue. The effects include degraded judgment, degraded situation awareness, degraded decision-making, memory; slowed reaction time, lack of concentration, fixation, and worsened mood. Other effects are decreased work efficiency, degraded crew coordination, reduced motivation, decreased vigilance, and increased variability of work performance.

Fatigue can lead to increased system errors, which can cause physical harm or loss of life. Humans have a biological clock that is approximately 24 hours 11 minutes[96]. It is reset with the presence of light.

Jet lag or circadian rhythm sleep disorders occur when there is a change in work schedules or time zones. Time is needed to transition from a day work shift to a night shift (sleep phase shift). The amount of time depends on the number of hours the schedule is shifted, and the direction of the shift. During this transition, the circadian rhythm disruption or jet lag can produce effects similar to sleep loss. The following guidelines should be used to minimize circadian rhythm sleep disorders.

- Minimize the frequency of shift changes and extend the weeks before changing a person's shift from day to night or vice versa.
- It takes time for the person to adjust to shift changes, from day to night, or midnight to day.
- A frequent change in schedule disrupts the circadian rhythm and results in chronic fatigue.
- One month or longer in one shift pattern is suggested and results in a lower chance of chronic fatigue.
- Direction of the shift change should rotate ahead. So an early day shift should go to a late day or early night shift, and late night shift to an early day shift.
- There should be at least 24 hours off from work before starting the next shift change.
- Jet lag recovery takes one day for every time zone crossed. The body clock gets out of synchronization with a rapid change in time zones; it encounters daylight or darkness contrary to the normal rhythm.

It is easier to reset the circadian rhythm (recover) from jet lag than from night shift work because the sun follows the person. Artificial methods have been used to try to reset the rhythm of night shift workers. The approach is to use high intensity lights at work and closed thick window curtains in the morning when the person is sleeping. Some suggest that the light intensity should be 10,000 LUX (bright sunlight) while others suggest 1000 LUX (overcast day) may be sufficient to reset the circadian rhythm. Office lighting is typically 300-500 LUX, which is similar to sunrise or sunset on a clear day.

When fatigue does set in due to circadian rhythm disruption and or progressive sleep loss, caffeine provides a temporary benefit. It affects the nervous system in 15

---

[96] Initial studies suggested the clock was closer to 25 hours, however those studies allowed participants to use incandescent room light. At the time it was believed the room light was not strong enough to affect the cycle. Later studies suggested that ordinary room light can shift the circadian cycle by more than 40 minutes.

to 20 minutes. The effects include increased heart rate and increased alertness. It lasts for approximately 4 to 5 hours, but can last up to 10 hours in especially sensitive individuals.

Table 7 Caffeine Levels[97]

Percolated Coffee	140 mg / 7 oz
Brewed Coffee	80-135 mg / 7 oz
Red Bull Energy Drink	115 mg / 12 oz
Jolt Cola	72 mg / 12 oz
Coca-Cola	34 mg / 12 oz
Tea	70 mg / 6 oz
Chocolate	5-35 mg / 1 oz
No-Doz or Vivarin	200 mg / tablet
Excedrine	65 mg / tablet
Dristan	30 mg / tablet

Regular caffeine use can lead to caffeine tolerance and undesirable side effects such as elevated blood pressure, stomach problems, insomnia, and disrupted sleep if taken too close to bedtime. Caffeine affects sleep and should not be consumed 4 to 5 hours prior to sleep. It makes falling asleep more difficult, reduces sleep length, and disrupts the quality of sleep. The follow situations are appropriate for caffeine:

- In the middle of a night shift especially on the first and second day of the work week when circadian disruption is most pronounced and alertness most compromised
- Mid-afternoon when post lunch alertness dips because of insufficient sleep
- Prior to an early morning commute following a night shift, but not within 4 hours of going to sleep because it prevents sleep

However, caffeine should not be part of the system. The system should be designed to respect the circadian cycles of humans. Ignoring this information and inducing stress on the humans in the system shifts long-term health costs to other stakeholders and still opens the system to possible circadian cycle fatigue based disasters[98].

There have been attempts to increase the circadian cycle from 24 hours to 28 hours. This would translate to a six-day week. However they have been

---

[97] Commercial Transportation Operator Fatigue Management Reference, Department of Transportation, Human Factors Coordinating Committee, Research and Special Programs Administration, July 2003.

[98] August 18, 1993 in Guantanamo Bay, Cuba involving a DC-8. NTSB report probable cause: "The impaired judgement, decision-making, and flying abilities of the captain and flightcrew due to the effects of fatigue".

unsuccessful and this suggests that life on Earth is based on the 24-hour cycle[99].

A big question is do we live to work or do we work to live. Are we slaves to our systems or do our systems serve us. These are profound cultural questions that get to the heart of quality of life.

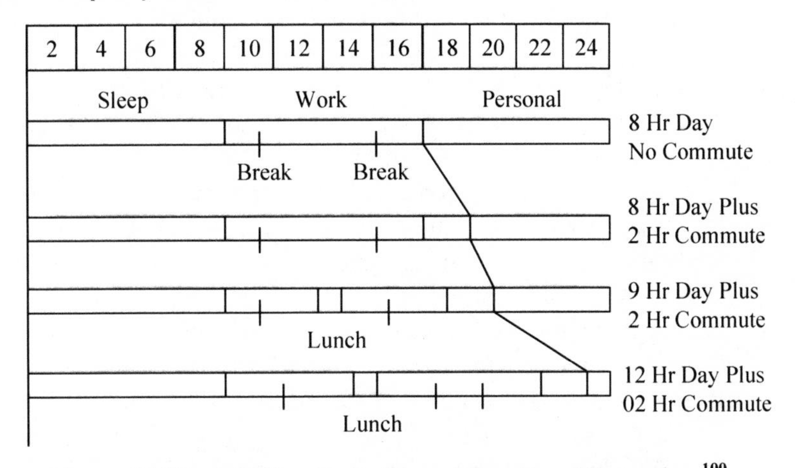

Figure 65 Work Shifts - Daily Time Allocation Alternatives[100]

There are some systems that need people 24/7. Examples are health care, air traffic control, police, fire fighters, etc. The question then arises, which work schedule is the best for maintaining healthy alert people. The 8-hour shift allows for shift changes to happen with minimum disruption to circadian cycles. The 12-hour shift has the worst impact when shifts change. Rather than 8-hour slides, it is a 180-degree flip between sleep and non-sleep time. The rate of shift changes (weekly, monthly, quarterly, semi annually, or random) can minimize the severe negative impacts of 12 hour shifts.

Personal time is a different state of being for a human. It is non-sleep time where people are engaged in other activities rather than the focused needs of work. It is unclear if this additional stimulation offers benefits in the short and long term. Removing this waking decompression time in the 12-hour shift is something to consider in such a system. For example offering comfortable pleasant lounges with different seating and private space for quick power naps during the 12 hour work shift may reduce fatigue and benefit the system and the people.

---

[99] Work performed at Harvard University by Charles Czeisler, M.D., Ph.D. and associates. William J. Cromie, Human Biological Clock Set Back an Hour, The Harvard University Gazzette, 1999.

[100] This time budget allocation is no different then allocating timing budgets in systems such as computers, communications, command and control, hospital emergency rooms, call processing centers, maintenance activities, etc.

10.14 Stress Sources

Understanding sources of stress and designing the system to minimize stress will not only maintain health of the people in the system but also minimize system errors that can lead to cost shifting to unsuspecting stakeholders, property damage, physical harm, or even loss of life. The sources of stress are:

- Poor display design with excessive field of view, characters and symbols that are tool small, use of blue colors, poor contrast, low brightness, excessive useless data or information
- Excessively loud noise or prolonged exposure to moderately loud settings
- Poor audio with poor frequency response, excessive digital quantization distortion and under sampling, digital dropouts, excessive static clicks pops, insufficient volume control
- Poor ergonomic design leading to repetitive motion disorders, restricted blood flow, excessive force strains or bone breaks, bumps, bruises, and cuts
- Poor environmental conditions of extreme heat, cold humidity, insufficient ventilation that leads to toxic fume exposure or reduced oxygen intake
- Ignoring circadian cycles and need for rest by not having breaks during work periods, excessively long work schedules, shifting work schedules with no time for circadian cycle readjustments
- Poor response time and inconsistent response time from the machines in the system increase stress and lead to fatigue and system errors

10.15 HMI Style Guide and HMI Design

The Human Machine Interface (HMI) needs to be captured in a style guide and a design document. The style guide feeds the design and is used to ensure consistency in the design. However a style guide is not enough to convey all the information. The design needs to be fully represented with pictures of all the HMI screens, controls, indicators, etc. The design should follow the mechanisms present in the existing technologies. For example forcing a Windows like style on a Browser Interface is inappropriate especially if there is no need to invent new HMI controls.

Ideally the HMI organization or specialist performs the HMI design. In the absence of an HMI specialist, then a systems team or person well versed in the Human Factors practices performs the design. It should never be left to the design or implementation team. That will always lead to serious omissions, poor design selections, and massive inconsistencies. The HMI design review should always ask:

- Where are <u>ALL</u> the screen layouts
- Is the HMI consistent
- Did you follow MIL-STD-1472
- Are the colors appropriate
- Is there maximum contrast
- Is test size sufficient
- Is field of view appropriate
- Is reach appropriate

The HMI architecture is like all other architectures. There should be alternatives. The alternatives are based on different technologies, which impact the architectures. These technologies and architectures need to be clearly understood from a function and performance point of view. They should follow the basic principals of great architectures such as consistency, simplicity, symmetry, intuitive, etc. Most of us are currently focused on computer settings but the following guidelines can be extrapolated to almost any MHI setting:

- Modularization or chunking based on logical groupings, symmetry, simplicity, expandability, is critical to a good design
- Backspace and delete keys should work consistently across all screens
- Menus should be short lists of 7 items and menu depth should be less than 3 if possible
- At no time should controls get hidden especially when windows are resized
- Press a button, provide an immediate clear response, never make the operator guess if they pressed a button
- Offer a consistent status indication if the operator needs to wait or processing is happening in the background for all conditions
- Provide a positive clear indication when processing is complete
- Have very good reasons to freeze a display and force the operator to wait, provide a way to instantly escape the wait cycle
- Context sensitive controls must be clearly identified and follow simple logical connections
- Touch entry is different from point and click is different from keyboard entry is different from voice entry, respect those differences
- Do not create an HMI control that is not offered by the technology, study all the available options in the technology, make the best selections, and use them consistently across the design

10.16 Ethics

Ethics is a major concern in all areas of a system. All information products should contain a section on ethical issues and considerations. This is especially true in human factors. Poor human factors decisions can lead to damaged eyes, ears, muscles, joints, bones, tendons, lungs, circulation, nerves, glands, skin, etc. The big issues are usually addressed but the small issues such as straining eyes tend to get overlooked except in the most sophisticated system settings. However these small issues when ignored lead to increased healthcare costs and degraded system performance as users struggle to interact with a bad system.

Author Comment: Even though we know all these human factors issues, new appliances such as televisions, audio, and video devices have recently started offering remote control devices that violate simple ideas of color use, text size, button grouping, button sizes, and tactile feedback. Using gray text over a black button is outrageous. Using buttons so small that multiple key press errors occur for a device that sits on a coffee table is wrong. Not using functional groupings and using same size buttons is a total disrespect of the user. Not standardizing remote controls for multiple home entertainment appliances so that a typical setting has a plethora of separate remote control devices is a sign of not appreciating technologies impact on the world and the user in particular.

10.17 Exercises

1. What is the resolution of film and how does it compare with current electronic displays? How do you feel about the findings?
2. What should you do if illegal on coming vehicle headlights or deregulation allows such headlights cause you to lose your night vision?
3. What should you do if your computer displays have excessive use of blue colors? What if you cannot modify the colors?
4. What should you do if your audio or video is poorly digitized and over compressed?
5. Describe the field of view characteristics of humans. Where is depth perception lost? Where is color lost? Where is the detailed vision? Does wide screen presentation lead to visual fatigue because of excessive eye movement?
6. What are examples of poor contrast? Are transparent displays an example of poor contrast?
7. What should the text size be for a display that is 12 to 18 inches away from your eyes?
8. Provide performance specifications for a high fidelity audio system.
9. Provide performance specifications for a high quality video system.
10. What techniques can be used to minimize sensory and cognitive overload?
11. What stress sources exist in a system and how are they minimized?
12. What are the implications of circadian cycles and quality of life?
13. Is texting a form of phraseology?
14. Is touch and tactile feedback an element of the cockpit design of your automobile?
15. Is ergonomics an element of the cockpit design of your automobile?
16. What elements can you add to a system to minimize information overload?
17. How can you prevent system instability in the human feedback loop?
18. Provide response specifications for a system that minimizes stress.
19. Provide examples of using vision versus sound as methods to minimize input time into a human in the system loop.

20. What does the human do with visual and audio noise? What happens if the noise becomes sporadic or bursts such as when images pixelate or the entire screen freezes or goes blank?
21. How does abusing the Circadian Cycle impact quality of life?
22. Is it ethical to allow users to shrink text size on their displays to such a level that it leads to eyestrain? Is it ethical to offer blue colors on computer displays that are offering text information?
23. What are some of the ethical issues you see in your environment given your new knowledge of human factors?

10.18 Additional Reading

1. Anthropometry of US Military Personnel, DOD-HDBK-743A, February 1991.
2. Discrete Time Systems, James A. Cadzow, Prentice Hall, 1973, ISBN 0132159961.
3. Ergonomics Program Management Guidelines For Meatpacking Plants, U.S. Department of Labor Occupational Safety and Health Administration, OSHA 3 123, 1993.
4. Guidelines for Nursing Homes Ergonomics for the Prevention of Musculoskeletal Disorders, U.S. Department of Labor, Occupational Safety and Health Administration, OSHA 3182-3R, 2009.
5. Guidelines for Poultry Processing Ergonomics for the Prevention of Musculoskeletal Disorders, U.S. Department of Labor Occupational Safety and Health Administration, OSHA 3213-09N, 2004.
6. Guidelines for Retail Grocery Stores Ergonomics for the Prevention of Musculoskeletal Disorders, U.S. Department of Labor, Occupational Safety and Health Administration, OSHA 3192-06N, 2004.
7. Guidelines for Shipyards Ergonomics for the Prevention of Musculoskeletal Disorders, United States Department of Labor, Occupational Safety and Health Administration, OSHA 3341-03N, 2008.
8. Handbook For Human Engineering Design Guidelines, DOD, MIL-HDBK-759C, 31 July 1995, MIL-HDBK-759B 30 October 1991.
9. Human Engineering Design Criteria, DOD, MIL-STD-1472C May 1991, MIL-STD-1472D March 1989, MIL-STD-1472E October 1996, MIL-STD-1472F August 1999.
10. Human Information Processing, Peter H. Lindsay, Donald A. Norman, Academic Press, 1977.
11. Man Systems Integration Standards, NASA, NASA-STD-3000, July 1995.
12. Noise Limits, DOD Design Criteria Standard, MIL-STD-1474B June 1979, MIL-STD-1474C March 1991, MIL-STD-1474D February 1997.

11 Flowcharts

The Flowchart shows sequence, flow, decision, and feedback. It is typically used in representing process and algorithms. Many believe the flowchart originated with software and is a software tool. The flowchart originated in 1921 and is traceable to Frank Gilbreth[101].

11.1 Flowchart Use

The flowchart is an extremely flexible formal method and can be found in studies, patent descriptions, plans, specifications, design documents, etc. It is found across disciplines such as science, engineering, management, etc. As the years have passed various practitioners have adapted the flowchart to their needs and named the new applications accordingly. Some flowcharts are:

Table 8 Flowcharts

Process flowcharts	General flowchart
Document flowcharts	Detailed flowchart
Data flowcharts	Decision flowcharts
System flowcharts	Logic flowcharts
Program flowchart	Product flowcharts

In many ways all of the modern methods such as the Unified Modeling Language (UML)[102] activity diagrams are traceable to the flowchart. This is the characteristic of a great process or method. It is formal and broad enough to be used across many fields of study. It is fundamental just like mathematics, logic, and set theory. Many formal methods may try to separate themselves from these roots, but then they start to fail in their purpose, which is to communication something important to the stakeholders. This topic is an extremely important because new systems typically require equally new break through processes and methods. But these processes and methods must be based on the scientific method.

[101] Frank Bunker Gilbreth, Sr. (July 7, 1868 - June 14, 1924) introduced the first structured method for documenting process flow in 1921 as a presentation "Process Charts - First Steps in Finding the One Best Way" at an American Society of Mechanical Engineers meeting. Purdue University houses his papers.

[102] Unified Modeling Language (UML) standardized general-purpose modeling language in software engineering.

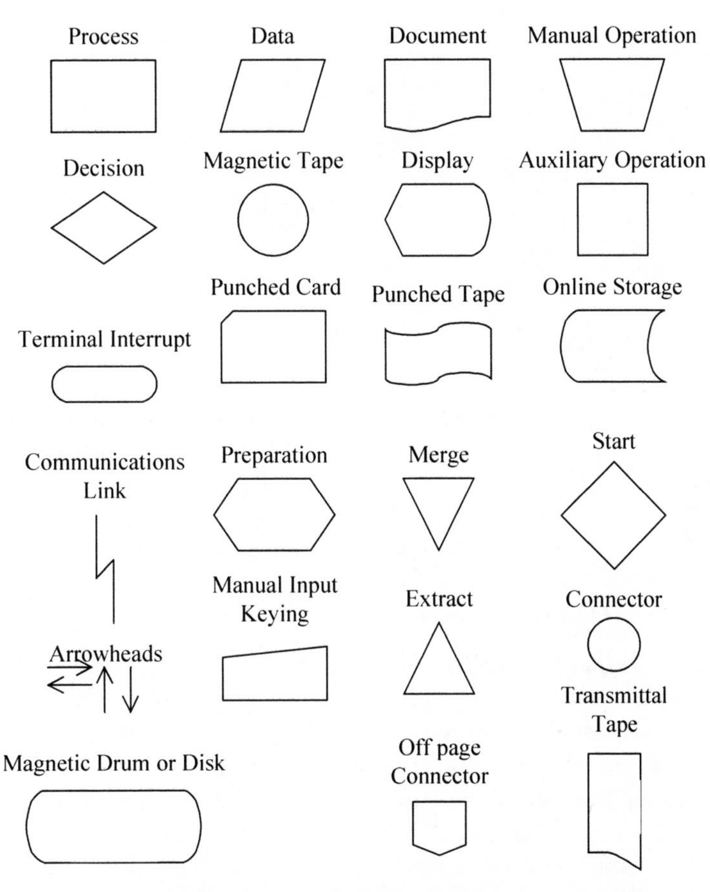

Figure 66 Flowchart Symbols - Software circa 1970

Herman Goldstine and John von Neumann adapted the process chart and used flowcharts to plan computer programs[103]. The symbols are arbitrary. They represent what needs to be communicated. For example in the late 1960's early 1970's computers used magnetic drums, magnetic tape, and paper tape. Today they use optical media and Universal Serial Bus (USB) flash stick storage.

---

[103] Programming flowcharts, Goldstine and von Neumann "Planning and coding of problems for an electronic computing instrument, Part II, Volume 1," 1947 unpublished, reproduced in von Neumann's collected works: Taub, Abraham (1963), John von Neumann Collected Works.

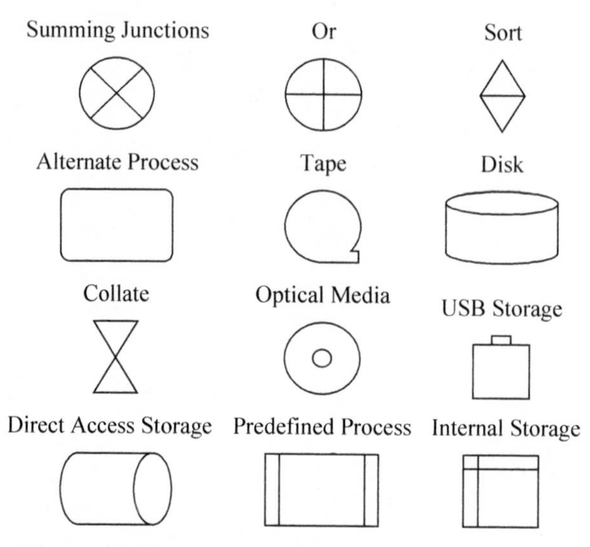

Figure 67 Flowchart - Additional System Symbols

Example 21 Flowchart - Emergency Health Care

When examining a system and looking for inefficiencies the flowchart communicates a great deal of information. For example how many steps are in the process? How many decisions are in the process? Is there a feedback loop that is pointless and leads to system instability, a positive feedback mechanism?

The less formal functional sequence diagram can show the same information, especially if it is annotated. Both the functional flow diagram and flowchart can show time allocations. Again simple annotations near each block can convey the information. The difference is the flowchart changes the practitioners thought patterns. They become more rooted in logic and an algorithm or process is forced from the exercise. It is this process which may show a new insight in a system. The insight could be significant and change the functional and performance view of the system.

Flowcharts normally flow from top to bottom, left to right. However if the sequence is very simple then 1 row with left to right orientation fits more appropriately in other information products such as a target document.

A flowchart should be laid out using the principles of information chunking. A flowchart template can be created with 10 rows and 5 columns. The stack of flowcharts should follow the principals of decomposition, which is based on logic, set theory, common sense, and reasonableness. Each individual flowchart and the entire stack representing a standalone (atomic) subject can be configuration managed.

The following are two possible examples of getting health care services in an emergency situation. There are significant differences and the implications are significant.

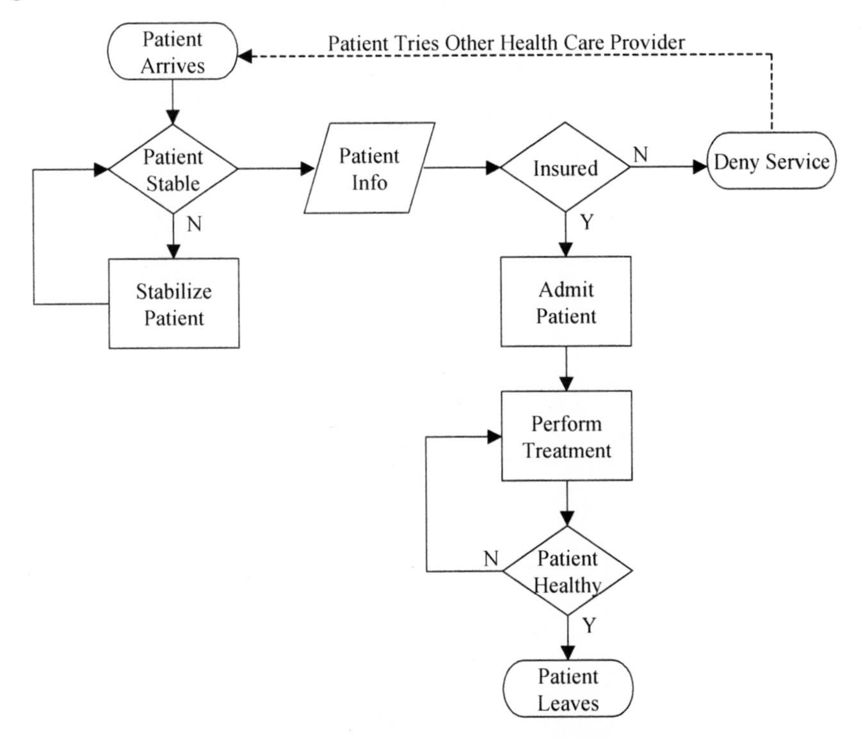

Figure 68 Current USA Emergency Health Care Process

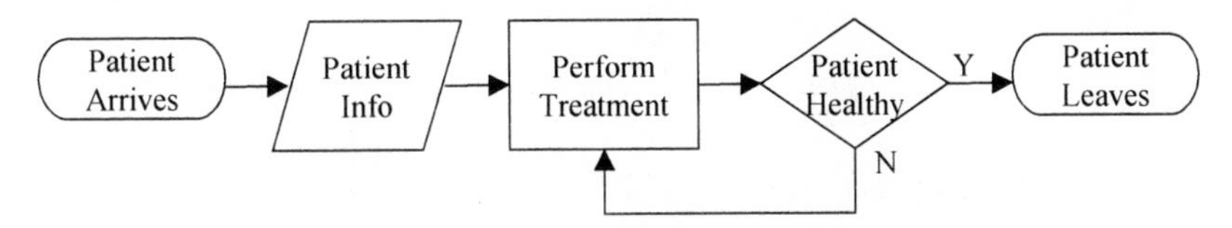

Figure 69 Ideal Emergency Health Care Process

Example 22 Flowchart - Troubleshooting

Flowcharts can be used in maintenance manuals to support troubleshooting. Typically the troubleshooting guide includes a family of flowcharts, organized by subsystems.

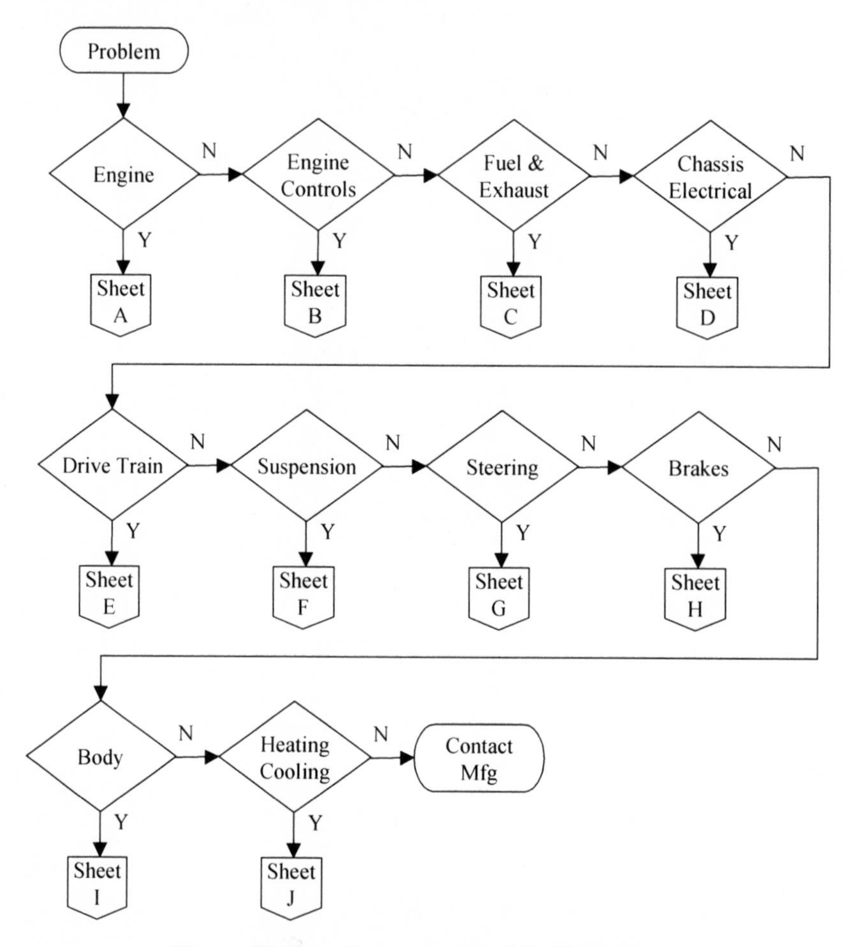

Figure 70 Flowchart - Automobile Troubleshooting

Example 23 Flowchart - Unintended Car Acceleration

Flowcharts also can be used to understand single points of failure and the impacts of inappropriate functional allocations. Typically failure analysis is performed for life dependent systems. The failure analysis includes the development of a fault tree. The fault tree is based on logic gates of AND / OR sequences. The flowchart analysis considers the major decisions in the system and how those decisions are processed from a single point of failure view.

A failure scenario can be identified, such as sudden uncontrolled acceleration. The operational sequence can then be shown as a series of decisions. The decisions are mapped to the relevant subsystem. In this case the decisions are mapped to a single computer subsystem. The implications are the failures within this subsystem should not propagate between major functions, such as the ability to disable the engine or place the transmission into neutral.

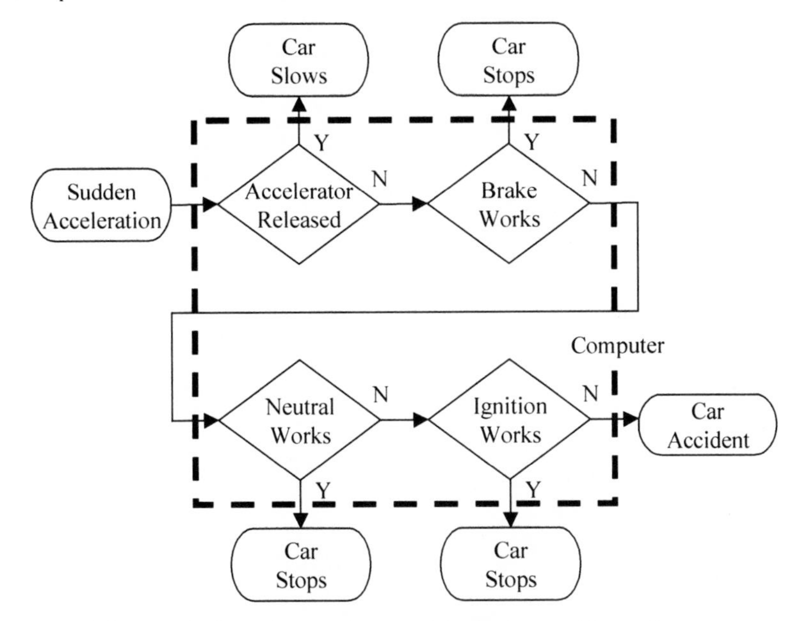

Figure 71 Flowchart - Automobile Drive by Wire Controls

The implications of a single computer subsystem are significant. For example, It must be fault tolerant such that services are always available, not make mistakes, and operate within the real world scenario of poor maintenance. If it does fail, it needs to fail in a safe manner. This is fail safe operation.

Mission critical systems are systems where loss of life is possible if they fail. Mission critical systems typically are able to support 3 single points of failure. This implies that there are 3 levels of redundancy. Further one of the redundant elements should be an orthogonal solution to avoid the possibility of a latent design defect appearing in each element simultaneously. In this particular case, the system might use 4 voting computers where 1 computer, the orthogonal element, is a different design and implemented by a different development team.

In a traditional automobile the control subsystems for ignition, acceleration, braking, and transmission gearshift are completely separate. Further, the technology, design, and implementation is different. So the risk of failure propagation between each critical control element is zero. In the case of the brake system we see 3 redundant elements, 2 of which are separate hydraulic elements, and the third is an orthogonal solution of a cable. In many ways this represents the evolution of the automobile. As new technology was added to the system, previous generations of the system were preserved and integrated into the system.

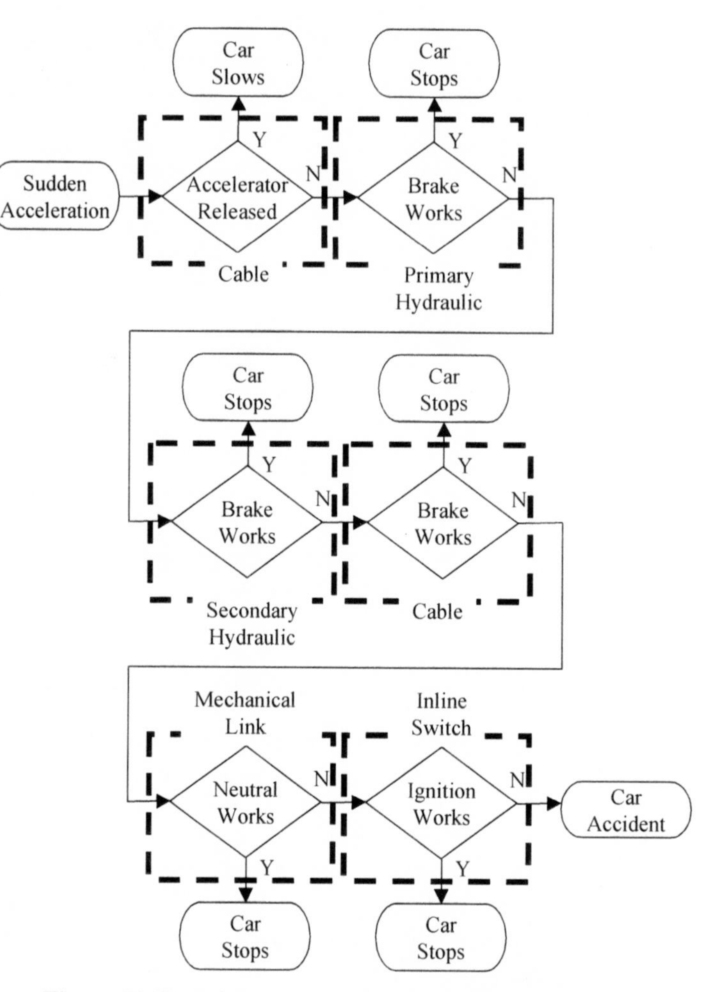

Figure 72 Flowchart - Traditional Automobile Controls

11.2 Key System Logical Operations

Many large mission critical systems exhibit the characteristic of preserving previous generation technologies and reintegrating them into the new system. The traditional automobile control subsystem is an example. Another example is Air Traffic Control (ATC). In the early days ATC was primarily based on Air Ground Communications. Then Navigational Aids were added. Then Broad RADAR was added. Then computer based processing for situational awareness was added, ala SAGE. So as failures occur in the ATC system, there are fallback mechanisms from computer to broad band, to radio with navigational aids, to navigational aids. Of course the desire is always to maintain the highest level of services, so the computer based systems employ all the known techniques for mission critical processing. That means redundancy for fault tolerance, orthogonal mechanisms to stop fault propagation, and other techniques. The flowchart with fault trees can be used to provide these system insights.

Flowcharts are extremely important because of the thinking process they enable in systems architects. The if-then-else sequence and ideas of control and feedback are critical to the design of good system architectures. After the practitioner develops a number of flowcharts in a number of different settings, this thinking process becomes second nature and issues surface on the fly from these practitioners rather than from formal activities using flowcharts. In many ways it is a right of passage.

11.3 Exercises

1. Identify an analysis effort where you use new flowchart symbols. Create an example of this flowchart and explain it to your classmates.
2. Identify commonality between the various flowcharts created by your classmates.
3. Why is a kill switch in a laboratory offered? What kind of kill switches might be available in a laboratory? Show the kill switch in a flowchart.
4. Should kill switches be part of all mission critical control systems? Can a flowchart be used to answer this question and if so, develop such a flowchart?
5. Develop a flowchart for a system engineering process, such as developing an automobile control subsystem.
6. Develop a flowchart showing the interaction of various information products on a project.
7. Develop a flowchart for an algorithm.

11.4 Additional Reading

1. Calculating Instruments and Machines, Douglas Hartree, University of Illinois Press, 1949.
2. IBM's Early Computers, Charles Bashe, MIT Press, 1986.
3. John von Neumann Collected Works, Abraham Taub, Macmillan, 1963.
4. The Computer from Pascal to Von Neumann, Herman Goldstine, Princeton University Press, 1972, ISBN 0-691-08104-2.

12 Data Flows and Structured Analysis

The Data Flow Diagram (DFD) became an established analysis technique in the late 70's coincident with the popular phrases structured system analysis, structured system design[104] and top down analysis. Aside from its approach to modeling a system from a processing and data point of view while showing the flow of data through the system, a method was developed to methodically capture the data flow diagrams and accompanying data dictionary and mini specifications[105].

Prior to this time analysts would stop their decomposition at some arbitrary level with no measurable rational. It did not matter if they were developing functional block diagrams or flowcharts. Many analysts began to notice that systems were not behaving as expected because the lower level details were never decomposed. So the data flow diagramming technique filled a known void of identifying in some measurable way when we are done.

A DFD is a logical or physical graphical representation of the flow of data through a system. The system is represented by functions and data stores. Data items flow from an external data source or an internal data store to an internal data store or an external data sink, via internal functions. The functions can be broadly categorized as data transport or data transformation. Transport functions are typically associated with moving data; they do not examine the data content. Transformation functions examine the data and perform operations on the data. The transformation operations are infinite in possibilities.

12.1 Syntax and Semantics

The components of the DFD syntax[106] are circles and lines with arrows, boxes, parallel lines, and diagrams. Circles traditionally represent processes. They can be used to represent activities, functions, transformations, organizational entities, etc. Arc lines with arrows represent data. Boxes represent external interfaces. Parallel lines represent data stores. Diagrams provide a format for representing the model

---

[104] Structured Analysis and Design Technique (SADT) developed by Douglas T. Ross and SofTech, Inc. SADT was developed and field-tested between 1969 to 1973.

[105] Structured Design circa 1975 with Larry Constantine, Ed Yourdon and Wayne Stevens. Tom De Marco applied decomposition to software in 1978 in his book Structured Analysis and System Specification published by Yourdon, ISBN 0917072073.

[106] The structural components and features of a formal method and the rules that define relationships among them are referred to as a method's syntax. Semantics refers to the meaning of syntactic components of a method and aids in correct interpretation.

both verbally and graphically. The format also provides the basis for configuration management. The following summarizes the syntax and semantics for DFD analysis:

- **Process**: A series of one or more steps that converts inputs to outputs. A process is named as a verb or verb phrase. It must have at least 1 input and 1 output. A primitive process or leaf process can be recognized if it has only one input and one output. A mini specification is written against primitive processes.
- **Data Store**: A data store is data at rest. In computer based systems it can be a file or database. It is named as a noun or a noun-phrase. Data must flow through a process it cannot move directly from one data store to another data store. Data cannot be moved directly from an outside source to a data store and data cannot be moved directly from a data store to an outside sink. It must first go through a process.
- **Source and Sink**: An external agent is a source or a sink of data. It lies outside the system boundary and is the fundamental element of the context diagram. Examples of sources and sinks are paper forms, printers, scanners, other systems, etc. A source or sink is named as a noun or noun-phrase[107]. Data cannot move directly from a source to a sink. A process must move the data from a source to a sink.
- **Data Flow or Connector**: A data flow is named with a noun or noun-phrase[107]. A data flow is represented with an arc line and has only one direction of flow. This is called a net flow. For example a read before an update will show one arrow for the update only. A fork shows a copy of the data is going to more than one location. A join shows data is being received from more than one process, data store and/or data sink/source. Both forks and joins should be avoided. Forks and joins suggest possible system issues. A data flow cannot loop back to itself. If it does need to loop back, it must flow through a process. A data flow to data store shows an update (add, delete, or modify). A data flow from a data store shows a read.
- **Control Flow or Connector**: In later incarnations of the DFD, practitioners showed control flow in parallel with the data flow[108]. Sometimes this was done on separate diagrams with the data replaced by control flow or on the same diagram as the data flows. The control flow is represented with dashed arc lines and has only one direction of flow. A control flow is named with a verb or verb-phrase.
- **Mini-Spec**: A mini spec is written against the process leaves. They are

---

[107] Modern Systems Analysis and Design, 3rd edition, Hoffer, Prentice Hall, 2002.

[108] Hughes Aircraft Fullerton used this technique on the Advanced Automation Program in the early 1980's for next generation air traffic control and air defense systems.

aggregated and form the body of a specification document of the system under analysis. Some practitioners attempt to write mini specs for each decomposition view. This will lead to duplicate requirements and inconsistencies. It also may lead to circumventing the process where the decomposition stops and the leaves are never reached, invalidating the strength of this analysis approach.

- **Data Dictionary**: The data dictionary lists all the data flows in the analysis. It should be organized by the final decomposition views. It should easily trace to each DFD diagram view. It must be consistent and complete with the DFD view. The data should be fully described in terms of use, type, format and other attributes needed for the system stakeholders.
- **Context Diagram**: The context diagram is the first diagram that is drawn in the analysis. It defines the entire system boundary. It has only one process, the system. The data flows from the system to the sinks and data flows from the sources to the system. The sources and sinks are external agents. There are no data stores.
- **Data Flow Diagram**: The data flow diagram is drawn on a single sheet of paper and traces to the upper level. The paper is titled using both the numbering convention and name of the next higher level process. This helps to enforce traceability. If it is the first diagram to decompose the context diagram, it is labeled "Level 0 - System Name". Lower layer diagrams are developed when the current view is getting too complex. The general rule is 7 +-2 processes.
- **Allocation Boundaries**: In later incarnations of the DFD, practitioners showed functional allocation to physical elements in parallel with the data flow[109]. This was accomplished by drawing dashed boundaries around the applicable data flow diagram elements. The dashed boundaries are then named using the actual physical architecture element responsible for those elements. Alternatively the process (circle) is color-coded or shaded or numbered or named to indicate the physical allocation.

12.2 Top Down, Bottom Up, Middle Out

There are different abstraction levels associated with understanding any system. Many refer to a top down approach, suggesting that it is superior to other approaches. But what is a top down view? Starting with a context diagram and continuing with successive decompositions is a top down approach. However when lower level DFDs start to impact the upper level DFDs that is a bottom up view. When the analysis is thought to be complete most analysts will plunge into the middle of the DFD decomposition stack and find they navigate up and down the

[109] Hughes Aircraft Fullerton used this approach in the early 1980's for next generation air traffic control (Advanced Automation System) and air defense systems.

abstraction level. The key being that the analysis is complete and consistent when viewed from the top down, bottom up, or middle out.

If a top down approach is based on starting from the top and decomposing down the abstraction level, what is a bottom up approach?

Initially Top-down and bottom-up were viewed as strategies of knowledge ordering involving software, mostly because of the work associated with structured systems analysis and the DFD. However, it applies everywhere and can be seen as a style of thinking and teaching. In many cases top-down is used as a synonym for analysis or decomposition, and bottom-up is used as a synonym for synthesis.

Bottom-up synthesis is not to be confused with design synthesis when creating something new and innovative. Instead it is more of an integration of lower level details to higher level details. In many ways it is set theory in practice where things are grouped and aggregated. When developing costs for a system the activity is typically bottom up, where the individual practitioners closest to the problem make the estimates. This is an aggregation process.

So a top-down approach is breaking down a system to gain insight into its sub-systems. In a top-down approach an overview of the system is first suggested, identifying but not detailing any first-level subsystems. Each subsystem is then refined in yet greater detail, until the entire system is reduced to atomic elements. Atomic elements refuse to decompose to a lower abstraction level. A top-down functional model uses "black boxes" rather than processes and data as in the DFD. This makes it easier to manipulate. However, black boxes fail to suggest lower level decompositions. There is nothing consistent to follow in the system like data as in the DFD case. It is also difficult to validate the model if there is nothing consistent to follow like data. **In many ways data is almost a universal constant present in almost any system.** In later incarnations of the DFD dashed lines would represent control, further aiding the decomposition and validation process.

A bottom-up approach is piecing together components and or subsystems to give rise to systems. Many practitioners reject this as a system approach because the emergent properties of the final system are not surfaced and understood. Also the vested interests of the individual components (manufacturers) override what is really needed for the system. It is for this reason that the top down approach is strongly embraced.

Piecing together a system from lower level elements is referred to as “the house that Jack built”[110]. Jacks house would be characterized as bizarre, with useless functions and very poor system characteristics such as poor maintainability, unpredictable behaviors, poor efficiencies, bad performance, little or no ability to grow or insert technology, etc. The art of systems practices is to surface the

---

[110] Many systems start out well, however after decades of upgrades and technology insertion they start to resemble Jacks house. This was a term used to describe the USA air traffic control system and justify the Advanced Automation System Program.

emergent properties when the whole is considered rather than the individual parts.

So, in a bottom-up approach the lower level components or subsystems are first specified in great detail. They are then linked together to form the system. This approach can start as a prototype or initial version. The initial system is small but eventually grows in complexity and completeness. However, this "organic strategy" will always result in a tangle of components and subsystems. Everything is developed in isolation and subject to local optimization as opposed to meeting a global purpose. In many ways this is an anti-systems perspective[111].

The reality is a system must be developed using a top down, bottom up, middle out approach. The goal is to surface all the emergent properties of the system. That means all views and analysis practices should be considered until a point of diminishing returns is reached. At some point placing more people on an effort, doing more analysis, doing different analysis yields no more insights about the system. This can and should be measured. When that point is discovered, it is the point of diminishing returns and the system has its lowest risk of unexpected negative emergence.

12.3 Structured Analysis

Structured analysis is the whole point of developing the DFDs. The output is the stack of paper with the data flow diagrams, data dictionary, and mini spec. The heart of the formal analysis yields a structured specification from these artifacts. It is a top down technique that is tuned when viewed from the bottom up. The bottom views change the top views. This parent child relationship is key to identifying a set of requirements that are complete and consistent. This traceability of parents and children became so important in succeeding decades that tools were created to manage requirements and show the parent child relationships.

"Structured analysis consists of a synthesis activity, which is what actually generates the data flow diagrams, mini specs and data dictionary, and an analysis activity, which checks that the resulting specification conforms to conventions intended to prevent inconsistency, incompleteness and ambiguity. Synthesis tends to be creative and often heuristic, while analysis is methodical and somewhat mechanical. The two proceed iteratively, with the goal of optimizing the structured specification"[112]. The tasks of synthesis are basically the rules for data flow diagrams and their elements[113]:

---

[111] Sustainable Development Possible with Creative System Engineering, Walter Sobkiw, 2008, ISBN 0615216307 suggests that we must return back to systems engineering if we are to build a sustainable world in this new century.

[112] Summarized in a doctoral dissertation and is partially duplicated herein: A Logical Approach To Requirements Analysis, Dr. Peter Crosby Scott, A Dissertation in Systems, Presented to the Faculties of the University of Pennsylvania in Partial Fulfillment of the Requirements for the Degree of Doctor of Philosophy, 1993.

[113] The list has been slightly modified from the original dissertation.

- Every data flow diagram must contain at least two processes and two data flows, and any number of data stores.
- Every process and data store must have at least one input data flow and one output data flow.
- Every source must have at least one output data flow.
- Every sink must have at least one input data flow.
- Every data flow must have a source and a destination.
- Every data flow must have a process as its source, destination or both. One end may be connected to a data store, source or sink.
- If both source and destination of a data flow are processes, they must be different processes.
- Any data flow may be decomposed into child data flows.
- Any non-primitive (leaf) process may have a mini spec but it is not encouraged. Instead list its key requirements.
- Each primitive or leaf must have a mini spec.
- Any description or reference to a description of what a process is to do is in its mini spec.
- All data flows and data stores must have a data dictionary entry.
- Every data flow, which is decomposed into two or more data flows (dictionary components) must have that decomposition defined in the data dictionary.
- Every data store, which is decomposed into two or more data stores, must have that decomposition defined in the data dictionary.
- Every data flow diagram, process, data flow, data store and external must have a unique name. However, a data flow or data store may appear on two or more data flow diagrams with a parent-child relationship.
- The data flows entering and leaving a lower data flow diagram must be equivalent to the data flows entering and leaving the parent process. Equivalence of data flows is defined by data dictionary entries.

12.4 Leveling and Optimization

Leveling is the process of ensuring that the top level data flows match the lower level data flows. They must have the same names. Optimization happens when the lower level data flows are examined and they change the upper level data flows. This is an iterative process where the children change the parents and the parents change the children until there is a convergence as evidenced by consistency and completeness.

There are various approaches for deciding when to decompose a given process group processes on a particular level and manage the mini spec requirements[114]:

- Minimize the number of data flows among processes.
- Group processes that manage external interfaces.
- Group processes that perform similar tasks or operate on related data flows.
- Distribute flows evenly between processes. Concentrate all processing of one input in one region of the model. Partition the system in ways that are familiar to the analysts, and that will divide the later analysis into meaningful tasks[115].
- Decompose a process when its mini spec exceeds a single page in length.
- Decompose a process with more than one data flow output.
- Decomposition cannot be done entirely top-down. Iterate in both directions.
- Minimize interconnection by combining children.
- For each process decomposition or merge identify additional requirements if necessary.
- Group the allocated requirements of a process into about seven groups.
- Allocate requirements of a parent process to its lower level decomposition DFD.
- Add data stores by detecting references to storage
- Add data stores based on the temporal relationships among data flows.
- For each process that cannot be decomposed, use its allocated requirements as a starting point to write a mini spec. It also should describe how to implement the leaf or primitive process.
- During decomposition consider coupling, cohesion, encapsulation, modularity, performance, overhead, test, duplication, maintenance, growth, technology insertion, producibility, etc. It is not possible to design a system without knowing something about how it will be implemented[116].

12.5 DFD Analysis Process

Traditional DFDs provide no information about timing or ordering of processes, control, or about whether processes will operate in sequence or in parallel.

This is different from flowcharts, functional flow diagrams, or IDEF0, which can show the flow of control through a system or an algorithm. These other

---

[114] A Logical Approach To Requirements Analysis, Dr. Peter Crosby Scott, A Dissertation in Systems, Presented to the Faculties of the University of Pennsylvania in Partial Fulfillment of the Requirements for the Degree of Doctor of Philosophy, 1993. Slightly modified for this description.

[115] Strategies for real-time system specification, Derek J. Hatley and Pirbhai A. Imtiaz, New York: Dorset House, 1987.

[116] Hughes Aircraft Fullerton used this approach in the early 1980's for next generation air traffic control (Advanced Automation System) and air defense systems.

practices allow analysts to determine what operations will be performed, in what order, and under what circumstances.

DFDs however show what kinds of data are input output to and from the system, where the data will come from and go to, and where the data will be stored. In addition, using this view of data in a system allows analysts to methodically decompose the system to its leaves and in most cases clearly shows the system primitives. This view is rarely possible using other practices, leaving the analysts to stop at arbitrary levels which are by definition incomplete and thus at risk of being wrong.

Because DFD analysis can offer a rigorous, methodical, and full decomposition of a system, many analysts augment traditional data flows with additional syntax and symbols to capture elements such as control, timing, sequence, allocation, etc. The traditional process begins with a clean stack of paper, pencil, and a template to draw a circle, like a quarter dollar coin. The process is summarized as follows:

- **Saturate**: The first step is to immerse the analysts into the new system. As part of saturation the analysts may have read various studies, interviewed stakeholders, observed similar operations, attended meetings, etc. The point is the analysts are familiar with the problem space and will now attempt to bound system. Once the initial system boundary is established they will then use a formal and somewhat rigorous technique to gain other insights into the system. Saturation continues during this formal analysis. Additional information about the system can come from any source and impact the analysis. The analysis should always keep up with the new information, as it is uncovered. In many cases the analysis itself will prompt for the search of new information and a discovery may impact the analysis.

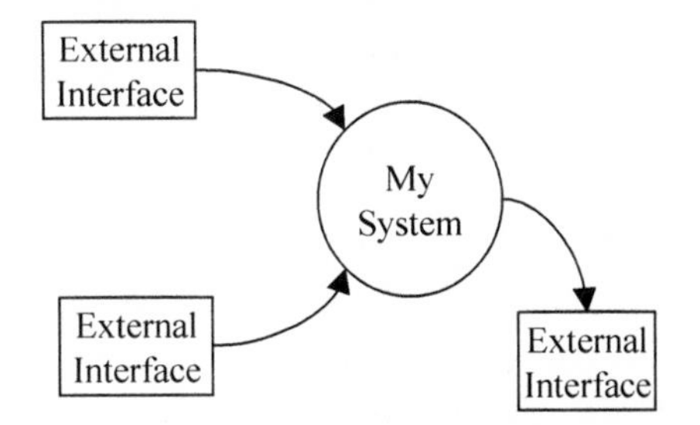

Figure 73 DFD Context Diagram

- **System Boundary**: Draw the context diagram as a single process on a single sheet of paper just like in functional analysis. Name the system and identify the external interfaces. Identify the nature of the data communicated. Use arc lines to represent the data flows between the system and the external interfaces. Only think about the whole system and its interfaces. Spend no more than 10 minutes on this diagram. Title the sheet at the top "Context Diagram". Identify and list the key requirements at this level.

- **Context Diagrams**: Many times a context diagram for a DFD is essentially the same as for a functional block diagram. That is because one body of work was built upon the other in the span of just a few years. While DFDs were created to support software, the analysis roots are found in functional decomposition. Conversely, many practitioners exposed to DFDs and this analysis technique modified it as needed to analyze non-software based system elements.

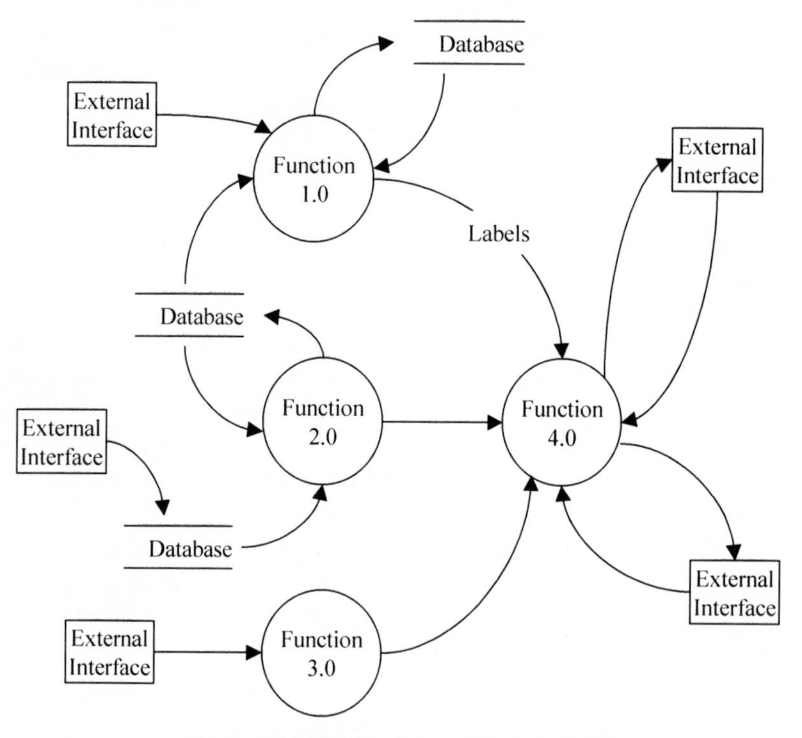

Figure 74 DFD First Level Decomposition

- **First Level Decomposition**: Take a new sheet of paper and draw what you think are the 7 plus or minus 2 top-level functions in the system. Think about how closely related these functions are to each other as you place the circles on the sheet of paper. Name and number the functions. Look at the context diagram external interfaces and add them to the functions on this level 0 diagram. Now start linking the functions with data flows. Label the data flows. If you are able to identify the high level stores, place them between the data flows. For each data flow generate a data dictionary entry. Identify and list the key requirements for this entire view and for the individual processes. Spend no more than 10 minutes on this diagram. Title this sheet "Level 0 Diagram".

- **Keep it Simple**: The time limitation is actually a key element to the analysis. The issue is to not over think the problem at any given level. Over thinking the problem will actually slow the process. The process is based on levels of understanding at different abstraction levels. As the abstraction levels progress from high to low or from less detail to more detail, understanding surfaces that impacts the previous upper levels. The analogy of pealing an onion applies and adds humor to the adventure. Notice that this data flow diagram violates some of the rules for data stores.
- **Next Level Decomposition**: While looking at the function labeled number 1.0 on the Level 0 Diagram take a new sheet of paper, and draw 7 plus or minus 2 top-level child functions originating from the parent function (a subset). Look at the parent function external interfaces and add them to the child functions on this level 1 diagram. Now start linking the child functions with data flows. Label the data flows. If you are able to identify the data stores, place them between the data flows. For each data flow generate a data dictionary entry. Identify and list the key requirements for this entire view and for the individual processes. Spend no more than 10 minutes on this diagram. Title this sheet "Level 1 Diagram".
- **Succeeding Level Decompositions**: You now have a choice. You can proceed to decompose the Level 1 diagram or you can go to the next function in the Level 0 diagram. Your decision should be based on your immediate reaction so you can start to decompose with the greatest ease and detail. What is happening is you are following your educated intuition that formed while you were saturating yourself on the system.
- **Layering**: The layering of the DFDs helps to manage complexity. Each lower layer traces to its higher layer. This is a bi-directional trace that must be complete and consistent in names and numbering conventions. Changing names even slightly will cause the process to collapse and you will need to re-start.
- **Update Higher Levels as Needed**: As you decompose the system in this top down structured fashion you are also learning about the system. Eventually you will discover an error at a higher level that is only visible when you are focused on the lower level. It may surface at the next level down or at the 4th level down. It is this discovery, ripple effect, and then the rework that is key to why this methodical method is so effective.
- **Reaching the Leaves**: At some point you reach a natural level where you are unable to decompose and find children of a parent function. These are called primitive functions. They are atomic. This may happen when you only have 1 input and output. When you think you have a primitive develop its mini-spec. If the mini-spec is larger than one page then you probably do

not have a primitive and some children are possible.

12.6 Structured System Analysis DFD Reach

In many ways data is almost a universal constant present in almost any system and was mentioned in the discussion of top down, bottom up, and middle out analysis. However systems have many elements that flow through them, such as: Electrons, Liquids, Gas, Money, etc. People also flow through systems. For example patients flow through health care systems and students flow through school systems, etc.

This idea of a universal medium, like data, to model a system and decompose it to its fundamental state is very powerful. It allows users to follow the structured system analysis techniques and decompose a system even if data is not the heart of the analysis. That is why many attempt to use the structured system analysis technique and modified "DFDs" to model their systems.

Example 24 DFD Decomposition - Air Traffic Control

Most people offer Air Traffic Control (ATC) as an example for many analysis techniques. It is a large public system where many people can relate to its functions. One of the major command and control centers of air traffic control is the Enroute Facility. It has been renamed to a less descriptive name, Area Control Facility.

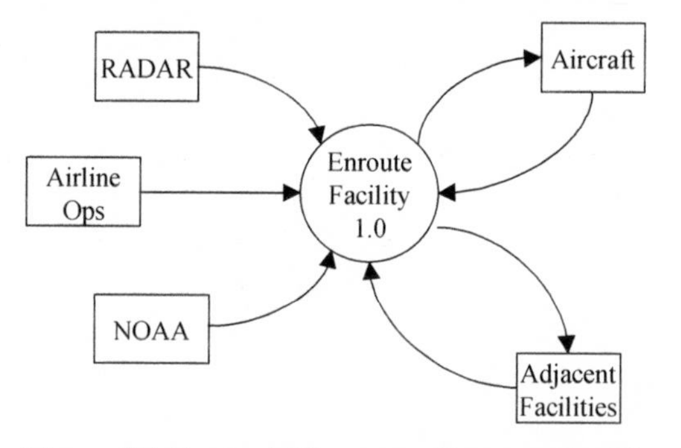

Figure 75 Context Diagram - Enroute Facility

The old Enroute name is very descriptive especially when viewed in the context of the other ATC command and control centers, the Tower and Terminal. The Tower is responsible for aircraft on the ground and approximately 1-mile after takeoff. The Terminal uses a **T**erminal **R**adar **A**pproach **Con**trol facility or TRACON. The TRACON is responsible for aircraft approximately 30 miles around and airport. The Enroute center is responsible for traffic that is en-route between TRACONS.

Author Comment: So in our analysis, it is important to be very descriptive in our names. We are not rock music stars inventing words that can mean anything.

For a rock star the more meaning a word has the more "cool" the word becomes in the collective. Obfuscation will lead to failed systems. Systems by their very nature have thousands of things that must be tracked and understood. Decomposition is a way to deal with the complexity. This complexity must be never made more complex, especially with bad naming conventions.

Initially the context diagram does not have labels for the data flows. Many times the data flows are not understood until the lower layer levels are identified. Notice the context diagram also starts with Enroute Facility rather than ATC. If the context diagram started with ATC then the Level 0 would have had Tower, TRACON, and Enroute as the first level processes or functions.

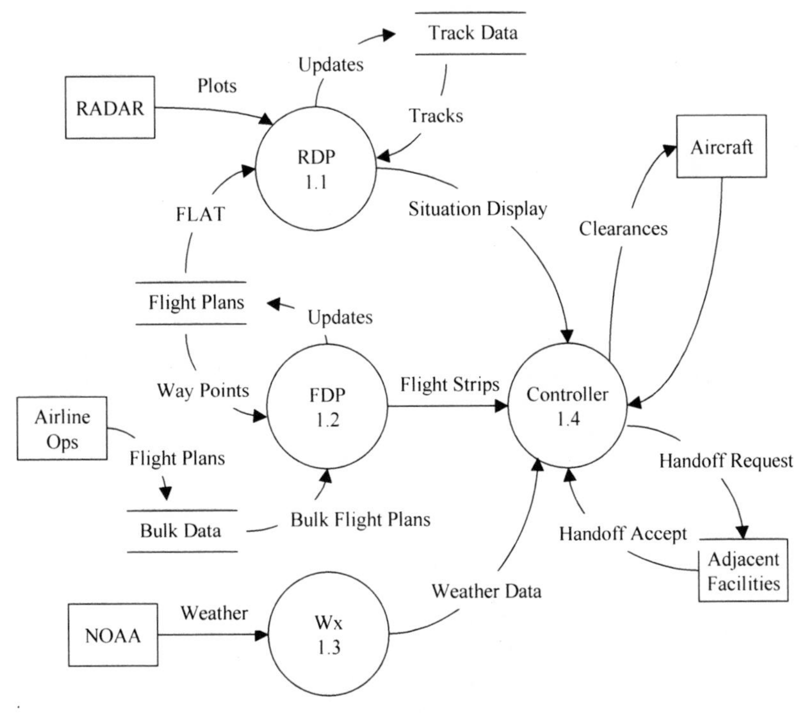

Figure 76 Level 0 Enroute Facility

Sometimes a DFD violates the rules, such as the rule for data stores where there is no process, the data store just magically appears. Is this a system issue? Perhaps it makes sense to leave DFD diagram as is until the issue is closed. How are the issues being tracked? Is there an action item list being maintained for this activity? Who is tracking the action items? Who is responsible for researching and closing out the issues as they surface?

As the analysis proceeds for this decomposition view, examine the parent. Update the parent and child views so that they are consistent. Update the data dictionary and list the key requirements for the DFD view and the individual processes. Are there key requirements associated with the data stores? Are there key requirements associated with the data flows? These key requirements will feed your next level decomposition. The context diagram is updated to reflect insights gained while developing the level 0 decomposition.

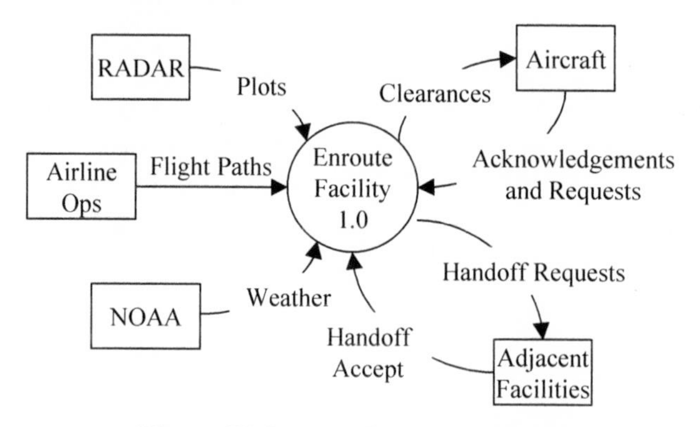

Figure 77 Context Diagram Updated

12.7 Allocations

Allocations and decomposition are closely related. Allocation also starts to enter the architecture and implementation phases of systems development. So what is it that we allocate our processes or functions to?

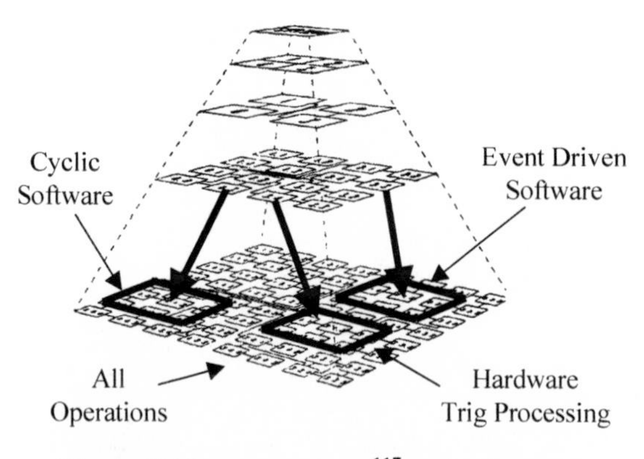

Figure 78 Decomposition[117] and Allocations

Allocations are multi dimensional and address various aspects of the systems view. We can start the allocation process from any perspective. Once the DFDs are allocated to possible physical entities, analysis can be performed to test the

[117] Basic pyramid showing decomposition is from functional analysis section in Systems Engineering Management Guide, Defense Systems Management College, January 1990.

effectiveness of the decomposition. The analysis can include availability, maintenance, performance, safety, etc[118]. The following is a list of possible allocation targets:

- **System**: This is the highest level of the system and an allocation to this level suggests that the process, function, and resulting requirements apply to all the system elements. An example is a maintenance philosophy that dictates user manuals be developed. Other examples include safety, fault tolerance, security, etc.
- **Segment**: A system segment is a physical location in a system that has multiple physical locations. For example the ATC system has four segments: Enroute, TRACON, Tower, and Support segments. At one time there were 23 Enroute facilities, 234 TRACONS and 400 Towers[119]. Other examples are ground, space, surface sea, under sea, and under ground segments typically found in very large defense systems. The segments will share common subsystems and lower level functions and their implementations.
- **Subsystem**: A subsystem is a physical entity that executes the processes or functions. In computer automation it is a computer with software that supports the functions. Subsystems can decompose into subsystems.
- **Hardware Configuration Item**: A Hardware Configuration Item (HWCI) is a hardware element that supports a function. For example an amplifier supports the functions of bass, treble, and loudness control. An HWCI also can be a computer, display, keyboard, mouse, cell phone, etc that are absent any software or firmware.
- **Software Configuration Item**: A Software Configuration Item (SWCI) is a software package that supports a group of functions. Examples include operating systems like UNIX, Windows, servers like Apache, applications like client web browser, applications that support functions like word processing, drawing, spreadsheets, etc.
- **Operations**: Operations are processes or functions performed by humans. Prior to computer automation, people in large organizational entities performed many processes or functions. Today people tend to be in maintenance roles and augment automation implemented on computer based systems. When systems are studied, automation tends to be the first approach to making a system better. However, not all system problems can be solved with technology. The system solutions might come from the social political economic sciences. For example, replacing teachers with a

[118] The idea of allocating the DFD functions to physical entities based on various analyses was part of Hughes Aircraft circa 1982.

[119] These are approximate numbers circa 1982.

computer screen may not yield an effective educational system.

- **Operating System User**: In a Multi User Operating System like UNIX, a process or group of processes can be allocated to users. For example in the ATC Level 0 DFD the RDP function can be allocated to each air traffic controller as a UNIX user account. If something should cause a failure for one user that failure would be isolated to that user, it will not propagate. An example is an overrun condition or divide by zero, which would cause processing to get stuck in an endless loop.
- **Operating System Task**: In a Multi Tasking Operating System like UNIX a process or group of processes can be allocated to a task. For example in the ATC Level 0 DFD the RDP function can be allocated to several tasks based on region of airspace. If something should cause a failure for a part of the airspace that failure would be isolated to that task, it will not propagate. An example is an overrun condition or divide by zero, which would cause processing to get stuck in an endless loop. UNIX can take advantage of its Multi Tasking and Multi User capabilities and use both mechanism to minimize failure propagation.
- **Directory**: The software that implements the processes or functions can be grouped into various file system computer directories. This grouping affects maintainability of the system.
- **Source File**: The software that implements the processes or functions can be grouped into various source code files. This grouping affects maintainability of the system.
- **Class**: The software that implements the processes or functions can be grouped into a class. The class can then be instantiated to support the system. For example the ATC RDP class can be instantiated for each air traffic controller. This is similar to the isolation achieved in UNIX. This grouping affects maintainability and performance of the system.
- **Process or Function Characteristics**: Processes or functions tend to behave in certain ways. As a staring point there are event driven, cyclic, real time, extremely important such that the function is always available, complex, simple, new, old and proven, etc. Categorizing the functions adds additional understanding of the system and helps in the decomposition and allocation to an architecture solution.

12.8 Threads and Analysis

Structured system analysis and the DFD stack of parents and children can be used to visualize the system from more than just a functional perspective. It is an information product that can feed other activities. Those activities will then offer further validation of the DFDs. The analysis may change the DFD results.

Analysts viewing the system from an operational perspective can identify key operational sequences, threads, and trace the operation through the decomposition.

This typically starts with selecting an external input and conceptually dropping a "ball" into the machine. The path is watched as it traverses through the DFD. Highlighting the DFD elements on a particular decomposition may show this view. In this way operators can validate the decomposition, analysts gain further insight into the decomposition, and specialty engineers can start to develop performance understandings.

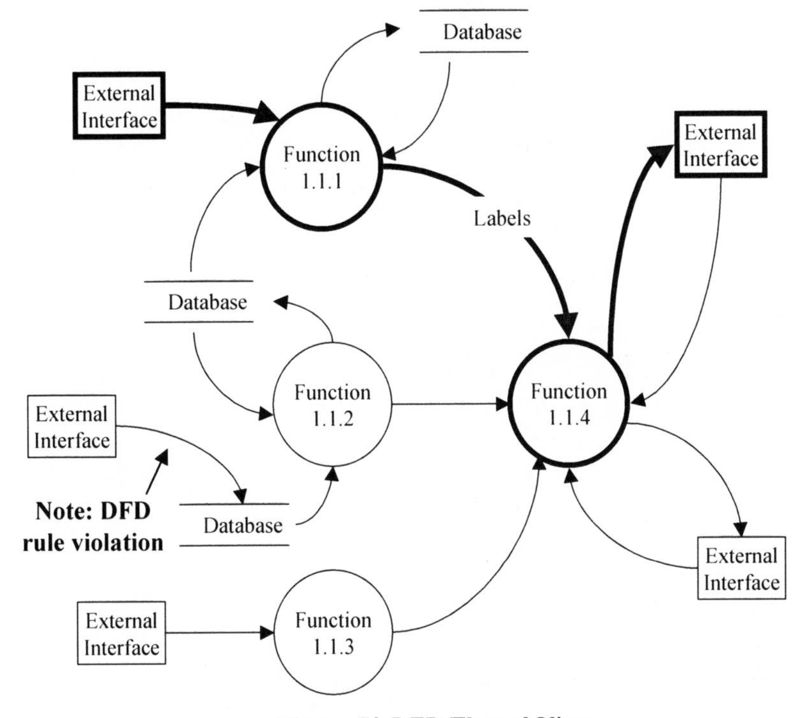

Figure 79 DFD Thread View

For example what is the maximum allowed time to traverse the path? What is the actual time when the functions are allocated to a possible architecture implementation? What is the sum of errors because of numerical precision? Since airplanes move what is the sum of errors because of time in storage? What is the reliability allocation? What is the actual expected availability when the functions are allocated to an architecture implementation? What is the expected input and output rate? Can these rates be supported at peak load and steady state level for an architecture implementation?

The threads and analysis can be visualized at a single individual decomposition view or at the aggregate decomposition views. The aggregate view may be offered first to get the big picture and then the individual DFD sheets can be offered as a thread is analyzed.

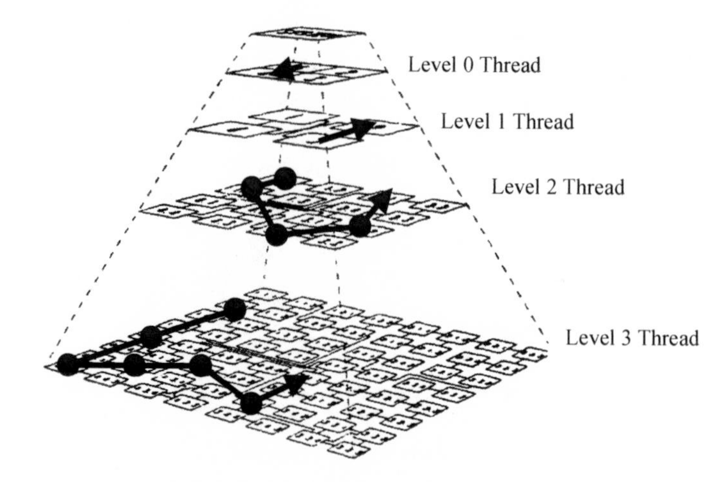

Figure 80 Decomposition Thread View

12.9 Hierarchical Input Process Output

Hierarchical Input Process Output (HIPO) is a technique for representing the modules of a system as a hierarchy and for documenting each module[120]. The HIPO also has been referred to as Input Processing Output (IPO). The HIPO has the following 4 elements where storage is optional and other elements are required:

- Input - Information, ideas, and resources to be processed
- Processing - Actions taken on the input or stored material
- Output - Results of the processing
- Storage - Material inside the process stored for later

As in the case of the DFD, the HIPO or IPO can be used to understand more than software in a system. Just like the DFD it can be organized into a parent child relationship. Child HIPOs of a system have their own set of inputs and outputs that trace to parent HIPOs. Outputs of a HIPO are either input for another HIPO or become part of the ultimate output of the system.

The HIPO can be coupled with the DFD where a HIPO can be drawn to represent a decomposition level. So it clearly shows the inputs, outputs, lists the processing and optionally shows storage. This is a different visual and captures the essence of a decomposition level rather than the details of the interaction or its context. In this way it is ideal for education, brainstorming, and preliminary investigation prior to full-scale decomposition analysis.

[120] Sandia Software Guidelines Volume 5; Tools, Techniques, and Methodologies; Sandia Report, SAND85–2348 I UC–32, Reprinted September 1992.

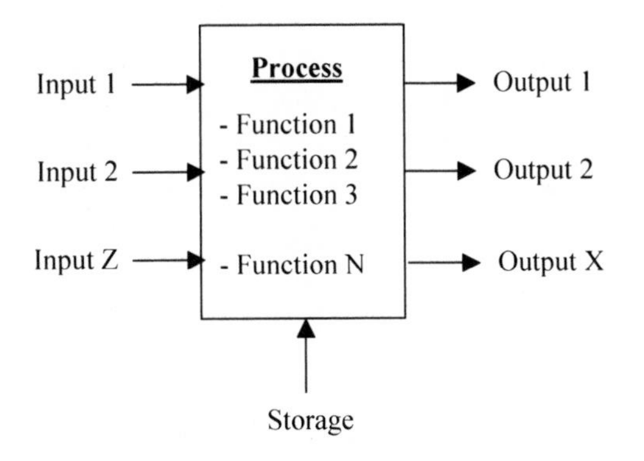

Figure 81 HIPO / IPO

Example 25 HIPO - RADAR Tracker

A HIPO can be used to consider a system question at the macro or micro level. This arbitrary level of detail is called scope. It is not unreasonable to create a HIPO that is not a complete fit with other analysis information products, such as data flow diagrams. The HIPO offers a specific view that the stakeholders need as they consider a problem. For example the introduction to a study capturing the results of the tracking function in a new air traffic control system might start with a HIPO and text to introduce the basic functions of the tracker.

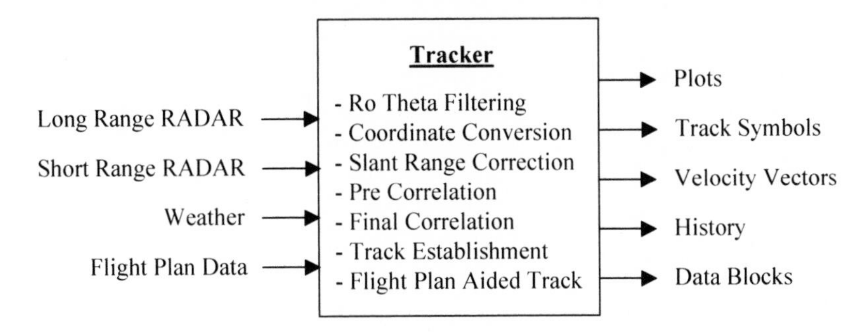

Figure 82 HIPO - ATC Tracker

A HIPO is typically captured on a single sheet of paper with fields such as input, processing, output, called by, calls, author, project, diagram number and name, module name, date. The sheets are placed in a book to represent a system. The book introduction lists the HIPO diagrams using an indentured list and also may include a graphical tree diagram.

12.10 Exercises

1. Using paper, pencil, and eraser, develop a DFD context diagram and Level 0 DFD for a good and a bad website.
2. Using paper, pencil, and eraser, develop a DFD context diagram and a Level 0 DFD for the US health care system and education system.
3. Try to take a process to the leaf level for any of the above context diagrams and Level 0.
4. Develop a data dictionary for any of the previous context diagrams and Level 0 diagrams.
5. Develop mini specs for some previous context diagrams and Level 0 diagrams. Develop a structured specification.
6. Characterize the functions. Based on the characteristics, allocate them to a potential architecture.
7. Using paper, pencil, and eraser, develop a top level HIPO of you personal computer. Pick an application and develop a HIPO or set of HIPO diagrams for the application.
8. Compare the DFD and HIPO diagrams that you developed. What are your findings? Was the architecture changed?
9. Try to develop DFD and HIPO diagrams using your computer. What have you detected?

12.11 Additional Reading

1. Specification Practices, Military Standard, MIL-STD-490 30 October 1968, MIL-STD-490A 4 June 1985.
2. Strategies for Real-Time System Specification, Derek J. Hatley, and Pirbhai A. Imtiaz, New York: Dorset House, 1987.
3. Structured Analysis and System Specification, Tom DeMarco, Englewood Cliffs, NJ: Yourdon Press, 1978.
4. Structured Design, Edward Yourdon and Larry L. Constantine, New York: Yourdon Press, 1978, ISBN 0917072073.
5. Systems Engineering Management Guide, Defense Systems Management College, January 1990.
6. System / Segment Specification, DOD Data Item Description DI-CMAN-80008A, June 1986.
7. System Software Development, Military Standard Defense, DOD-STD-2167 4 June 1985, DOD-STD-2167A 29 February 1988.
8. The Magical Number Seven, Plus or Minus Two: Some Limits on Our Capacity for Processing Information, G.A Miller, The Psychological Review, 63, 2 (March): 81-97, 1956.

13 IDEF0

Integration Definition for Function Modeling (IDEF0) is a formal graphical method to represent a system. It was matured while performing analysis for development, re-engineering, and integration of information systems, organizational processes, and software engineering. While the functional block diagram shows functional relationships, the more formal Functional Flow Block Diagram (FFBD) shows the functional flow of a system, and the DFD shows data flow through a system, the IDEF0 shows system control and the mechanism that implements the function. It really is a reaction to DFDs where many analysts would ask questions about control and allocations when DFDs were presented. It also conveys data or objects related to functions to be performed [121].

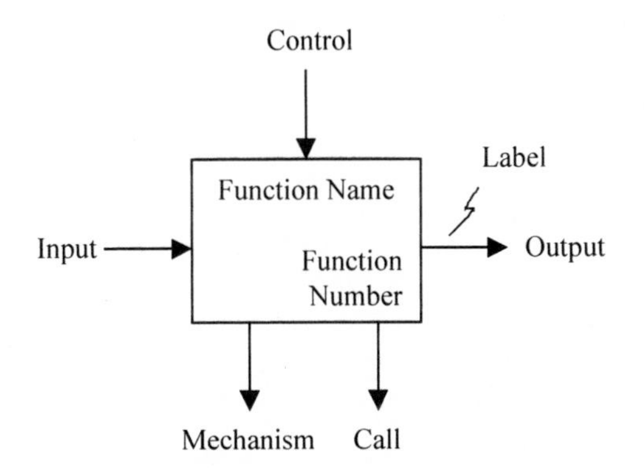

Figure 83 IDEF0 Elements

IDEF0 also has more rules and as such is more formal than the FFBD. This formality can offer a rigorous and precise system description that facilitates consistency of usage and interpretation. However the rules also may alienate some stakeholders who will not take the time to understand the rules. This is typical of all formal methods and must be understood when the results are embedded in information products that should effectively summarize the results of the IDEF0 activity.

---

[121] IDEF came from the graphic modeling language Structured Analysis and Design Technique (SADT) developed by Douglas T. Ross and SofTech, Inc. SADT was developed and field-tested between 1969 to 1973. SADT was used in the MIT Automatic Programming Tool (APT) project. It received extensive use starting in 1973 by the US Air Force Integrated Computer Aided Manufacturing program.

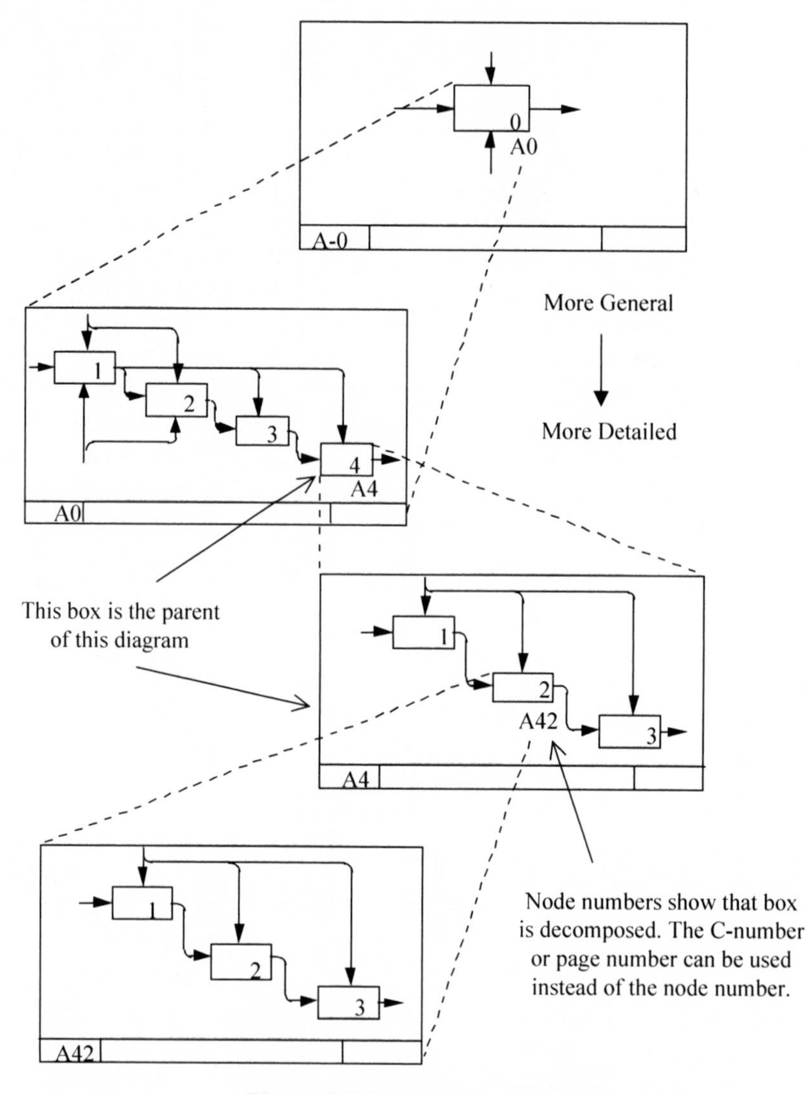

Figure 84 IDEF0 Decomposition

IDEF0 is a series of cross-referenced hierarchical diagrams, text, and glossary. The primary elements are functions represented as boxes, and data or objects represented by arrows that connect the functions. The position where an arrow attaches to a box represents a specific role. The controls enter the top of the box. The inputs handled by the function enter the box from the left. The outputs of the function leave the right-hand side of the box. Mechanism arrows for performing the function enter the bottom of the box.

The activity starts with the identification of the highest level IDEF0 function to be decomposed. This function is identified on a Top Level Context Diagram. It defines the scope of the particular IDEF0 analysis.

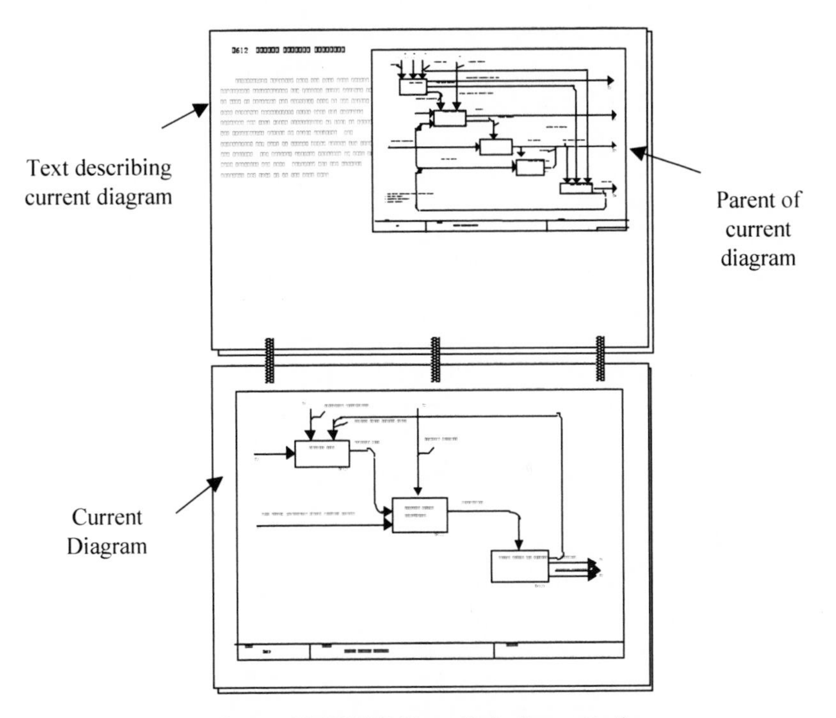

Figure 85 IDEF0 Page Pair Presentation

This diagram is then successively decomposed to lower-level diagrams. This is just like functional decomposition described in the Functions Performance Subsystems section. So there are parent child relationships between the diagrams.

During the 1970s there was great interest in productivity improvements, especially in the engineering and manufacturing domains. It was also a time where computers were becoming more accessible as costs were dropping. The U.S. Air Force Program for Integrated Computer Aided Manufacturing (ICAM) was established to increase manufacturing productivity through systematic application of computer technology. The ICAM program identified the need for better analysis and communication techniques for people involved in improving manufacturing productivity. They developed a series of techniques known as the IDEF[122] (ICAM Definition) techniques, which included the following:

[122] IDEF0 standard, Federal Information Processing Standards Publication 183 (FIPS 183), and IDEF1x standard (FIPS 184). Maintained by the National Institute of Standards and Technology (NIST).

- IDEF0, is used to produce a "function model". A function model is a structured representation of the functions, activities or processes within the modeled system or subject area[123].
- IDEF1, is used to produce an "information model". An information model represents the structure and semantics of information within the modeled system or subject area.
- IDEF2, is used to produce a "dynamics model". A dynamics model represents the time-varying behavioral characteristics of the modeled system or subject area.

In 1983, the U.S. Air Force Integrated Information Support System program enhanced the IDEF1 information modeling technique to form IDEF1X (IDEF1 Extended), a semantic data modeling technique.

The IDEF has significantly more information than a functional block diagram. Using the magical number 7 plus or minus 2 concepts developed in the functional analysis section, we see that with 5 functions on a page, the human also needs to track control elements and mechanism elements. That triples the amount of information on a single view and we exceed our magical number.

13.1 Syntax

The structural components and features of a language and the rules that define relationships among them are referred to as the language's syntax. The components of the IDEF0 syntax are boxes and arrows, rules, and diagrams. Boxes represent functions, defined as activities, processes or transformations. Arrows represent data or objects related to functions. Rules define how the components are used, and the diagrams provide a format for depicting models both verbally and graphically. The format also provides the basis for model configuration management.

A box provides a description of what happens in a function. Each box has a name and number inside its boundaries. The name is an active verb or verb phrase that describes the function. Each box contains a number located in the lower right corner. The box numbers are used to trace to the associated text. Boxes represent functions that show what must be accomplished. The box syntax rules are:

- Sufficient in size to insert box name
- Rectangular in shape, with square corners
- Drawn with solid lines

---

[123] Structured Methods: Structured programming circa 1967, Edsger W. Dijkstra; Structured Design circa 1975, Larry Constantine and Ed Yourdon; Structured Analysis circa 1978, Tom DeMarco, Yourdon, Gane & Sarson, McMenamin & Palmer; Information Engineering circa 1990, James Martin. In 1981 IDEF0 published, based on SADT.

Arrows are composed of one or more line segments, with a terminal arrowhead at one end. Arrow segments may be straight or curved (with a 90° arc connecting horizontal and vertical parts), and may have branching (forking or joining). Arrows do not represent flow or sequence as in traditional functional flow diagrams. Arrows convey data or objects related to functions to be performed. The functions receiving data or objects are constrained by the data or objects made available. Today it is difficult to draw a 90° arc using a computer presentation tool and so many practitioners draw 90° line segments without the arcs. The arrow syntax rules are:

- Bends are curved using only 90 degree arcs
- Drawn in solid line segments
- Drawn vertically or horizontally, not diagonally
- Ends touch outer perimeter of function box and do not cross into the box
- Attach at box sides, top, bottom, not at corners

13.2 Semantics

Semantics refers to the meaning of syntactic components of a language and aids in correct interpretation. Interpretation addresses items such as box and arrow notation and functional relationship interfaces.

The box name is a verb or verb phrase, such as "Perform Inspection", that is descriptive of the function that the box represents. The example "Perform Inspection" function transforms un-inspected parts into inspected parts. The definitive step beyond the phrase naming of the box is the use of arrows (matching the orientation of the box sides) that complement and complete the expressive power (as distinguished from the representational aspect) of the IDEF0 box.

Standard terminology is used to ensure precise communication. Box meanings are named descriptively with verbs or verb phrases and are split and clustered in decomposition diagrams. Examples of function name are: process parts, plan resources, conduct review, monitor performance, design system, provide maintenance, develop detailed design, fabricate component, inspect part, etc.

Arrow meanings are bundled and unbundled in diagramming and the arrow segments are labeled with nouns or noun phrases to express meanings. Arrow segment labels apply exclusively to the particular data or objects that the arrow segment graphically represents. Arrow meanings are further expressed through fork and join syntax. Arrows identify data or objects needed or produced by the function. Each arrow is labeled with a noun or noun phrase, For example: specifications, test report, budget, design requirements, detailed design, directive, design engineer, board assembly, requirements, etc.

Arrows entering the left side of the box are inputs. Inputs are transformed or consumed by the function to produce outputs. Arrows leaving a box on the right side are outputs. Outputs are the data or objects produced by the function.

Arrows entering the box on the top are controls. Controls specify the conditions required for the function to produce correct outputs.

Arrows connected to the bottom of the box represent mechanisms. Upward pointing arrows identify the means that support the execution of the function. Other means may be inherited from the parent box. Mechanism arrows that point downward are call arrows. Call arrows enable the sharing of detail between models (linking them together) or between portions of the same model. The called box provides detail for the caller box.

Supporting information concerning the function and its purpose is addressed in the text associated with the diagram. A diagram may or may not have associated text. When acronyms, abbreviations, key words, or phrases are used, the fully defined terms are provided in the glossary. The box arrow semantic rules are:

- A box is named with an active verb or verb phrase
- Input arrows interface with the left side of a box
- Control arrows interface with the top side of a box
- Output arrows interface with the right side of the box
- Mechanism arrows (except call arrows) point upward and connect to the bottom side of the box
- Mechanism call arrows point downward, connect to the bottom side of the box, and are labeled with the reference expression for the box which details the subject box
- Arrow segments, except for call arrows, are labeled with a noun or noun phrase unless a single arrow label clearly applies to the arrow as a whole
- A “squiggle” is used to link an arrow with its associated label, unless the arrow label relationship is obvious
- Arrow labels do not consist solely of any of the following terms: function, input, control, output, mechanism, or call

13.3 Diagrams

IDEF0 models are composed of graphic diagrams, text, and a glossary that are cross-referenced. The graphic diagram is the major component, containing boxes, arrows, box/arrow interconnections and associated relationships. Boxes represent functions of a subject area. These functions are decomposed into more detailed diagrams, until the subject is described at a level necessary to support the goals of the analysis. The top-level graphic diagram is the most general or abstract description of the subject. This diagram is followed by a series of child diagrams providing more detail about the subject.

Context Diagram - Each model has a top-level context diagram represented by a single box with its bounding arrows. This is called the A-0 diagram (pronounced A minus zero). The arrows on this diagram interface with functions outside the subject area. This bounds the system. Since a single box represents the system, the descriptive name written in the box is general. The same is true of the interface arrows since they also represent the complete set of external interfaces to the system. The A-0 diagram sets the model scope or boundary and orientation.

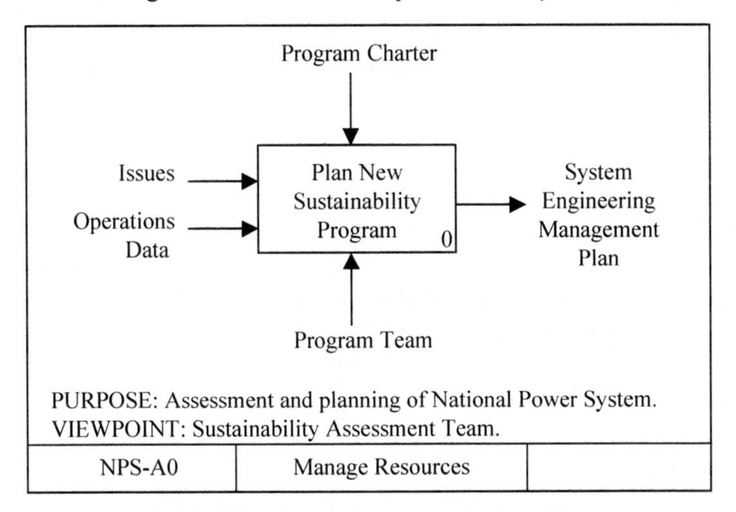

Figure 86 A-0 Diagram - Context Diagram

The A-0 context diagram has brief statements identifying the model's viewpoint and purpose. This helps to guide and constrain the creation of the model. The viewpoint suggests what is within the model and offers the perspective or slant of the analysis. Depending on the audience, different statements of viewpoint may be used that emphasize different aspects of the analysis. Elements that are important in one viewpoint may be irrelevant in another viewpoint and not appear in that model presented view.

The purpose suggests why the model is created and determines the structure of the model. The most important elements come first in the hierarchy, as the whole top-level parent functions are decomposed into child sub-functions. The decomposition continues until the viewpoint is adequately expressed.

Child Diagram - A child diagram is the result of decomposing a single parent function box. Each of the child sub-functions may be decomposed, creating other, lower-level child diagrams. The parent child relationship continues down the decomposition tree. On a parent diagram, some of the functions, none of the functions or all of the functions may be decomposed. Each child diagram contains the boxes and arrows that provide additional detail about the parent box. The child diagram covers the same scope as the parent box it details. Thus, a child diagram may be thought of as being the inside or subset of its parent box.

Parent Diagram - Except for the context diagram, every diagram is also a child diagram. They all detail a parent box. So a diagram may be both a parent diagram (containing parent boxes) and a child diagram (detailing its own parent box). A box may be both a parent box (detailed by a child diagram) and a child box (appearing on a child diagram). The hierarchical relationship is between a parent box and the child diagram.

Text and Glossary - A diagram may have text, which is used to provide a concise overview of the diagram. Text is used to highlight features, flows, and inter-box connections that clarify the intent of items and patterns. Text is not used to describe, redundantly, the meaning of boxes and arrows.

The glossary is used to define acronyms, key words, and phrases used in the diagram graphics. The glossary defines words in the model to convey a common understanding so analysts can correctly interpret the model content.

For Exposition Only Diagrams - For Exposition Only (FEO, pronounced fee-oh) diagrams are used where additional knowledge is needed to understand specific model areas. Additional detail is limited to what is needed to achieve the stated purpose for a knowledgeable audience. A FEO diagram does need not to follow IDEF0 syntax rules.

13.4 Definitions

1. **A-0 Diagram:** The special case of a one-box IDEF0 context diagram, containing the top-level function being modeled and its inputs, controls, outputs and mechanisms, along with statements of model purpose and viewpoint.
2. **Arrow:** A directed line, composed of one or more arrow segments, that models an open channel or conduit conveying data or objects from source (no arrowhead) to use (with arrowhead). There are 4 arrow classes: Input Arrow, Output Arrow, Control Arrow, and Mechanism Arrow (includes Call Arrow). See Arrow Segment, Boundary Arrow, and Internal Arrow.
3. **Arrow Label:** A noun or noun phrase associated with an IDEF0 arrow or arrow segment, specifying its meaning.
4. **Arrow Segment:** A line segment that originates or terminates at a box side, a branch (fork or join), or a boundary (unconnected end).
5. **Boundary Arrow:** An arrow with one end (source or use) not connected to any box on a diagram. Contrast with Internal Arrow.
6. **Box:** A rectangle, containing a name and number, used to represent a function.
7. **Box Name:** The verb or verb phrase placed inside an IDEF0 box to describe the modeled function.
8. **Box Number:** The number (0 to 6) placed inside the lower right corner of an IDEF0 box to uniquely identify the box on a diagram.
9. **Branch:** A junction (fork or join) of two or more arrow segments.

10. **Bundling/Unbundling:** The combining of arrow meanings into a composite meaning (bundling), or the separation of arrow meanings (unbundling), expressed by arrow join and fork syntax.
11. **C-Number:** A chronological creation number that may be used to uniquely identify a diagram and to trace its history; may be used as a Detail Reference Expression to specify a particular version of a diagram.
12. **Call Arrow:** A type of mechanism arrow that enables the sharing of detail between models (linking them together) or within a model.
13. **Child Box:** A box on a child diagram.
14. **Child Diagram:** The diagram that details a parent box.
15. **Context:** The immediate environment in which a function (or set of functions on a diagram) operates.
16. **Context Diagram:** A diagram that presents the context of a model, whose node number is A-n (n greater than or equal to zero). The one-box A-0 diagram is a required context diagram; those with node numbers A-1, A-2, ... are optional context diagrams.
17. **Control Arrow:** The class of arrows that express IDEF0 Control, i.e., conditions required to produce correct output. Data or objects modeled as controls may be transformed by the function, creating output. Control arrows are associated with the topside of an IDEF0 box.
18. **Decomposition:** The partitioning of a modeled function into its component functions.
19. **Detail Reference Expression (DRE):** A reference (e.g., node number, C-number, page number) written beneath the lower right corner of an IDEF0 box to show that it is detailed and to indicate which diagram details it.
20. **Diagram:** A single unit of an IDEF0 model that presents the details of a box.
21. **Diagram Node Number:** That part of a diagram's node references that corresponds to its parent box node number.
22. **For Exposition Only (FEO) Diagram:** A graphic description used to expose or highlight specific facts about an IDEF0 diagram. Unlike an IDEF0 graphic diagram, a FEO diagram need not comply with IDEF0 rules.
23. **Fork:** The junction at which an IDEF0 arrow segment (going from source to use) divides into two or more arrow segments. May denote unbundling of meaning.
24. **Function:** An activity, process, or transformation (modeled by an IDEF0 box) identified by a verb or verb phrase that describes what must be accomplished.
25. **Function Name:** Same as Box Name.
26. **Glossary:** A listing of definitions for key words, phrases and acronyms used in conjunction with an IDEF0 node or model as a whole.
27. **ICOM Code:** The acronym of Input, Control, Output, and Mechanism. A code that associates the boundary arrows of a child diagram with the arrows of its parent box; also used for reference purposes.

28. **IDEF0 Model:** A graphic description of a system or subject that is developed for a specific purpose and from a selected viewpoint. A set of one or more IDEF0 diagrams that depict the functions of a system or subject area with graphics, text and glossary.
29. **Input Arrow:** The class of arrows that express IDEF0 Input, i.e., the data or objects that are transformed by the function into output. Input arrows are associated with the left side of an IDEF0 box.
30. **Interface:** A shared boundary across which data or objects are passed; the connection between two or more model components for the purpose of passing data or objects from one to the other.
31. **Internal Arrow:** An input, control or output arrow connected at both ends (source and use) to a box on a diagram. Contrast with Boundary Arrow.
32. **Join:** The junction at which an IDEF0 arrow segment (going from source to use) merges with one or more other arrow segments to form a single arrow segment. May denote bundling of arrow segment meanings.
33. **Mechanism Arrow:** The class of arrows that express IDEF0 Mechanism, i.e., the means used to perform a function; includes the special case of Call Arrow. Mechanism arrows are associated with the bottom side of an IDEF0 box.
34. **Model Note:** A textual comment that is part of an IDEF0 diagram, used to record a fact not otherwise depicted.
35. **Node:** A box from which child boxes originate; a parent box. See Node Index, Node Tree, Node Number, Node Reference, and Diagram Node Number.
36. **Node Index:** A listing, often indented, showing nodes in an IDEF0 model in "outline" order. Same meaning and content as Node Tree.
37. **Node Number:** A code assigned to a box to specify its position in the model hierarchy; may be used as a Detail Reference Expression.
38. **Node Reference:** A code assigned to a diagram to identify it and specify its position in the model hierarchy; composed of the model name (abbreviated) and the diagram node number, with optional extensions.
39. **Node Tree:** The graphical representation of the parent-child relationships between the nodes of an IDEF0 model, in the form of a graphical tree. Same meaning and content as Node Index.
40. **Output Arrow:** The class of arrows that express IDEF0 Output, i.e., the data or objects produced by a function. Output arrows are associated with the right side of an IDEF0 box.
41. **Parent Box:** A box that is detailed by a child diagram.
42. **Parent Diagram:** A diagram that contains a parent box.
43. **Purpose:** A brief statement of the reason for a model's existence.
44. **Semantics:** The meaning of the syntactic components of a language.
45. **Squiggle:** A small jagged line that may be used to associate a label with a particular arrow segment or to associate a model note with a component of a diagram.

46. **Syntax:** Structural components or features of a language and the rules that define relationships among them.
47. **Text:** An overall textual (non-graphical) comment about an IDEF0 graphic diagram.
48. **Title:** A verb or verb phrase that describes the overall function presented on an IDEF0 diagram; the title of a child diagram corresponds to its parent box name.
49. **Tunneled Arrow:** An arrow (with special notation) that does not follow the normal requirement that each arrow on a diagram must correspond to arrows on related parent and child diagrams.
50. **Viewpoint:** A brief statement of the perspective of the model.

13.5 Exercises

1. Develop an IDEF0 model for a system or subject of interest.
2. Contrast IDEF0 with Structured Systems Analysis and The DFD described in this book.
3. Which analysis technique (IDEF0, DFD, FBD, FFBD) would you use and why.

13.6 Additional Reading

1. Integration Definition for Function Modeling (IDEF0), Federal Information Processing Standards Publication FIPS 183, 21 December 1993.
2. Integration Definition for Information Modeling (IDEF1X), Federal Information Processing Standards Publication FIPS 184, 21 December 1993.

14 State Machine

Many systems have multiple states and modes. Examples of states and modes include: operational, idle, ready, active, training, degraded, emergency, backup, failsafe, maintenance, online, offline, reset, diagnostic, standby, off, etc[124]. The system developers establish the distinction between modes and states. A system can have states only, modes only, states within modes, modes within states[125]. Different modes tend to be associated with operations, as in operational modes[126]. Usually a mode is a subset of a state. There may be several modes in the operational state. This discussion will use the term, state.

States aggregate requirements. So requirements can be allocated to states. States also can be an attribute of requirements. State is yet another view of the system in operation. It can be represented with a state diagram or a state table.

A state is a response to an event or stimulus. The response varies depending on the current state, conditions, and set of rules. A state transition diagram or table capture this information in a formal deterministic manner. When an event or transition condition happens, the next state depends on the event and the current state. This is called a state change or state transition.

For simple state machines it is easy to get confused with a flowchart. A state machine is an aggregator of large collections of functionality or operational scenarios while a flowchart is associated more with an algorithm. For example a traffic light is an example of a simple finite state machine that can be represented and analyzed with a state diagram or state table, but why bother. This is a misuse of process or technology and just introduces noise to the activity rather than understanding.

State machine analysis is very important but it needs to be used when and where it adds to the discussion and understanding of the system. Creating system states and modes means that requirements eventually need to be allocated to the states and modes and that adds complexity to the system description that may be unneeded. The following are some general rules that can be applied to determine if states and modes are needed:

---

[124] System / Subsystem Specification (SSS), Data Item Description, DI-IPSC-81431, December 1994.

[125] Software Development And Documentation, MIL-STD-498, 5 December 1994; Superseding DOD-STD-2167A 29 February 1988, DOD-STD-7935A, 31 October 1988 DOD-STD-1703 12 February 1987.

[126] Specification Practices, Military Standard, MIL-STD-490 30 October 1968, MIL-STD-490A4 June 1985

- Are there safety considerations between the states and modes
- Does the system need to fail safely in a failsafe state or mode
- Are there security considerations between the states and modes
- Does the system need to fail securely in a secured state or mode
- Is the system certified daily
- Is the system used for training while supporting normal operations
- Will maintenance disrupt operations

Obviously if a system does have states and modes then a state diagram should be developed and if needed a state table that traces to the diagram. If there are inconsistencies in the diagram or disagreements within the team on the diagram, then there is an issue that must be resolved.

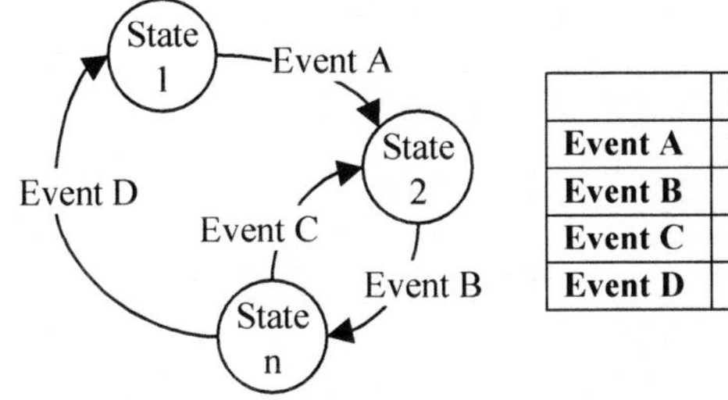

	State 1	**State 2**	**State n**
Event A	State 2		
Event B		State n	
Event C			State 2
Event D			State 1

Figure 87 State Diagram and Table

A state diagram uses circles to represent the states. Each state has a name and number. A line represents an event that comes from a source state and results in a destination state. The line includes a label that represents the event.

A state table lists the events in rows and the states in columns. There are always fewer states than events, thus the row column allocation. An event in combination with the column that represents the source-state yields the resulting state that is named in the intersecting cell.

When a system has states and modes the requirement description in the specification becomes challenging because requirements tend to get duplicated in various sections. This is difficult to detect if there are multiple contributors each assigned to a different section. Typically the requirements are partially duplicated or conflicting. In this case the specification review process is extremely critical and needs to be focused on finding and reconciling these requirements. As always there only should be one requirement in one place. This can be accomplished by using references back to the states and mode section, For example, if there is a failsafe state, the failsafe state requirements sections should be referenced as needed. This simple concept will prevent the invention and proliferation of multiple failsafe approaches within the system.

Example 26 State Diagram

Does a personal computer have states? What are those states? Does it make sense to show the states? The following is a possible state diagram for a personal computer being used to access the Internet to check for information and email. Is it possible to power off from the logged in state?

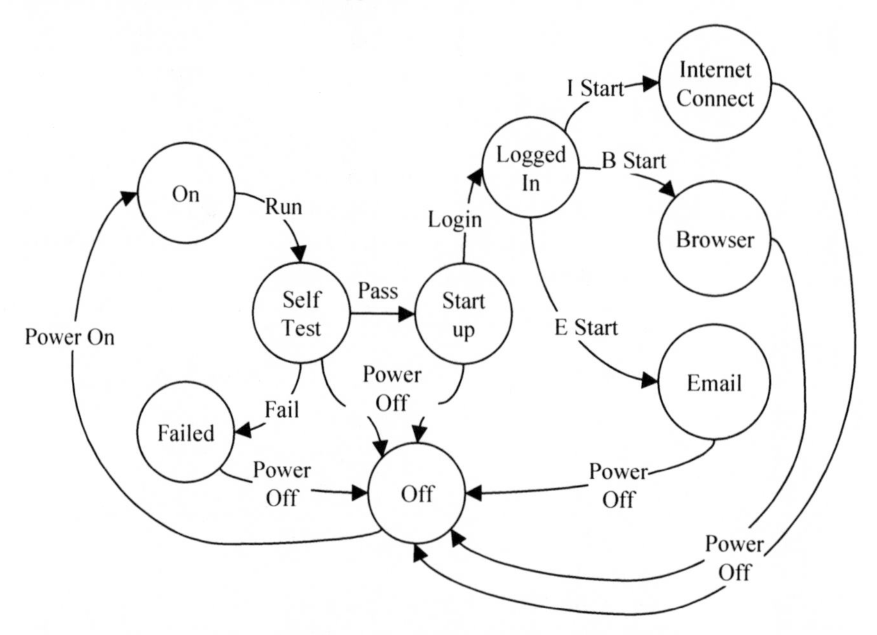

Figure 88 State Diagram - Personal Computer

After self-test and before startup there is usually another state where the user can access computer settings before the operating system is loaded. The diagram does not show the operating system being loaded prior to the logged in state.

14.1 Exercises

1. Does an automobile have states and modes and if so what are they?
2. Draw a state diagram for a reel-to-reel tape recorder, videocassette recorder, digital videodisk (DVD) recorder, digital video recorder (DVR), digital MP3 recorder. How do they compare and differ?
3. Do you have states and modes and if so draw a state diagram of you.

14.2 Additional Reading

1. Specification Practices, Military Standard, MIL-STD-490 30 October 1968, MIL-STD-490A 4 June 1985.

15 Interfaces and Interactions

Systems have internal and external interactions, which are accomplished with interfaces. There are manual interfaces that require humans and automated interfaces that are handled by machines. The interfaces transport data or physical items using various media. Occasionally when data is transported a translation needs to happen so the final receiving entity can understand the information. Occasionally when data is transported it needs to be secured via data encryption. Physical items also may need to be secured using tamper proof packaging and other techniques such as trusted couriers with tracking receipts. When considering interfaces and interactions there are various mechanisms and approaches.

15.1 Interface Types

There are four broad interface types that should be considered when structuring a system. The interfaces exist within the system linking the various subsystems. These are internal interfaces. There are also interfaces to other systems, these are external interfaces. The interface types are:

- Person to person
- Machine to machine
- Person to machine
- Machine to person

A system may use any combination of these interface types. Their characteristics are driven by the media and protocol used in the interface mechanism. Many will assume that a person to person interface is the slowest, however a simple voice communication from one person to the next person may be the fastest and most effective interface approach. For example there is no faster or more effective interface to provide an alert than a screaming passenger in an automobile with a distracted driver as brake lights are quickly approaching the front on the vehicle.

15.2 Interface Media

Interfaces can be implemented using different media. When most people think of interface media they think in term of computers. However there are many different types of machines each with their own interface media approaches. The media are as varied as the interface types. The same is true of person to person interfaces. The following is a sampling of interface media:

- Co-located conversation, smoke signals, hand carried documents
- Voice and or text communications via electronic media
- Machine hydraulic or physical linkage interface
- Telephone wire, fiber optics
- Terrestrial analog or digital radio
- Satellite analog or digital
- Fasteners attaching similar or different structures, screws, nails, glue, weld

15.3 Interface Mechanisms

An interface mechanism or interface protocol is a set of rules, a mechanism that is used to support an exchange between a sender and a receiver. The protocol description may include message formats, rules for exchange, signaling, synchronization, error detection and correction, authentication, timing, etc. When selecting or developing a protocol there are a number of different interface elements to consider as described below.

15.3.1 Message Transfer

A transfer is one way. Something is just sent assuming that it successfully arrived and was correctly processed.

Example 27 Internet Associate Program

When a user clicks an associate program link[127] on a website, they are taken to another website, the vendor. As part of the associate program, the vendor promises to offer a percentage of the sale to the originating website if a user buys something on the vendor website. Even though there is a specially encoded link that is used to credit an account with the transfer and a possible purchase, there really is no way to verify if the purchase happened. The originating website is able to access their web logs and see how many users clicked a link. They are also able to access the vendors associate program account and see how many visitors were sent to the vendor. However, full correlation is difficult and not part of the interface interaction. Essentially the associate must trust the vendor to credit your account.

This was a problem with the early Internet and many websites removed associate links because they believed they were not being credited for sales. There were some attempts to establish transaction based Internet links but they never passed into the infrastructure[128].

---

[127] Amazon.com associate program patent 6,029,141 Internet-based customer referral system. In 2002 they had 600,000 associates suggesting the associates are one of the primary reasons for Amazons success.

[128] In the late 1990s the World Wide Web Consortium (W3C) worked on HTML micropayments but eventually stopped efforts in this area.

15.3.2 Acknowledgement

An Acknowledgement or ACK is used to signal a sender that the receiver has successfully received something. This can be a computer message or a physical item. It does not mean that the something is intact or that the something has been processed. It is typically associated with inter-machine communications.

15.3.3 Non Acknowledgement

The Non Acknowledgement or NAK is used to signal a sender that the receiver has rejected the sent item. This can be a computer message or a physical item. It can be because of an error in a message or the receiver is busy. It is typically associated with inter-machine communications.

15.3.4 Interrupt

An Interrupt originates from a sender. It is an asynchronous or random signal indicating the need for attention. Once the attention is acknowledged then transfer takes place. It is possible that an interrupt may hog the resources and not allow other elements to interrupt and offer their information. To deal with this issue there can be different types of interrupts, different priority levels of interrupts, and timeouts.

15.3.5 Polling

The receiver drives polling. It periodically queries the sender non-stop. If the sender has something to offer, it is placed in the polling area (buffer) and extracted by the receiver with the next polling sequence. A computer sound card is an example where polling is used. The audio is periodically sampled at some rate and its digital equivalent is placed into memory. Analog sampling is the conceptual equivalent to digital polling. A trackball or mouse buffer is polled at some rate and its position is placed on the display.

15.3.6 Round Robin

Round Robin is a form of scheduling where each element has an equal share of time. During a time slice an interface element assigned to a time slice can initiate a communication transfer. The time slice slot is assigned with placement relative to other devices on the round robin mechanism. It is circular and runs indefinitely. Because each element has an equal slice of time the mechanism is starvation free. It is the simplest form of scheduling.

15.3.7 Queue

A queue is a way of organizing things that need to be done. It is a waiting list or line. It can be a series of transactions in a buffer, a list of tasks in a task list, a line of people waiting to be seated at a restaurant, etc. Things can be removed from the queue using a last in first out (LIFO) approach, occasionally called a stack, or using a first in first out (FIFO) approach. Waiting lines or queues have been extensively studied using mathematics under the broad topic of queuing theory and have been applied in telecom, traffic engineering, computing, factories, shops, offices, hospitals, etc. The system can include high, medium, and low priority queues and other schemes to maximize performance.

15.3.8 Event Driven

Event driven processing is associated with processing that is triggered by some event such as an interrupt. The event can come from any source. The issue is to deal with the event before it goes away without losing current on going activities and their states.

15.3.9 Cyclic

A cyclic process periodically checks something associated with the system. For example the display of information is performed with a cyclic process that updates a screen 30, 60, 120, etc times per second. It also can be associated with receiving information from a buffer such as a client email application that checks for new email every 5, 10, 15, etc minutes.

15.3.10 Buffer

A buffer is a holding area. When data needs to be sent it is placed into a buffer until it can be fetched. When data arrives it may be placed into a buffer until it can be processed. A buffer is an isolation area. It allows two areas to share such that one area does not negatively impact another area. This is a transition zone. Examples include a buffer amplifier in electronics and a buffer zone for separating spaces or groups of people, etc.

15.4 Transaction

A transaction goes beyond the boundary of a communication protocol and includes the successful processing and the generation of a receipt acknowledging the successful processing of the items sent by the sender. A transaction is an end to end transfer that is not complete unless there is a receipt showing that the processing of the received item is complete.

Example 28 Banking Transaction

When a customer goes to a bank to deposit money the customer fills out a form and hands it to the teller. The teller takes the form and cash then proceeds to verify the form cash and ID information is correct. This is message validation. The teller then proceeds to enter the deposit into the account. The system accesses the customer account record and updates the account balance. When this process is complete a receipt is printed. The teller hands the receipt to the customer and the customer verifies the change in the account balance. From the customers point of view the transaction is complete when they get the receipt and they verified it is correct.

A transaction guarantees that the exchange is complete and all parties have been properly addressed. A transaction is not complete if a receipt is not provided showing correct processing and full completion.

15.5 Exercises

1. What are your computer interface types?
2. What are your computer interface media?
3. What are your computer interface mechanisms?
4. Are there any transaction-based interfaces with your computer?
5. Can you list the interfaces between you and your automobile? Can you characterize the interfaces? Did you include your interface with the car radio, horn, steering wheel, etc?

15.6 Additional Reading

1. Specification Practices, Military Standard, MIL-STD-490 30 October 1968, MIL-STD-490A 4 June 1985.
2. System Software Development, Military Standard Defense, DOD-STD-2167 4 June 1985, DOD-STD-2167A 29 February 1988.

16 Reliability Maintainability Availability

Reliability Maintainability and Availability (RMA) merged into a holistic view of the system when systems became mission critical and possible loss of life surfaced. To achieve this level of performance it was realized that preplanned maintenance could prevent critical failures in the field. An example of this is aircraft and the approach of replacing critical parts before the end of their expected lifetime or failure time. Additionally new systems surfaced that must operate 24 hours per day 7 days per week 365 days per year (24/365). To achieve this level of service preplanned maintenance would need to occur while the system is operational. An example of this is a national Air Traffic Control system. The measure of operational time is referred to as availability and the goal is to maximize availability.

There are three ways to achieve high availability. The first is to build something that never fails. This is a case where the reliability is spectacular.

The second approach is to structure the system so that hot preventative maintenance can be performed while the system is operational. However this requires excellent empirical data of the reliability data and expected life of the components needing periodic maintenance. It also assumes that since failures happen randomly after burn in, the system can tolerate a failure. For example the pilot and crew of an airplane have parachutes and will jump if a wing falls off the airplane.

The third approach is to build in redundancy, such that when a failure happens, a redundant path can be used until the primary path is fixed. The reality is that the second approach is not possible in a 24/365 system without redundancy. So redundancy is key to boosting availability. However redundancy adds components, which reduces reliability. So it is possible to have a high availability system that is constantly breaking and being fixed. However, a system that is constantly breaking and being fixed is vulnerable to wear and degradation, which reduces reliability. Constant maintenance also increases the chance of human error, which can bring the whole system down leading to a catastrophic event.

16.1 Reliability Prediction

Reliability is a measure of how quickly a component will fail. It can be expressed as failure rate lambda (λ) or the inverse of failure rate, which is Mean Time Between Failure (MTBF). In a system with millions of components it is a measure of how many failures will occur in the system during a selected time frame.

$$\mathbf{MTBF = 1/\lambda}$$
$$\mathbf{\lambda\ (system) = \lambda 1 + \lambda 2 + \lambda n}$$
$$\mathbf{MTBF\ (system) = 1/\lambda(system)}$$

Some failures are irrelevant and allow the system to continue to operate. For example the loss of a panel back light might make it hard for an operator to read a control label, but it will probably not lead to a catastrophic system failure. Other failures are critical and cannot be tolerated. For example the loss of break fluid in an automobile break system because of a failed hose is critical. These are referred to as Mean Time Between Critical Failure (MTBCF). Measuring the failure rate of each component, adding the failure rates together, and then taking the inverse of the result determines the MTBF.

$$\mathbf{MTBF = 1/\lambda 1 + 1/\lambda 2 + 1/\lambda n}$$

Where: λ = failures / unit time

Unit time:
failures / hour
failures / 1 million hours or 10^6
failures / 1 billion hours or 10^9

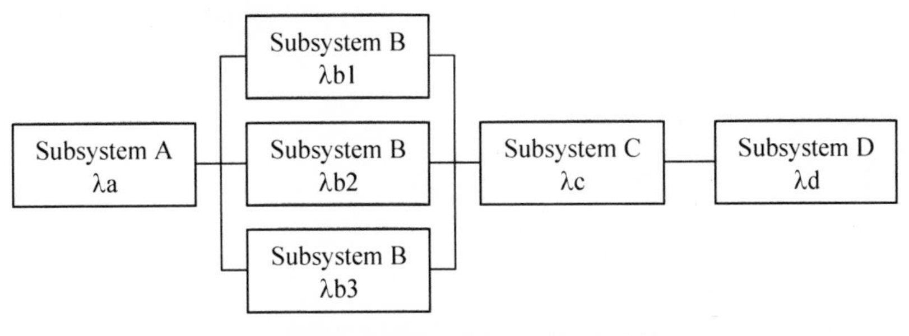

Figure 89 Reliability Block Diagram

Reliability is modeled using a reliability block diagram. The diagram uses series blocks to represent cases where if any one of those blocks fails the system fails. It uses parallel blocks to show redundancy. For example a failure in subsystems A, B, C or D will bring down the entire system. However since subsystem B has three parallel elements, all three must fail for subsystem B to bring down the total system. The sum of failures for the system is $\lambda = \lambda a + \lambda b1 + \lambda b2 + \lambda b3 + \lambda c + \lambda d$. However for Subsystem B to fail in the system all three subsystem elements must fail at the same time. So λ represents the total failures that must be corrected but not all those failures will bring the system down. MTBF for the system is MTBF = $1/\lambda$, where is $\lambda = \lambda a + \lambda b + \lambda c + \lambda d$. An easier way to understand this is from the availability point of view, which is described in the availability section.

16.2 Maintainability Prediction

Maintainability is the maintenance philosophy associated with the system. In the early stages of system development where there are alternative system architectures being considered the maintenance philosophy could be different with each architecture alternative. In a maintainability analysis the primary outputs are the lowest replaceable unit definition and the expected Mean Time to Repair (MTTR). Other outputs include numbers of maintainers, their skill levels, and dispatching approaches. For example travel time and on the spot learning may be part of the MTTR. The following are items to consider in the time line:

- Travel to suspected failure location
- Diagnose cause of failure
- Procure or deliver parts for repair
- Gain access to the failed parts
- Remove and replace failed parts
- Bring system back up to operational status
- Run diagnostics and verify system is working properly
- Closing up the system and returning it to normal status

In the case of software, if there is a condition where the software crashes freezes, or is not operating as expected then the MTTR is potentially the sum of:

- Detect the software problem
- Reboot
- Applications startup
- Database synchronization
- Resume to last known point

This can take from a few seconds to days if database synchronization is extensive. It also does not consider what happens if the software needs to be reinstalled because the failure damaged the software image.

Table 9 Maintenance Philosophy Assumptions

Spares Philosophy	Staffing Philosophy	Estimated MTTR
Onsite	24 hrs/day	30 minutes
Onsite	On call 24 hrs/day	02 hours
Onsite	Regular work hours, week, weekend, and holidays	14 hours
Onsite	Weekday only	24 hours

Table 9 Maintenance Philosophy Assumptions

Spares Philosophy	Staffing Philosophy	Estimated MTTR
Offsite shipped	Called when fault happens	1 week
Offsite warehouse	Remote location air transport needed	2 weeks

16.3 Availability Prediction

Availability considers the MTBF and MTTR to identify the probability that the system is operational in the expected operational time interval. If the system is operational 24/365 then that is 8760 hours or 31,536,000 seconds of operation. The operational availability for an air traffic control center in an emergency mode is .9999995 or 99.99995% or 6-nines five in slang terms. That is an operational availability of 31,535,984 seconds or 16 seconds of unavailability per year. Just as in the case of reliability, availability can be modeled. The availability can be calculated using the following equations:

A = Uptime / (Uptime + Downtime)
A = MTBF / (MTBF+MTTR)

Subsystems are shown in series if failure of any subsystem results in a system failure. The series is calculated as:

A = A1 * A2 * An

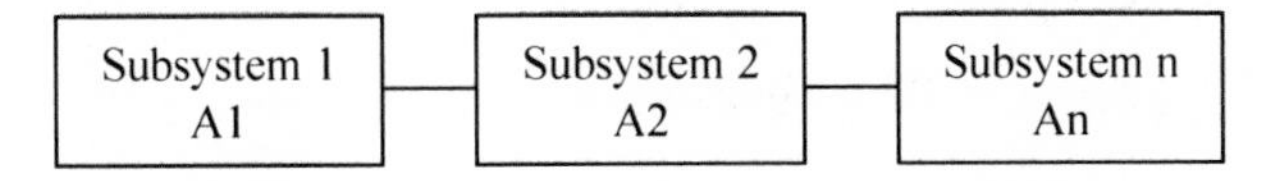

Figure 90 System Availability - Series Subsystems

The combined availability of the elements in series is always lower than the availability of any individual subsystem. The overall availability is pulled down by the lowest availability subsystem. It is the weakest link in the system. This concept of the weakest link is critical when synthesizing high availability architectures.

Table 10 Series Subsystems Availability Impact

Subsystem	Availability		Downtime
1 string (where each string is 99%)	99%	2-nines	3.65 days/year
2 strings (fault tolerant)	99.99%	4-nines	52 minutes/year
2 strings (fault tolerant) + 1 string in series	98.99%	1-nine	3.69 days/year

Subsystems are shown in parallel if the combination is redundant and that is considered failed when all redundant paths fail. The system is available if any string

/ leg / redundant path is available. The availability in this case is 1 - (all paths are unavailable).

So using redundant paths can significantly increase the overall availability. The redundant path can be a duplicate of another subsystem or a different subsystem. If a redundant path is the same as an existing subsystem there is the risk of a common latent defect that can bring both subsystems down simultaneously. This is especially true if the subsystems are software intensive.

A = 1 - [(1-A1) * (1-A2) * (1-An)] where: unavailability = 1-A

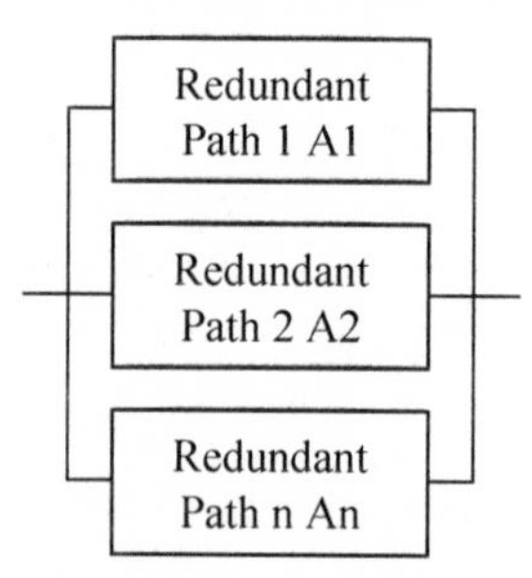

Figure 91 System Availability - Redundant Subsystems

If the redundant paths are different, they are considered orthogonal and will more likely avoid a common latent defect, especially if the subsystems are significantly different. For example a mechanical parking break as a backup to the hydraulic based primary path has no common latent defects that can surface at the same time. However, there are common points that they share such as the tires, wheels wells, break pads, and rotors or drums.

Table 11 Redundant Subsystems Availability Impact

Redundancy Level	Availability		Down Time
1 String	99%	2-nines	3.65 days/year
2 Strings	99.99%	4-nines	52 minutes/year
3 Strings	99.9999%	6-nine	31 seconds/year

Table 12 Availability and Downtime

Availability		Downtime
90%	1-nine	36.5 days/year
99%	2-nines	3.65 days/year
99.9%	3-nines	8.76 hours/year
99.99%	4-nines	52 minutes/year
99.999%	5-nines	5 minutes/year
99.9999%	6-nines	31 seconds/year

16.4 Reliability Growth

Sometimes the only way to achieve high reliability and availability[129] is through a reliability growth program. There are several elements in a reliability growth program:

- Environmental stress testing including automated temperature, humidity, shock, vibration cycling
- Tracking of all problems on all devices through out the lifecycle including prototypes in lab environments
- Introduction of parts screening
- An effective failure reporting and corrective action system (FRACAS)
- New technology injection

The first step is to predict the expected reliability growth for each possible change. The second step is to track the actual reliability results of each introduction. The predicted reliability growth curve for each new higher reliability introduction is always updated based on actual results.

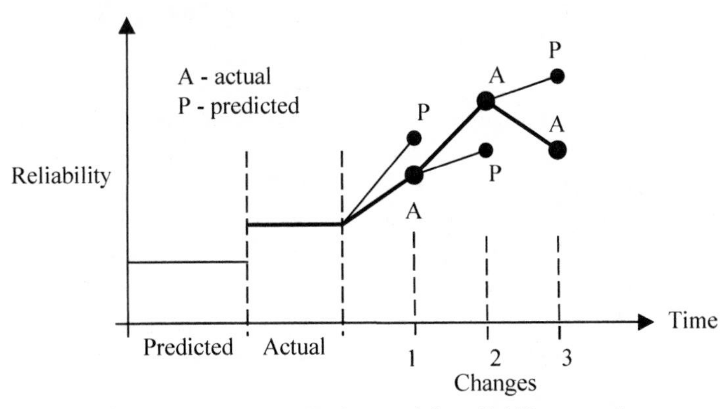

Note: Change 3 actually damaged the reliability growth curve

Figure 92 Reliability Growth Curve

The reliability growth should be tracked for each element change. Ideally elements leading to the greatest reliability growth should be selected first, to accelerate the reliability growth curve. However other factors such as cost, support, training, logistics, etc change priorities and lead to slower reliability growth.

As each change is made a new reliability number is predicted and then tracked.

---

[129] Availability growth is especially true with software since software redundancy is rare in most system. Software redundancy suggests multiple design and implementation teams working in parallel to off the same functionality without sharing their work.

It is possible that changes in field operations may not match the lab and analysis reliability predictions. They can be changed by environmental factors such as heat, vibration, humidity, etc and by operational factors such as user interaction and maintenance activities. A change may actually hurt the reliability of the system once fully implemented in the field.

16.5 FRACAS

A Failure Reporting Analysis and Corrective Action System (FRACAS) is used to increase reliability and safety. The FRACAS documents, categorizes, and analyzes failures or incidents[130] and determines corrective actions and tracks the results of the corrective actions[131]. The FRACAS includes or is closely coupled to the field support system. The FRACAS starts during development and is pre-populated with data gathered during development. So it is also closely coupled to the problem reporting and tracking system[132] used by engineering and manufacturing. The FRACAS has a process, procedures, tools, and staff for:

- Initiating failure and incident reports
- Independently analyzing the failures and incidents
- Providing feed back for consideration to design, manufacturing, and test

The analysis is formal and uses flow diagrams, block diagrams, schematics, drawings, and other information products to show the failure or incident and its impact. The analysis finds the root cause of the failures or incidents and categorizes them so trends can be detected. The FRACAS also has mechanisms to ensure that effective corrective actions are taken and not delayed or deferred.

The FRACAS includes a reporting mechanism so that data can be visualized, understood, and managed. The reporting includes follow-up audits with affected organizations. The information includes all open failure and incidents, analysis, corrective action, suspense dates, and delinquencies. The root failure or incident cause should be clearly stated.

16.6 Exercises

1. If you are planning to ship 10 million units and offer a 1 year warranty what is more important, reliability or availability?
2. If you are designing a command center to achieve 6-nines availability, what is the possible system impact if there are 30 large command centers and 234 smaller command centers?

---

[130] The term incident is used to represent safety, security, returned items, etc. FRACAS is also a mechanism for quality improvement, discussed in the quality section of this text.

[131] Reliability Program For System And Equipment, MIL-STD-785B, 15 September 1980

[132] Discussed in the configuration management section of this text.

3. When identifying architecture alternatives and trying to make a selection should RMA numbers be calculated?
4. When working with new technologies with little history what can you do to calculate RMA numbers?
5. At what point in the project or program should you stop RMA calculations?
6. How can you calculate the RMA numbers for a software intensive system?

16.7 Additional Reading

1. Designing And Developing Maintainable Products And Systems, DOD Handbook, MIL-HDBK-470A August 1997, MIL-HDBK-470 June 1995, MIL-HDBK-471 June 1995.
2. Electronic Reliability Design Handbook, Military Handbook, MIL-HDBK-338B, October 1998.
3. Definitions Of Terms For Reliability And Maintainability, DOD Standard, MIL-STD-721 25 August 1966, MIL-STD-721C 12 June 1981.
4. Guide For Achieving Reliability Availability And Maintainability, DOD, August 3 2005.
5. Reliability Program For System And Equipment, Military Standard, MIL-STD-785B, 15 September 1980
6. Software Safety Standard, NASA-STD-8719.13B, July 2004.
7. Test and Evaluation of System Reliability Availability Maintainability, DOD 3235.1-H, March 1982.

17 Impact of Failure

Some systems cannot fail or make subtle errors. These systems need to be subjected to failure analysis at all stages of development from the early concept stage to the final operational stage. The traditional approach has been to prepare Fault Trees and or perform a Failure Mode and Effects Analysis (FMEA). However these techniques apply once the architecture has been selected and there are some implementation details. A technique called Impact of Failure (IOF) can be applied in the very early stages of architecture exploration[133].

IOF analysis is performed on the early architecture alternatives. The approach is to take an architecture block diagram and sequentially break each element and connection in the architecture and then determine the impact on operations. This break or failure is a result of some fault that can cause loss of service or make subtle errors not detectable by operators. It can be a hard fault that is persistent or a transient fault that comes and goes at random. These faults should point to previous experience of similar systems and architectures.

The goal is to find the worst case IOF for each architecture alternative. An architecture (Arch A) may have a huge IOF but the failing item is known to rarely fail either because of simplicity or extreme maturity. On the other hand another architecture (Arch C) may have a similar huge IOF and is known to fail regularly either because of complexity or extreme technological or operational immaturity. Obviously Arch A has a lower exposure to IOF than Arch C. Additionally Arch A may have a medium IOF for an element that is known to fail often and so this medium rating is the final IOF assessment for Arch A.

In all cases an architecture has a High, Medium, or Low IOF rating. The goal is to modify the architectures that have high IOF ratings while preserving the spirit of the original architectures. The quest to lower the IOF in all the architectures will eventually lead to a final differentiation of the IOF between all the architectures.

Table 13 Impact of Failure

Alternatives	Impact	Worst Case Failures
Arch A	Low	One user display freezes
Arch B	Medium	Some user displays freeze
Arch C	High	All users displays go black

---

[133] Impact of Failure was developed at Hughes Aircraft by Walter Sobkiw during the FAA Advanced Automation System pre-proposal to help characterize and select an architecture.

The analysis can include a count of the number of different failure types in an architecture alternative. The number of High, Medium, and Low impacts of failure (IOF) can be tabulated. The goal is to search for a pattern that might lead to further insights and improvements to each architecture alternative.

Table 14 Impact of Failure Counts

Alternatives	Failure Types	High IOF	Med IOF	Low IOF
Arch A	5	0	0	**5**
Arch B	3	0	**2**	1
Arch C	2	**1**	1	0

Unlike other techniques, the IOF is a high level qualitative assessment of an architecture. There is less data gathered in this approach than other approaches because the goal is to quickly improve the architecture or discard the architecture because the IOF is unacceptable.

Example 29 Impact of Failure Concept Application

Impact of failure is an interesting concept and can be applied to any system using any concept of failure. For example the loss of a critical resource for a system is a failure and it has an impact. Some examples of impact of failure concept being applied in other areas are loss of oil on food production, transportation, and electrical systems; loss of clean water on food production, sanitation and health care systems.

17.1 Exercises

1. Draw an architecture diagram of a Thanksgiving Dinner and perform an IOF analysis on the architecture. Modify your Thanksgiving Dinner architecture to improve the IOF rating and update the architecture diagram.
2. Draw an architecture diagram of your computer and perform an IOF analysis on the architecture. Modify your computer architecture to improve the IOF rating and update the architecture diagram.
3. What is the impact of failure in a car if the engine, a tire, the steering, headlight, or suspension fails at high and low speed? Use an IOF table.

17.2 Additional Reading

1. Flight Assurance Procedure Performing, A Failure Mode and Effects Analysis, NASA, GSFC-431-REF-000370, Number: P-302-720.
2. Procedures For Performing A Failure Mode, Effects And Criticality Analysis, Military Standard, MIL-STD-1629a, 24 November 1980.

18 Failure Mode and Effects Analysis

Failure Mode and Effects Analysis (FMEA)[134] is a qualitative analysis that systematically identifies failure modes and assigns a severity level to each failure mode. The goal is to extensively study the failure modes with the highest severity levels. The results are used to improve the system by identifying and initiating corrective actions.

FMEA is a bottom up methodology. An element at the lowest level of interest is selected. The possible failure modes for each item of the selected level are tabulated. The failure effect at the next higher level is then interpreted. This effect on each higher level is interpreted until the highest level of the system is reached[135].

The FMEA starts as soon as preliminary design information is available. It continues throughout the design process. Although FMEA in the past was typically applied to electrical circuit hardware elements and their aggregations, it can be applied in other physical and non-physical areas (e.g. software). It also can be applied before a design is available such as during the conceptual phase of a system development. In the non-physical areas the functions can be used as the analysis elements. Failures are postulated for the minor lowest level functions, which are aggregated to major functions, then allocated to assemblies, subsystems, and systems. So the bottom up approach is still maintained but the starting point is a non-physical element.

18.1 Failure Modes

Systems can fail in various ways. FMEA identifies all the ways a failure may happen at a designated level. This includes historical failure mechanisms and failures based on sound engineering.

The failure modes are based on the knowledge of the physical component, functional specification, interface requirements, schematics, and other relevant information products. Failure can be attributed to the element or its interface. Failure modes that happen are possible under the following conditions:

1. Premature operation

[134] FMEA was specifically developed in the early days of electronic hardware, which was prone to frequent failure, circa 1940 to 1970. The technique is however very applicable and useful outside the hardware domain. Today, hardware is very reliable because it has moved to high-density integrated circuits eliminating many sources of previous failures.

[135] This is called inductive synthesis.

2. Failure to operate at prescribed time
3. Failure to stop operation when required
4. Failure during operation
5. Loss of output
6. Intermittent operation
7. Degraded output or operational capability
8. Other unique failure conditions

Example 30 Electronic Hardware Interface Failure Modes

Hardware failure modes at the interfaces involve connectors and cables. Failures within the unit appear as short to ground, short to voltage, open, or short to signal for both signal and power lines.

Hardware and software interfaces can be examined from two perspectives. The first is hardware effects on the software where failures of the hardware result in improper or lack of response to the software. The second is software effects on the hardware where the software causes improper operation of the hardware. Examples of software failures that affect the hardware are:

1. Commands that come to early
2. Commands that come to late
3. Failure to command
4. Erroneous commands

18.2 Failure Effect Severity

Each failure mode is assigned a severity category. Safety and impact to other systems are reflected in the severity category. The failure effect is first assigned at the lowest level being analyzed. This is rolled up to the next higher level, the subsystem level, and so on to the system or mission level. In selecting the severity category, the worst case consequence at any level is used for the failure mode. The severity categories[136] are:

Table 15 Failure Effect Severity Levels

Level	Failure Effect	Description
1	Catastrophic	A failure that could result in serious injury or death or system loss.
2	Critical	A failure that could result in severe injury, major property damage, or major system damage which will result in mission loss.
3	Marginal	A failure that could result in minor injury, minor property

[136] MIL-STD-1629a, 24 November 1980, Military Standard Procedures For Performing A Failure Mode, Effects And Criticality Analysis.

Table 15 Failure Effect Severity Levels

Level	Failure Effect	Description
		damage, or minor system damage which will result in delay or loss of availability or mission degradation.
4	Minor	A failure not serious enough to cause injury, property damages, or system damage, but which will result in unscheduled maintenance or repair.

18.3 FMEA Process

The FMEA process can be broken down into a series of steps. The goal is to populate a worksheet or database that contains several fields. The primary fields are Failure Mode Number, Function or Item, Failure Mode and Cause, Failure Effect, Severity Level, and Comments. The worksheets form the primary body of a report, which includes the following:

1. **Identify the ground rules** assumptions and mission phases. Identify the subdivision level for the analysis.
2. **Obtain block diagrams** or construct block diagrams. Functional and reliability block diagrams illustrate the operation, interrelationships, and interdependencies of functional entities for each item involved in the system's use. All system interfaces should be identified.
3. **Describe the system** and requirements to be analyzed. This includes identification of internal functions, internal interface, and expected performance at all indenture levels, system restraints, and failure definitions. Describe each mission in terms of functions, which identify tasks to be performed for each mission, mission phase, and operational mode. Describe the environmental profiles, expected mission times and equipment utilization, the functions and outputs.
4. **Identify problem areas**. Identify all potential failure modes (including interface) and define their effect on the immediate function or item, on the system, and on the mission. Identify failure detection methods and compensating provisions for each failure mode.
5. **Assign severity categories**. Evaluate each failure mode in terms of the worst potential consequences at any level.
6. **Recommend corrective actions**. Identify corrective design or other actions needed to eliminate the failure or minimize its impact. Identify effects of corrective actions or other system attributes, such as requirements for logistics support.
7. **Complete work sheets**. Show the failure modes, failure effects, severity classification, and corrective measures.
8. **Document the analysis**. Summarize the problems which could not be

corrected by design and identify the special controls which are necessary to reduce failure risk.

18.4 FMEA Worksheet

The FMEA worksheet can be built on paper or a computer using a database. The elements or attributes of the worksheet are:

1. **Identification Number**: This can be a serial number or other unique reference number. It is used for traceability and provides a reference to each failure mode.
2. **Item / Function Identification**: This is the name of the physical item or function being analyzed for failure modes and effects. Information products such as official schematics, block diagrams, drawings, etc can be used to identify the item or function. Names are typically nouns.
3. **Function**: A succinct statement of the function performed by the item or function. Functions are typically noun verb phrases.
4. **Failure Modes and Causes**: Failure modes are postulated by examining the outputs shown in block diagrams and schematics. Failure modes are also postulated based on the requirements and the failure definitions included in the ground rules. The most probable causes associated with the postulated failure mode are identified and described. Since a failure mode may have more than one cause, all causes for each failure mode should be identified and described. The failure causes within the adjacent indenture levels are considered. For example, failure causes at the third indenture level is considered when conducting a second indenture level analysis. Examine each failure mode in relation to:

    - Premature operation
    - Failure to operate at a prescribed time
    - Intermittent operation
    - Failure to cease operation at a prescribed time
    - Loss of output or failure during operation
    - Degraded output or operational capability
    - Other unique failure conditions

5. **Mission Phase / Operational Mode**. A succinct statement of the mission phase and operational mode in which the failure happens. The most definitive timing information should be entered for the assumed time of failure occurrence.
6. **Failure effect**. The consequences of each failure mode on operation, function, or status is identified, evaluated, and recorded. Failure effects

focus on the specific block diagram element. The failure may impact several indenture levels; the local, next higher level, and end effects is evaluated. Failure effects should consider the mission objectives, maintenance requirements and personnel, and system safety.

- **Local effects**. Local effects concentrate on the impact an assumed failure mode has on the operation and function in the indenture level under consideration. The consequences of each postulated failure affecting the item is described along with any second order effects. The purpose of defining local effects is to provide a basis for evaluating compensating provisions and for recommending corrective actions. It is possible for the local effect to be the failure mode itself.
- **Next higher level**. Next higher level effects concentrate on the impact an assumed failure mode has on the operation and function in the next higher indenture level above the indenture level under consideration. The results of each postulated failure affecting the next higher indenture level is described.
- **End effects**. End effects evaluate and describe the total effect a failure has on the operation, function, or status of the uppermost system. The end effect may be the result of a double failure. For example, failure of a safety device may result in a catastrophic end effect only in the event that both the prime element and the safety element fail. Note end effects resulting from a double failure.

7. **Failure Detection Method**. A description of the failure mode detection method. The failure is ultimately detected by the operator using visual or audible warning devices, automatic sensing devices, sensing instrumentation, etc. Alternatively there may be no indications provided to the operator.

    - **Other indications**. Descriptions of operator indicators showing that the system has malfunctioned or failed. Proper detection of a system malfunction or failure may need identification of normal indicators as well as abnormal indicators. If no indicators exist, identify if the undetected failure will jeopardize the mission objectives or personnel safety. If the undetected failure allows the system to remain in a safe state, a second failure situation should be explored to determine if that indicator is evident to an operator.

        a) **Normal**. An indicator that is evident to an operator when the system or equipment is operating normally.
        b) **Abnormal**. An indicator that is evident to an operator when

the system has malfunctioned or failed.

c) **Incorrect**. An erroneous indicator to an operator due to the malfunction or failure of an indicator.

- **Isolation**. Describe the most direct procedure that allows an operator to isolate the malfunction or failure, such as running built-in-test (BIT). The failure may be of lesser importance or likelihood than another failure that could produce the same symptoms. Fault isolation procedures should describe the actions by an operator, followed by a check or-cross reference either to instruments, control devices, circuit breakers, or combinations.

8. **Compensating provisions**. Design provisions or operator actions, which circumvent or mitigate the effect of the failure are identified and evaluated. This step records the true behavior of the malfunctioning or failed item.

- **Design provisions**. Describe the compensating elements of the design at any indenture level that will nullify the effects of a malfunction or failure, control, or deactivate system items to halt generation or propagation of failure effects, or activate backup or standby items or systems. This includes:

  a) Redundant elements that allow continued and safe operation.
  b) Safety or relief devices such as monitoring or alarms that permit effective operation or limits damage.
  c) Alternative modes of operation such as backup or standby items or systems.

- **Operator actions**. Describe the compensating actions of the operator to circumvent or mitigate the effect of the failure. Determine the compensating action that best satisfies the observed indicators when the failure happens. If needed investigate the interface to determine the most correct operator actions. The consequences of incorrect operator actions in response to an abnormal indication should be considered and the effects recorded.

9. **Severity classification**. Assign a severity classification category to the failure mode. Identify the effect on the item under analysis caused by the loss or degradation of output so the failure mode effect is properly categorized.
10. **Remarks**. Clarify any other fields in the worksheet. Note recommendations for design improvements. Identify unusual conditions, failure effects of

redundant items, recognition of particularly critical design features or any other remarks. Category 1 and Category 2 failure modes that cannot be designed out, provide a mitigation strategy or reasonable actions and considerations to reduce the occurrence of the failure mode and provide a rationale for accepting the design. The rationale for accepting of Category 1 and Category 2 failure modes should address:

- **Design**. Features of the design that minimize the occurrence of the failure mode; i.e., safety factors, parts de-rating criteria, etc.
- **Test**. Tests that verify the design features at acceptance or during maintenance that would detect the failure mode.
- **Inspection**. Inspections to ensure that the item is being built to the design requirements and inspections during maintenance that would detect the failure mode or evidence of conditions that could cause the failure mode.
- **History**. History related to this particular design or a similar design.

Table 16 FMEA Worksheet or Database Record

Field	Content
Identification Number	25
Nomenclature	Wireless Router
Function	Access to Internet
Failure Mode	Unable to access Internet
Causes	Internet attack
Mission Phase	Daily web updates
Failure Effect - Local	Unable to access Internet
Failure Effect - Next Higher	Unable to FTP to web server
Failure Effect - End	Unable to update website
Failure Detection Method	Broken URLs, failed email access, and network indicator shows red X
Failure Isolation Method	Bypass the router and connect directly to the DSL modem. Note that this is dangerous and may infect the connecting computer. Alternatively cycle power on router and see if connectivity is re-established.
Compensation	Use dialup modem
Severity	4
Remarks	Cycle power on router to reset from Internet attack. If unable to restart router use dial up modem. Consider other router products that may be immune to Internet attacks.
System	Website: www.cassbeth.com
Indenture level	2

Table 16 FMEA Worksheet or Database Record

Field	Content
Reference Drawing	1
Mission	Education
Date	10/11/2010
Sheet	25
Compiled By	Walt
Approved By	Claudia

18.5 FMEA Worksheet Summary

Obviously the FMEA database can become very large. The findings need to be summarized in an information product (document). It should list the Category 1 and 2 failure modes and their mitigation strategies.

18.6 Exercises

1. Prepare a FMEA worksheet for a category 1 failure mode.
2. Prepare a FMEA worksheet for a category 2 failure mode.
3. Prepare a FMEA worksheet for a category 3 failure mode.
4. Prepare a FMEA worksheet for a category 4 failure mode.
5. How would you sort the fields in the FMEA worksheet?

18.7 Additional Reading

1. Flight Assurance Procedure Performing, A Failure Mode and Effects Analysis, NASA, GSFC-431-REF-000370, Number: P-302-720.
2. Procedures For Performing A Failure Mode Effects And Criticality Analysis, Military Standard, MIL-STD-1629A, 24 November 1980.

19 Fault Tree Analysis

Fault Tree Analysis (FTA) is used to identify how an undesired event could happen in the system. It is a risk analysis technique used in safety and security analysis or when there is a need to understand how critical failures might surface in a system. FTA is a top down deductive reasoning technique where the analyst moves from general to specific. It is qualitative, but probabilities can be assigned to different elements allowing comparisons between different areas in the fault tree. This quantitative comparison between the fault tree areas allows the analyst to work on an area of the represented system and strengthen or fix the system area contributing to the undesired event.

Since FTA is a top down approach and FMEA is a bottom up inductive synthesis[137] approach, they are both typically performed on a system. So the two analysis efforts are complementary and together provide a more comprehensive and accurate result as one-analysis impacts the other analysis.

19.1 Identification of Undesired Events

FTA begins with the identification of a list of undesired events. A fault tree is then developed for each undesired event. This is not a model of all possible system failures or all possible causes for loss of safety or security. So it is critical to identify all the undesired events and state them at an appropriate level. The events should be a completely catastrophic, safety, security event or massive failure rather than a drift type failure.

If the top-level event is too broad the analysis becomes unmanageable. If it is too specific, the analysis does not provide a sufficiently broad view of the system and interactions between events may be lost. It is best to error on the side of being too broad and use a top level OR gate to decompose the event and then start the detailed fault tree for each input of the OR gate.

19.2 Fault Trees

FTA should be started in the early architecture phase and refined and updated as the design evolves. The simple act of identifying unauthorized events feeds the entire system engineering effort as the team considers the key requirements,

[137] Inductive reasoning is used to detect similar and dissimilar patterns while synthesis is putting things together to form a whole.

architecture alternatives, design and implementation. Fault trees[138] use the following symbols:

- **AND gate:** "A" exists if B1, B2 Bn simultaneously exist.
- **OR gate:** "A" exists if any B1, B2, Bn exist.
- **Undesired Event:** input or output of an AND or an OR gate.
- **Basic Event:** A failure at the lowest level that cannot be decomposed further.
- **Undeveloped Event:** An event not developed because of lack of information or because of insignificant consequence.
- **Connecting Symbol:** Points to another part of the fault tree on a different page or view.
- **Inhibit Gate:** Used to describe the relationship between one fault and another. The input fault directly produces the output fault if the indicated conditions are satisfied.

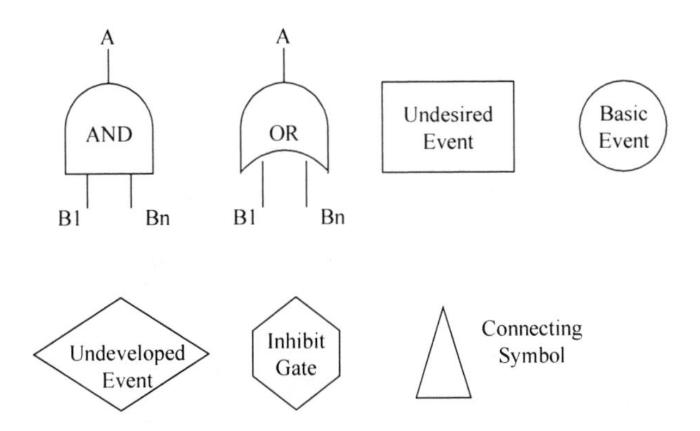

Figure 93 Fault Tree Symbols

Typically FTA is performed on mission critical systems. These systems use redundancy and other techniques to prevent undesired events. A blanket statement like: it requires three failures before an undesired event can be permitted to surface, provides a specification the team can use for the system. In this way reliability numbers are removed from the analysis. Instead the analysis lists the undesired events and the number of failures needed for the undesired event. This is referred to as a cut set. The system meets the three-failure requirement if the cut set shows that the unauthorized event needs three or more failures in all cases before the

[138] Although fault trees can be developed with drawing tools, it is strongly recommended that a fault tree tool be used. The amount of information and editing that is needed is significant and hand drawing the trees will lose all the details and rigor needed for an effective analysis. This is the opposite of many analysis approaches I recommend in this text.

unauthorized event surfaces.

19.3 Cut Sets

It is interesting to see the final FTA output. There are pages of fault tree diagrams and then a table listing the cut sets for all the undesired events. The table may have multiple rows, cut sets, where each row identifies the failures that lead to an unauthorized event. In this case if there are less than three failures shown in any row then there is a problem with the system. Some cut set rows may list 3, 6, 10, 20, etc failures needed before the unauthorized event surfaces.

Minimal cut sets are all the unique combination of basic events that can lead to the top level undesired event. So using the fault tree in Figure 94 as an example, if basic events 1,2,3,4 or 1,2,3 or 1,2,4 happen, then the undesired event will happen. If the criteria for the system is to have 4 failures before allowing the event then we see we need to work on the leg triggered by event 1,2,3 or 1,2,4.

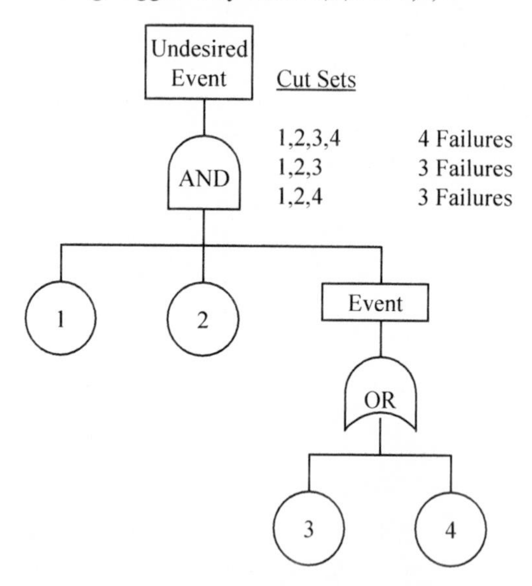

Figure 94 Fault Tree and Cut Set - 3 Failures

An undesired event is caused by one or more faults. An undesired event can be a major system failure. The analyst starts with the top-level event (e.g. technician exposed to RADAR microwaves). The analyst then postulates how that event might occur at the highest abstraction level possible. If it is based on more than one lower level event the analyst must decide if an AND gate or an OR gate should be used to capture the event. The next step is to postulate how the lower level events might occur, again picking the highest abstraction level possible. This continues until the analyst reaches a basic event. A basic event cannot be decomposed[139]. The

[139] This is just like DFD decomposition where at some point the DFD reaches a primitive

development of the tree is a decomposition exercise where lower level events trigger higher level events until the unauthorized event is reached.

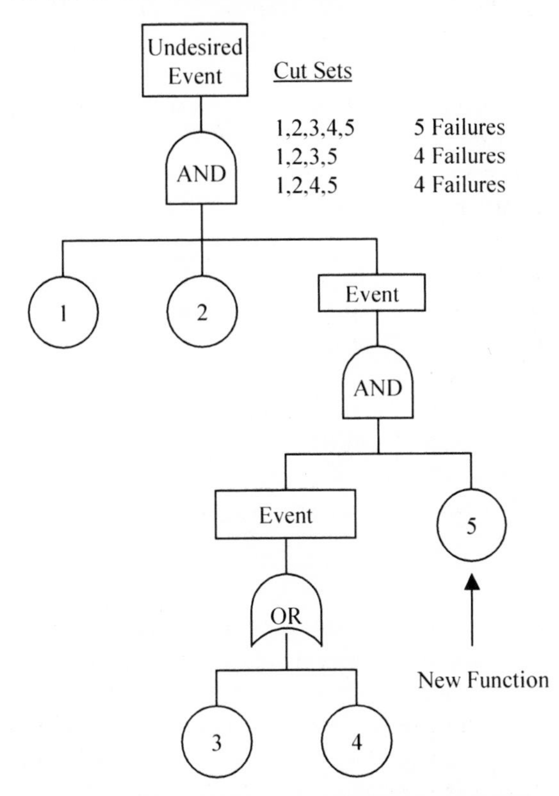

Figure 95 Fault Tree and Cut Set - 4 Failures

Early in a project the fault tree is function based. As the system unfolds the fault tree is subsystem based. In later stages of the system development, the fault tree is a physical representation of the system components. Some might suggest that as the fault tree moves closer to an actual physical abstraction of the system, there is greater confidence in the analysis and thus the system. However it all starts with the unauthorized events. For example, no one thought of a cabin fire on the ground for Apollo 1[140]. Later at congressional hearings the most powerful statement was given by astronaut Frank Borman who when asked, "What caused the fire?" responded with:

---

and cannot be decomposed.

[140] Apollo 1 was to be the first manned mission to the moon. On January 27, 1967 a fatal fire killed all three crewmembers (Command Pilot Virgil "Gus" Grissom, Senior Pilot Edward H. White, and Pilot Roger B. Chaffee). This happened during a pre-launch test of the spacecraft on Launch Pad 34 at Cape Canaveral.

"**A failure of imagination.** We've always known there was the possibility of fire in a spacecraft. But the fear was that it would happen in space, when you're 180 miles from terra firma and the nearest fire station. That was the worry. No one ever imagined it could happen on the ground. If anyone had thought of it, the test would've been classified as hazardous. But it wasn't. **We just didn't think of it.** Now who's fault is that? Well, it's North American's fault. It's NASA's fault. It's the fault of every person who ever worked on Apollo. It's my fault. I didn't think the test was hazardous. No one did. I wish to God we had."

Lack of imagination is one of the fundamental flaws in any systems effort and it is especially true in FTA. All the information and evidence can be right there staring everyone in the face and everyone will miss it. That is why systems are analyzed from multiple dimensions or points of view using formal, semi formal, and ad hoc methods. FMEA and FTA are semi formal methods. FMEA comes from the bottom and FTA comes from the top. Other analysis is performed and captured in a myriad of information products. All for the sole purpose of understanding the system, trying to surface the emergent behavior, and not being blindsided by the unexpected.

19.4 Exercises

1. Identify unsafe events associated with your automobile. State how many failures you will accept before any unsafe event surfaces. Create a fault tree and identify the cut sets.
2. Identify un-secure events associated with your computer. State how many failures you will accept before any un-secure event surfaces. Create a fault tree and identify the cut sets.

19.5 Additional Reading

1. Electronic Reliability Design Handbook, Military Handbook, MIL-HDBK-338B, October 1998.
2. Reliability Analyses Handbook, Prepared By Project Reliability Group, JET Propulsion Laboratory, JPL-D-5703_JUL1990, July 1990.

20 Fault Tolerance

Fault tolerance is a broad subject that branches into all aspects of faults, errors, and failures. The broad approaches to dealing with system failures are fault avoidance, fault removal, and finally fault tolerance. When it is realized that a system needs fault tolerance then all the other approaches to mitigate faults should have been used in the system. It is similar to a layering approach where the final layer is fault tolerance. So before we can begin discussing fault tolerance we need to appreciate all aspects of faults, errors, failures and the approaches to dealing with them in a system.

20.1 Faults Errors and Failures

Faults lead to failures. A fault is the result of either age or stressing something beyond its tolerable limits such that it breaks and the break results in a failure. A failure is the inability to support a function, performance, or characteristic requirement. From a user perspective it is the inability to provide a service or a service with its expected level of performance.

Faults can be either because of a part that goes bad or it can be the result of a bad design. Design faults have the additional characteristic of usually being hidden or latent. Obviously if the fault were known, then the design would change.

A latent fault is a fault that was always present but never detected and removed via test. Latent design faults are a big issue in complex software systems where test may not exercise all the functional sequences and data. These are typically called data dependent latent faults, latent design faults, or design faults.

Faults can be hard or intermittent. An intermittent fault surfaces at random. Typically as mechanical parts fail they have intermittent faults. Eventually the faults surface more frequently and then become a permanent fault. Strategies can be developed to detect intermittent faults and replace a failing part before a hard system failure occurs. The implication is that a hard failure is more serious than an intermittent failure.

Some faults may not lead to failure but can be used as indicators of a final fault that will soon surface. These are symptoms prior to the final fault. For example an automobile water pump makes noise and leaks long before a fault leads to a failure.

Errors are typically associated with software but they can apply to anything that offers a signal or interfaces to another element in a system. An error is the wrong signal. The wrong signal results in a failure. For example an oxygen sensor in an automobile is sending the wrong signal and the result is that the engine exhibits very poor performance at idle or high speed. Just like with faults the errors can be latent, intermittent, hard, or symptoms suggesting a more catastrophic error in the future.

20.2 Well Check or Certification Diagnostics

Most people equate diagnostics with maintenance and trying to isolate a fault so that it can be repaired. However diagnostics are also used to certify that a system is ready to be used. Well check or certification diagnostics are used to surface faults in an operational system. The idea is that the diagnostics stimulate all the critical points in the system that normal operations may not immediately stimulate.

The diagnostics can be manual or automated. Most people think in terms of automated diagnostics especially in computer based systems. However manual diagnostics are just as important. A manual diagnostic might be a visual inspection using an approved or certified checklist. For example those who fly small personal airplanes check the airplane using an approved checklist prior to starting the engine and proceeding to taxi. The checklist includes visual inspection of the tires, landing gear, rudder, tail, ailerons, wings, flaps, propeller, fuselage, etc. When the automobile was new it was not uncommon for people to check the belts, tires, fluids, brakes, lights, and filters prior to a long trip[141].

20.3 Fault Avoidance

The goal of fault avoidance is to reduce the likelihood of failure. This is usually accomplished by over designing and using high reliability components. Over designing is taking a part and de-rating it by significant levels or by taking the performance numbers and adding significant margins.

Author Comment: With the introduction of computers and extremely accurate measurement equipment it is now possible to design systems very precisely. The issue is that this precision can lead to disaster as all the requirements are met including the expected lifetime. As the expected end of life approaches, all the components fail at the same time as predicted. However stakeholders rarely acknowledge the end of life prediction form the original design criteria and they are rarely ready to let go of a system even as it is collapsing because there is no replacement. This is a big issue in this new century.

Fault avoidance is also accomplished by structuring the system so that there are a minimum number of parts and interconnections. As the number of parts increase the chance of failure increases as predicted by the reliability models. The same is true for the number of connections. Connections also have the attribute of being impacted by the surrounding environment, as they become loose, dirty, and broken from the effects of dust, dirt, vibration, shock, temperature expansion contractions, maintenance actions, etc.

---

[141] I still tell my kids to check their automobile prior to a long trip.

20.4 Fault Removal

Fault removal acknowledges that there are design faults and that they can be surfaced via testing. The testing program is usually very large and intensive in this setting. It begins with unit level tests that stimulate components that otherwise may be untested when in a full system. It continues with integration and system tests. The tests are as varied as the systems. Testing is described elsewhere in this text.

20.5 Failure Propagation

Failure propagation is the result of a failure, which causes other system elements to fail. For example loss of a single support beam in a physical structure should not cause other beams to fail, leading to a collapse of a structure. It also should not impact other subsystems. For example an automobile tire failure should not damage the wheel well, cause the fender to fly off, continue propagation into the engine compartment, causing the engine to fail, causing engine pieces to be thrown into passenger compartment, causing physical harm.

A system needs to be structured so that failure propagation is minimized or eliminated. The failure propagation is stopped or limited based on the system structure and its strength in various areas. Failure propagation characteristics tend to be associated with different architecture approaches and once the architecture is selected it may be difficult to minimize some failure propagation scenarios.

20.6 Broad Fault Tolerant Concepts

Fault tolerance acknowledges that even with the best components significant over engineering (margins) and massive test programs there is still the possibility of a known or unknown fault that must be mitigated and prevented from causing a failure. Fault tolerance is added to a system so that it can continue to operate in the presence of one or more failures.

Fault tolerance is accomplished through replication. The replication can be accomplished by using additional physical elements or by duplicating a process in time using the same physical elements. In some cases both strategies may be used.

There are many different aspects of fault tolerance that need to be understood from the perspective of the system, and then selectively applied. These different aspects of fault tolerance can be grouped into three different approaches to fault tolerance: Inherent Fault Tolerance, Passive Fault Tolerance, and Reconfiguration Fault Tolerance.

20.6.1 Inherent Fault Tolerance

Inherent Fault Tolerance is accomplished with an architecture that has no special fault tolerant mechanisms. It is inherently fault tolerant and nothing special is needed beyond the main system functionality. This is important because special fault tolerant mechanisms are needed to support a crisis situation. However special fault tolerant mechanisms may fail just when they are needed in a critical time. In fact they may be broken and the breakage my not be detected because they are normally not used.

An example of an inherently fault tolerant architecture is a beehive. Physical damage to a part of the beehive does not stop the hive from providing its fundamental service of storing honey. The honeycombs offer structural integrity yet minimize propagation of failure to other compartments.

20.6.2 Passive Fault Tolerance

Passive fault tolerance offers fault tolerance that is part of the normal function of the architecture. No special fault detection or reconfiguration mechanisms are needed to accomplish the fault tolerance. Usually redundant elements are added that work in parallel with the primary elements to provide services. When an element fails the service is still provided but at reduced performance levels that may or may not be detected by the user.

An example of this is the brake system in an automobile. There are front and rear brakes that work off a primary and secondary hydraulic system. The primary hydraulic system feeds both the front and rear brake. If there should be a failure the secondary hydraulic system still engages the rear brakes. If both hydraulic systems fail, the brakes can be engaged via the parking brake, which is typically a mechanical linkage[142] and like the regular breaks is used consistently.

20.6.3 Reconfiguration Fault Tolerance

This fault tolerant approach relies on fault detection and then a reconfiguration mechanism. If the fault is not detected then there is no reconfiguration and the system fails. Since the process begins with fault detection there is usually several fault detection mechanisms in place. The final detection mechanism in many systems is a human in the loop. However, by the time the human detects the fault it may be too late and there may be massive damage.

Once a fault is detected, then action is taken to reconfigure the system to remove the faulty element. The reconfiguration may mean switching to a complete duplicate string, a partial duplicate string, or a shut down of the offending subsystem.

Fault detection and reconfiguration take time. They also have a probability of success. For example mechanisms may be put in place to detect 95% of the possible system faults. However 5% will be undetected. Although fault detection may be relatively fast, reconfiguration may be long and result in a system outage. Also

---

[142] Hydraulic control was invented at Hughes Aircraft to control aircraft surfaces. They act as amplifiers reducing the manual stiffness of the controls. The mechanical wires and links were previously used and it is easy to see why they were maintained for safety just in case the hydraulic hoses burst or are physically damaged.

since reconfiguration is rarely performed the reconfiguration mechanisms may fail.

20.7 Fault Detection Approaches

There are two broad approaches to fault detection: (1) manual and (2) automated. A system might use one or both approaches to fault detection. One of the issues associated with fault detection mechanisms is what detects faults in the fault detection mechanism. So there may be a chain of fault detection mechanisms which include fault detectors of the fault detection functions. Once a fault is detected an action needs to be taken. More advanced fault detection mechanisms include an inherent approach to dealing with the fault. Examples include cold standby, hot standby, dual string, and voting architectures. These are typically associated with computing but can apply to any processing mechanism.

20.7.1 Check Lists and Inspections

Manual fault detection is based on using people to detect faults. People are used to periodically inspect parts of the system. The inspection can be methodical and rigorous or ad hoc. A methodical rigorous inspection uses a checklist. The checklist can form a legal record of the process and can be signed at the conclusion of the inspection. An ad hoc inspection is a physical walk through a system looking for unusual items. In many cases a checklist-based inspection also includes an ad hoc walkthrough the system. The observations can be fed into a computer database for post analysis and statistical predictions.

For example counting the number of rusty bolts on a physical structure may be used to determine when maintenance should be performed. Inspection of a post operation checklist list after major surgery may result in a different surgical process.

20.7.2 Sensors and Instrumentation

Automated fault detection approaches use instrumentation to detect faults. The instrumentation can include physical, visual, sound, temperature, gas, smoke, humidity, liquid and other detectors and sensors. The sensor information can be fed to visual and audible alarms so that people in the loop can detect the faults. The sensor data can be fed to machines, which can process the information and decide if there is a fault. Once a fault is detected the machine can notify people in the loop and take action to deal with the fault.

20.7.3 Inverse or Reverse Processing

Inverse or reverse processing is a fault detection mechanism used in a processing function. For example an accountant may check the addition of numbers in a column by subtracting the numbers from the total. When complete the result should equal zero or the value of the last number to be subtracted. Changing the order of the subtraction sequence from the original addition sequence adds further strength to the fault detection results. For example if the numbers are added in sequence, subtract them in a random order. Conversely add the numbers at random but subtract them using a sequence such as top to bottom or bottom to top. The same can be done with multiplication and division. Complex processes based on logic and sequence also can be performed in some inverse or reverse order with some randomization.

This approach is interesting when applied to large logic processing mechanisms. For example early mainframes used error detection logic rather than full-scale duplication of processing blocks to detect errors. This allowed more resources to be applied to the processing and yet provide very high levels of error detection. More logic circuits can be applied in a computer central processing unit to speed up the math functions rather than be wasted as a parallel string used only in comparison for error detection.

20.7.4 Parity

Parity is a mathematical technique applied to binary data in computing to detect errors. It is used primarily during data transport especially in high noise environments where the signal can be compromised and bit errors can be generated.

There are two approaches to parity processing: even parity bit and odd parity bit. In even parity, if the number of ones in a given set of bits is odd, the parity bit is set to 1 making the entire set of bits even. In odd parity, if the number of ones in a given set of bits is even, the parity bit is set to 1 keeping the entire set of bits odd.

Table 17 Parity Error Detection

Data (7 bits)	number of 1s	8 bits with even parity	8 bits with odd parity
0000000	0	0000000 **0**	0000000 **1**
1010001	3	1010001 **1**	1010001 **0**
1101001	4	1101001 **0**	1101001 **1**
1111111	7	1111111 **1**	1111111 **0**

The parity is calculated and appended as part of a process. Once the data is transferred to another process the parity is recalculated and compared with the parity bit setting. If the sent parity does not match the received parity an error is declared or detected.

20.7.5 Cyclic Redundancy Check

Cyclic Redundancy Check (CRC) is a mathematical technique applied to binary data in computing to detect and correct errors. It is used primarily during data transport especially in high noise environments where the signal can be compromised and bit errors can be generated. The CRC is an outgrowth of the parity concept and takes error detection to the next level where errors can be corrected. Most CRC mechanisms can detect multiple bit errors and correct single bit errors. Some CRC mechanisms can correct two bit errors. There are different CRC calculations that have been developed with different characteristics. The CRC-32-IEEE is an example of attempting to use a standard CRC.

CRC is considered to be a stronger fault detection approach than parity. For example multiples of two bit errors in the parity approach will not be detected while CRC will be able to detect many of these types of errors. So CRC tends to be used to check the integrity of large blocks of data, such as a program or a database. A CRC can be calculated and then checked prior to executing a program or accessing data. If the CRC is incorrect the program or data is considered corrupted.

20.7.6 Sequence Counter

A sequence counter is a monotonically increasing counter (a counter that increases by 1 with each new element) that is associated with something of interest in the system. If the counter jumps, then an error is suggested. The jumping counter value also represents loss of data (the item being monitored) in the system. For example an audit log may use a monotonically increasing counter. If the counter jumps in value, an audit entry is missing. If the audit entry is missing other items in the system also may have been lost. This suggests a system overload.

The sequence counter can be monitored online during system operations using a separate mechanism. When a jump is detected an alarm can be generated alerting an operator. Also system action can be taken such as controlled shutdown. The sequence counter also can be analyzed offline as part of periodic diagnostics, maintenance, or daily certification.

Table 18 Sequence Counter

No	Time Stamp	Events	Status
4	02142011 14:01:01	Event 8	
7	02142011 14:02:05	Event 5	**Warning:** Count Jumped by 2
8	02142011 14:03:00	Event 6	
20	02142011 14:04:45	Event 5	**Error:** Jumped by 12 system shutting down

The action can be a function of the level of count loss and the actual event. Some events are not as serious as other events. Also the level of lost count may impact the decision for a particular event. For example a small count loss with a particular event may lead to a warning while a large count loss for the same event may translate to a system shutdown.

20.7.7 Progress Monitor

A progress monitor is similar to the sequence counter concept. There is a known good view of the processing sequence. A number represents the processing sequence. The number can be calculated on the fly or looked up by the process and used to represent the sequence. If the sequence deviates, an error is detected.

Table 19 Progress Monitor

From	To	Expect From	Expect To	Status
7462	9845	7462	9845	
9845	**6235**	9845	**9951**	**Error**: Wrong destination restarting
9951	6236	9951	6236	
6236	7567	**7236**	7567	**Error**: Wrong source shutting down

Depending on the detected error different system actions can be taken The error can be noted, the system restarted, system shutdown, etc. If the progress monitor has a different starting value than the expected value from sequence, that may suggest a problem in the sequence monitor rather than the system being monitored.

20.7.8 Keep Alive Signals

Keep Alive signals or heart beat signals are exchanged between system elements. There is an element sending a signal and an element expecting a signal. They are sent periodically such as once per second. If some predefined number of signals is missed then a fault is assumed. The number of missed signals is usually set at three. If one or two signals are lost, the information can be logged, but a failure is not declared. This data can be for a transient fault detection mechanism.

20.7.9 Diagnostics

Diagnostics are used to detect and isolate faults. There are different diagnostics for different system needs. Maintenance diagnostics are used to fix broken equipment and other diagnostics are used by the system to increase levels of fault tolerance. There are also diagnostics used to certify that a system is ready for operational use. In some cases the diagnostics between each of these areas may overlap and even may be the same design and implementation.

- **Maintenance Diagnostics**: Are used by maintainers to detect and isolate faults. They are part of a troubleshooting tool set used by maintainers to fix the system and ensure it is performing as expected.
- **Certification Diagnostics**: Are used by licensed maintainers to certify that a system is ready for operational use. The diagnostics tend to be executed on a daily basis or when there is a shift change. The licensed maintainers sign off that the system can be used after the certification diagnostics successfully complete.
- **Power Up Diagnostics**. Are forms of certification diagnostics but are not associated with any certification. However they perform the same function of making sure the system is ready for operation.

- **Online Diagnostics**: Execute periodically while the system is not busy with operational functions. They are used to boost the level of error detection in a system so that failure propagation or operation in a failed condition is minimized. A failed condition can result in erroneous outputs not detected by humans in the loop. The online diagnostics also stimulate parts of the system that are rarely used so that when they are accessed and used there is some confidence the system will behave as expected. As system utilization increases online diagnostic activity decreases. There eventually may be a point where the online diagnostics may not execute, thus reducing the fault detection coverage level. Various schemes have been developed to run online diagnostics without perturbing the main operational processing. They include the addition of specialized hardware and even stand alone maintenance processors that do not in any way use any resources associated with the operational processes.
- **Periodic Diagnostics**: Are online diagnostics. They run periodically no matter what the operational load may be in the system. Examples include computer instruction, register, and memory tests.
- **Continuous Diagnostics**: Are the special case of online diagnostics that run continuously. This requires special hardware so that the main operational processing is not impacted by the continuous diagnostics. Examples include computer interface loop back tests, sensor tests, continuity tests, power monitoring, and temperature checks.

20.7.10 Data Checking

Data checking is a very easy fault detection mechanism to implement. It can be applied to all inputs, all outputs, or all inputs and outputs. If applied to inputs and outputs the assumption is the transport mechanism may introduce a fault into the data. This may be probable for external interfaces but it may be highly unlikely for internal interfaces such as between software functions running on the same computer. However if the computer is in a hostile environment such as outer space it even may make sense to check data being passed between functions on the same computer. Data can be checked from different perspectives.

- **Type**: Type checking determines if the data is a integer number, text string, binary number, floating point number, etc. The best way to view this is from an operator point of view. When entering a phone number, if the entry includes a non-numeric other than dash or parenthesis it should be rejected.
- **Range**: Many entries have a reasonable range of values. For example entering a negative value for altitude in a flight plan update of an air traffic control system should be rejected because of a range error. The same is true when an operator enters a stock trade request. Selling one billion shares is

not reasonable and should be rejected based on a range check.

- **Validity**: Validity checking involves accessing internal data and comparing it against incoming data. For example a user may want to sell 1000 shares of a stock, but they only own 100 shares. The request should be subjected to a validity check and rejected based on different rules.

Passing internal data in a computer system is a significant source of latent design errors especially with poorly typed languages like C. Some software languages are inherently immune to data size and type errors, such as PERL.

In the 1980's there was a strong movement to create a computer language that would offer extreme flexibility in defining data to minimize memory size (like C offered) and yet be immune to data design errors. The language that surfaced was Ada and is used in mission critical systems. Ada forces data type and range checking as part of the coding process. On the other hand C allows the data to be incompatible and may only generate a warning during compile time, which the coders usually ignore, especially in immature organizations.

An example of a data error is passing a floating-point value into an integer. This is called type recasting and may result in the wrong value as the data is truncated. Type recasting is a source of many latent software design errors in systems today as the software revolution has led to people in positions of authority not understanding the issues. The results can be catastrophic.

20.7.11 Statistical Transient Fault Tracking

Many faults manifest themselves as transient faults prior to becoming hard faults. These transient faults may not even impact the system. For example a daily diagnostic may detect a fault, rerun, and the fault may not be present with multiple reruns. A message might be sent across an interface, a fault is detected, and the message is either corrected via CRC or a retransmit. These occurrences can be counted and maintained in a database. Statistical analysis and historical information can be used to remove a component before it becomes a problem in the system. So the system can include a Statistical Transient Fault Tracking function.

20.8 Redundancy and Replication

The approach to fault tolerance is to add redundancy to a system through replication. If the main part should break the redundant element can support the function. The redundancy can be inherent in the structure of the system or a special consideration of the system. For example a wood frame house uses walls made of 2x4 wood studs placed some distance apart from each other. A typical wall will have several 2x4 studs and if one or two studs should break the wall will remain intact. It is inherently fault tolerant and redundant.

In the case of physical structures, replication can only be accomplished with the addition of physical items. The replicated physical items can be of the same type or a different type. The redundant structure can duplicate the main structure or be a completely different structure. For example a bridge can be replicated by using two side-by-side structures. Alternatively a bridge can be backed up by another bridge a few miles away using the same materials and structure or different materials and structure. One or more ferries also can back up a bridge.

If the same materials and structure are used for redundancy, then the backup is a replication. If different materials and or structures are used then the back up is a redundant element but not a replication. Fault tolerance tends to be stronger if replication is not used for redundancy because replication may have an unknown single point of failure. For example termites or fire can compromise wood studs.

In the case of a process the replication or redundancy can be physical and or time displaced. A physical structure implements the process so the same physical redundancy approaches can be used. However processes rarely use the physical structure 100% of the time. So the process can be duplicated in different time slots and compared against previous process runs. The duplicate process can follow the same design and algorithms or use a different design or algorithms. It can be designed and implemented by the same team or a different team.

20.9 Single Points of Failure

A single point of failure is a place or condition in the system, which if failed will cause the loss of function, performance, or the complete system. If the whole system is lost it is a catastrophic failure or catastrophic single point of failure. The goal is to always identify the single points of failure and remove them from the system. This can be accomplished via redundancy by structuring the system such that the single point of failure is inherently not present in (removed from) the system.

As different architecture approaches are considered different single points of failure are analyzed. Some architecture approaches have fewer catastrophic single points of failure. Some architecture single points of failure cannot be effectively or reasonably mitigated with replication. Some may be mitigated with replication while other architectures inherently just do not have catastrophic single points of failure. Yet others may have large numbers of single points of failure and others may have few single points of failure. These analysis findings all need to be tracked for the different architectures and considered in the tradeoff selection process.

20.10 Fault Tolerant Structures

There are different structures that can be used to provide fault tolerance. They vary in complexity, cost, and degree of fault tolerance. In many ways they evolved through the years. The simplest exhibit deficiencies that the next steps address. The Architecture section offers additional discussion of fault tolerant structures.

20.10.1 Cold Standby

In a cold standby architecture the error detection is internal to each string. The error detection can be automated at any level but in the end the human in the loop is the final error detection mechanism. When an error is detected another string not in use is bought into being so that it can replace the failed string. The mechanism to bring the cold string can be established using either automated, manual, or a combination of mechanisms.

For example an automobile can have a tire blowout or go flat while the vehicle is in motion. Alternatively the tire can have very low tire pressure detected by a sensor or by a human while the automobile is parked. In all cases once the failed tire is detected the user can remove the bad tire and replace it with a spare tire, the cold standby. The problem is the spare tire also may be flat.

The same cold standby approach can be used when a computer fails or a cell phone fails. The user just moves to another computer or cell phone. Unless the other computer or cell phone is in use, they also can be in a failed state, such as a dead battery condition. Processing devices also have the issue of reestablishing the data and configuration from the failed device to the backup device. The full recovery of data and configuration my not be possible and will take time.

20.10.2 Hot Standby

The hot standby addresses the primary deficiency of the cold stand by approach by quickly and fully reestablishing operations. The error detection is still in the operating string. However the back up string is running in parallel and is able to immediately take on the operations performed by the primary string.

For example a truck trailer uses two side by side tires on each end of an axle. If one tire fails the other tire will allow the truck to safely continue driving. Once the truck stops the failed tire is detected by inspection and the operator can have it replaced before continuing. The decision is based on current load and distance to the final destination.

The same hot standby approach can be applied to processing and communications. The system can be structured so that there is a primary and backup string. The backup string shadow processes the primary string so that the data and configuration are current. Once a failure is detected, the backup string is switched online and continues to offer services. There is delay associated with the failure detection and the reconfiguration to the secondary string. This delay may lead to lost data and configuration differences. For example 10 seconds can lead to lost video, sound, RADAR data etc.

20.10.3 Multiprocessing

Multiprocessing is similar to a hot standby except the backup string is adding to the main system function and performance. For example additional tires on truck axles increase the weight safety margin. In a processing or communications system the additional resources reduce processing time or provide less important functions that can be shed in the presence of a failure. Since the multiprocessing nodes are always in operation contributing to the system there is no reconfiguration time, it is instantaneous. Also there is no risk of failing to properly reconfigure and start adding services. The backup nodes are always online.

20.10.4 Dual String

A dual string addresses the deficiency of fault detection in cold standby, hot standby, and multiprocessing fault tolerant structures. Although a great deal of automated mechanisms can be added to increase levels of fault detection, the reality is many unanticipated faults surface during years of operation and the human becomes the final fault detection mechanism. In these cases many times the fault is detected too late and damage happens to the system, its users, or the surrounding community. Statistical mechanisms of measuring transient faults are the last stages of fault detect, but they are also insufficient.

In a dual string configuration the single primary and single secondary string work in parallel and compare outputs. They together are now called a dual string because they compare their outputs[143]. If a difference is detected, the process can be rerun a number of times (3 times) before a failure is declared and processing is transferred from the failed dual string to a working dual string.

The replication level is four in a dual string configuration. Two elements are used in the primary dual string and two elements are used in the secondary dual string. The level of fault detection can be 100% if the comparison is at the output interface level. As the comparison moves further back into the system the level of fault detection drops significantly. So there is some art that needs to be applied so that the comparison can happen at the last possible location and time in the configuration.

Dual processors can run in lock step mode or asynchronously. The advantage of running lock step is that the fault can be detected very early and thus minimize fault propagation. It is excellent for detecting hardware failures. The problem is that a data dependent transient failure will bring down both processors in a lock step configuration. An asynchronously running mode may offer enough change between the two processors so that the data dependent transient failure only impacts one of the processors.

[143] Tandem computers studied by the FAA for Flight Service Stations offered comparison at the transaction level. It was used by a telephone text system a precursor to the Internet in France in the 1980's. Stratus Computer developed a four-processor system that offered comparison at the interface level and challenged Tandem.

20.10.5 Voting

Voting reduces the number of total elements from four to three. Rather than have two elements compare outputs and then switch to another set of two elements that are not being used, the voting approach uses three elements that vote all the time. If all three elements have matching outputs the system is fully operational. If two elements match then that is declared as the correct answer and the system continues to operate. If no elements match then the system is declared failed.

The availability calculations will show that two element failures are extremely unlikely. However, if needed the number of processors can be increased (e.g. 3 out of 4, 3 out of 5, etc). Also the processors can be orthogonal and use different designs and algorithms to arrive at an answer that is subjected to voting.

A major advantage of a voting system is that there is no separate reconfiguration mechanism. So there is zero fault detection time and zero reconfiguration time. The term Triple Modular Redundancy (TMR)[144] has been used in the past but a voting system can use more than three elements[145].

20.10.6 Path Isolation

Just because a system may have backup mechanisms it does not mean the system does not have single points of failure. One of the more critical sources of a single point of failure when redundant elements are used is the path that delivers power, signals, structural support, hydraulic pressure, etc.

So when structuring a system, paths need to be examined to make sure there are no sources of a single point of failure. For example hydraulic fluid pipes and hoses should be routed at opposite ends of an aircraft to prevent loss of hydraulic control in the event of structural failure of part of the airplane. The same concepts apply when developing message formats. The message formats should be structured so that significantly different behavior is significantly differentiated in the format. For example using a digital bit to represent one state and another digital bit right next to it that represents an opposite state opens the system to a false indication if the two bits are internally shorted[146]. This is not unlike the loss of hydraulic fluid in tubes that are side by side in a redundant system. Similar situations can arise in a form such as a medical record. There should be some degree of separation between items

---

[144] August Systems was a vendor that offered a TMR system for the process control industry in the early 1980's. It was also considered for the Hughes Aircraft Advanced Automation System (AAS) to support AERA processing.

[145] The Space Shuttle uses four identical computers running in lock step mode the same software, and a fifth computer running different software developed by a second contractor. The fifth computer is only invoked if the voting four-computer set fails to offer a consistent answer during the voting process, attributed to a common software latent defect.

[146] Hamming distance is used to determine how different two messages are in a system. Hamming distances of two or more are used in mission critical messages and flags.

that if confused can lead to serious repercussions.

Example 31 Single Point of Failure

On November 24, 1961 all communication links went down between Strategic Air Command (SAC) Headquarters (HQ) and North American Air Defense Command (NORAD). Communications was lost from three Ballistic Missile Early Warning Sites (BMEWS) at Thule Greenland, Clear Alaska, and Fillingdales England. The communication systems used redundant and independent routes. However the redundant communications routes all ran through one relay station in Colorado. A motor overheated and caused interruption of all the lines.

All SAC bases in the United States were alerted, and B-52 bomber crews started their engines, with instructions not to take off without further orders. An on the fly ad hoc approach was used to determine system status. A B-52 near Thule was contacted by radio and it was able to communicate with the BMEWS stations by radio and report that no attack had taken place. An orthogonal backup mechanism was used in this case (radio versus redundant ground communications).

20.11 Graceful Degradation

When a fault happens in the system, the system behaves in some way. Ideally its behavior minimizes the impact of the failure. Also the fault may actually cascade and become a series of faults. Graceful degradation is a term used to convey the concept that the system should slowly shed function and performance as failures start to surface in the system. A minor fault or series of minor faults should not lead to a catastrophic system failure.

20.12 Fault Tolerance Measurement

There are different approaches that can be used to compare the degree of fault tolerance between different architectures. The first is to develop an availability number for each architecture approach. The second is to determine the reconfiguration risk. The third is to identify the number of single points of failure and estimate their impacts on the system for each architecture approach.

Availability calculation is described in the reliability section of this text. Reconfiguration risk is easily described in terms of whether the system uses or does not use reconfiguration. In both cases as the architectures become mature the availability and reconfiguration risks start to become similar between the architecture approaches. However they should be determined and captured in a tradeoff table.

What will separate the architecture alternatives is the third technique of identifying the number of single points of failure and their impacts. This statement seems to be in conflict with the fundamental concept of fault tolerance, which suggests that there are no single points of failure. The reality is that fault tolerance is always approached in a system and rarely achieved. There are usually one or more single points of failure in most systems. The issue is how many single points of failure remain, and what is the impact if they fail. Failure severity levels are used to assess each single point of failure.

Table 20 Single Point of Failure Severity Levels[147]

SPFL	Impact	Description
5	Catastrophic	Complete system outage all users loose all services.
4	Devastating	Significant system outage most users loose all services.
3	Serious	System operational but many services are unavailable to all users.
2	Marginal	System operational but some services are unavailable to some users.
1	Insignificant	System operational performance degraded but still acceptable all services present.

The single point of failure level (SPFL) for each single point of failure (SPF) is used to determine a total SPFL, the lower the SPFL the better the approach. For example assume a system has the following single points of failure and SPFL assessments: SPFL = SPF(5) x SPFL(5) + SPF(4) x SPFL(4) ... + SPF(1) x SPFL(1). The following is a list of failures and the impact rating for each failure:

Local Area Network Cable Plant 5
Power 5
Data Processor 3
External Communications Processor 2
Recording System 1
The resulting SPFL is: 2 x 5 + 3 + 2 + 1 = 16

Table 21 Fault Tolerance Measurement

Criteria	Arch 1	Arch 2	Arch n
Availability	.999 999	.999 999	.999 999
Reconfiguration	Yes	None	Yes
SPFL	16	21	26

Architecture 1 has the lowest SPFL so it is considered to be the most fault tolerant. However it does use an active reconfiguration mechanism. Architecture 2 has the next best SPFL but it does not use any active reconfiguration mechanisms. That may actually significantly outweigh the low the SPFL in architecture 1 thus architecture 2 may actually be the best fault tolerant architecture approach.

---

[147] The practice of assigning levels and defining them is a recurring practice that surfaces in safety analysis, security analysis, defect tracking, and general risk assessment.

20.13 Exercises

1. What are the fault tolerant elements in your health care system? What are the single points of failure in your health care system?
2. What are the fault tolerant elements in your educational system? What are the single points of failure in your educational system?
3. What are the fault tolerant elements in your city? What are the single points of failure in your city?
4. What are the fault tolerant elements in your automobile? What are the single points of failure in your automobile?
5. What are the fault tolerant elements in your house? What are the single points of failure in your house?
6. What are the fault tolerant elements in your body? What are the single points of failure in your body?
7. Identify systems where checklists can increase the level of fault tolerance.

20.14 Additional Reading

1. Electronic Reliability Design Handbook, Military Handbook, MIL-HDBK-338B, October 1998.
2. Definitions Of Terms For Reliability And Maintainability, DOD Standard, MIL-STD-721 25 August 1966, MIL-STD-721C 12 June 1981.
3. Guide For Achieving Reliability Availability And Maintainability, DOD, August 3 2005.
4. Practice For System Safety, DOD Standard, MIL-STD-882D, 10 February 2000.
5. Reliability Program For System and Equipment, Military Standard, MIL-STD-785B, 15 September 1980.
6. System Engineering Glossary - INCOSE SE Terms Glossary Document; October 1998; File: Glossary Definitions of Terms 1998-10 TWG INCOSE.doc; Prepared by: INCOSE Concepts and Terms WG, International Council on Systems Engineering (INCOSE).

21 Safety

System safety is the hunt for unexpected hazards. If the unexpected can be discovered then a system mechanism can be provided. The problem is one of finding the unexpected. Fault tree analysis and Failure Mode Effects Analysis (FMEA) are formal techniques for finding the unexpected. However safety is generally addressed through past experience. These are hard painful lessons learned from history. Each domain has long lists of safety hazards that have resulted in unsafe events. These lists need to be accessed and reviewed so that the team can focus on finding and addressing new safety issues related to their system.

21.1 Safety Hazards

Hazards are conditions that can lead to an unsafe situation that can result in or contribute to a mishap or accident. Are there general hazards that can apply to all systems? Can these general hazards be decomposed in the same way functions are decomposed and mitigated in the system? The short answer is yes. We are human biological systems with various subsystems that we can easily relate to and find hazards. The following is a list of our bodies subsystems and possible hazards:

- **Light Hazards (Eyes)**: Mixing blue with other colors causes the eyes to constantly refocus thus potentially leading to myopia. Poor contrast ratio such as using gray letters on black buttons leads to eyestrain. Poor ambient light for documents or machine markings leads to eyestrain. Loss of night vision due to on coming bright lights or non-use of red lights in cabin instrumentation. Small letters forcing eye strain. Small objects with close placement forcing excessive crossing of eyes.
- **Noise Hazards (Ears)**: Excessive noise for short periods of time or low level noise for long periods of time leading to hearing loss. Quick air pressure changes that may burst eardrums. Talking over loud ambient noise leading to strained vocal cords.
- **Physical Hazards (Body)**: Sharp edges that can cut. Poor placement causing frequent bumps and bruising. Excessive physical weight, shock (traumatic movement), or vibration that can pull muscles, damage tendons, or break bones. Repeated motion causing damaged tendons. Loud vocalizations for long periods of time.
- **Environmental Hazards (Body)**: Excessive heat or cold. Exposure to high voltage or radiation. Poor, polluted, or poisoned air. Dangerous substances

or poisons ingested or absorbed by the skin. All of these can seriously damage the body.

- **Cognitive Processing Hazards (Brain or Mind)**: Shift changes that do not follow natural circadian cycles. Frequent shift changes with no time to transition internal body clock. Long periods of work without rest. Long periods before sleep cycles. Visual, sound, and or tactile sensory overload. Light and noise hazards, which result in unnecessary cognitive processing to detect images. Trying to detect garbled speech in high noise static settings. Trying to detect objects in high noise low contrast light settings. All of these lead to impair thinking, judgement, and cognitive abilities.
- **Physical and Emotional Stress Hazards**: Can lead to temporary or permanent physical or cognitive impairment. Sources can be poor machine, social, and economic systems. For example constant reorganization and workforce churn in a company or industry is a significant stress source with future indirect health care costs not paid by the offending system.

21.2 Formal Safety Analysis

A formal safety analysis follows the same principles as FMEA and fault tree analysis. It is just adapted to match the safety domain. Fault trees are developed for different safety hazards. The fault tree logic represents the safety architecture elements in place to stop the hazards from surfacing. The FMEA is a formal decomposition and assessment of the safety hazards. The following paragraphs will focus on a FMEA like method for safety analysis.

21.2.1 Hazard Identification

Hazards are listed and reviewed by the stakeholders. The more people and perspectives involved in identifying the hazards list the greater the probability of surfacing hazards. The starting points for identifying the hazards are other FMEA studies such as for the system hardware and history of previous system safety failures and their hazards.

21.2.2 Hazard Assessment

Some hazards are more serious than others are and so they need to be understood and assessed in some consistent manner. When a hazard results in a mishap there are different impacts or levels of severity.

Just like in FMEA and problem tracking the concept of severity categories is used and different categories or severity levels are named and described that are used in the analysis.

Table 22 Hazard Severity Levels

LVL	Severity	Description 1[148]	Description 2[149]
1	Catastrophic	Death or system loss.	Could result in death, permanent total disability, loss exceeding $1M, or irreversible severe environmental damage that violates law or regulation.
2	Critical	Severe injury, occupational illness, major system or environmental damage.	Could result in permanent partial disability, injuries or occupational illness that may result in hospitalization of at least three personnel, loss exceeding $200K but less than $1M, or reversible environmental damage causing a violation of law or regulation.
3	Marginal	Minor injury, occupational illness, minor system or environmental damage.	Could result in injury or occupational illness resulting in one or more lost work days, loss exceeding $10K but less than $200K, or can mitigate environmental damage without violation of law or regulation where restoration activities can be accomplished.
4	Negligible	Less than minor injury, illness, system damage or environmental damage.	Could result in injury or illness not resulting in a lost workday, loss exceeding $2K but less than $10K, or minimal environmental damage not violating law or regulation.

Once the hazards are identified they are assigned a severity level. Some hazards may have multiple mishaps with different severity levels. The hazards and mishap ratings are maintained in a hazard assessment table.

Table 23 Hazard Assessment Table

Hazard	Mishap	Severity
Hazards 1	Mishap 1.1	1 Catastrophic
Hazards 1	Mishap 1.2	3 Marginal
Hazards 2	Mishap 2	2 Critical

21.2.3 Hazard Assessment Probability

Analysis can be performed in the hazard or mishap domain. For this discussion the hazard domain is elected. Different hazards and mishaps have different

[148] Software System Safety Handbook, Joint Services Computer Resources Management Group, US Navy, US Army, And US Air Force, December 1999.

[149] Practice For System Safety, Department of Defense Standard, MIL-STD-882D, 10 February 2000.

probabilities of occurring. These probabilities are described in terms of potential occurrences per unit of time, events, population, items, or activities. The ideal approach is to assign a quantitative hazard probability but that may be impossible, especially if all hazards are consistently treated. A qualitative approach based on history of similar systems is offered. There are suggested hazard mishap probability levels that can be used for the analysis.

Table 24 Hazard Probability Levels

LVL	Description	Prob.	Comments
A	Frequent	1 in 100	Likely to occur often in the life of an item. Continuously experienced. Will constantly occur.
B	Probable	1 in 1000	Will occur several times in the life of an item. Will occur frequently. Will occur.
C	Occasional	1 in 10,000	Likely to occur some time in the life of an item. Will occur several times. Likely to occur.
D	Remote	1 in 100,000	Unlikely but can reasonably be expected to occur. Unlikely to occur.
E	Improbable	1 in 1,000,000	Very unlikely to occur, but possible. Assume it will not occur.

21.2.4 Hazard Risk Index

Risk assessment is performed using the hazard severity and its probability. A risk assessment matrix is used to assign a Hazard Risk Index (HRI). The values in the risk assessment matrix are arbitrary but follow a logical pattern.

Table 25 Risk Assignment Matrix

Probability	Catastrophic		Critical		Marginal		Negligible	
Frequent	**1**	**very high**	**3**	**very high**	7	high	13	medium
Probable	**2**	**very high**	**5**	**very high**	9	high	16	medium
Occasional	**4**	**very high**	6	high	11	medium	18	low
Remote	8	high	10	medium	14	medium	19	low
Improbable	12	medium	15	medium	17	medium	20	low

Finding the intersection of the severity and probability yields an HRI number. The lower the HRI the greater the hazard risks. The HRI numbers are assigned to levels, which require different action.

Table 26 HRI Hazard Risk and Acceptance Levels

HRI	Hazard Risk	Hazard Risk Acceptance
1 to 5	Very high	Unacceptable risk. If not reduced, needs acquisition executive approval.
6 to 9	High	Unacceptable risk. If not reduced, needs program executive.
10 to 17	Medium	If not reduced, needs program manager approval.

Table 26 HRI Hazard Risk and Acceptance Levels

HRI	Hazard Risk	Hazard Risk Acceptance
18 to 20	Low	As directed by program policy.

So each hazard is listed in a table. The hazard has its associated mishaps. In many cases there is only one mishap per hazard. Severity and probability are assigned. This then results in an HRI and the HRI yields an overall risk and actions to be taken.

The safety analysis is an iterative process where hazard risk is identified then reduced for all hazards until there are no hazards or hazard risks. However at the end of the system design there will be residual risk. These are hazard risks not lowered or eliminated. These remaining hazards need to be clearly re-stated to all impacted stakeholders and incorporated into documents, training, refresher courses, and warning labels.

21.3 Safety Mitigation Approaches

There are various safety mitigation approaches. They each have their levels of effectiveness. The approaches for developing the safest system starting with the most effective and ending with the least effective are:

- **Inherently Safe**: Use design selection to make the architecture, design, and implementation inherently safe by eliminating the hazard. If the hazard cannot be removed try to make the mishap of minimal consequence via an inherently safe approach exclusive of safety mechanisms.
- **Use Safety Devices**: If hazards cannot be eliminated or their risk not reduced through design selection use fixed, automatic, or other protective design features or devices. Consider fault tolerances and make provisions for periodic safety checks of the safety devices.
- **Warning Devices**: When Inherent Safety or Safety Devices are not possible, use devices to detect an unsafe condition and produce an adequate alarm signal to alert personnel of the hazard. The signals are designed to minimize incorrect personnel reaction. Standards when present are followed. Use a warning device for each Safety Device just in case it fails.
- **Procedures and Training**: When Inherent Safety, Safety Devices, or Warning Devices are not possible, use procedures and training to mitigate the risk. Procedures and training can include personal protective equipment, checklists, and two or more people checking each other's actions in a fault tolerant process. Certification or personnel proficiency training is used for safety-critical tasks and activities. For Level 1 and 2 hazards using just warning, caution, or other forms of write advisories is unacceptable. They can be used but the hazard remains and appropriate approvals are needed as

part of the hazard risk acceptance rules. Use procedures and training as a backup mechanism for other mitigation approaches just in case they fail.

21.4 Safety Architecture

So the safety architecture is a combination of safety mitigation approaches. The safest architecture uses inherent safety as the strongest mitigation approach for all hazards. Strong architectures use the next highest-level mitigation approach followed by the lower levels as backups. The weakest use a mitigation approach with no lower level backup. For example no procedures or training is offered.

Table 27 Safety Mitigation Architectures

Mitigation	**1**	**2**	**3**	**4**	**5**	**6**	**7**	**8**
Inherently Safe	X							
Safety Device		X	X	X				
Warning Device		X	X		X	X		
Procedures and Training		X			X		X	

Arch 1: Most Safe and Arch 8: Least Safe

21.5 Failsafe

Failsafe is acknowledging that if there is any unexpected failure, the system fails in a safe manner. Many times this means a graceful degradation or loss of functions so that the system can continue to operate but at a significantly reduced performance and or functional level otherwise the loss of the system might lead to an unsafe condition.

21.6 Safe Practices

The following is a list of practices that attempt to mitigate safety hazards. They are safe practices:

- Minimize potential of human error
- Unobstructed clear safety labels such as for high voltage, heat, pressure, radiation exposure
- Rounded corners of shelves, tabletops, chairs, controls
- Proper use of ambient light
- Proper use of color in displays
- Sufficient size for display text and symbols
- Sufficient size for physical label lettering and symbols
- Sufficient display size to allow for proper distance to prevent crossed eyes
- Normal reach and field of view not exceeded
- Weight of objects less than 10 pounds
- Size of objects well within typical arm length

- Proper work shifts respecting normal biological circadian cycles, sleep, and breaks

21.7 Exercises

1. Provide several examples of a failsafe condition.
2. Identify hazards associated with using an elevator.
3. Identify a failsafe mechanism for an elevator.
4. Using a fault tree show how a car can become unsafe. How does this compare to a fault tree used for failure analysis? Are unsafe conditions system failures?
5. How does Formal Safety Analysis compare to Impact of Failure?
6. How does Formal Safety Analysis compare to FMEA?

21.8 Additional Reading

1. Human Engineering Design Criteria, DOD, MIL-STD-1472C May 1981, MIL-STD-1472D March 1989, MIL-STD-1472E October 1996, MIL-STD-1472F August 1999.
2. Practice For System Safety, DOD Standard, MIL-STD-882D, 10 February 2000.
3. Software Safety Standard, NASA-STD-8719.13B, July 2004.
4. Software System Safety Handbook, Joint Services Computer Resources Management Group, US Navy, US Army, And US Air Force, December 1999.

22 Security

Security is a systems practice that everyone can relate to because it is fundamental like eating, breathing, drinking, being safe, etc. There are instinctive responses that guide us in the area of security. However there are basic elements of security that have been established and are used to help us develop secure systems.

Security is about protecting something of value from a threat. All things have value and all things have threats. The issues are do the things have sufficient value and are the threats likely to happen such that protection is needed. For example your life has great value and if you live in a high crime rate area or war zone you have real threats that you acknowledge and try to avoid.

Protecting life and property are physical things that physical security addresses. However security also applies to non-physical things. There are virtual things, which may manifest in a physical form such as a document, storage media, Internet cloud, etc. Examples include your identity, bank account, personal information, employee performance appraisals, your image (picture), ownership records, court records, etc. This is typically referred to as information security.

There are security techniques and mechanisms that can be used to assess and protect physical and or information items. The initial physical and information security elements to consider are:

- **Threat Assessment**: Identify the things that must be protected or secured and rate their importance or value.
- **Security Perimeter**: Establish security perimeters or boundaries and place the most important or valuable items in the inner most security boundaries.
- **Penetration Analysis**: Identify the threats, determine the changes of each threat associated with an item penetrating the boundary. Once the boundary is penetrated determine the impact (loss of life, physical harm, significant financial harm, etc)

As the security analysis and architecture are further refined additional security techniques and mechanisms need to be considered. Those that tend to be associated more with information security are need to know, authentication, and non-repudiation. Other security elements to consider are security abuse, simplicity, logging auditing, overhead impacts.

22.1 Threat Assessment

Threat assessment starts by building a list of items to be protected. The next step is to assign importance or value to each of the items in the list. The items in the list should be grouped and re-grouped based on their relationships to each other and their importance or value. At this time there may be a desire to start to discuss the security boundary or perimeter. Allow the discussions and thoughts to freely flow. For example a company has many databases. What are the most valuable company databases that need to be protected from a threat?

- Financial
- Contracts
- Employee Performance Appraisals
- Product Documents
- Intellectual Property Documents
- Internal Process Documents
- Payroll
- Employee Retirement
- Employee 401K
- Employee Health Selection

It is hard to separate the threat from the list of secured items. For example some will view the external threat of theft by a competitor. Others view the internal threat of lost data due to broken equipment or operator error. Others may view the internal threat from an employee just given a termination notice. It does not matter on the view, make a full list and place a threat assessment against each item in the list.

Table 28 Security Threat Assessment Levels

Level	Threat Assessment	Description
1	Grave	Loss of life or Physical Harm
2	Financially Catastrophic	Financially destroys an organization or and individual, no chance of recovery
3	Financially Devastating	Financial losses causing reorganization or change in life style
4	Financially Serious	Financial losses leading to setbacks in plans, investments, or savings / profits
5	Financially Insignificant	No impact on organization or individual

Example 32 Irresponsible Security Offloading

It is critical that all stakeholders be considered not just the entity needing security. For example, an organization pushing responsibility to employees to not divulge proprietary information on a company tool that is implemented on an Internet cloud system is irresponsible and violates common sense security. The company should host the tool on its own computer network inside its own security boundary or inspect and validate all information before it is transferred to the Internet cloud computing service.

22.2 Security Perimeter

A security perimeter is a boundary that prevents possible threats from entering and causing harm. For example a barn for chickens protects them from wolves. There can be multiple security perimeters or layers such that the least secure area is the outside perimeter and the most secure is the inside perimeter. For example a chicken farm may have a fence around the barn area, a barn, and a room with closing doors in the barn housing new baby chickens. Breaking into the most secure area requires penetrating several security boundaries, each becoming more challenging to penetrate.

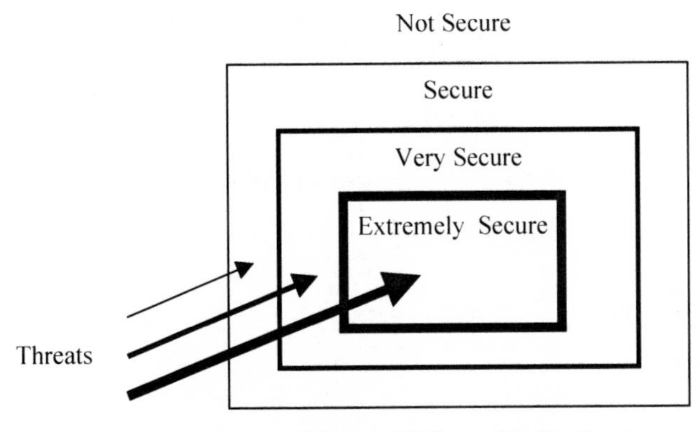

Figure 96 Security Perimeter

22.3 Inherent Security No Security Perimeters

It there a way to secure a system without having one or more security boundaries? Is such a system more secure than one that does have security boundaries? Such a system is inherently secure. No security mechanisms are needed to offer security. There is no need for a security boundary.

Example 33 Inherently Secure No Security Perimeters

For example you can implement a software tool on a computer network that is completely isolated. There are no connections anywhere. The application might be to prepare your tax returns and store your financial information using an isolated standalone computer. Alternatively you can implement the tax tool on a computer network connected to the Internet. The computer network may be very secure with multiple security boundary levels but it will always be less secure than the totally isolated computer. The tax application also might be to prepare your tax return using an Internet cloud computing service.

22.4 Penetration Analysis

Penetration analysis is a study of the effectiveness of the security architecture or approach. The system is subjected to threats and an assessment is made. An approach is to use fault tree analysis to security. The fault in this case is loss of security for a class of items. The basic events represent the threats and the logic gates represent the security architecture. There may or may not be redundancies in the security architecture. For example a perimeter fence with dogs and cats roaming the barn area to protect the farm animals are examples of redundancy that can be captured in the fault tree analysis.

Once a threat has been shown to penetrate a boundary it needs to be assessed. This assessment was performed as part of threat assessment but it needs to be transferred or copied to the penetration analysis. The number of security failures then characterizes it. In general the more serious the threat, the more security elements and backup mechanisms should be broken.

Table 29 Penetration Analysis Criteria

Level	Penetration Result	Minimum
1	Grave	Three backup security mechanisms fail.
2	Financially Catastrophic	Two backup security mechanisms fail.
3	Financially Devastating	Multiple security mechanisms fail.
4	Financially Serious	Security mechanism failed.
5	Financially Insignificant	No security mechanisms triggered.

For example if a grave threat surfaces and three or more backup mechanisms failed then the architecture is considered acceptable[150]. However if the grave threat surfaces with just the breakage of two backup mechanisms then the security architecture needs to be improved until there are a minimum of three failed backup mechanisms to allow the grave threat to penetrate.

22.5 Compartmentalization or Need to Know

Compartmentalization or need to know is fundamental in information security and is similar in concept to Impact of Failure (IOF) described in this text. The goal is to reduce the impact of information loss by limiting access to the information. Only the people who need private information to perform their jobs have access to the private information. This is called need to know.

People who have access to private information with no need to know are potential threats just like people who have a need to know, but they increase the size of the threat pool. The concept is that the more people that know about a private information database the greater the threat. So the approach is to reduce the size of

---

[150] Loss of automobile breaks is a threat and can lead to loss of life. There are two hydraulic breaks (primary and secondary) and one manual cable based parking brake.

the pool and remove access from those with no need to know the information.

22.6 Authentication

Security authentication is confirming the identity of an individual or machine. There are four approaches to authenticating an individual:

- **Special Knowledge**: such as a user name and password, employee number, social security number, account number, etc.
- **Physical Token**: such as a badge with name an picture, automobile key, house key, electronic device with changing alpha numeric pattern, electronic key stick, etc.
- **Biological Evidence**: such as finger print, retinal pattern, hand pattern, signature, face, voice, etc.
- **Third Party**: such as an authenticated employee signing in an employee that forgot their badge, a bouncer at the door of a knight club screening potential customers for entry, etc

Authentication can use one or more authentication approaches. Many times in very secure computer settings special knowledge in the form of a user name password and a computer security token is offered. When accessing very secure facilities a token such as a badge and either special knowledge in the form of access code or biological evidence such as a signature is offered.

22.7 Access Control

Access control is a mechanism that allows authorities to control access to physical areas and information. In order to gain access the entity requesting access needs to be authenticated. If proper authentication is not provided the entity is denied access. This can be in the form of a security guard or guards protecting a perimeter or a machine denying access to a physical location or its services.

22.8 Privileges

Once access is granted to an area, certain privileges are associated with the access. These privileges for physical security are unlimited. You can walk within the entire security perimeter. However in the case of information security access may be associated with different privilege attributes. These privilege attributes are a function of the domain. For example in banking they can be associated with deposit, withdraw, and transaction records privilege access attributes. In the case of a computer they can be read, write, and program execute privilege attributes.

Privileges also exist at different levels. For example someone can allow their information to be shared with a group. The sharing can be one or more of the privilege attribute settings. There is also the concept of a super authority, which can access everything at anytime. Examples of super authorities are government via court actions, potentially internal system maintainers, and potentially trusted internal system operating specialists.

22.9 Non Repudiation

Non repudiation is a mechanism where the source or recipient cannot deny that the source is not the originating source or that a third party cannot claim that the originating source is not the source. This accomplished using a trusted third party person, a trust third party machine, or mutually exchanged receipts.

When a contract is signed a third party also can sign the contract stating that they witnessed the signature of one or both parties. A notary public with an official number and stamp is used to witness many contract signatures such as for the transfer of real estate property. The third party signature is the mechanism that supports non-repudiation in a contract. A notary will check the identity of the individual, maintain a log of the transaction, and complete the signature with a note identifying the type of identity credential checked (e.g. driver license).

Another approach to non-repudiation is to issue a receipt to both parties and have both parties acknowledge the mutual receipt exchange. For example in a transaction involving the transfer of goods and services in exchange for currency, the exchange is completed only when a receipt is generated and given to both parties. There is also a log generated, which may be in the form of another copy of the paper receipt or it may be an electronic entry in a computer file. The transaction is not complete until both parties acknowledge and receive the receipt. So a transaction is a mechanism that supports non-repudiation via the receipts. There is no third party but the source and recipient acknowledge both receipts (copy to buyer and copies to seller).

Non repudiation in the digital Internet domain is accomplished with third party electronic certificates coming from trusted electronic certificate authorities. The certificates are used in the public key infrastructure to establish an encrypted trusted communication mechanism on an open network. The certificates attest that the identity and public key belong together in the web of trust. The web servers log the various server requests.

The strongest non-repudiation mechanism is a trusted third party person and the weakest is the mutual exchange of receipts. Third party witnessing and transaction receipts have an extremely long history. Trusted third party machines is a relatively new development that arose with the proliferation of the Internet and online business transactions.

22.10 Basic Security Systems

At this point there is sufficient information to discuss the fundamentals of security. This section is a combination of thought experiments or what-ifs and knowledge transfer.

When you go to a bank you open an account and list yourself as the account holder. You are given an account number and you provide your name, personal information, and a hand written signature. Until the bank tellers get to know you, you need to provide your account number, some form of identification, usually with a picture, and a signature. At that point you deposit or withdraw from your account. You are given a transaction receipt. You are able to add other people to your account. They may be only authorized to make deposits but no withdraws. Or they may be authorized to make deposits and withdraws but no access to transaction records. These people are now part of your group and you assign appropriate account privilege attributes for your group (deposit or deposit and withdraw). If you break the law and are found guilty the government may gain a court order to access your account. They may decide the entire bank location has a problem and they may request access to multiple accounts. As part of this process they can review the deposit and withdraw records and confiscate the account balances. In this simple bank thought experiment are the fundamentals of security: authentication, access control, privileges, and non-repudiation.

Visualize a time when people thought that one massive computer could provide processing services for thousands of people with very different needs. That time existed in the 1960's. The thought process was that the central computer was extremely expensive and needed to be centralized to keep costs to a minimum. People would use terminals that offered a display screen, keyboard, pointing device, and perhaps voice recognition and synthesis. How would people keep their work separate such that an error from one user would not propagate to thousands of users? How would they ensure their data was private and secure? What was born was the multi-user multi-tasking operating system and one implementation was called UNIX.

A UNIX[151] account is accessed via a user name and a password that is set by the user. The password is encrypted so no one knows its pattern. If forgotten the account needs to be reset with a new temporary password. Once logged in the user is assigned virtual memory for program use and a computer time slice to run programs. The user is also given storage space to house data and programs. The virtual memory is completely private. No one can access it. Further the current users programs cannot go beyond the users assigned virtual memory. If the user creates a broken program with an endless loop the computer time slice cannot be exceeded so the other users are not impacted. Once logged in the user decides the file and program execution privileges. The privilege attributes are read, write, and execute. The privilege levels are user, group, and everyone. The defaults are set by your

[151] I had the privilege of using a Systems Engineering Labs (SEL) 3277 with the MPX-32 operating system circa 1980. A few dozen Hazeltine monochrome terminals provided multi-user access and I never used a keypunch again.

account and start with only the user having read, write, and execute privileges. These privileges are clearly shown when listing the directory content using three groups of three letters. From left to right the first three letters are the privileges for all, followed by the next three letters which are the privileges for a group, followed by the last three letters which are the privileges for the user.

- rwx rwx rwx   Everyone has read write execute access
- r-- rw- rwx   Everyone has read access, group has read write access

When an account is created it is either a user account, a user account associated with one or more groups, or a super user or root account with unlimited privileges.

So the UNIX security approach follows the manual security approach used for hundreds of years in domains such as banking. The idea is to attach authentication, access control, privileges, and non-repudiation to a person, group of people, or a super account.

Recently security mechanisms have surfaced which attach security attributes and levels to data elements. This obviously translates to enormous complexity, because in the end someone in the form of a person needs to assign the security settings. This becomes interesting because people tend to make mistakes and the more they have to do the greater the number of mistakes.

22.11 Security and Simplicity

Inherent in security is simplicity. It is like fault tolerance. The simpler the mechanisms the more secure. The less people have to do the more secure. When establishing a security boundary simplicity also translates to smaller size. That is why a reasonable threat assessment needs to be performed. If everything in the system is tagged at the same high threat level then there is high probability nothing will be secure as the system struggles with traditional system load and capacity issues. So simplicity also translates to making sure the things that need protecting are protected and not surrounded by less critical system elements that burden the security-processing core.

22.12 Security Overhead Impacts

Security always translates to system overhead. The overhead results in reduced response time and more processing needs. This translates to cost. If the system is subjected to excessive security processing because of a poor threat assessment, security effectiveness will decrease.

22.13 Security Logging and Auditing

Security logging tracks certain security related events. For example access logs for visitors into a facility can be used which record the visitors identity, association, contact information, escort name, time in, and time out. A computer can log users attempting to log into a system. Successful and unsuccessful attempts can be logged with information such as user account, time stamp, and number of failed login attempts. A website may have logging enabled, which shows the visiting domain or Internet Protocol (IP) address, what link they went to on the site, what link they came from on the Internet, and their software configuration (operating system and browser).

Once logs are generated they can be audited to look for security concerns. Repeated attempts to break into user accounts may cause changes to the user account visibility or access to the login screen. Review of web server logs may show repeated attempts to deny service and the decision might be to block one or more IP addresses at the router. Auditing is also a review of the security mechanism and its records. The audit may involve physical inspections of facilities, computer logs, reports, actions taken, etc.

22.14 Fail Secure

Fail secure acknowledges that if there is any unexpected system failure, the system fails in a secure state. A secure state is achieved if no security compromise is possible. For example a broken access control mechanism should fail such that access is prevented under all but the most extreme trusted conditions.

22.15 Security Monitor

A security monitor is a passive mechanism that searches for insecure patterns. A camera in a bank or department store coupled to some form of processing to detect insecure patterns is an example. A separate computer program that scans all inputs, outputs, and process start-ups shutdowns is another example of a security monitor. This is generally referred to as virus detection software. A monitor may or may not take action if a security issue is detected. For example a security monitor may detect three unsuccessful account login attempts and lock the account requiring the user to be re-established in the system. Re-establishing the account is accomplished with a more secure mechanism.

Example 34 Appropriate Inappropriate Security Monitoring

A security monitor takes resources away from the system. It becomes less and less effective as more elements are monitored more often. If everything is monitored, it is not a very effective approach to security. A count of the incorrect number of logins mechanism (three tries and account is locked) is an effective security monitor. It is also better than forcing users to have long obscure passwords that they will just write down. Alternatively scanning every piece of data entering a computer is not an effective security monitor as evidenced by the constant release of virus scanning software and patterns. In fact this approach always results in some number of users being compromised before the new security (virus) detection pattern and or software is released. A good analogy is that you cannot monitor every blade of grass and then tell each blade of grass which direction it should grow. It eventually becomes silly and futile.

22.16 Additional Information Security Elements

Information security protects information from unauthorized access, viewing, recording, disclosure, modification, use, disruption, and destruction. Information security uses all traditional security elements discussed thus far and includes new elements of confidentiality, integrity, and availability.

- **Integrity**: Data cannot be modified without detection. There are many mechanisms such as error detection codes and electronic signatures that can be used to ensure the data is not modified.
- **Confidentiality**: Information is not disclosed to unauthorized individuals or systems. This can be accomplished by encrypting the data transmissions, access control, and physical layouts that prevent over the shoulder computer screen views.
- **Availability**: Service is always available when needed. The system can become unavailable because of hardware, software, and operator failures. It also can become unavailable because of an attack such as denial of service.

22.17 Secure Architectures

As in all architecture approaches, the best security architecture is the simplest least complex approach. As complexity increases so does the possibility of failure.

Complexity can be measured by the number of functions, interfaces, and items subjected to security. Complexity also can be measured by the number of people in the loop. They are in the loop because of flexibility that does not lend itself well to automation. The problem is that people make mistakes especially under stressful repetitive conditions. As with all architectures there are collections of trick and lessons learned that should be considered:

- **Changing computer passwords too frequently**. Forcing the use of very obscure passwords. Too many passwords. These will lower security as people violate policy and write them down. A better approach might be to lock the account after three unsuccessful login attempts.
- **Applying the same tight security screening measures equally** to all threat elements even those that are obviously very low security threat levels. This wastes precious resources that can be applied to the real potential threats.
- **Not logging security related events**. Not reviewing the logs. Not using automation techniques to review the logs.

- **Ignoring audit findings**. Not modifying or improving the system when there are security breaches.

22.18 Security Abuse

Security abuse is an interesting concept. Everyone is familiar with the concept of people given special privileges in the name of security and then abusing these privileges. History is full of horrific events in this regard. Electronic security abuse is new. The following is a sampling of recent security abuses:

- Blocking access to legitimate websites because of security software filters thus infringing or breaking restraint of free trade laws in the U.S.
- Forcing users to frequently change or have obscure passwords rather than unsuccessful login account activity detection and block.
- Allowing computer operating systems that are inherently insecure to proliferate through an infrastructure.
- Forcing automated downloads and installation without user knowledge or understanding.
- In large systems there is a simple concept of freezing a proven stable baseline. Serious damage is done with moving baselines and unstable releases.

22.19 Sources of Security Failure

There are many sources of security failure. Obviously understanding the domain with a history of known issues and applying Fault Tree Analysis and Failure Mode Effect Analysis (FMEA) can surface security failures. When people think of security they automatically think of threats. However security can break down in the absence of any threat. Security can break down as part of normal system operations. The following is a list of various security failures:

- **Non Security Failure**: The system normally breaks and security information is left vulnerable. This does not mean security is compromised. It is only compromised if a threat is able to capitalize on the situation.
- **Untested Latent Defect**: Some aspect of the system is untested and an internal set of conditions results in security information left vulnerable. Again this does not mean security is compromised. It is only compromised if a threat is able to capitalize on the situation.
- **Latent Defect Exploit**: Some aspect of the system is untested and an adversary exploits it to gain access to the system and cause a security failure.
- **Security Failure**: A clever threat breaks through the security mechanisms

and is able to gain access to the system and cause a security failure.

- **Trusted People Mistakes**: Trusted people make undetected human mistakes, which leaves information vulnerable.
- **Trusted People Compromised**: Trusted people are compromised and willingly compromise security either by disrupting security mechanisms and or transferring trusted information to unauthorized users.

22.20 Exercises

1. Describe a situation where a trusted person is compromised and security is violated.
2. Describe a situation where a trusted person makes a mistake and security is violated.
3. Using a fault tree show how an authentication scheme can break down.
4. Identify several threats and characterize them.
5. How do security analysis techniques (threat assessment and penetration) compare to Impact or Failure?
6. How do security analysis techniques (threat assessment and penetration) compare to FMEA?
7. Use the Formal Safety Analysis techniques to assess security on a system of your choice. Should you call it Formal Security Analysis?

22.21 Additional Reading

1. Trusted Computer System Evaluation Criteria, DOD 5200.28-STD, December 26, 1985.

23 Modeling Simulation Prototypes

A model is not about fancy equations and complex math. A model is about assumptions and validating those assumptions. You build models because it is too hard to build the real thing and throw it away if it does not work. You build models if you need to accelerate time and see what happens. You build models if you want to see all the places where a thing breaks.

Models are the first step in understanding some aspect of a system or the system in total. As details surface the need for simulators and prototypes starts to surface. As these simulators and prototypes are developed they may find uses outside understanding system behavior and include training, potential backup mechanisms, and research platforms.

23.1 Mental Models Scenarios Assumptions

A mental model is someone's view of the system or some aspect of the system. All stakeholders have mental models. The mental models are converted to scenarios and their view of the expected results.

Scenarios and assumptions are the most important aspects of a model. The stakeholders and other analysts easily verify the algorithms and equations. However stakeholders have different views and perceptions which affect the scenarios and assumptions. In all cases the scenarios and assumptions should be clearly delineated and described. There should be no confusion on a scenario or assumption even though some may not agree with a scenario or assumption. The scenarios and assumptions should include the following cases:

- Expected average
- Expected peak
- Expected worst case
- Expected continuous peak
- Expected continuous worst case
- Breakage points 1-n

The scenarios and assumptions also should include the important system life cycle stages. Dates should be attached to the life cycle points in time. As a minimum the following life cycle stages should be considered:

- Initial delivery (year 1)

- Midlife operations (year 5)
- End of life operations (year 20)

23.2 Static Model

A static model is like an accounting ledger with entries. So you can use a spreadsheet to create a static model. It is a worst case analysis of a situation. It does not show behavior over time. You start with identifying your worst case scenario and translating it into a worst case set of inputs. You identify the key elements and what parameters they accept. These elements are then stimulated by your inputs. For example you can identify the amount of memory in a computer by identifying all the computer functions and their sizes. You also can identify the processing load by taking the functions, determining their processing path length, then multiplying them by the inputs.

Example 35 Sum of RADAR Errors

Sometimes a static model is the most effective approach. Track position error is not only associated with the algorithm mathematics such as rounding errors and trigonometric approximations, but also other error sources such as time and physical placement:

- RADAR registration error which is error in knowing the actual physical location of the RADAR and its alignment to true North.
- Wind jitter, which causes the RADAR dish to vibrate and change the reported position of the return.
- Time in transit associated with communications transfer and its varying times due to communications buffering and queue characteristics.
- Time in storage within the computer before the software starts processing the data.
- Display update time associated with actually getting the track symbol on the display console.

23.3 Dynamic Model

A dynamic model is able to accept inputs over time. It accepts a load scenario rather than a snapshot of the worst case set of inputs. The model internals are partitioned in the same way as a static model, but the behavior is modeled by using an algorithm representing the characteristics of the modeled elements and their interactions.

23.4 Simulation

A simulation is a dynamic model. It may use equipment similar to the actual thing being modeled, but not the actual thing because of prohibitive costs. In the worst case simulation it just might be software attempting to mimic the thing being modeled. In a better simulation it might be a cheaper version of the actual thing.

Example 36 Tracking Algorithm Modeling

Sometimes models are developed to not only understand dynamic behavior but also to test a major element in a system. For example a RADAR tracking algorithm in air traffic control or air defense must be fully understood and it must be fully tested. The alternative is to either approximate the tracker in a dynamic model or implement the actual tracking code in a simulation test bed environment. If it is implemented in a test bed environment the final version of the code can be subjected to unit testing prior to the integration into the final system.

The purpose is to stress the algorithm and capture its performance. The stress testing is based on multiple scenarios that would investigate performance such as:

- Track drop probability because of fast maneuver
- Track swap probability between two very close planes
- Track establishment time or number of RADAR scans
- Track position error

Obviously this will require different test scenarios. Each scenario has its own number of aircraft and flight path profiles. When testing drop track probabilities the planes need to fly different maneuvers and speeds. When testing track swap probabilities the planes need to fly different crossing patterns. When testing track establishment time the planes need to fly different speeds and maneuvers in noisy environments. In all cases track position error needs to be assessed.

Using the real software in computing hardware that is the same or close to the final target hardware is ideal. Not only could the tracking algorithm be studied, optimized, and tuned, but it also can be tested without using live aircraft in dangerous maneuvers. It also would not tie up an entire command and control center, an expensive and difficult option during the early stages of refinement.

23.5 Emulation

Emulation uses the real thing to accept various inputs and determine behavior. This is the most faithful form of modeling, but it is extremely expensive and may be unable to support studies like accelerated time or beyond normal stress analysis.

Example 37 Automobile Crash Tests

Automobile crash testing is used to determine how a vehicle behaves in various crash scenarios. A real automobile is instrumented along with crash dummies. The instrumentation is used to gather the effects on the automobile occupants and the automobile. The data is then used to rate the performance of each automobile and offer data to engineering for possible changes to the automobile design.

23.6 Prototypes

The team needs to determine if a prototype can be created using the same resources as a model. This is especially true when trying to surface qualitative measures and characteristics like functional requirements. A prototype offers significantly more functional and performance fidelity but is less flexible in studying the internals of a system. There are different prototypes:

- Research and Development prototypes
- Proof of concept prototypes
- Simulation labs that emulate possible solutions
- Engineering models
- Pre-production prototypes
- Production prototypes
- Market analysis prototypes

It is not uncommon for a system to evolve from different types and levels of prototypes. This is the traditional definition of synthesis. A system is assembled from existing known elements. The danger is that this synthesis may have undesirable emergent properties that may have been surfaced with a full system analysis that was not performed. So its ok to synthesize a system from prototypes, but it also should include the underlying system analysis and resulting glue or changes to allow these elements to predictably and dependably operate.

23.7 System Dynamics

System Dynamics is an approach that models a systems behavior as a function of time. Professor Jay W. Forrester[152] of the Massachusetts Institute of Technology (MIT) created System Dynamics in the 1950s to help corporate managers improve their understanding of industrial processes. It is obvious that this work was built from his previous experience including WHIRLWIND and SAGE[153].

---

[152] His first research assistantship was under Professor Gordon Brown, the founder of MIT's Servomechanism Laboratory. The Servomechanism Laboratory conducted pioneering research in feedback control mechanisms.

[153] In 1947, the MIT Digital Computer Laboratory was founded and placed under the direction of Jay Forrester. The Lab developed WHIRLWIND I, MIT's first general-purpose digital computer. It was also an environment for testing digital computers for control combat systems. After WHIRLWIND I, Forrester lead a division of MIT's Lincoln Laboratory to develop computers for the North American Semi-Automatic Ground Environment (SAGE) air defense system.

System Dynamics[154] is not a static snapshot of the system. It is this time view that allows the analysts to surface and start to understand unexpected behavior. In many ways system dynamics is a broad generalization of other models. For example, circuit analysis uses feedback loops and the properties of components understood in the physics of electricity and magnetism to model behavior as a function of time. With the introduction of computers and communications, understanding message buffers and processing queues were key to optimizing communications throughput and processing delays. It is within this environment that system dynamics emerged as a very general modeling approach that can be used on almost any system. It uses fundamental elements that are found in all systems. These elements are:

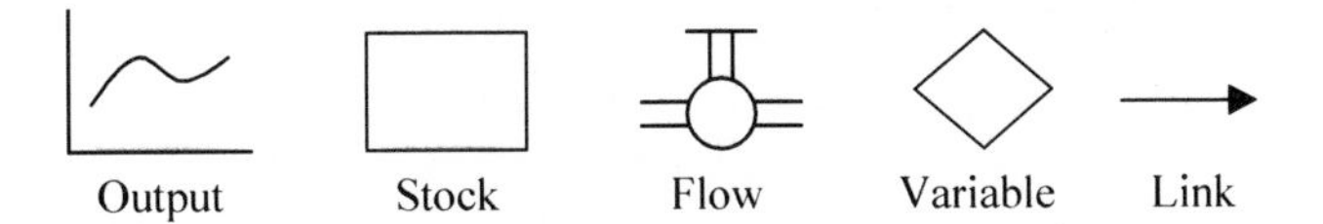

Figure 97 System Dynamics Symbols

- **Output Time Paths**: This is the output of the model. There are four broad categories of outputs possible from any system. They are linear, exponential, goal seeking, oscillation and s-shaped.
- **Stocks**: A stock is a collection of anything, an aggregate. For example, a Stock can represent a population of sheep, the water in a lake, number of widgets in a factory, or the amount of energy reserves, food reserves, etc.
- **Flows**: A Flow brings things into or out of a Stock. Flows look like pipes with a faucet because the faucet controls how much of something passes through the pipe. Think in terms of a bathtub or sink. The basin is the Stock, the faucet is a Flow, and the drain is a Flow.
- **Variables**: A Variable can be an equation that depends on other Variables, or it can be a constant.
- **Links**: A Link makes a value from one part of the diagram available to another. A link transmits a number from a Variable or a Stock into a Stock or a Flow.

23.7.1 Linear Output Time Paths

The simplest output time path is linear response. It can be linear growth, decline, or equilibrium. In this situation there is no feedback in the system. This is an oversimplified condition, but may be a starting point for building the model. The equilibrium response is a system in perfect balance. It may be a goal but it is rarely

[154] NetLogo modeling tool available from Center for Connected Learning and Computer-Based Modeling, Northwestern University, Evanston, IL, Wilensky, U, 1999.

achieved. Although the model is typically started with the system in the Equilibrium State its actual response will not be an equilibrium response. If the output is equilibrium the model should be closely examined for problems.

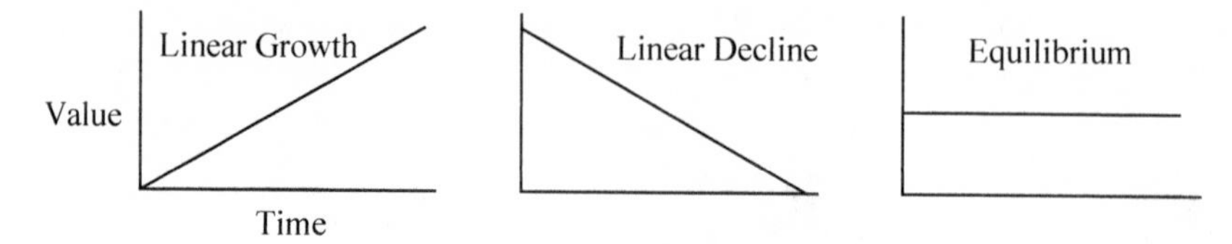

Figure 98 Linear Time Paths

23.7.2 Exponential Output Time Paths

Most systems are non-linear. Non linear situations happen when there is feedback in the system. The exponential time paths grow or decay exponentially.

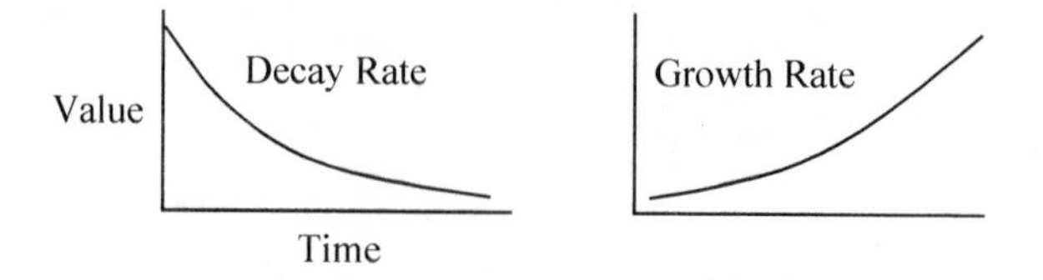

Figure 99 Exponential Time Paths

They are the result of some feedback in the system. However, as the system model gets more detailed the response will become more complex. Usually simple components have exponential response characteristics. For example capacitors and inductors as electrical components have exponential responses but when they are integrated into larger circuits the response becomes more complex.

23.7.3 Goal Seeking Output Time Paths

Many systems are structured to seek a goal. An example is a cruise control in an automobile. Living systems exhibit goal-seeking behavior. Goal-seeking behavior is similar to exponential decay except instead of reaching zero a value is asymptotically approached. The value can be either a maximum or minimum value.

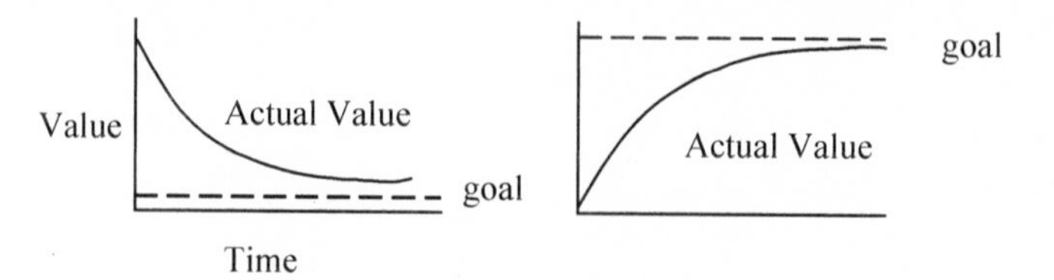

Figure 100 Goal Seeking Time Paths

23.7.4 Oscillation Output Time Paths

Oscillation occurs around an idealized equilibrium. The oscillation can be chaotic, exploding, sustained, or damped. Most systems exhibit stable oscillation.

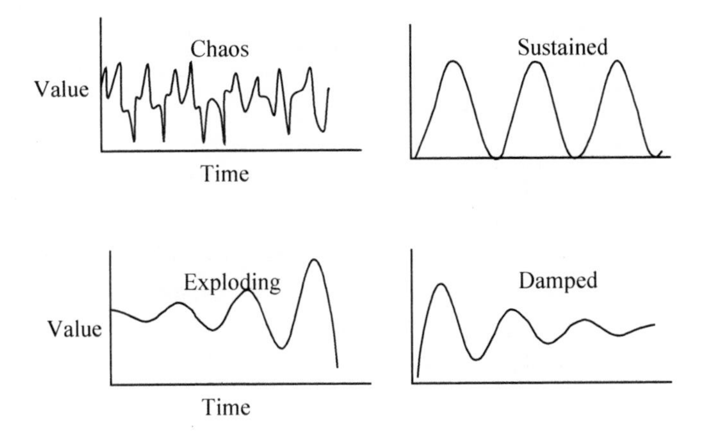

Figure 101 Oscillation Time Paths

Stable oscillation is either sustained with very small amplitude or damped with a small starting amplitude and short decay time. Sustained oscillation is periodic with constant amplitude. Damped oscillations exhibit reducing amplitude with time. This can be accomplished with negative feedback or low pass filter mechanisms. They both add stability to the system.

Exploding oscillations exhibit increasing amplitude until a maximum is reached or the system self-destructs. This is usually the result of an unchecked system using a positive feedback mechanism.

23.7.5 S-Shaped Output Time Paths

The s-shaped system response is an exponential growth that shifts to goal seeking behavior as the system approaches its limits. There are four possible scenarios if the system limit is exceeded.

S-Shaped Growth: If the system limit is exceeded and its Carrying Capacity (CC) is not completely destroyed the system will operate around its limit. An example would be driving a car in first gear. The engine revs to its maximum rotations per minute (RPM) and just stays at that limit making a great deal of noise as the car moves relatively slowly.

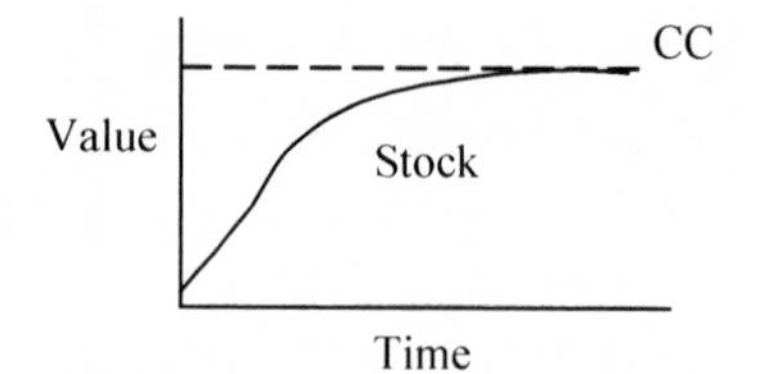

Figure 102 S-Shaped Growth

Overshoot and Collapse: If the system significantly overshoots its limit and its carrying capacity is damaged, the system collapses. This is an overshoot and collapse or burnout system response. An example would be driving a car in first gear with a modified ignition system that does not limit the engine spark timing. The engine revs to its maximum RPM then exceeds its maximum design limit RPM. Eventually the engine is spinning so fast oil starts to foam and the engine loses lubrication. As the parts lose lubrication the engine starts to slow down.

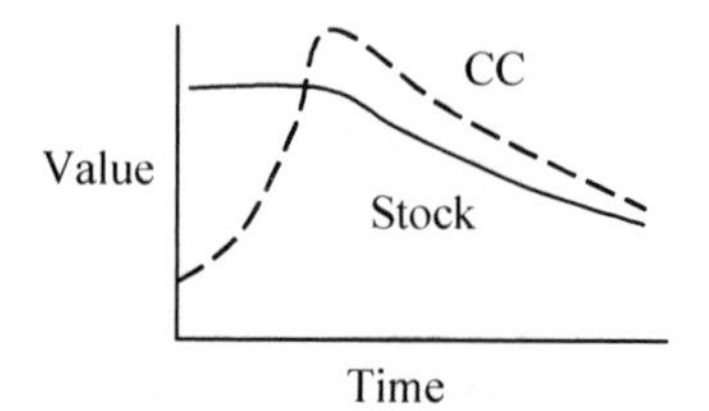

Figure 103 Overshoot and Collapse

Overshoot and Oscillate: If the system limits are exceeded and the system does not collapse, but does not maintain the limit response, the system may overshoot and oscillate. This is just like the pervious example except as the RPM approaches the design limits the valve springs let the valves float. The floating valves cause the engine to oscillate around its maximum RPM leading to a jerking car movement while in the first gear setting. This is actually an example of a failsafe design. The floating valve springs cannot be easily manipulated so if someone does modify the engine to exceed its design RPM the engine will not self-destruct.

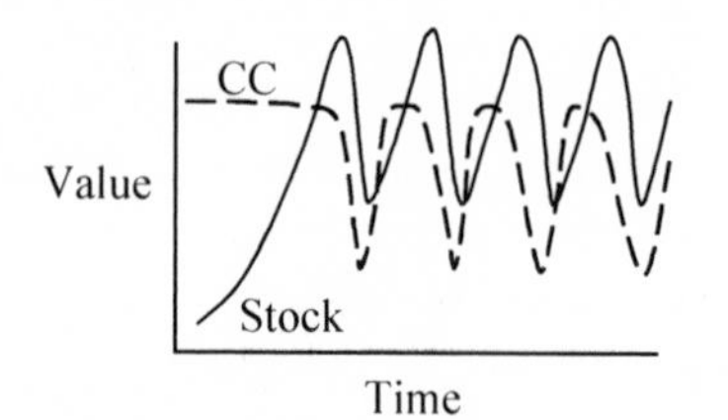

Figure 104 Overshoot and Oscillate

Reverse S-Shaped Decline: If the system limits are exceeded and the system does not collapse, but does not maintain the limit response, the system reverses direction in a reverse s-shaped pattern. This is just like the pervious example except that as the RPM approaches the design limits the engine starts to overheat. The excess heat causes the piston rings to over expand and slow the engine down until it stops.

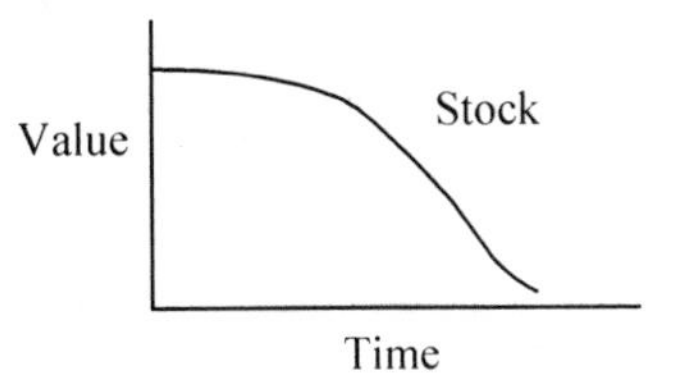

Figure 105 Reverse S-Shaped Decline

23.7.6 Stocks and Flows

A stock is a holder or container. It is like a bathtub holding water. Dynamic behavior happens when there is inflow and outflow. The flows can be enabled, disabled, and change their rates over time. If the inflow exceeds the outflow the stock will reach its limit then overflow. If the outflow exceeds the inflow then the stock will eventually empty. When inflow matches the outflow then the system is in equilibrium. However, that equilibrium can be at the system limit.

Just as in the case of functional analysis the magical number 7 +/- 2 applies when associating flows with a stock. So less than four to six inflows and or outflows should be the goal for a single stock.

It can be challenging to detect the difference between a stock and a flow. The following guidelines are offered to identify stocks and flows:

- Stocks are nouns flows are verbs
- Stocks do not disappear if time stops
- Flows disappear (stop flowing) when time stops
- Stocks send state information to the rest of the system

Stocks have memory, persistence, or inertial. If the inflow to a stock is shutoff the stock will not decrease unless there is an outflow. If there is no outflow it stays at the level it is at when the inflow stops. This is important because people believe that shutting off an inflow to a stock representing a problem will cure the problem. However the stock has an accumulated backlog of the problem. For example in 1990 steps were taken to stop the release of ozone depleting chemicals, however there is a buildup of these chemicals in the atmosphere that will continue to cause damage at some rate even though no new chemicals are being released.

23.8 Physical Models and Mockups

Physical models and mockups are models of some aspect of the system. They are usually built to scale but not always. The detail in the model is a function of the issues and the teams desire to understand various physical and aesthetic aspects of the system. Examples of scale models are:

- Buildings, houses, structures such as bridges
- Large machines such as satellites, automobiles, airplanes
- Power plants, factories, processing factories, industrial parks
- Towns, country sides, celestial maps

In some cases a physical model is built to full scale using low cost lightweight materials. The materials can be Styrofoam backed paper, cardboard, balsa wood, composites, fiberglass, wax, clay, etc Examples of mockups include:

- Operator workstations
- Appliances such as computers, toasters, televisions, cell phones
- Theme park structures[155]

23.9 Virtual Models

Virtual models use computer technology to model a physical system. In some cases a computer based virtual model can be converted to a physical model using automated manufacturing.

The computer model can use simple two-dimensional presentations or three-dimensional visualization. The two-dimensional presentation can be on a small screen in front of a user or on a large screen in an auditorium. The three-dimensional visualization can use the same display technology as two-dimensional presentations with the addition of glasses worn by a user that enable and disable light entry into the left and right eye while synchronized to the display on the screen[156]. Lightweight glasses with internal displays also can be offered.

The immersion in a virtual model can continue beyond sight and include sound[157]. Sound processing techniques can be used to move electronic sound to a

---

[155] Cinderella Castle at Disney theme parks: Magic Kingdom at Walt Disney World Resort, and Tokyo Disneyland at Tokyo Disney Resort. Cinderella Castle is more than twice the size of Sleeping Beauty Castle at Disneyland in Anaheim, California. Forced perspective makes the Castle appear larger. Its proportions get smaller as height increases. Major elements of the Castle are scaled and angled to give the illusion of distance and height.

[156] Previous techniques used light polarization to offer a three-dimension perception.

[157] In the 1990's Air Traffic Control Tower trainers used large screen displays and extremely high quality multi channel sound to simulate aircraft noise. Source: this Author.

level so that it could be perceived as a live situation[158]. The visualization and other sensory processing can become extremely sophisticated and moves into the realm of a simulation of a system rather that just a simple virtual model of some aspect of the system.

23.10 Simulator

A simulator attempts to faithfully reproduce the whole system. It can be implemented with the same equipment as the operational system augmented with simulation functions or with lower cost equipment and less complex operational internals. Usually a simulator is part of a larger facility, a simulation lab that includes multiple simulators interconnected to support full-scale operational simulations and concurrent studies. Examples of a simulator are:

- Airplane simulator
- Air Traffic Control / Air Defense workstation simulator
- Air Traffic Control / Air Defense tower simulator
- Tank simulator, police car simulator, fire fighting simulator

23.11 Simulation Test Lab or Facility

When new functionality is added to an existing system it can be operationally validated at a live site. However to reduce risk of disaster, such as in a mission critical system[159], the validation can start in a simulation test lab or facility. Once the functionality is shown to be useful and not harmful to operations, it can be rolled out to a selected live site.

The live site can then validate the new functionality in an isolated simulation like setting. This can be accomplished by operating during low use time (midnight) while mixing live and small sets of simulated data on a backup system string. So a simulation test lab can take various forms and be located at a dedicated facility, an operational site, or both.

The simulation testing can become progressively more real, until the simulation becomes live and operational suitability is determined (movement from similar equipment to backup strings to live strings). Once the initial live site signs off on the new functionality then the simulation can be considered complete and successful.

23.12 Simulation Training Lab or Facility

The simulation training lab can be located in a dedicated training facility, a test

---

[158] Hughes Sound Retrieval System (SRS) is an example that modeled the behavior of the ears on a human head to capture all the phase and time shifts present in a live setting.

[159] Mission critical systems can cause loss of life if something goes terribly wrong. An example of a mission critical system is Air Traffic Control.

and evaluation facility, at operational sites, or any combination. Because simulation test labs are extremely faithful representations they can be used for training. This causes scheduling conflicts between training and other organizations trying to use the same lab resources. An approach is to build a simulation lab that can be segmented to support training as needed.

An operational system can be segmented to offer training during low workload times. This is accomplished by using spare resources and redundant elements that are not needed for short periods of time. Simulation onsite training using spares and back up strings is the most accurate form of simulation.

23.13 Simulation Research Lab or Facility

A simulation research lab or facility is typically appended to test and training simulators. The research elements are new functionality implemented in either new products or one of kind devices that are attached or integrated to an existing simulation test lab. The research simulation lab offers maximum flexibility to allow for various studies, while not being burdened by working with the internals of real system elements or high fidelity trainers.

23.14 Model Verification Validation Accreditation

How can you trust a model or simulation? The approach is to establish confidence in the model or simulation through formal Verification, Validation, and Accreditation (VV&A)[160].

- Model verification is the process of determining that a model implementation accurately represents the developer's conceptual description and specifications.
- Model validation is the process of determining the manner and degree to which a model is an accurate representation of the real world from the perspective of the intended uses of the model and of establishing the level of confidence that should be placed on this assessment.
- Model accreditation is the formal certification that a model or simulation is acceptable for use for a specific purpose. Accreditation is given by the organization best positioned to make the judgment that the model or simulation in question is acceptable. That organization may be an operational user, the sponsor, an independent organization, and an industry body, depending on the intended purpose.

---

[160] Systems Engineering Fundamentals, Supplementary Text, Defense Acquisition University Press, January 2001.

23.14.1 Model Verification

Model verification ensures that the model correctly implements the requirement specification and the algorithms are correctly represented and implemented. It also ensures that there are no errors, oversights, or problems. Verification does not ensure the model accurately represents the problem to be solved or correctly reflects the real world setting.

Verification is accomplished by running test cases and examining the results against the expectations established by the specification and reasonableness. This includes running extreme cases not typically found in real world scenarios. All expected points of inflection and extreme case should be executed. For example zeroing out parameters one at a time. Setting parameters to "relative infinity" one at a time. As the number of test cases increases confidence grows that the model is implemented correctly.

23.14.2 Model Validation

Model validation builds confidence that the model addresses the correct issues, provides accurate representations of the real world settings, and is actually used because of its value and correctness in representing some aspect of the system.

A model can be validated using historical data, however there is no guarantee that history will repeat unless a logical analysis can prove the repetition. For example a RADAR tracking algorithm is deterministic and various flight patterns can be duplicated to both verify and validate the model.

When a model is non-deterministic then validation using historical data is of limited use. Validation becomes challenging under the following conditions:

- There is no historical data
- Controlled experiments on the live system cannot be performed to gather real world data and compare with the model
- The system does not exist there are no similar systems to compare against
- The model is fundamentally non deterministic
- The model is extremely non linear with many points of inflection and state changes

The approach to validate a model is to use various mental model[161] inputs to break the model. These mental models translate into scenarios and expected outputs. If the model breaks then it needs to be updated to capture the mental model and its scenario. The danger is that the scenarios derived from various stakeholder mental models may not capture unexpected and hidden non-linearity's, points of inflection,

---

[161] A mental model is a person's intuitive perception or thought process about how something works in the real world.

and state changes. So initial incorrect model outputs must be clearly explained in a precise accurate unambiguous manner or the mental model must be invalidated using the same approach.

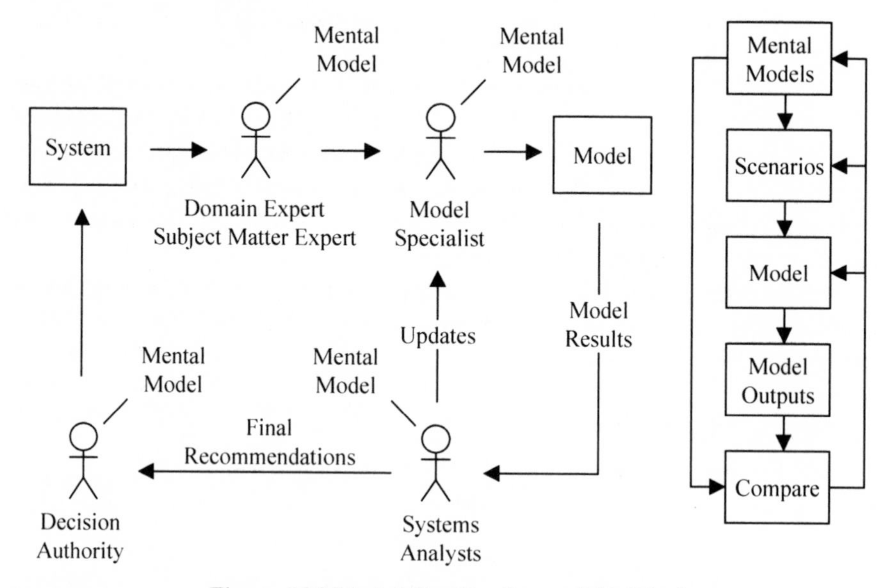

Figure 106 Model Verification and Validation

Once the mental model scenarios have been executed then those scenarios for the model have been validated. The model is technically not validated but there is a high degree of confidence that the model will properly execute similar scenarios and the model should be considered validated. The process and logic is similar to a legal case where there is a large body of evidence showing why the model is valid for its intended use.

At the end of the validation process everyone has a better understanding of the system and its behavior. During the iterative validation process of running different mental model scenarios various discussions surface and validation can be stated:

- Assumptions and requirements validation
- Calibration validation
- Process validation
- Face validity
- Data, Theory, and algorithm validation
- Emergent (unexpected) behavior validation

The timeline between model development, verification, and validation quickly blurs. This suggests that model implementation should be started early and expanded as new insights surface. It is not unreasonable to spend a majority of the time in model validation as each new scenario results in major changes, minor changes, and then tweaks to the model.

23.14.3 Model Certification and Accreditation

A certified model is a model that meets an industry standard. These models tend to be used in simulators to test new products. Examples include environmental test, chambers, communication modem testers, product safety test devices, etc.

Accreditation is applied to an organization and represents a certain minimal level of performance that has been achieved by the organization. So accredited test labs run certified models to test new products.

Certification is a process where all information products (plans, specifications, tests, etc) are examined and a confidence level is achieved so that the item can be certified or accredited. Certificates are normally issued to people and devices while accreditation is issued to organizations or institutions. Certification uses people engaged in the review of the evidence. A separate body of people based on the certification evidence approves accreditation.

A major consideration when seeking model certification and accreditation is the preparation of the evidence. A model in a development setting needs to support the developers, suggesting limited documentation and non-independent verification and validation. A model submitted for certification needs to be appropriately documented and independently verified and validated. The analogy is one of moving from an R&D prototype to a full production version.

23.15 Exercises

1. If models are about assumptions and validating those assumptions why are so many models so complex no one can understand them or their results?
2. You start a new project and spend the next 10 years of your career on this project. What is the sequence of the types of models that you may have developed?
3. Why are mental models and scenarios so important?
4. What types of models are needed to produce a consumer product?
5. What types of models and simulation facilities are needed to study and produce a sustainable system such as a wind turbine farm?

23.16 Additional Reading

1. Systems Engineering Fundamentals, Supplementary Text, Defense Acquisition University Press, January 2001.
2. Sustainable Development Possible with Creative System Engineering, Walter Sobkiw, 2008, ISBN 0615216307.

24 Technology Assessment and Maturity

There are two ways to view technology. The first is as an observer trying to understand its maturity and adoption rate. The second is from a technology management viewpoint.

In systems that are driven by goals, objectives, and resulting requirements, new technologies may need to be developed, matured, and infused. Infusion of mature technologies from other systems into a different architecture and or into a different operational setting also requires methodical technological advancement otherwise key steps in the development process will be missed.

Technology assessment, advancement, and maturity are not new ideas and have been practiced by generations of people. There is a formal technology assessment and advancement approach that has been codified by NASA. It started in the 1970's and progressed through the years. There are also informal approaches and various observations of technology, such as maturity, that are worthy of reviewing.

24.1 Formal Technology Assessment

Technology assessment has always been a part of systems engineering. However it was more of an ad hoc process closely related to research and development cultures rather than development or production process oriented cultures. That process started to be formalized by NASA in 1974[162]. Some organizations included the activity as part of feasibility analysis and studies[163]. Many project problems and failures have been attributed to poor requirements. However history keeps repeating itself and there is evidence to suggest that in some of these cases there is a lack of understanding the maturity of the technologies for a successful project[164]. Understanding the maturity of the key technologies and identifying the needed technological advancement will help projects to understand their challenges and avoid technology surprises.

---

[162] NASA researcher, Stan Sadin, conceived the first scale in 1974. It had seven levels, which were not formally. In the 1990s NASA adopted a scale with nine levels which gained widespread acceptance across industry: Technology Readiness Levels Demystified, NASA, August 2010.

[163] Feasibility Analysis is a result of technology assessments and trade studies to justify system design approach. From Systems Engineering Fundamentals, Supplementary Text, Defense Acquisition University Press, January 2001.

[164] NASA Appendix G Technology Assessment, NASA Systems Engineering Handbook, NASA/SP-2007-6105 Rev1, National Aeronautics and Space Administration NASA Headquarters Washington, D.C. 20546, December 2007.

Technology assessment consists of performing a Technology Maturity Assessment (TMA) and an Advancement Degree of Difficulty Assessment (AD2).

Technology assessment should start very early in the concept phase, continue as part of the architecture development and selection, and proceed during design and implementation. The initial TMA is a baseline maturity of the systems technologies. It allows for monitoring progress as the system unfolds.

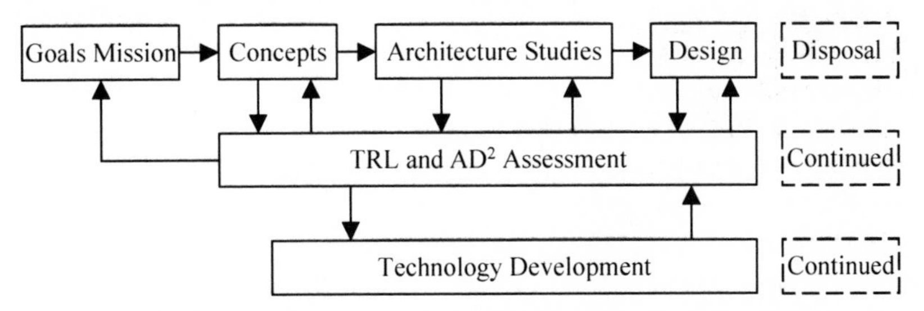

Figure 107 System Life Cycle and Technology Development

Some suggest that the final TMA be performed as part of the preliminary design review. However, key technologies associated with implementation, verification, validation, maintenance, training, installation, switch on, decommissioning, and disposal may need to be understood throughout the systems life. For example disposal may have significant sustainability needs and the technologies may need to be tracked and matured as the system ages but before its disposal and abandonment.

There is a two-way relationship between technology assessment and other system engineering practices. As a practice offers a straw man[165] for technology assessment, the assessment offers its findings, which then alters the straw man approach. This relationship is especially important in the concept and architecture stages. A poor foundation will only lead to serious problems as a system unfolds.

The technology assessment needs to be done against something tangible and traceable to the project. The problem is that at the early concept stage a system block diagram, architecture, or product break down may be unavailable. A suggestion is to try to list the technologies along side the context diagram or concept diagram. Eventually the technology assessment should be organized by systems, subsystems, and components, which should be traceable to the work break down structure, so that reporting is in a form that facilitates the project cost and schedule tracking mechanisms including Earned Value Management System (EVMS).

Although technology assessment is performed through out the life cycle it is extremely important when considering alternative system architectures. Many times

[165] Straw man is an approach intended to stir debate and discussion to surface disadvantages so that a stronger approach can surface. As an approach becomes more mature and difficult to knock down it is a given name suggesting more mass and strength such as stone-man and iron-man.

architecture selection is based on the underlying technologies.

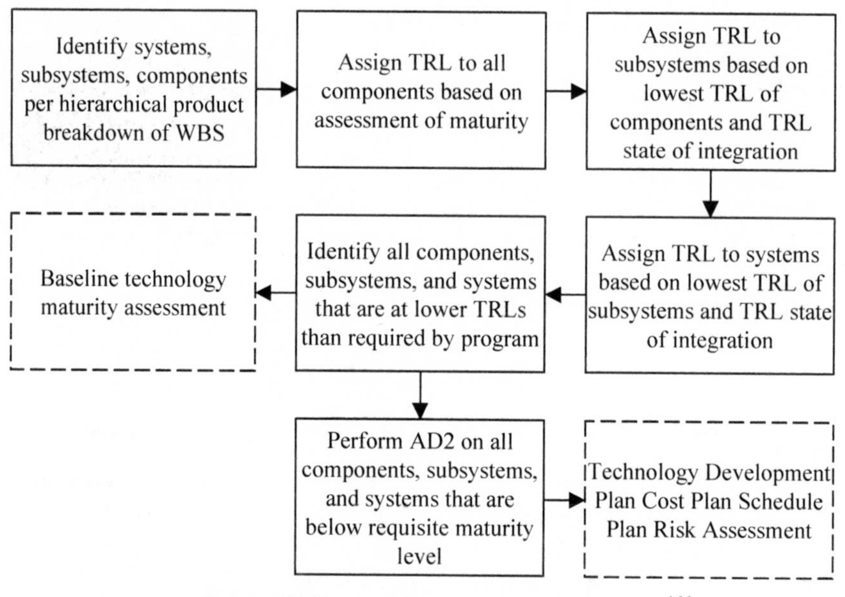

Figure 108 Technology Assessment Process[166]

24.1.1 Technology Readiness Levels

Technology assessment is used to determine the need to develop or inject technological advances into a system. The first step is to determine the current technological maturity of the system in terms of technology Readiness Levels (TRL). The second is to determine the difficulty with moving a technology from one TRL to the next using an Advancement Degree of Difficulty Assessment (AD2).

TRL is a gage that measures the state of the art of a technology. There are different TRL scales associated with different organizations. There is overlap between the different scales and some translation may be needed as part of establishing the TRL scale for an organization. The following is a list of different organizations offering TRL scales:

- National Aeronautics and Space Administration (NASA)
- Department of Defense (DOD)
- North Atlantic Treaty Organization (NATO)
- European Space Agency (ESA)

[166] NASA/SP-2007-6105 Rev1, NASA Systems Engineering Handbook, December 2007.

- Federal Aviation Administration (FAA)

24.1.1.1 NASA Technology Readiness Levels

Table 30 Technology Readiness Levels - NASA[167]

NASA TRL	Description
1. Basic principles observed and reported.	Lowest level of technology readiness. Scientific research begins to be translated into applied research and development. Examples might include paper studies of a technology's basic properties.
2. Technology concept and / or application formulated.	Invention begins. Once basic principles are observed, practical applications can be invented. The application is speculative, and there is no proof or detailed analysis to support the assumption. Examples are still limited to paper studies.
3. Analytical and experimental critical function and / or characteristic **proof of concept**.	At this step in the maturation process, active research and development (R&D) is initiated. This must include both analytical studies to set the technology into an appropriate context and laboratory-based studies to physically validate that the analytical predictions are correct. These studies and experiments should constitute "proof-of-concept" validation of the applications / concepts formulated at TRL 2.
4. Component and / or **breadboard** validation in **laboratory environment**.	Following successful "proof-of-concept" work, basic technological elements must be integrated to establish that the pieces will work together to achieve concept-enabling levels of performance for a component and / or breadboard. This validation is devised to support the concept that was formulated earlier and also should be consistent with the requirements of potential system applications. The validation is relatively "low-fidelity" compared to the eventual system: it could be composed of ad hoc discrete components in a laboratory.
5. Component and / or **breadboard** validation in **relevant environment**.	At this level, the fidelity of the component and / or breadboard being tested has to increase significantly. The basic technological elements must be integrated with reasonably realistic supporting elements so that the total applications (component-level, subsystem-level, or system-level) can be tested in a "simulated" or somewhat realistic environment.
6. System / subsystem model or **prototype** demonstration in an	A major step in the level of fidelity of the technology demonstration follows the completion of TRL 5. At

[167] Generic TRL descriptions found in NPR 7123.1, NASA Systems Engineering Processes and Requirements, Table G-19.

Table 30 Technology Readiness Levels - NASA[167]

NASA TRL	Description
operation environment.	TRL 6, a representative model or prototype system or system, which would go well beyond ad hoc, "patch-cord," or discrete component level breadboarding, would be tested in a relevant environment. At this level, if the only relevant environment is the environment of space, then the model or prototype must be demonstrated in space.
7. System **prototype** demonstration in an operational environment.	Prototype near or at planned operational system. TRL 7 is a significant step beyond TRL 6, requiring an actual system prototype demonstration in a space environment. The prototype should be near or at the scale of the planned operational system, and the demonstration must take place in space. Examples include testing the prototype in a test bed.
8. Actual system competed and "flight qualified" through test and demonstration.	Technology has been proven to work in its final form and under expected conditions. In almost all cases, this level is the end of true system development for most technology elements. This might include integration of new technology into an existing system.
9. Actual system flight proven through successful mission operations	Actual application of the technology in its final form and under mission conditions, such as those encountered in operational test and evaluation. In almost all cases, this is the end of the last "bug fixing" aspects of true system development. This TRL does not include planned product improvement of ongoing or reusable systems.

Although the TRL descriptions appear to be straightforward it is important to define the terms. Moving the domain from Space to a different domain such as Air Traffic Control will add even more confusion further necessitating clear definitions. Without these definitions trying to assign levels even in the space domain is a challenge. Many will think they can define a breadboard, but not everyone will have the same definition and the alternative definitions may be appropriate. This is a classic communications and perception problem that surfaces when a team starts to work on any problem. A relevant environment in one application may be irrelevant to another. Many of these terms cross engineering fields and had, at one time, very specific meanings to a particular field. The following TRL related definitions and terminology come from NASA[168] and can be used as a starting point:

[168] Lists hardware and software TRLs and offers a description of the terminology: Appendix J. Technology Readiness Levels (TRLs), NPR 7120.8, NASA Procedural Requirements, NASA Research and Technology Program and Project Management Requirements, February 05, 2008.

- **Proof of Concept (TRL 3)**: Analytical and experimental demonstration of hardware / software concepts that may or may not be incorporated into subsequent development and / or operational units.
- **Breadboard Laboratory (TRL 4)**: A low fidelity unit that demonstrates function only, without respect to form or fit in the case of hardware, or platform in the case of software. It often uses commercial and / or ad hoc components and is not intended to provide definitive information regarding operational performance.
- **Breadboard Relevant Environment (TRL 5 - 6)**: Sometimes called a Brassboard, a medium fidelity functional unit that typically tries to make use of as much operational hardware / software as possible and begins to address scaling issues associated with the operational system. It does not have the engineering pedigree in all aspects, but is structured to be able to operate in simulated operational environments in order to assess performance of critical functions.
- **Prototype Unit (TRL 6 - 7)**: The prototype unit demonstrates form, fit, and function at a scale deemed to be representative of the final product operating in its operational environment. A subscale test article provides fidelity sufficient to permit validation of analytical models capable of predicting the behavior of full-scale systems in an operational environment.
- **Engineering Unit (TRL 6 - 8)**: A high fidelity unit that demonstrates critical aspects of the engineering processes involved in the development of the operational unit. Engineering test units are intended to closely resemble the final product (hardware / software) to the maximum extent possible and are built and tested so as to establish confidence that the design will function in the expected environments. In some cases, the engineering unit will become the final product, assuming proper traceability has been exercised over the components and hardware handling.
- **Mission Configuration (TRL 9)**: Final architecture / system design of the product that will be used in the operational environment. If the product is a subsystem / component, then it is embedded in the actual system in the actual configuration used in operation.
- **Laboratory Environment (TRL 1 - 4)**: An environment that does not address in any manner the environment to be encountered by the system, subsystem, or component (hardware or software) during its intended operation. Tests in a laboratory environment are solely for the purpose of demonstrating the underlying principles of technical performance (functions), without respect to the impact of environment.
- **Relevant Environment (nominally TRL 5)**: Not all systems, subsystems, and / or components need to be operated in the operational environment in

order to satisfactorily address performance margin requirements. Consequently, the relevant environment is the specific subset of the operational environment that is required to demonstrate critical "at risk" aspects of the final product performance in an operational environment. It is an environment that focuses specifically on "stressing" the technology advance in question.

- **Operational Environment (TRL 6 - 9)**: The environment in which the final product will be operated. In the case of space flight hardware / software, it is space. In the case of ground-based or airborne systems that are not directed toward space flight, it will be the environments defined by the scope of operations. For software, the environment will be defined by the operational platform.

Past experience is needed to make judgment calls on the actual TRL. Even with clear definitions, deciding if a prototype is really a prototype or an engineering breadboard can be challenging. Describing elements in terms of form, fit, and function based on design intent and subsequent performance helps in assessing the TRL.

A team should perform the TRL assessment. Team members can be systems engineers, users, and other stakeholders. They do not have to be discipline experts but they must understand the current state of the art technology. Some feel it should be well balanced and experienced. I feel the team should include young engineers, mathematicians, or scientists who will challenge the status quo. This will obviously require an open non-threatening environment. The questions, especially the obvious questions will stimulate learning and offer different perspectives.

There will be significant discussion when attempting to assign TRL numbers to the various system elements. Sometimes it helps to view the TRL elements from a different perspective. For example a series of questions that frame the TRL can be developed with the accepted set of definitions and terms.

Table 31 Assigning TRL Strategy

NASA TRL Assessment Questions		TRL
1. Has an identical unit been successfully operated / launched in identical configuration / environment?	If Yes	9
If NO then		
2. Has an identical unit in a different configuration / system architecture been successfully operated in space or the target environment or launched? If so, then this initially drops to TRL 5 until differences are evaluated.	If Yes	5
If NO then		
3. Has an identical unit been flight qualified but not yet operated in space or the target environment or launched?	If Yes	8
If NO then		

Table 31 Assigning TRL Strategy

NASA TRL Assessment Questions		**TRL**
4. Has a prototype unit (or one similar enough to be considered a prototype) been successfully operated in space or the target environment or launched?	If Yes	7
If NO then		
5. Has a prototype unit (or one similar enough to be considered a prototype) been demonstrated in a relevant environment?	If Yes	6
If NO then		
6. Has a breadboard unit been demonstrated in a relevant environment?	If Yes	5
If NO then		
7. Has a breadboard unit been demonstrated in a laboratory environment?	If Yes	4
If NO then		
8. Has analytical and experimental proof-of-concept been demonstrated?	If Yes	3
If NO then		
9. Has concept or application been formulated?	If Yes	2
If NO then		
10. Have basic principles been observed and reported?	If Yes	1
If NO then		
Rethink position regarding this technology.		

Question number two refers to reuse from heritage or previous systems. Apparently, from NASA's point of view, using a proven technology in a new environment or architecture is considered risky and the TRL rating drops significantly. Additional testing is needed for the new use or new environment. If the new environment is sufficiently close to the old environment or the new architecture sufficiently close to the old architecture then the resulting evaluation could be a TRL 6 or 7, but it is no longer at a TRL 9 rating.

To summarize the TRL assessment findings, a matrix is developed which lists the systems, subsystems, and components and then associates TRL findings with each of the items in the list. The columns identify the categories that are used to determine the TRL ratings. The columns are grouped into the maturity of the units, the environment the units have been operating in, and an assessment of the form fit and function.

The TRL ratings are rolled up in the hierarchy of the list. The TRL of a subsystem is driven by the lowest TRL rating of its components. The TRL of a system is driven by the lowest TRL rating of its subsystems. So the TRL of the system is determined by the lowest TRL present in the system even if it is a single element in a subsystem. Multiple elements at low TRL ratings are addressed in the AD2 process.

Integration affects the TRL of every system, subsystem, and component. All of the elements can be at a higher TRL, but if they have never been integrated as a unit, the TRL will be lower for the unit. How much lower depends on the complexity of the integration.

TRL Assessment Summary															
R Red = Below TRL 3 Y Yellow = TRL 3,4 & 5 G Green = TRL 6 and above W White = Unknown X Exists		Demonstration Units					Environment				Unit Description				
	Concept	Breadboard	Brassboard	Developmental Model	Prototype	Operation Qualified	Laboratory Environment	Relevant Environment	Operational Environment	Live Operations	Form	Fit	Function	Appropriate Scale	Overall TRL
1.0 System															
1.1 Common Console															R
1.1.1 FDDI Interface															
1.1.2 Sector Processor															
1.1.3 Display Processor					X				X		X	X	X	X	G
1.1.4 Display Monitor															
1.1.5 Touch Entry															
1.1.6 Voice Recognition							X				X	X			Y
1.1.7 Voice Synthesis		X													R
1.1.8 Sector Software															
1.1.9 Display Software															
1.1.10 Composite Console															
1.1.11 VSCS Panel	X														R
1.1.12 Overhead Maps															
1.2 Display Record Playback															
1.3 Radar Gateway															
1.4 Communications Gateway															
1.5 Mass Storage Device															
1.6 AERA Processors															

Figure 109 TRL Assessment Summary

Applying this process at the system level and then proceeding to lower levels of subsystem and component identifies those elements that require development and sets the stage for the subsequent phase, determining the AD2 levels.

24.1.1.2 DOD Technology Readiness Levels

The DOD TRL descriptions and ratings are very similar to the NASA TRL descriptions and ratings. The process used to determine the TRL rating is the same as described for the NASA TRL assessment[169].

---

[169] Technology Readiness Assessment (TRA) Deskbook, Department Of Defense, May 2005, July 2009. Systems Engineering Fundamentals, Supplementary Text, Defense

Table 32 Technology Readiness Levels - DOD

DOD TRL	Description
1. Basic principles observed and reported	Lowest level of technology readiness. Scientific research begins to be translated into applied research and development. Examples might include paper studies of a technology's basic properties.
2. Technology concept and/or application formulated	Invention begins. Once basic principles are observed, practical applications can be invented. Applications are speculative and there may be no proof or detailed analysis to support the assumptions. Examples are limited to analytic studies.
3. Analytical and experimental critical function and/or characteristic proof of concept	Active research and development is initiated. This includes analytical studies and laboratory studies to physically validate analytical predictions of separate elements of the technology. Examples include components that are not yet integrated or representative.
4. Component and/or breadboard validation in laboratory environment	Basic technological components are integrated to establish that they will work together. This is relatively "low fidelity" compared to the eventual system. Examples include integration of "ad hoc" hardware in the laboratory.
5. Component and/or breadboard validation in relevant environment	Fidelity of breadboard technology increases significantly. The basic technological components are integrated with reasonably realistic supporting elements so it can be tested in a simulated environment. Examples include "high fidelity" laboratory integration of components.
6. System/subsystem model or prototype demonstration in a relevant environment	Representative model or prototype system, which is well beyond that of TRL 5, is tested in a relevant environment. Represents a major step up in a technology's demonstrated readiness. Examples include testing a prototype in a high-fidelity laboratory environment or in simulated operational environment.
7. System prototype demonstration in an operational environment	Prototype near, or at, planned operational system. Represents a major step up from TRL 6, requiring demonstration of an actual system prototype in an operational environment such as an aircraft, vehicle, or space. Examples include testing the prototype in a test bed aircraft.
8. Actual system completed and	Technology has been proven to work in its final form

Acquisition University Press, January 2001. Defense Acquisition Guidebook, Defense Acquisition University, August 2010.

Table 32 Technology Readiness Levels - DOD

DOD TRL	Description
'flight qualified' through test and demonstration	and under expected conditions. In almost all cases, this TRL represents the end of true system development. Examples include developmental test and evaluation of the system in its intended weapon system to determine if it meets design specifications.
9. Actual system 'flight proven' through successful mission operations	Actual application of the technology in its final form and under mission conditions, such as those encountered in operational test and evaluation. Examples include using the system under operational mission conditions.

24.1.1.3 NATO Technology Readiness Levels

The North Atlantic Treaty Organization (NATO) TRL descriptions and ratings are very similar to the NASA TRL descriptions and ratings except that there is a new level added for basic research. So there is a conscious distinction between basic and applied research. Applied research is worthy of a TRL rating while basic research, although acknowledged, has no TRL value (i.e. the TRL is 0).

Table 33 Technology Readiness Levels - NATO[170]

NATO TRL	Description
0. Basic Research with future Military Capability in mind	Systematic study directed toward greater knowledge or understanding of the fundamental aspects of phenomena and /or observable facts with only a general notion of military applications or military products in mind. Many levels of scientific activity are included here but share the attribute that the technology readiness is not yet achieved.
1. Basic Principles Observed and Reported in context of a Military Capability Shortfall	Lowest level of technology readiness. Scientific research begins to be evaluated for military applications. Examples of Research and Technology (R&T) outputs might include paper studies of a technologys basic properties and potential for specific utility.
2. Technology Concept and / or Application Formulated	Invention begins. Once basic principles are observed, practical applications can be postulated. The application is speculative and there is no proof or detailed analysis to support the assumptions. Example R&T outputs are still mostly paper studies.
3. Analytical and Experimental Critical Function and/or Characteristic Proof of Concept	Analytical studies and laboratory/field studies to physically validate analytical predictions of separate elements of the technology are undertaken. Example

[170] Website, http://www.nurc.nato.int/research/trl.htm, Year 2010.

Table 33 Technology Readiness Levels - NATO[170]

NATO TRL	Description
	R&T outputs include software or hardware components that are not yet integrated or representative of final capability or system.
4. Component and/or Breadboard Validation in Laboratory / Field (e.g. ocean) Environment	Basic technology components are integrated. This is relatively low fidelity compared to the eventual system. Examples of R&T results include integration and testing of ad hoc hardware in a laboratory/field setting. Often the last stage for R&T (funded) activity.
5. Component and/or Breadboard Validation in a Relevant (operating) Environment	Fidelity of sub-system representation increases significantly. The basic technological components are integrated with realistic supporting elements so that the technology can be tested in a simulated operational environment. Examples include high fidelity laboratory/field integration of components. Rarely an R&T (funded) activity if it is a hardware system of any magnitude or system complexity.
6. System / Subsystem Model or Prototype Demonstration in a Realistic (operating) Environment or Context	Representative model or prototype system, which is well beyond the representation tested for TRL 5, is tested in a more realistic operational environment. Represents a major step up in a technologys demonstrated readiness. Examples include testing a prototype in a high fidelity laboratory/field environment or in simulated operational environment. Rarely an R&T (funded) activity if it is a hardware system of any magnitude or of significant system complexity.
7. System Prototype Demonstration in an Operational Environment or Context (e.g. exercise)	Prototype near or at planned operational system level. Represents a major step up from TRL 6, requiring the demonstration of an actual system prototype in an operational environment, such as in a relevant platform or in a system-of-systems. Information to allow supportability assessments is obtained. Examples include extensive testing of a prototype in a test bed vehicle or use in a military exercise. Not R&T funded although R&T experts may well be involved.
8. Actual System Completed and Qualified through Test and Demonstration	Technology has been proven to work in its final form and under expected conditions. In almost all cases, this TRL represents the end of Demonstration. Examples include test and evaluation of the system in its intended weapon system to determine if it meets design specifications, including those relating to supportability. Not R&T funded although R&T experts may well be involved.
9. Actual System Operationally	Application of the technology in its final form and under

Table 33 Technology Readiness Levels - NATO[170]

NATO TRL	Description
Proven through Successful Mission Operations	mission conditions, such as those encountered in operational test and evaluation and reliability trials. Examples include using the final system under operational mission conditions.

24.1.1.4 ESA Technology Readiness Levels

The European Space Agency (ESA) TRL descriptions and ratings are very similar to the NASA TRL descriptions and ratings. Instruments and spacecraft sub-systems are classified according to a TRL on a scale of 1 to 9. Levels 1 to 4 relate to creative, innovative technologies before or during mission assessment phase. Levels 5 to 9 relate to existing technologies and to missions in definition phase. If the TRL is too low then a mission risks being jeopardized by delays or cost over-runs.

Table 34 Technology Readiness Levels - ESA[171]

ESA TRL	Level description
1	Basic principles observed and reported
2	Technology concept and/or application formulated
3	Analytical & experimental critical function and/or characteristic proof-of-concept
4	Component and/or breadboard validation in laboratory environment
5	Component and/or breadboard validation in relevant environment
6	System/subsystem model or prototype demonstration in a relevant environment (ground or space)
7	System prototype demonstration in a space environment
8	Actual system completed and "Flight qualified" through test and demonstration (ground or space)
9	Actual system "Flight proven" through successful mission operations

24.1.1.5 FAA Technology Readiness Levels

The FAA has taken a slightly different path for the TRL ratings. The first observation is that there are fewer levels. The second is that they have been closely tied to their process, in this case it is associated with human factors. Also there is a concept that to move from one level to another, research products must meet a number of exit criteria.

[171] Website, http://sci.esa.int/science-e/www/object/index.cfm?fobjectid =37710, Year 2010.

Table 35 Technology Readiness Levels - FAA[172]

FAA TRL Description	Exit Criteria
1. Basic Principles Observed / Reported	Initial concept description is provided and is consistent with top-level Concept of Operations; benefits, risks, and research issues are identified.
2. Technology Concept and/or Application Formulated	Research management plan is delivered and FAA Research Management plan is delivered if applicable. Single year benefits assessment showing performance and economic benefits, preliminary safety risk assessment, and preliminary human factors assessment and research plan must be completed.
3. Analytical / Experimental Critical Function or Characteristic Proof-of-Concept	Initial Feasibility report is submitted showing capability is feasible from technical, benefits, safety, and human factors perspectives. Initial analytic or experimental quantification of technical performance metrics shows improvement over baseline.
4. Component and / or Integrated Components Tested in a Laboratory Environment	Research demonstrates capability is feasible from safety, human factors, and development perspectives, and expected benefits outweigh costs based upon human-in-the-loop testing with representative potential users. A FAA baseline Concept of Use for the capability is developed.
5. Components and / or Subsystems Verified in a Relevant Environment	Pre-development prototype is developed and evaluated in a high fidelity environment. This could involve a full mission simulation in a laboratory or a demonstration or test in a field setting. Specifications and design documentation are updated based upon lessons learned in testing. An updated report documents capability feasibility from safety, human factors, and development perspectives and summarizes what has been learned to date. R&D organization continues research on as-built prototype while FAA begins acquisition program baseline definition.
6. System Demonstrated / Validated in a Relevant Environment	Field evaluations demonstrate technical functionality of prototypes, benefits, and resolution of human factors issues. FAA and research organization review capability to determine its readiness to transfer to development organization. An acquisition strategy is required and a development contractor is engaged.

[172] This table and text following this table comes from DOT/FAA/AR-03/43, FAA/NASA Human Factors for Evolving Environments: Human Factors Attributes and Technology Readiness Levels, Human Factors Research and Engineering Division, FAA, April 2003.

TRL 1: Basic Principles Observed / Reported is the stage at which an Air Traffic Management (ATM) concept is initially identified and described. The appropriate development group analyzes a deficiency or need in the National Airspace System (NAS) for which the capability may be a solution. During this TRL phase, the following should occur: development of an initial operational concept, completion of a trade/risk/benefit analysis, and identification of research issues. A key concern during this stage is the viability of the operational concept.

The Free Flight Research Program Plan notes that the research organization proceeds relatively independently and coordinates research activities with the FAA during TRLs 1-3 via the Interagency Air Traffic Management Integrated Product Team (IAIPT)[173].

TRL 2: Technology Concept and / or Application Formulated is the stage at which a detailed research plan is developed that provides a definition of the technical solution to the deficiency and identifies critical feasibility issues. The plan describes activities, schedule, likely facilities, and resources required to address research issues in tool development. Human factors research issues including human effectiveness are also identified in this research plan as well as resources necessary to resolve these human factors issues.

TRL 3: Analytical / Experimental Critical Function or Characteristic Proof-of-Concept is the stage during which a conceptual prototype of the tool is developed with initial requirements defined. Initial laboratory evaluations may include part-task computer-human interface (CHI) evaluations, preliminary procedures development, functionality testing, and performance evaluations. The proof-of-concept conceptual design should consider the relationships of roles and responsibilities with the conceptual design and architecture.

To successfully exit from this TRL, initial research should show the tool to be feasible from technical, benefits, safety, and human factors perspectives based on research to date. Also, initial quantification of technical performance metrics should reflect an improvement over the baseline, or if improvement cannot be demonstrated, the cause is understood and improvement is expected at some point.

TRL 4: Component and/or Integrated Components Tested in a Laboratory is the stage during which a research prototype is developed and evaluated by representative potential users. Evaluations may consist of medium-fidelity human-system interface evaluations, procedures evaluation, human performance evaluations, functionality testing, and performance evaluations. The laboratory real time simulation environment is at a higher fidelity level than at TRL 3 using standalone or integrated components. The research organization and the FAA

[173] In September 1995, FAA and NASA formed the IAIPT to plan and conduct integrated research related to air-based and ground-based Air Traffic Control (ATC) and Air Traffic Management (ATM) decision support tools and procedures. FAA, NASA, and research partners selected a model of Technology Readiness Levels (TRL) for coordinating various activities, roles, and expectations.

participate in the laboratory and, if appropriate, site evaluations such as a shadow mode or back room test at a field site with user teams. Exiting from this TRL requires development and baselining of an initial FAA Concept of Use for the capability. The cost/benefit analysis is updated, as appropriate.

When TRL 4 evaluations are complete, the research team generates an updated feasibility report describing technical progress, life-cycle cost/benefits indicated, current safety and human factors status, and issues that might require a "return to the drawing board." The updated feasibility report is intended as an executive summary of what the team has learned to date and should be more reflective of the operational environment in which the capability is expected to operate.

Formal acquisition involvement begins at TRL 4 as design and architecture requirements are refined. Assessments of procedures along with roles and responsibilities become more robust relative to the range of conditions afforded in the laboratory environment.

TRL 5: Components / Subsystems Verified in a Relevant Environment is the stage during which a pre-development prototype of the tool is developed and evaluated. The evaluation environment should be at a high fidelity such as can be achieved through a full-mission simulation platform or through demonstration in the field with an integrated architecture for representative normal and off-normal traffic conditions. Based upon lessons learned, specifications and design documentation are updated during this TRL.

Also during this TRL, the FAA focuses on activities to prepare for acquisition. The FAA acquisition office forms user teams to address user inputs, specifications, maintenance concepts, concept of use issues, and human factors concerns. FAA will begin to develop contractual documentation such as statements of work and contract data requirements lists.

The research organization will continue on the as-built prototype system while the FAA begins its acquisition program baseline definition. It is FAA policy that research prototype development and evaluation should adhere to the same fundamental paradigm as a full-scale system acquisition that includes early user involvement. The FAA uses prototyping activities in this vein to assist in assessing alternative solutions to an identified mission need.

TRL 6: System Demonstrated / Validated in a Relevant Environment is the stage at which an operational demonstration of the pre-production prototype system is conducted in a FAA field facility if deemed feasible and necessary. Field evaluations may include a substantial demonstration of the prototype's functionality, and could involve a daily use version of the pre-development operational software application. Comprehensive human factors assessment for the prototype capability should be completed to exit TRL 6. The research organization focuses on completing the technology transfer and cost/benefits activities based on data obtained during the field test. Final system engineering documents are produced including system specifications, interface requirements, and design descriptions. It is critical that when operational demonstrations and field evaluations are conducted they involve FAA operational personnel.

During TRL 6, the final high fidelity, integrated system demonstration of the transfer prototype is accomplished, using a large variety of traffic nominal and off-nominal conditions. Documents produced at earlier TRLs will be finalized for formal transfer to the FAA.

Table 36 TRL Human Factors Considerations - FAA[174]

TRL	Pipeline Output	HF Component
1. Basic Principles Observed / Reported	ATM concept initially developed and described	Initial identification of human factors issues
2. Technology Concept and / or Application Formulated	Detailed research plan developed	Preliminary human factors assessment and research plan: - Prioritize human factors research issues - Identify the activities, schedule, and resources required to resolve identified human factors issues
3. Analytical / Experimental Critical Function or Characteristic Proof-of-Concept	Conceptual prototype of tool developed and evaluated; initial feasibility report developed	Address human factors issues identified in TRL 2
4. Component or Integrated Components Tested in a Laboratory Environment	Research prototype developed and evaluated; initial FAA Operational Concept of Use developed and baselined	Assess human factors issues associated with the concept to show how they have been resolved; document human factors research; update human factors plan
5. Components / Subsystems Verified in a Relevant Environment	Pre-development prototype of tool developed and evaluated	Resolve human factors issues from the TRL 4 update
6. System Demonstrated / Validated in a Relevant Environment	Operational demonstration of the pre-production prototype system	Collect human factors data to show that all issues have been addressed and that operations are practicable in nominal and off-nominal conditions

24.1.2 Advancement Degree of Difficulty Assessment

Once there is a TRL assessment the next step is to determine how to move the TRL to the next levels. The Advancement Degree of Difficulty Assessment (AD2)

[174] DOT/FAA/AR-03/43, FAA/NASA Human Factors for Evolving Environments: Human Factors Attributes and Technology Readiness Levels, Human Factors Research and Engineering Division, FAA, 2003.

has been suggested[175].

Initial AD2 provides information needed to develop preliminary cost and schedule plans and offer preliminary risk assessments. Detailed AD2 provides information to build a technology development plan in a process that identifies alternative paths, fallback positions, and performance de-scope options. Once an effort is established, the AD2 information is also used to prepare milestones and metrics for subsequent Earned Value Management (EVM) for the project or program.

AD2 is a very tricky subject. On the one hand there is the desire to create a scale and somehow equate that scale to time, cost and risk. On the other hand we know that some technologies can move from TRL 1 to TRL 9 very quickly. The following is an example of a technology that moved from Level 1 to Level 9 in the span of 12 months with little cost.

Example 38 Traceability to the Line of Code

In the late 1990's a process and tool was developed to trace software requirements to the line of code and output the requirements as log data during live testing[176]. This was done to support certification of mission critical systems. The original technology was developed for an E-commerce website to track how and why visitors arrive.

The technology was quickly modified adapted and matured for a pilot program. When it was clear that the technology significantly benefited the pilot program it was successfully rolled out to other programs. This was accomplished within 12 months with little cost.

Each program had their own unique needs and even though a "product model" was used for this technology, the technology or product model was expanded to include new needs. For example on the pilot program the largest test generated a maximum of approximately 1000 log entries representing requirements allocated to the line of code. The second program had tests yielding in excess of 30,000 log entries. The third program had tests yielding in excess of 20 million log entries.

Obviously the technology needed to advance to address the correlation and visualization challenges of the new programs. Without this advancement the verification strategy would have failed with significant cost and schedule impacts.

The big picture was to create a mechanism that would transparently add traceability down to the line of code in software. However no one wants to pay for technology using their budget. In this case there was a strong need to show this traceability to shorten the traditional certification cycle. So there was a desire to investigate the technology and try to make it work.

The details were in the implementation. They included using the new Internet

---

[175] NASA/SP-2007-6105 Rev1, NASA Systems Engineering Handbook, December 2007.

[176] Walt Sobkiw developed this, your Author.

technologies of a web server, browser, and PERL to parse and process the software text. They also include the details of the System Requirements Database and the interface to the mechanism that implemented the traceability to the line of code. The details also included modifications to the process such as code walk through to ensure the traceability was accurate and the test environment to allow for the automated output of the log data.

It was estimated that it took approximately 500 hours of computer execution time to mature the PERL implementation. This was in a time when 10 MIP personal computers just surfaced. Just one year earlier and this approach would have proved unfeasible because the 500-hour execution time would have converted to 5000 hours. Perhaps the 5000 hours could have been shortened with the addition of more computers and people however the program would have then rejected the idea and just done things the old way, even though it would have required significant time and resources. So there are many lessons in this example:

- No technology before its time. Some technology's take long times to mature others are very fast.
- Shifting from one domain or application may be less risky than some might suggest.
- It is important to constantly consider new technologies and not be afraid of the initial low TRL rating.
- New technologies work only when the details are addressed but it is important to watch the key gates and decide to cut the losses if the technology gets stuck and is unable to mature.

The AD2 descriptions and rating are the inverse of the TRL ratings. They can be stated as generic tasks that must be performed to move from the current TRL to the next TRL. Obviously the resulting technology plan will contain the details specific to the technology. Each of the AD2 ratings can be given a number and an arbitrary development risk can be applied to each AD2 level.

We see that different organizations have slightly different TRL ratings. In this text the AD2 ratings are numbered 1 - 9 to track to the corresponding TRL ratings. Other approaches may use a reverse AD2 rating scale or a shortened AD2 scale[177].

What is proposed is that the AD2 ratings should be the inverse of the TRL ratings that are eventually adopted by the organization. In other words if a TRL item addresses a deficiency the AD2 item describes the generic activity to address the deficiency.

Some organizations have more TRL ratings and others have less. What matters

---

[177] John C. Mankins proposed 5 degrees of difficulty in a White Paper Research & Development Degree Of Difficulty (R&D3) March 10, 1998, Advanced Projects Office, Office of Space Flight NASA Headquarters.

is that each organization adopts a standard scale and tries to adhere to the scale so that meaningful comparisons can be made as the portfolio increases. If the scale changes then a translation map should be offered until the transition to the new scales are completed and the organization has internalized the new scales.

There are general maturing activities that are associated with each AD2 level[178]. The idea is that you perform these activities to move to the next AD2 level. These activities translate into a potential risk level.

The risk associated with an AD2 item and its activity is arbitrary until the organization develops a significant history at which time a non-linear distribution might be assigned to the AD2 ratings. In this text the risk is uniformly applied across the AD2 items with a jump at AD2 rating 4.

Table 37 AD2 Level and General Maturing Activities

AD2 Level and General Maturing Activities	Status	Risk
1. Requires new development outside of any existing experience base. No viable approaches exist that can be pursued with any degree of confidence. Basic research in key areas needed before feasible approaches can be defined.	Chaos RED	90%+
2. Requires new development where similarity to existing experience base can be defined only in the broadest sense. Multiple development routes must be pursued.	Unknown Unknowns RED	80%
3. Requires new development but similarity to existing experience is sufficient to warrant comparison in only a subset of critical areas. Multiple development routes must be pursued.	Unknown Unknowns RED	70%
4. Requires new development but similarity to existing experience is sufficient to warrant comparison on only a subset of critical areas. Dual development approaches should be pursued in order to achieve a moderate degree of confidence for success. (desired performance can be achieved in subsequent block upgrades with high degree of confidence.	Unknown Unknowns RED	60% normally 50%
5. Requires new development but similarity to existing experience is sufficient to warrant comparison in all critical areas. Dual development approaches should be pursued to provide a high degree of confidence for success.	Known Unknowns Yellow	40%
6. Requires new development but similarity to existing experience is sufficient to warrant comparison across the	Well Understood	30%

178 Based on white paper Systematic Assessment of the Program / Project Impacts of Technological Advancement and Insertion, James W. Bilbro George C. Marshall Space Flight Center, December 2006, with acknowledgements: Dale Thomas, Jack Stocky, Dave Harris, Jay Dryer, Bill Nolte, James Cannon, Uwe Hueter, Mike May, Joel Best, Steve Newton, Richard Stutts, Wendel Coberg, Pravin Aggarwal, Endwell Daso, and Steve Pearson.

Table 37 AD2 Level and General Maturing Activities

AD2 Level and General Maturing Activities	Status	Risk
board. A single development approach can be taken with a high degree of confidence for success.	GREEN	
7. Requires new development well within the experience base. A single development approach is adequate.	Well Understood GREEN	20%
8. Exists but requires major modifications. A single development approach is adequate.	Well Understood GREEN	10%
9. Exists with no or only minor modifications being required. A single development approach is adequate.	Well Understood GREEN	00%

24.1.3 The Big Picture

It looks like no technology is ready until it flies in space or in a real operational setting. Makes sense, it is a reasonable expectation. It also looks like there are stepping stones that need to be touched as a technology is matured. That also makes sense. Bypass one of the stepping stones and you may not capture a cost schedule box that is needed to make the next step work. It also makes sense from the point of view of small goals small cost schedule risk. If you can make it through a gate or a stepping stone then you should try the next stepping stone or gate. If however you try to leapfrog, you may never know that the first stepping stone contains showstoppers that will not allow the technology to mature without decades of work.

There is another perspective that might make planning easier. Think in terms of prototypes that evolve to a final real world solution:

1. Proof Of Concept Prototypes
2. Research And Development Prototypes
3. Engineering Prototypes
4. Pre-Production System
5. Staging Area Production System
6. First Field Production System
7. Field Proven System

Each prototype has their cost and schedule estimates. If progress is good, then all is well. If the team gets stuck at one of the levels then a decision needs to be made, stop the effort, or change direction. The trick is to not lie about the function performance of each stepping stone. In the end the field will reject the system, after great cost and time, if the team was in denial and missed the true technology needs of the system.

The devil is in the details. In the end a technology will only surface and mature because people with great perseverance focus on the details to make everything work elegantly. If no one is focused on the details, or if they do not have the authority to focus on the details, the technology development will fail. At the same time there needs to be people looking at the big picture and willing to think outside the box to determine if a new direction is needed. It's possible the current approach where everyone is focused on the details may be a dead end.

24.2 Informal Technology Assessment

The following is an example of an informal approach to technology assessment that was used to support the upgrade of an Air Traffic Control Simulation Lab.

Example 39 Informal Technology Assessment ATC Simulation

In 1979 the Air Traffic Control Simulation Facility (ATCSF) at the National Aviation Facilities Technical Center (NAFEC)[179] was interested in replacing its old computers and Air Traffic Control (ATC) displays[180]. The computers were Xerox Sigma 5 and Sigma 7 models. The displays were ITT with a custom in-house display processor.

The original approach was to focus on the Xerox computers and try to salvage the huge investment in the simulation software available from the rather large tape library room. Telefile advertised a plug-compatible computer that would execute the existing Xerox sigma software. So that was the desired approach by many in the lab. As you will see when you review the section on Technology Readiness Levels (TRL), the Telefile approach was obviously a relatively reasonable TRL. What unfolded was something very interesting.

A computer study was initiated. It included a literature search, vender search, and consulting from SRI[181]. This immersed a small team tasked with the upgrade in the latest computing technology. Some of the featured computers and architectures were the Xerox Star system, various mainframe computer vendors, a few mini computer vendors, and a new collection of companies offering a new category of computers called super mini computers. Eventually the System Engineering Labs (SEL) 32 series computers were selected to replace the Xerox computers.

This opened the door for addressing the ATC display consoles. They needed to be interfaced to the new computers, not a trivial task. Also they really did not represent the ATC workstations as found in the field. A similar activity was started to investigate various display alternatives. One of the alternatives was to piggyback

[179] Now referred to as FAA Technical Center (FAATC).

[180] This is experience is recalled by Walt Sobkiw, your author.

[181] Stanford University established Stanford Research Institute in 1946. SRI became independent of the university in 1970, and changed its name to SRI International in 1977. SRI is a non-profit organization.

on the much larger ETABS[182] program. ETABS was a new distributed approach to display processing where each workstation was allocated its own display processing subsystem. Eventually the Sanders Graphic-7 Display Processor was selected as the display processing alternative. The thought was that ETABS had already paid for much of the development, which included the maturing of the Graphic-7, and the sheet metal associated with the console physical structure.

For the ATCSF, the displays would be interfaced to the SEL computers using either RS-232/422 point to point interfaces, a custom bus from Sanders, or Ethernet from the Xerox Star investigations. The initial approach was to use the low cost RS-232/422 point to point interfaces.

Obviously this pushed the lab TRL ratings into the red areas. However within 18 months a lab had appeared with the new displays interfaced to the new computers and test data was passing across the interfaces[183].

This is an important lesson when performing technology assessment. At the time the staff knew they needed to move forward. There was significant technology movement in computers, software, and ATC displays. The old system was using keypunch cards. The new system used multi-user multitasking programming consoles with an early version of UNIX. This was a huge leap. The same was true of the ATC workstations and the distributed Graphic-7 Display Processors.

The approach taken was to use the term "proven state-of-the-art" technology. State-of-the-art implied new while proven was defined by the team as something that was commercially available for a year. Interesting logic because something can be commercially available for a year with no sales. The idea of selling a few items was factored into the final decisions.

So a unique TRL rating was defined. That TRL gate once compared to the current TRL ratings resulted in a system approach that today might be considered too high of a risk.

24.3 Technology Maturity and Adoption

Every existing system consists of different technologies. Every new system will consist of different technologies. Technologies have a life cycle that represents technology maturity.

In the beginning new technologies are complex, difficult to manage, and have few practitioners. In the end a technology is old overcome by other technologies. Some might assume that because of its age it is easy to manage. However, as many practitioners have moved on to newer technologies there are fewer practitioners

---

[182] Electronic Tabular Display System (ETABS), which would electronically display the paper flight strips used in the current system.

[183] At this point I left for Hughes Aircraft. I am not sure if the new hardware was ever populated with new simulation software. I assume that the system eventually did start to offer simulation services.

making it difficult to manage and complex because of missing information.

Technology adoption occurs in a modified S curve[184] as suggested by the logistics curve and the diffusion and innovation theory[185]. Rogers using more of a bell shape groups the technology curve by Innovators, Early Adopters, Early Majority, Late Majority, and Laggards. These adopters are associated with five stages of technological maturity starting with Bleeding Edge, Leading Edge, State-of the-Art, Dated, and Obsolete.

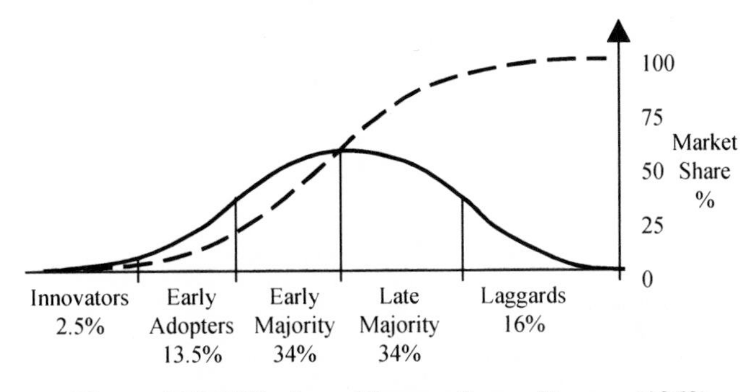

Figure 110 Diffusion of Innovations - Rogers (1962)

Table 38 Technology Adopters and Maturity Levels

Adopters	Numbers	Maturity
1. Innovators	2.5%	Bleeding Edge
2. Early Adopters	13.5%	Leading Edge
3. Early Majority	34%	State of The Art
4. Late Majority	34%	Dated
5. Laggards	16%	Obsolete

24.4 Technology Assessment and Sustainability

There is a huge caution that needs to be addressed with technology assessment. Organizations will try to minimize risk and in the process no new technology will be introduced or considered. In this case it is thought that the competition will remove these organizations as the technology maturity and adoption rates follow their normal course.

However what happens if the adoption rate drops across the board in an industry

[184] Pierre François Verhulst (1845) studied the logistic curve in relation to population growth. The S-shaped curve of the initial stage of growth is approximately exponential, as saturation begins the growth slows, and at maturity growth stops.

[185] Technology not initially exponentially adopted, Textbook: Diffusion of Innovations, Everett Rogers, 1962.

or a nation state?

In many ways organizations have an obligation to push the technology envelopes, just in case there are competitors, but more importantly for a sustainable company, industry, world. So what is a reasonable distribution of TRLs within an organization and how do they relate to different project types?

Just because the TRL is very low it does not mean that the organization is engaged in a healthy research and development activity. The organization might actually be attempting to adopt technologies that other organizations and industries may have abandoned because of obsolescence. One way to prevent this false view of technology is to link the technology maturity level to the TRL. If the TRL reasonably tracks the maturity then the activity is reasonable. There is nothing worse than working with dead technology.

Table 39 Project Types and TRL

Project Types	TRL	Maturity
Basic Research	0 or 1	1
Applied Research	2 or 3	1
Advanced Research	4 or 5	1
Advanced Technology Demonstration	6 or 7	1 or 2
Development Program	8 or 9	1, 2, or 3

24.5 Technology Prediction

Systems are in development and used for several years even if the initial capability might be offered in 6 months. These systems use technologies that are moving. Some are in their infancy, some are maturing, others are being overcome and becoming obsolete. The issue is to understand the maturity of these technologies and make reasonable predictions of where various technologies will be with the initial system delivery, its mid life operations, and its final stages of operations. This requires developing strong relationships[186] with the technology leaders to gain visibility into where they think they will be in 1, 3, 5, and 10 years.

Example 40 Technology Prediction - ATC Workstation

In the early 1980's Hughes Aircraft was developing the Common Console for the Advanced Automation System (AAS) air traffic control workstation. This work was also closely tied to their existing air defense systems. There were three technologies being tracked and evaluated[187] circa 1982:

- **Embedded Computer Processing:** Bit slice processing, Intel 80186,

[186] Non disclosure agreements enable communications between entities.

[187] The author worked on various aspects of AAS including the common console.

Motorola 68010. The choices after visiting the appropriate organizations moved towards 80286 and 68020.

- **Display:** Stroker monochrome, beam penetration color stroker, raster, plasma, liquid crystal display (LCD), three CRT projection, and liquid crystal light valve projection.
- **Human Interaction:** Touch entry (capacitive, resistive, and perimeter light emitting diode light interruption with pen or finger), variable function keys using touch entry, voice synthesis, and voice recognition.

The following summarizes the results of the technology prediction and management:

- **Embedded Computer Processing:** The choice eventually narrowed on Motorola because of processor performance and memory access.
- **Display:** The user goal was to push the display to a 20-inch square 2000 X 2000 monitor to match what was perceived using the existing round 19-inch stroker technology. The choice eventually narrowed on the highest risk color raster approach. This pushed several technologies but satisfied the user goal of a 20-inch square 2000 X 2000 color display. The manufacturing process needed to support the building and mounting of a 2000 X 2000 shadow mask. The current approach caused the shadow mask to evaporate from high heat as a laser burned the holes in the shadow mask (i.e. too many holes). There were also challenges associated with physically stabilizing the large shadow mask in the glass tube. The large glass CRT bottle originally weighted 2 tons, because of the large amount of glass needed for the corners. The deflection electronics needed to move the beam across a very large area and at a very large resolution, leading to large power and high frequency deflection amplifiers. These technologies were pushed and stabilized and a 20-inch color raster display with a 2000 X 2000 resolution was developed[188].
- **Human Interaction:** Touch entry was always plagued with grease smeared screens or lost pointing devices. Voice recognition and synthesis was found to be in channel conflict with the operators normal voice communications with pilots and other controllers. So these technologies were used only on a case by case basis.

---

[188] This work can be credited to Augie and Bob at Hughes Aircraft Fullerton, Ca, they were determined to make this happen, visiting Corning Glass, Radio Corporation of America (RCA), and other relevant companies. Art created the first deflection amplifiers for the new 20-inch square monitors.

24.6 Function and Performance Allocation

At some point functional requirements and performance budgets need to be allocated to the various system elements. These allocations need to be based on sound analysis. It makes no sense to allocate a budget such as weight to an element if that allocation exceeds the known available products or worse the technologies for that element, and that portion of the weight budget can be easily accommodated by another element.

So technology impact is a key issue that needs to be considered when decomposing and allocating function and performance. This means that someone must always be cognizant of the technology impacts throughout the system engineering effort especially during the architecture development phases. As architecture based on an imbalance of technology maturity or worse needlessly excessive technology immaturity must be avoided.

24.7 Misuse of Technology

Misuse of technology is an interesting topic. There are different dimensions that should be known to the stakeholders. Most will focus on the simple philosophical case of technology being used for good or bad activities. If it is used for bad activities then it is a misuse of technology. However there are other dimensions to misuse of technology. The following are other perspectives on technology misuse. At what point is technology in a system being misused?

- The technology adds no value to a system
- The technology actually hurts the system but the stakeholders are unaware of the damage
- The technology hurts the system, the stakeholders are aware of the damage, but are powerless to stop using it

- The technology is prematurely abandoned
- The technology is stopped because of status quo
- The technology is viewed as a commodity

- The technology only appeals to the ego
- The technology is poorly understood
- The technology is not sustainable

At what point does misuse of technology translate into deviant design practices? By definition systems practices should prevent deviant design practices and misuse of technology. Some deviant design practices and misuse of technology are:

- Wasting peoples time by forcing useless actions especially for ulterior motives

- Fooling or deceiving people into taking undesired actions especially when the user is harmed
- Manipulating biological systems to mask poor system performance while doing harm to the biological system (e.g. using blue background to fuzz out poor resolution displays causing eye fatigue as it continuously focuses and defocuses)
- Manipulating biological systems to mask poor system performance even though there is no harm
- Consciously reducing performance even though the technology supports much higher levels of performance at no additional cost
- Reducing performance to allow large quantity but undesired low quality content when performance can match technology with a reasonable quantity level of high quality content
- Ignoring scientific knowledge and previous experience especially as related to human factors
- Using technology to compromise inalienable rights

Example 41 Misuse of Technology

When we look at a system from a functional or requirements point of view we see that there are root functions and requirements. From the roots other elements are identified via decomposition until we reach a natural limit where no more decomposition is possible. These are the leaves or primitives. We also see that there are core functions and requirements. Without this core, the system does not exist.

So if inalienable[189] rights is a core root requirement where does the decomposition lead? The most famous first level decomposition is "life, liberty, and the pursuit of happiness" from the United States Declaration of Independence. Are there other decompositions from the key requirement of technology not compromising inalienable right? The following is a possible list:

- Technology that does not violate peoples inalienable rights to privacy. This is interesting because inalienable or natural rights are not dependent on cultural or government laws, customs, or beliefs.
- Technology that does not knowingly harm people. Knowingly harm is based on some human in some setting knowing the technology can cause harm to the uninformed or misinformed.
- Technology that does not place a significant burden on a people or society. This is where people become slaves to the technology regardless of benefit.

[189] Inalienable rights are self evident and universal.

It is most egregious if the benefit is small or nonexistent.

Author Comment: Much of world has shifted to digital machines. In the beginning digital quality was viewed as the primary driver to displace analog alternatives. So maximum digital technology was applied to offer services approaching analog technology.

Once analog was displaced, quality became irrelevant and revenue became the primary driver. The revenue is a function of transmission bandwidth and storage needs. So sound and images are compressed to such as high level that they are almost intolerable.

The digital movement has devastated television images produced on spectacular equipment of the last century. Fast moving images pixilate until the movement stops. Stationary images have compression artifacts surrounding facial features like the human eye. All facial detail such as subtle color, texture, lines and bumps that are natural and expected on a human face are lost. Even the image format has changed where striping or black lines are added to the top and bottom of visual content even though the original content contained the top and bottom images. There is also a false case made against home viewing, which is significantly different than theatre viewing, so that even though pan and scan alternatives exist they are not used to reduce the bandwidth, or worse they are striped.

Music recorded in spectacular recording studios is compressed to such a level as to lose all soul moving content. Even traditional analog voice grade telephony has been compressed on cell phones to such a level that words must be repeated.

This is not a technology limitation nor is it even a cost issue. The increased revenue streams associated with cutting the bandwidth and storage needs are insignificant compared to other costs[190]. However because the system permits this to happen, system managers have done this egregious act. It is an example of a gross misuse of technology. It is also unethical, as new generations of users not aware of the alternatives and tradeoffs are forced to use these systems in oligopoly settings.

There is an old axiom: those that do not understand a technology cannot manage it and those that did not develop a technology cannot understand it. So this is a complex problem moving forward. If we do not have people that understand a technology to effectively manage it, how will we sustain our civilization?

24.8 Technology Infusion - Push or Pull

At one time technology infusion was not a difficult political task. It was

---

[190] This has happened in other settings such as aviation where airlines planned to charge a fee for using toilet facilities circa 2009. This revenue stream is insignificant and only adds significant overhead and brutality to the society. These system managers should be removed because they are wasting everyone's time and distracting from what actually needs to be done to keep a complex system sustainable.

accepted that technology was good and more technology would lead to more good. It was also thought that technology was precious and difficult to mature and apply where needed. However, in the last two decades of the previous century some have come to think that technology is a commodity to be enabled and disabled when it makes financial sense. The reality is technology can and has been lost and that form of thinking will lead to an unsustainable world. Our world relies on technology for growth and stability. Technology is very precious.

Technology either can be pushed into an organization or pulled by an organization. So technology has a characteristic associated with user acceptably.

If the originators of the mission concept, the system stakeholders, identify the technology as needed, then it is pull.

If technologists surface a technology that is a potential solution to a system or enables an entirely new concept then it is push.

Once system stakeholders accept it, it becomes pull. It remains pull until it is either successfully integrated into the mission architecture or rejected as inapplicable or unsuccessful. So the trick is to convert technology acceptance into a pull activity.

Example 42 Traceability to the Line of Code Revisited

The previous example of Traceability to the Line of Code was initially a push technology effort. After the success of the first program it immediately changed to a pull technology effort. All future programs faced with the same challenge requested the technology and did what was needed to make it fit their needs. However the first program was an early adopter with members not afraid of risk and driven by vision and innovation. The other programs were more risk averse and less idealistic.

Example 43 Technology Infusion - FAA and AAS

In the late 1970's the FAA had a number of technologies that were ready to move from the laboratory into operational settings. They were associated with the ground and cockpit environment:

- Discrete Address Beacon System (DABS)
- Electronic Tabular Display System (ETABS)
- Color Stroker Plan View Displays

All three technologies were rejected as a new program called the Advanced Automation System (AAS) was created. AAS was supposed to incorporate the ETABS and color display work. The problem is AAS failed and with it the introduction of color displays and electronic tabular display of information was delayed by years. The DABS program eventually resurfaced as Mode-s and was eventually fielded.

So the big questions are why the delay in introducing these new technologies into Air Traffic Control (ATC) and what impact did these delays have? The simple answer is the user community was able to keep the current system operational within a reasonable performance envelope. The old system just continued to work. The more complex answer is politics, money, and entrenched interests.

New technologies are very disruptive to the status quo. That was true in this case but the status quo were in a better position to influence decisions in Washington D.C. than the smaller companies closer to the new technologies. As long as the current system was not leading to a visible crisis, there was no need to change as perceived by the non-technologist stakeholders.

If we believe that technology is not a commodity that can be created at anytime, then the delay of introducing these technologies to the ATC community probably caused irreparable harm. There was a technology growth path that was naturally evolving and it was disrupted and stopped for the air traffic control application.

However something interesting happened with the AAS program. The FAA is one of three government organizations in the U.S. that have taken the time to develop technology readiness levels. Research and development is part of the FAA as stated in the Federal Aviation Act of 1958 (see Appendix).

Some have suggested AAS was a failure because it was never fielded in the U.S. They suggest that AAS was a pure technology effort when technology was not the current need. However, Hughes Aircraft fielded AAS derivatives in Canada, the Pacific Rim, and other countries outside the US. AAS also pioneered many of the technologies we take for granted in our world today. They are:

- Local Area Networks
- Distributed Computing and Workstations
- Graphic User Interface (GUI) Windows Based Displays
- UNIX which hosted the first Internet Web Servers

It's unclear what level of funding is attributed to AAS for maturing these technologies that were adopted around the world, but there is no question AAS was stimulating these areas. Its possible that without AAS the Internet may have been delayed by several years as other funding sources would have been needed to help mature these technologies.

So other countries received ATC systems with newer technology than the US and the entire computing world benefited from the spin-offs of AAS[191]. Interesting turn of events.

---

[191] Hughes Aircraft viewed themselves as a technology company rather than a product company. They worked with industry to mature technology as needed. Their air defense and ATC systems were worldwide. These systems benefited from knowledge gained on AAS. This included the new workstation display, distributed processing, and LAN capabilities.

24.9 Organization Longevity and Research & Development

Organizations need to have some level of research and development to feed the pipeline so that in later year's systems can be fielded. Just because innovators arrive, it does not mean that there always will be early adopters. The solution could be abandoned by another approach or overcome by other technologies. There could be a failure at each stage of the maturity curve where a solution is displaced. The maturity curve in many ways represents only the most successful solutions, products, or systems. It is an idealized representation. A solution can even fail at the Innovators stage.

Table 40 Research and Development Success Scenarios

Adopters	No	Maturity	Scenarios			
1. Innovators	2.5%	Bleeding Edge	S	S	S	S
2. Early Adopters	13.5%	Leading Edge	F	S	S	S
3. Early Majority	34%	State of The Art	-	F	S	S
4. Late Majority	34%	Dated	-	-	F	S
5. Laggards	16%	Obsolete	-	-	-	F
		Total Market Reached	2.5%	16%	50%	84%

An organization can be evaluated for technology maturity. The technology maturity evaluation consists of the process, tools, and product for each product, product line, or system. This information can be used to visualize the remaining market and compare it with assessments from marketing staff. At some point a product, product line, or system needs technology infusion for it to become viable from a business point of view.

Table 41 Organizational Technology Maturity Assessment

Business	Product	Process	Tool	Product	Remaining Market
Area 1	product 1	Leading	Dated	Bleeding	100%
	product 2	Bleeding	Leading	Leading	84%
	product n	Obsolete	Obsolete	Obsolete	16%
Area 2	product 1	Dated	Obsolete	State of Art	50%
	product 2	Bleeding	Leading	Dated	16%

A product, product line, or system can be evaluated for Technology Readiness Levels (TRL) to support market expansion goals. The TRL evaluation consists of determining the TRL of each product for different domains such as air, land, sea, space or military, industrial, commercial. This can be used to determine the level of difficulty of moving to a new customer base.

For example a product may be considered obsolete with one customer base but state of the art for another customer base. For example a product that is obsolete for a Space based system may be considered state-of-the-art for a commercial application. Alternatively a commercial product may be considered state-of-the-art but leading edge for a military based system. The trick is to determine how to commercialize a Space based product or to military harden a commercial product.

When assessing a product for application in a new domain there are always key requirements that need to be addressed. For example sea applications need to protect against salt fog. Space applications need to protect against radiation. Air applications need to protect against vibration. Land applications need to protect against mud and rain. Air traffic control needs to be very fault tolerant, while military and industrial applications may be fault tolerant. Removing fault tolerance is easier than building in new levels of fault tolerance.

Table 42 Organizational Technology Readiness Levels Assessment

Domain	Prod 1	Prod n
Space	5	9
Sea	9	8-5
Air	8-5	8-5
Land	8-5	8-5

Domain	Prod 1	Prod n
Military	9	8-5
Air Traffic Control	9-5	9
Industrial	8-5	9-5
Commercial	8-5	8-5

24.10 Exercises

1. What do you think is a reasonable TRL distribution in a company and why?
2. What do you think is a reasonable technology maturity levels distribution in a company and why?
3. What can a company do if it has had enormous success with previous systems to ensure that its' internal status quo does not put it out of business?
4. What do you think a nation state can do if it has had enormous success with previous systems to ensure that the status quo does not move it into an unsustainable condition?
5. What is your new business development process? How does it compare with other students? How does it compare with systems engineering or development processes? How do these answers compare with your answers in the early part of this text?
6. What is your research and development process? How does it compare with other students? How does it compare with systems engineering or development processes? How do these answers compare with your answers in the early part of this text?
7. Is it appropriate for a technology or product to be in search of a system or is it always a system in search of a technology or product? Can you give examples of each? If not why not?

8. Is it appropriate for a system to be in search of a need or is it always a need in search of a system? Can you give examples of each? If not why not?

24.11 Additional Reading

1. Defense Acquisition Guidebook, Defense Acquisition University, August 2010.
2. Human Factors for Evolving Environments: Human Factors Attributes and Technology Readiness Levels, FAA/NASA Human Factors Research and Engineering Division, FAA, DOT/FAA/AR-03/43, April 2003.
3. Managing a Technology Development Program, James W. Bilbro & Robert L. Sackheim, Office of the Director George C. Marshall Space Flight Center, augments NASA Procedures and Guidelines, NPG 7120.5, NASA Program and Project Management Processes and Requirements.
4. NASA Procedural Requirements NASA Systems Engineering Processes and Requirements w/Change 1 (11/04/09), NPR 7123.1A, March 26, 2007.
5. NASA Research and Technology Program and Project Management Requirements, NPR 7120.8, NASA Procedural Requirements, February 05, 2008.
6. NASA Systems Engineering Handbook, NASA/SP-2007-6105 Rev1, December 2007.
7. Research & Development Degree Of Difficulty (R&D3) A White Paper, John C. Mankins, Advanced Projects Office of Space Flight NASA Headquarters, March 10, 1998.
8. Systems Engineering Fundamentals, Supplementary Text, Defense Acquisition University Press, January 2001.
9. Technology Readiness Assessment (TRA) Deskbook, DOD, May 2005, July 2009
10. Technology Readiness Levels A White Paper, John C. Mankins, Advanced Concepts Office, Office of Space Access and Technology, April 6, 1995, NASA.
11. White Paper, Systematic Assessment of the Program / Project Impacts of Technological Advancement and Insertion, James W. Bilbro George C. Marshall Space Flight Center, December 2006, with acknowledgements: Dale Thomas, Jack Stocky, Dave Harris, Jay Dryer, Bill Nolte, James Cannon, Uwe Hueter, Mike May, Joel Best, Steve Newton, Richard Stutts, Wendel Coberg, Pravin Aggarwal, Endwell Daso, and Steve Pearson.

25 Analysis and Requirements

If we examine all the analysis techniques we see that there are common elements. There are inputs and outputs that bound the system. There are operations between the inputs and outputs. There is layering to represent different abstraction levels so humans can focus attention on a layer as needed to understand some aspect of the system. There are words to describe each abstraction level that are qualitative and quantitative. The quantitative words become performance requirements and the qualitative words become the remaining requirements. The requirements are eventually moved to specifications.

There is a tight relationship between studies, requirements, specifications, and tests. Studies help to surface requirements. Requirements are housed in specifications and specification requirements are verified by tests.

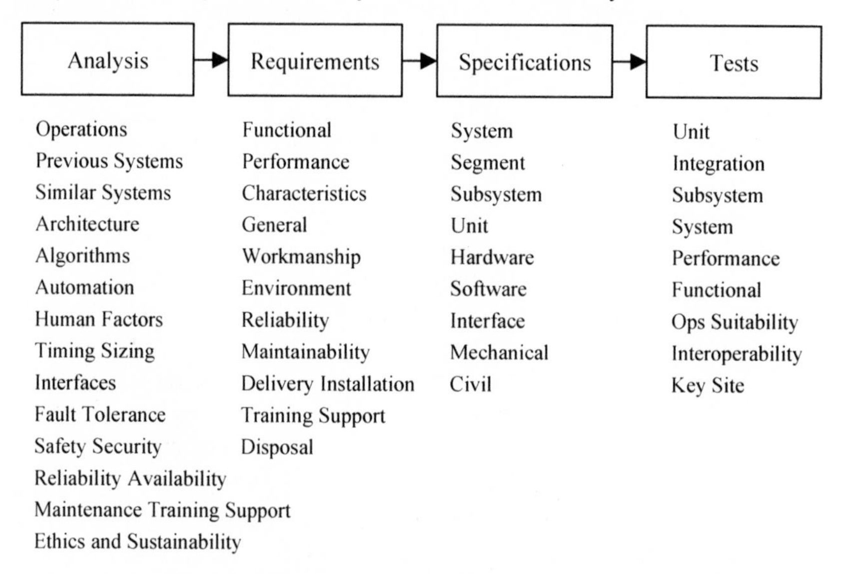

Figure 111 Studies Requirements Specifications Tests

The front-end systems effort is focused on the architecture and so the requirements are at a high level. As the development effort continues the emphasis moves towards identifying the detailed requirements so that the architecture can be designed and implemented. Obviously this is an iterative process and there is always the possibility that the architecture may be modified as the details are surfaced. If the front-end systems effort was strong then necessary details were uncovered at that time and the architecture will survive. For example, performance budgets are allocated and those allocations should not result in any subsystem being forced to meet an impossible requirement number.

It is reasonable to separate the front-end requirement effort, which is used to develop the architecture from the back end requirement effort, which is used to design and implement the architecture or system. The emphasis is different. The front-end effort is big picture driven while the back end is very detail oriented. The tools for housing and tracking the requirements also will be different. In the back-end effort a relational database is used to track and manage thousands of requirements. This may be unneeded for the front-end effort unless there is a desire to prepare for the back end effort.

The steps for identifying and allocating the requirements are not complex. The complexity is in the performance of the steps. They apply to both the front end and back end system activities. The following is a possible sequence that can be used for requirement and specification development:

- Elicit and gather the requirements
- Identify the key requirements
- Group requirements by functions, performance, and specialty areas
- Allocate requirements to the system architecture
- Allocate requirements to the subsystems
- Create a specification tree based on the architecture and subsystems
- Place requirements in the proper specification tree level
- Allocate requirements to test methods

25.1 Key Requirements

Identifying and tracking the key requirements and issues may mean the difference between success and failure. Not all requirements and issues are the same. Some are make or break situations. Key requirements are drivers of the system solutions. They can be technology, performance, cost, schedule, production, maintenance, logistics, support, and other drivers. These drivers may exist because the state-of-the-art is being pushed in an area or previous knowledge shows these requirements and issues are make or break items.

Author Comment: Everyone should know the key requirements. They should never be hidden. It is possible that some requirements and issues are so significant that management will try to hide them, but then no one will be able to use the magic of their abilities to tackle and address these key requirements. In the end the key requirements and issues don't go away, they always remain, but the successful team has addressed and effectively satisfied each of the key requirements and knows how the key issues were closed. It is the essence of the solution.

Key requirements and issues come in layers. There are key requirements and issues at the highest level and are parts of the architecture dialog. There are key requirements and issues at the subsystem levels. There are key requirements and issues at the component levels. There are key requirements and issues at the abstract levels like maintenance, training, support, and other areas that do not make up the physical solution but are part of the system solution. In many ways they are the kickoff of an activity for each new layer of the system development.

25.2 Requirement Types

These are some ideas for types of requirements you might consider. Think in terms of form, fit, and function (there is no order to the list):

- Operation, function, interface, transition, growth, upgrade, performance (typically numbers)
- Physical characteristics (weight, size, power, ingress, egress, tie downs, electromagnetic radiation, security, vulnerability), durability, ruggedness, health and safety
- Design, construction, fabrication, environment (temperature, humidity, dust, fungus, mold, vibration, altitude)
- Human engineering, safety, workmanship, interchangeability
- Reliability, maintainability, maintenance, repair, personnel
- Logistics, materials, parts, support, supply, training
- Transportability, shipping, delivery, installation
- Quality, test (inspection analysis demonstration that you will levy on the contractors that provide the subsystems)
- Product markings / labeling, nameplates
- Chemical, electrical, and physical requirements, dimensions, weight , color, protective coating
- Special tools, work stands, fixtures, dollies, and brackets
- Structural, architectural or operational features
- Raw material (chemical compound), mixtures (cleaning agents, paints), or semi-fabricated material (electrical cable, copper tubing) which are used in the fabrication of a product, toxic products and formulations
- Restraints/constraints, restriction of use of certain materials due to toxicity, dimensional or functional restrictions to assure compatibility with associated equipment

25.3 Eliciting Gathering Requirements

Eliciting and gathering requirements is the practice of surfacing requirements while interacting with stakeholders and performing system analysis. Many narrow eliciting requirements to just users, customers, maintainers, managers, and other stakeholders. However they only provide a limited view of the possible requirements. In many instances this view is bounded by their own experiences and limited exposure to the challenge.

So part of the practice is to build a team of stakeholders that is able to learn and evolve as the system is being conceived. These stakeholders are frequently called Subject Matter Experts (SME). They are different from the other team members because they must maintain the ultimate vision of the system mission and its peculiarities. As the technologists run down their particular paths the SME must understand the ramifications of these paths and provide appropriate checks and balances. The elicitation practice includes:

- Interviews
- Questionnaires
- User Observations
- Workshops
- Working Groups
- Brainstorming
- Operational Examples
- Role Playing
- Prototyping
- Site Surveys
- Previous Systems
- Similar Systems

The information is extracted from the SMEs but the entire team participates. There is a two-way exchange between the SMEs and the rest of the team.

The non-SME portion of the team focuses on their respective disciplines and offers requirements based on their research and studies. The introspective research and studies include:

- Site surveys, previous systems, similar systems
- Operational and functional analysis
- Maintenance training and support analysis
- Various performance analysis such as timing, sizing, capacity, thruput, load, power, cooling, heating, structural, electro-magnetic interference, environment, etc analysis
- Reliability, maintainability, safety, human factors analysis
- Specialized algorithms analysis
- Manufacturing, production, transport, installation, shutdown, decommissioning, disposal analysis
- Transition and interpretability analysis
- Environmental impact analysis
- Sustainability analysis including cost shifting
- Life cycle cost analysis
- Architecture alternatives and selection
- Models, simulations, mockups, prototypes analysis

Science and engineering are based on connections. A literature search is performed before new science is offered. Current solutions are understood before new engineering approaches are proposed. These simple ideas also apply to surfacing potential requirements. So the team also looks to external sources and performs research and studies from an external perspective to surface requirements:

- Study the past and understand historical connections
- Literature searches
- Industry surveys
- Vendor searches, requests for information
- Technology searches, forecasting, and prediction
- Marketing surveys
- Emerging technologies tracking
- Emerging products tracking

All these analysis and study efforts result in multiple requirements. They need to be vetted by the stakeholders and moved into the baseline as appropriate.

Although the list of requirement elicitation and gathering suggestions is long, do not discount simple reading of related documents[192] and interacting with the team members. This is a very low cost approach but can result in significant understanding of the system and its requirements.

Many times people need to take on different roles. This is especially true for small teams. The technologists need to read literature describing the needs of the stakeholders and act on their behalf when they are not physically part of the team. At the same time others may need to enter a specialist role that is new for them and they need to learn by reading literature so that they can act on behalf of the missing specialist. This is how true systems practitioners are born. They are forced by necessity to take on these other roles so that an effective system solution can be offered when the team is deficient or broken.

Author Comment: As the years click by and an organization practices this broad approach to requirement elicitation and gathering, the organization starts to change to conform to the activity. For example subject matter experts tend to group together. Senior staff with new members tends to group together and look outside the organization being less introspective. New members that previously were part of the less introspective activities are pulled into the introspective group and eventually they themselves become senior staff. So there are three groups that hone their craft and interact with each other to elicit and gather requirements.

---

[192] Do not discount the information offered on the Internet.

As the discussions unfold and requirements surface there is a constant culling activity where some requirements are accepted and others are rejected. The process should have the mechanisms to preserve this body of work. Some requirements become goals and others get moved to the dream category.

Author Comment: Goals are tomorrow's requirements. Dreams are next week's requirements. When we stop making goals and loose our dreams we stop progress. The challenge is to constantly challenge ourselves and every so often greatness happens when least expected. So always keep track of the Requirements, Goals, and Dreams. You now have the technology with System Requirements Databases (SRDB). Write specifications to house your requirements and figure out how to make your dreams come true.

25.4 Characteristics of Goo d Requirements

The authoring of good specification text is not a trivial exercise and has been documented for many years[193]. The following is considered to be a reasonable list of characteristics for good requirements.

- **Complete**: The requirement is fully stated in one place with no missing information. It is atomic it is standalone.
- **Clear**: The requirement is clearly stated using strong unambiguous words. There is no jargon or useless words or phrases. Vague subjects, adjectives, prepositions, verbs and subjective phrases are avoided. Figures and tables can be vague and should be converted to standalone requirement objects.
- **Consistent**: The requirement does not contradict any other requirement in the specification or other specifications associated with the system. For example software requirements specifications must be consistent with the hardware specifications.
- **Accurate**: The requirement is fully vetted and accepted by the stakeholders. The requirement is not obsolete or becomes obsolete. This suggests a requirement management mechanism to change requirements as the system unfolds.
- **Feasible**: The requirement can be implemented. Technology is not exceeded, the teams' abilities are not exceeded, and the project cost or schedule is not exceeded. This suggests a requirement management mechanism to track risks associated with key requirements. This also builds the case for identifying the key requirements in the system.
- **Testable**: Do not use negative statements they are not testable. Do not use

[193] In 2005 Carnegie Mellon performed a study, Report CMU/SEI-2005-TR-014, and once again visited the dilemma of poor requirement text.

compound statements, they lead to test confusion and complexity. When a requirement is stated, identify one or more of the traditional test methods: inspection, analysis, demonstration, test, or certification. The stronger methods are test or demonstration and they should be used when practical. Inspection or analysis are weaker test methods and only should be done in special cases.

- **Necessary**: The requirement is fully vetted and all stakeholders agree its absence will result in a deficiency that cannot be ameliorated. This may lead to multiple releases where requirements are pushed to future releases. This suggests a requirement management mechanism where requirements are associated with different releases.
- **Traceable**: The requirement addresses only one thing. Compound requirements lead to traceability and test issues. Figures and tables can house many compound requirements open to interpretation and should be converted to standalone requirement objects.
- **Black Box**: Some believe all requirements should be black box oriented where the requirement specifies externally observable events based on the input and output. Requirements that specify elements within the black box such as internal architecture, algorithms, design, implementation, testing, or other decisions are constraints, and should be articulated in the Constraints section of the requirement specification document. I strongly disagree with this approach. If the details are critical to make sure a particular design and implementation is followed, it should not be buried in the constraint section.

25.5 Requirement Style Guide

The need for clear consistent testable specifications is obvious. One approach to help in this area is to develop a requirements style guide[194] that offers guidance to staff when writing the requirements. The style guide offers example sentence syntax and states the preferred imperative. The imperative is the heart of a specification. It is what must be implemented and tested. The imperative is signaled with a key word like shall or must.

25.5.1 Requirement Imperatives

Imperatives are words and phrases that command something must be provided. Imperatives are used to signal the presence of a requirement. A requirement is something that must be implemented and tested. The following are different views of imperatives in a specification: NASA, FAA, MIL-STD-490, MIL-STD-463, and my view or Walt's view.

---

[194] For Example: CMS Requirements Writer's Guide, Centers for Medicare & Medicaid Services Integrated IT Investment & System Life Cycle Framework, Department of Health and Human Services, V4.11 August 31, 2009.

NASA View[195]

Imperatives are words and phrases that command something must be provided. They are in descending order of strength as a forceful statement of a requirement. The order from strongest to weakest is Shall, Must, Must Not, Is Required To, Are Applicable, Responsible For, Will, Should. The NASA requirement documents judged most explicit have a majority of imperative counts associated with the strongest words.

- **Shall**: Dictates provision of a functional capability.
- **Must or must not**: Establishes performance requirements or constraints.
- **Is required to**: Specification statements written in passive voice.
- **Are applicable**: Includes by reference, standards or other documentation as additions to specified requirements.
- **Responsible for**: Systems with predefined architectures e.g. "The XYS function of the ABC subsystem is responsible for responding to PDQ inputs."
- **Will**: Cites operational or development environment things provided to capability being specified e.g. "The building's electrical system will power the XYZ system" In a few instances "shall" and "will" are used interchangeably containing both requirements and descriptions of operational environment system boundaries not always sharply defined.
- **Should**: Not used frequently as imperative. When used, statement is always very weak, e.g. "Within reason, data files should have same time span to facilitate ease of use and data comparison".

FAA View

There was a movement within the FAA to use plain language techniques within the organization. The suggestion was to drop the use of "shall" and replace it with "must".

MIL-STD-490 View[196]

Use "shall" whenever a specification expresses a provision that is binding. Use "should" and "may" wherever it is necessary to express non-mandatory provisions. "Will" may be used to express a declaration of purpose on the part of the contracting agency. It may be necessary to use "will" in cases where the simple future tense is required, i.e., power for the motor will be supplied by the ship.

---

[195] Extracted from Automated Requirement Measurement (ARM) tool description. This is an internal tool used to review specification text.

[196] This text is from MIL-STD-490 describing the use of imperatives. Specification Practices, Military Standard, MIL-STD-490 30 October 1968, MIL-STD-490A4 June 1985

MIL-STD-963B View[197]

Use "shall" when an instruction is mandatory. "Will" may be used to indicate the Government will do something. Avoid the use of "should" and "may" since Data Item Descriptions (DIDs) normally contain only mandatory instructions.

Walt's View

Keep it Simple. Remove all special cases that need to be remembered. There is no time for fine distinctions of English syntax. Use "shall" for all requirements. Use System Requirements Database (SRDB) attributes like "future" to categorize the requirement statements. Don't try to specify other systems, specify your interface to other systems. Minimize descriptive text to minimize noise in the specification and avoid missing important requirements. Use analysis and design documents to offer descriptive text of the system.

25.5.2 Requirement Sentence Structure Suggestions

Obviously there are no absolutes. The following are general rules for writing requirement statements.

- Avoid compound sentences.
- Software processing requirements are stand-alone statements with trigger / action / response as appropriate.
- Requirements are on a single line; use carriage return or line feed consistently to separate requirements. It will simplify exports to other tools.
- Do not embed requirement project unique identifier (PUI) number in requirement statement. Automated tools automatically assign PUI numbers.

Enumerated lists capture sequence. They can be used to capture algorithmic information in English statements rather than flowcharts or other forms of non-text communication.

Upon power up failure the <device> shall:

1. step 1
2. step 2
3. step n

The numbers are part of the statement. The statement in the list does not need to include the "shall" imperative. An alternative is shown below.

---

[197] This text is from MIL-STD-963B describing the use of imperatives. Standard Practice Data Item Descriptions (DIDs), Department Of Defense MIL-STD-963B 31 August 1997, DOD-STD-963A 15 August 1986.

Upon power up failure the <device>:

1. shall step 1
2. shall step 2
3. shall step n

The following are example software requirements:

Request Validation

a. FDP Processing shall reject an <X-request> with an <X-response>, if ... (FDP performs validation checks)

Request Reporting

a. Upon receipt of a valid <X-request>, FDP shall … (no condition)

b. Upon receipt of a valid <X-request>, FDP shall forward <SS Commands> to the selected SS to ...(if a SS is used)

c. Upon expiration of [X-Timer], FDP shall ... (timed event)

d. Upon receipt of <SS Command Responses> resulting from the successful SS processing of an <X-request>, FDP shall ..

e. Upon receipt of <SS Command Responses> resulting from the successful SS processing of an <X-request>, FDP shall reject the <X-request> with an <X-response> if ... (FDP detects problem that SS can not)

f. Upon detecting an error condition reported in a <SS Command Response> resulting from processing an <X-request>, FDP shall ... (SS detects error FDP can't)

Results Reporting

a. FDP shall return the results of processing an "X" request to the ATC operator. (Operator notification)

b. FDP shall return the results of processing an <X-request> in a <X-response>. (Computer notification)

Audit Reporting

a. FDP shall store an <FDP Update Msg> in response to processing an <X-request>. (Unconditional reporting)

b. FDP shall store an <FDP Update Msg> in response to processing an <X-request>, if X was selected as an audit event. (Conditional reporting, depending on operator selection)

c. FDP shall store an <FDP Update Msg>, if any errors were encountered in processing an <X-request>. (Conditional reporting, depending on errors)

25.5.3 Requirement Words and Phrases to Avoid

There are words and phrases that should be avoided in requirement text. They typically suggest problems with the requirements. They are:

- **Buzz Words**: fault tolerant, high fidelity, user friendly
- **Compound**: and, or, with, also
- **Incomplete**: TBD, TBS, TBE, TBC, TBR, ???, huh, not defined, not determined, but not limited to, as a minimum
- **Options**: may, might, could, should, can, optionally, ought, perhaps, probably
- **Unbounded**: all, totally, never, fully
- **Undefined**: any, approximately, near, far, close, back, in front, useful, significant, adequate, fast, slow, versatile, a minimum, as applicable, as appropriate, old, new, future, recent, past, today's, normal, timely, clear, well, easy, strong, weak, good, bad, efficient, low, user-friendly, flexible, effective, usually, generally, often, normally, typically, but, except, unless, eventually, although
- **Unsure**: possible, possibly, eventually, if possible, if needed
- **Weak Words 1**: normal, effective, timely, similar, flexible, adaptable, rapid, fast, adequate, support, maximize, minimize, etc, clear, easy , useful , adequate , good , bad
- **Weak Words 2**: this, these, that, those, it, they, above, below, previous, next, following, last, first
- **Weak Phrases 1**: as appropriate, be able to, be capable of, capability of, capability to, capability of, as required, provide for, easy to, having in mind, taking into account, as fast as possible
- **Weak Phrases 2**: according to, on the basis of, relative to, compliant with, conform to, but not limited to
- **Untestable**: not

25.5.4 Requirement Document Structure

It is important that the document structure is considered and decisions made on where to place certain requirements. The goals are:

- Place logically grouped requirements in the same section
- Make sure there are no duplicate or partial duplicate requirements

The rule is one requirement one place, no exceptions. This can get tricky if the team is very thread oriented. In this case someone must methodically go through the document and remove the duplicates. It is also difficult if a system has multiple modes and states. The mode and state requirements can get duplicated in the functional requirements if the team is not careful.

A requirement document does capture the decomposition characteristics of a system. By counting the number of requirement statements (children) under a paragraph heading it is possible to get some insight into the level of completeness of the requirements.

The number of children at a particular level translates to a document shape. There are different document shapes and each shape has implications. The document shapes are random, rectangle, pyramid, inverted pyramid, trapezoid and diamond.

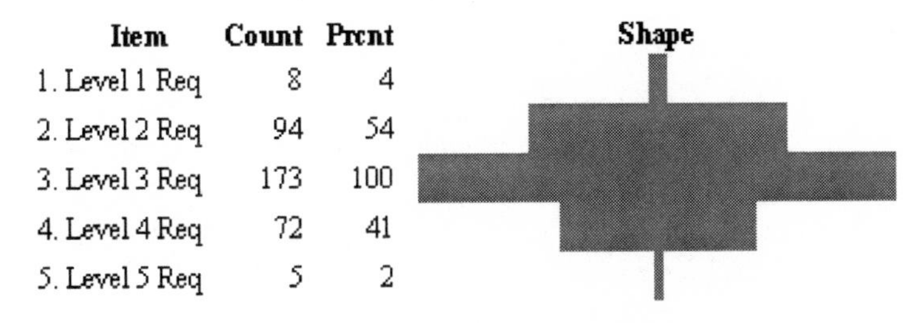

Item Count Prcnt
1. Level 1 Req 8 4
2. Level 2 Req 94 54
3. Level 3 Req 173 100
4. Level 4 Req 72 41
5. Level 5 Req 5 2
Shape

Figure 112 Document Shape

If there are insufficient paragraph levels, either the system is extremely small or there is a problem. The magic rule of 7 +/-2 should be considered when grouping requirements into paragraphs and paragraph levels.

If there are more children at lower document paragraph levels than higher levels, the document shape will be a pyramid, suggesting reasonable decomposition and leveling of requirements. However, it also suggests the team may not have completed their decomposition or they ventured into design[198].

A diamond shape also suggests the decomposition is reasonable and at the lower levels the team did not venture into too much detail suggesting design.

---

[198] When you look up it is requirements, when you look down it is design. In this case the team may have over achieved.

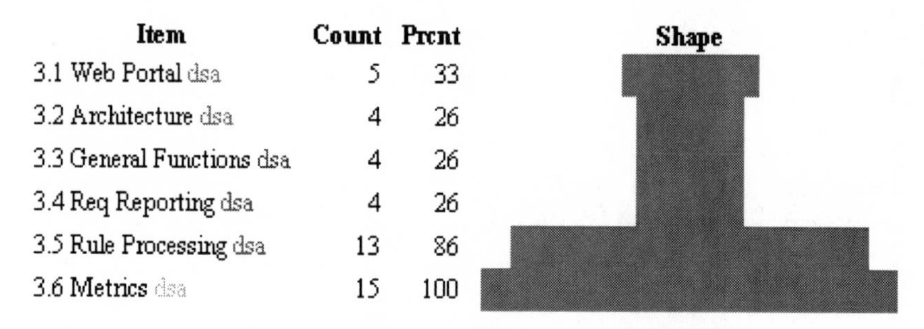

Item	Count	Prcnt
3.1 Web Portal dsa	5	33
3.2 Architecture dsa	4	26
3.3 General Functions dsa	4	26
3.4 Req Reporting dsa	4	26
3.5 Rule Processing dsa	13	86
3.6 Metrics dsa	15	100

Figure 113 Domain Specific Document Shape

The technique of counting the number of children under a paragraph heading can be expanded to include other views of a document. For example domain specific requirements can be counted. This offers a system view and stimulates dialog on the emphasis in the system. If the domain specific document shape is unexpected, then the team should try to address the issue. Either the team gained new insight or there is work to be done that they are now aware of and can execute.

25.6 Review Elements

Typically experts, novices, or some combination produce a specification document. They may or may not use a requirements style guide. At some point the staff decides if they will have a review. If they do not have a review, then they have to deal with the ramifications. If there is a review, then the review is a manual effort needing input from experts. The review includes domain and non-domain experts.

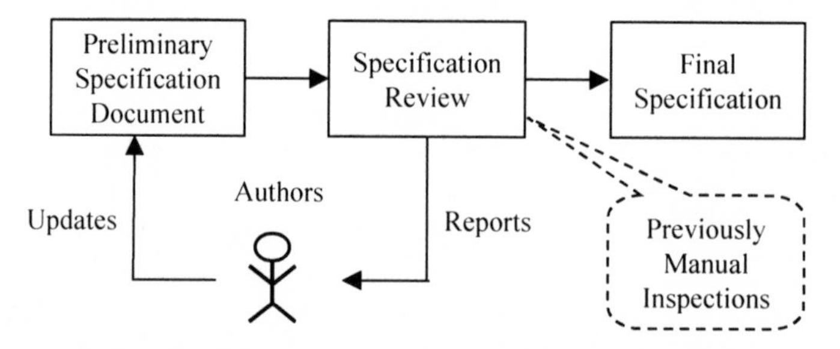

Figure 114 Requirements and Specification Review

The review can be manual or a tool[199] can be used to augment the manual review. As with all tools, the manual review should not be bypassed by just pressing a few buttons on a tool. Instead the manual and automated review should complement each other to form a full picture not possible with just one approach.

Using a tool does not shorten the review time. The time tends to be arbitrarily

---

[199] The Specification Analysis Tool (SAT) was developed as a direct result of the Carnegie Mellon, Report CMU/SEI-2005-TR-014 findings.

set. People will look for possible issues until they get bored or exhausted. The findings from a tool are more consistent. People tend to miss instances in categories while a tool tends to find all problems of a certain type. People tend to miss a whole category of a problem but are better at finding domain specific problems.

When venturing into a tool based review approach let the machines do what they do well such as search, count, filter, categorize, profile, and visualize. Let the humans do what they do well such as creativity, critical thinking, inspiration, and intuition. Also realize there are more findings when a tool is used because the people and machines tend to complement each other.

25.7 Exercises

1. Identify 5-7 key requirements of your laptop computer and compare them with your classmates. Are they different?
2. Identify 5-7 key requirements of your cell phone and compare them with your classmates. Are they different?
3. Identify 5-7 key requirements for a new cell phone or laptop and compare them with the previous generation product key requirements. Compare them with your classmates. Are they different?
4. Identify 25 requirements for a system you select (laptop, cell phone, house, car, etc) and group them into different categories.
5. Identify the possible analysis or study activities that may have led to your list of 25 requirements.
6. Review your 25 requirements and identify any problems with the requirements.
7. Move your 25 requirements into a possible document structure with different paragraph numbers and heading text.

25.8 Additional Reading

1. CMS Requirements Writer's Guide Version 4.11, Department of Health and Human Services, Centers for Medicare & Medicaid Services, August 31, 2009.

26 Specifications and Design Artifacts

Specifications are information products that house the requirements. Requirements at the highest abstraction level precisely describe what a system is suppose to do and at the lowest abstraction level describe how the system does it. The lowest level requirements are design and implementation details.

Specifications can take many forms. For example a flyer can list features that are important to a customer. These features can be any abstraction level that someone thinks separates this system or product from another system or product. They can be blue prints, which show design details of a house under construction. They can be text documents, which describe what the system is supposed to do. Low level specifications can be product brochures or even line items in a table of a book listing hundreds of items like pipe fittings.

So requirements and specifications can take on many forms and capture many levels of abstractions. Ideally the requirements should be grouped by abstraction level.

26.1 Guiding Principles

Many times in immature settings the team will ignore the requirements and not even produce specifications. This is a serious mistake. Even if a miracle happens and a product is developed, a list of features for marketing and customers is needed. The list is also needed for planning the next generation product. As the product is fielded and customers find problems, the problems are associated with the feature list. Eventually nature will take its natural course. If the product is marginal or fails it does not matter[200]. If the product is successful there will be a reasonable set of specifications housing the requirements following these guiding principals:

- **Precise**: The requirements are written very precisely. The level of precision must be respected and understood. The precise requirements form the foundation for a requirement driven design that yields a solution, which meets the requirements and only the requirements. This precision follows through to the test program where requirements are allocated to test cases that are clearly delineated and easily understood.
- **Accurate**: All the requirements must very accurate because they form the basis of a "proof like" environment that is established during test.

[200] Study after study has shown project failures are attributed to poor requirements. Report CMU/SEI-2005-TR-014, Carnegie Mellon SEI Report September 2005

Inaccurate requirements will yield an unexpected design. The allocation of requirements must be accurate or the test program will fail to provide a "proof like" environment showing that the design meets the requirements and only the requirements.

- **Rigor**: To establish high levels of precision and accuracy for a design solution a great deal of rigor is needed. For example the introduction of a Systems Requirements Database (SRDB) forces a level of rigor on the organization and allocation of the requirements not possible with a spreadsheet or word processing table. The SRDB is only one example of rigor. Another is the allocation of software requirements to source code.
- **Methodical**: The only way to get to an accurate and precise solution that has the level of rigor to provide for a truly effective system solution is to be very methodical and establish rules that are clear and consistently executed. Simple things like naming conventions become critical to the process. For example, if you use automated test scripts, ensure that the test script names are consistent between the SRDB, test script file names, and test procedure document.

26.2 Specifications and Requirement Leveling

There are various approaches to organizing requirements and the internals of a specification. Examples are organization by functional area, operational threads, functional area and operational threads, or a different approach as defined by the team[201]. It is important to realize that different abstraction levels may exist and so different specification documents may be needed that represent those abstraction levels. Many think in terms of upper, middle, and lower levels. However there may be separate groups of upper, middle, and lower level requirements.

If the specification is organized by operational thread, eventually after a few thread requirements are documented, duplicate requirements are created because many operational threads traverse many of the same functions as part of their normal sequence. It is ok to have duplicate requirements in a study such as an operational concept. In fact it is encouraged in studies to spur discussion. However specifications draw the line in the sand, commitment is made and a clear baseline must be articulated.

Duplicate requirements in the specification will lead to serious problems. Duplicate requirements are rarely stated the same way. This can lead to major or minor differences in the requirements. Also as the system evolves and the requirements change, keeping duplicates consistent is a maintenance nightmare. While various requirement alternatives are welcomed in studies, only one

---

[201] Although the team may think they may have invented a new breakthrough approach to specifications, before they go down that path they need to study the past including MIL-STD-

requirement clearly stated in one place in the specification is critical to success.

At the highest conceptual level, the possible requirement paragraph numbering scheme can be tree based where the requirements are organized such that logically grouped requirements are stated just once. In the past requirements were written as paragraphs under a logical paragraph title. However with the introduction of requirement management and traceability tools the paragraphs should be parsed into standalone sentences so that they can be easily imported into the tools. These standalone objects are still grouped by paragraph using a number and heading name.

3. REQUIREMENTS
3.1 First Paragraph
    3.1.1 First Subparagraph
3.2 Second Paragraph
    3.2.1 First Subparagraph
    3.2.2 Second Subparagraph
3.3 Third Paragraph
3.4 Fourth Paragraph

As we dig deeper into what may be needed for a specification other text surfaces to support the requirement story[202]. In its simplest form a specification has an introduction, requirements organized by functional area[203], statements associated with test methods, and a verification cross-reference matrix (VCRM) with test methods.

Later versions of the specification show traceability to test procedures and possibly other lower level specifications. This traceability is shown using tables in the child specification documents and child test procedure documents. The parent is the source specification document. So there can be several layers of parent child relationships. This is discussed further in the System Requirements Database (SRDB) section of this text.

26.2.1 Test Methods

There are several test methods, sometimes called verification methods, which can be used to verify the requirements. The test methods are typically identified after the requirement section in a specification using a new standalone section usually called Qualification. The test methods are as follows[204]:

---

490 and MIL-STD-1521B.

[202] MIL-STD-490 identified various specifications used to house requirements for different kinds of subsystems and different abstraction levels. It is probably the broadest and deepest description of the topic.

[203] Functional analysis uses cohesion and coupling concepts to group similar items. This removes duplicates from the model or conceptual view.

[204] Systems Engineering Fundamentals, Supplementary Text, Defense Acquisition

- **Analysis**: Analysis uses mathematical modeling and recognized analytical techniques including computer models to interpret or explain the behavior and performance of a system element and to predict the compliance of a design to its requirements based on calculated data or data derived from lower level component or subsystem testing. It is used when a physical prototype or product is not available or not cost effective. Analysis includes the use of both modeling and simulation.
- **Inspection**: Visual inspection of equipment and evaluation of drawings and other pertinent design data and processes is used to verify physical, material, part and product marking, manufacturer identification, and workmanship requirements.
- **Demonstration**: Demonstration executes a system, subsystem, or component operation to show that a requirement is satisfied. Usually performed with no or minimal instrumentation. Visual observations are the primary means of verification. It is used for a basic confirmation of performance capability and is differentiated from testing by the lack of detailed data gathering.
- **Test**: Test executes a system, subsystem, or component operation to obtain detailed data that is used verify or to provide sufficient information to verify performance requirements through further analysis. Usually uses special test equipment or instrumentation to obtain very accurate quantitative data for analysis. Testing is the detailed quantifying method of verification.
- **Certification**: This is verification against legal, industrial or other standards by an outside authority, which issues a certificate.

26.2.2 Test Verification Cross Reference Matrix

The Verification Cross-Reference Matrix (VCRM) lists each requirement and identifies one or more test methods for each requirement. Preparing the VCRM forces the team to examine the requirements for testability and reasonableness.

Obviously the VCRM must be consistent with the body of the specification. There is no excuse for any disconnect and the specification must be rejected if the Requirement column does not clearly follow the requirement specification paragraphs.

In the past, before the use of a System Requirements Database (SRDB), each specification included a section on Qualification, which had a description of the verification methods. The verification methods still apply however the need to identify them coincident with the initial specification is not as strong as in the past.

University Press, January 2001. Defense Acquisition Guidebook, Defense Acquisition University, August 2010.

This is because an SRDB is used in the process where each requirement is linked to one or more tests. The automation and process ensures that each requirement is testable.

Table 43 Test Verification Cross Reference Matrix (VCRM)

Requirement	I	A	D	T	C
3.1 Functional Area 1	-	-	-	-	-
Requirement 1			X		
Requirement 2			X		
Requirement 3				X	X
3.2 Functional Area 2	-	-	-	-	-
Requirement 1	X				
Requirement 2	X	X			

Before the SRDB, the specification also included a VCRM, which showed what test methods or methods are used to verify each requirement. This suggested that each requirement was testable even if the actual test procedure was unknown. Today the VCRM may remain blank until the test planning is started. At that point actual test procedures in the form of test paragraph Headings are shown in the matrix. This tends to make the allocation to test methods overcome by events. However most practitioners still track the allocation to the test methods, along with links to different kinds of tests[205] and the actual test procedure heading number, test PUI[206] and name in the SRDB. This view is then offered, as needed using an SRDB report export.

Although the specification baseline is very strong before test planning is initiated and completed, it is not perfect and changes will happen. Some of the changes include the impacts of test and the final test procedures. Ideally there will be no specification changes once test dry runs are completed. However, if dry run testing surfaces changes, those changes must be clearly and methodically addressed in the specification. Change tracking is discussed in the Configuration Management section of this text.

26.3 Specification Types and Practices

MIL-STD-490 is an example of grouping requirements at certain abstraction levels and offering a suggestion for placing them in an information product known for that abstraction level. Buried within the standard is not only the notion of abstraction levels but also examples of types of requirements to consider. This list broadens and deepens the thought processes of the stakeholders. It is a significant

---

[205] Different kinds of tests include functional, performance, environmental, reliability, safety, power, durability, etc.

[206] PUI – Project Unique Identifier is a unique number associated with database row or object.

starting point for making sure things are not missed in the system development[207].

Table 44 MIL-STD-490 Specifications[208]

Type	Title
Type A	System/Segment Specification
Type B	Development Specifications
Bl	Prime Item
B2	Critical Item
B3	Non-Complex Item
B4	Facility or Ship
B5	Software
Type C	Product Specifications
Cla	Prime Item Product Function
Clb	Prime Item Product Fabrication
C2a	Critical Item Product Function
C2b	Critical Item Product Fabrication
C3	Non-Complex Item Product Fabrication
C4	Inventory Item
C5	Software
Type D	Process Specification
Type E	Material Specification

These specifications use common numbered sections. The common sections are Scope, Applicable Documents, Requirements, Qualification Requirements (Software), Quality Assurance Provisions (Hardware), Preparation For Delivery, Notes, and Appendix. For this new generation I have taken the liberty to add Ethics and Sustainability to each information product, including specifications.

26.3.1 A Level System Segment Specification

The Type A System Segment Specification states the technical and mission requirements for a system or segment as an entity, allocates requirements to functional areas, documents design constraints, and defines the interfaces between or among the functional areas. Normally, the initial version is based on parameters developed during the Concept Exploration[209] phase.

---

[207] Much of this text in this section comes directly from MIL-STD-490. The goal was to extract the most salient points of the standard while minimizing interpretation. The intent is to use this as a universal example of specification practices.

[208] This example is selected because it is the original starting point dating back to October 30, 1968.

[209] This description is based on the approach of Concept Exploration, Demonstration, Validation, Development, and Production phases in a procurement cycle.

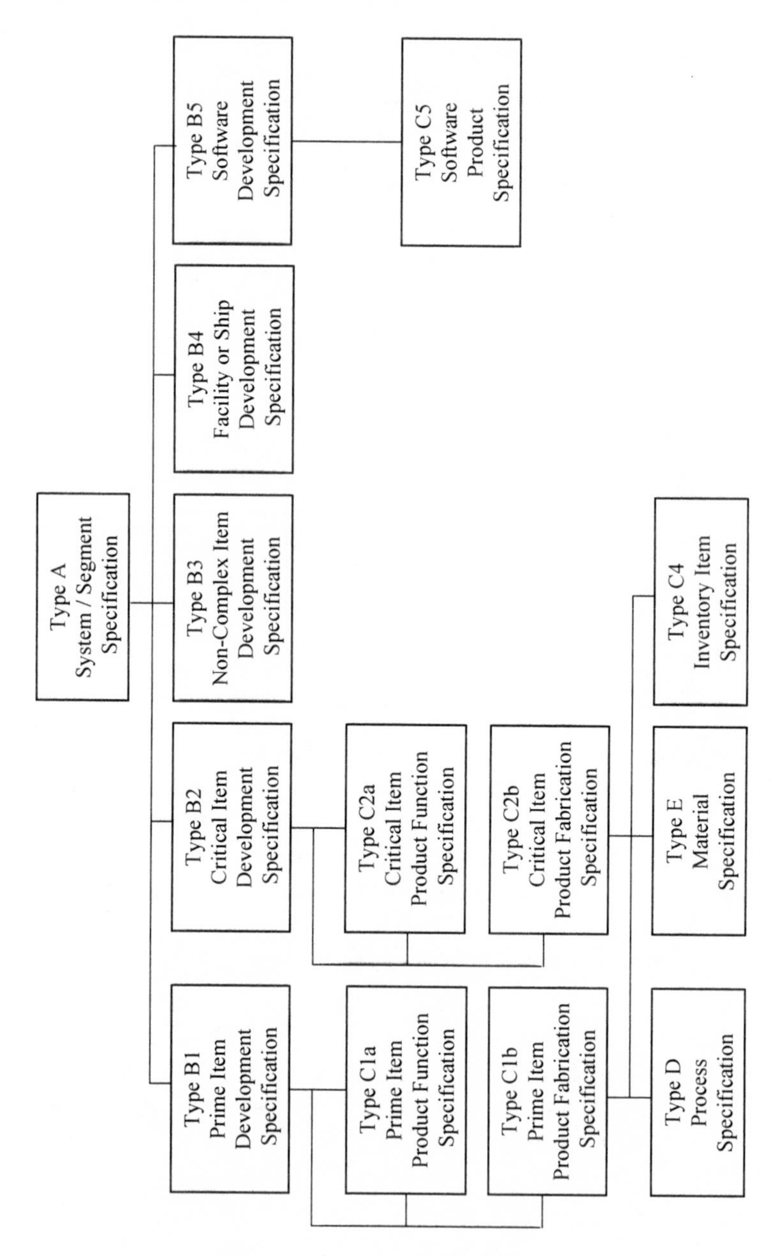

Figure 115 Generic Specification Tree[210]

[210] Specification Practices, Military Standard, MIL-STD-490 30 October 1968, MIL-STD-490A4 June 1985.

The initial version is used to establish the general nature of the system that is to be further defined and finalized during the concept Demonstration and Validation[211] phases. This specification is maintained current during the concept Demonstration and Validation phase, culminating in a revision that forms the future performance base for the Development and Production[212] of the prime items and configuration items[213].

1.0 SCOPE
1.1 Scope
1.2 Overview and Definitions
2.0 Applicable Documents

3.0 Requirements
3.1 System (Item) Definition
3.1.1 Functional Layouts
3.1.2 Interfaces
3.1.3 Major Components
3.1.4 Furnished Equipment
3.1.5 States and Modes

3.2 Performance
3.2.1 Operational Performance
3.2.2 Physical
3.2.3 Reliability/Availability
3.2.4 Maintainability

3.2.5 Environment
3.2.5.1 Natural Environments
3.2.5.1.1 Operating
3.2.5.1.1.1 Ambient Pressure
3.2.5.1.1.2 Ambient Air Temperature
3.2.5.1.1.3 Humidity
3.2.5.1.1.4 Fungus
3.2.5.1.1.5 Lightning
3.2.5.1.1.6 Precipitation
3.2.5.1.1.7 Sand and Dust
3.2.5.1.2.9 Solar Radiation
3.2.5.1.2.10 Corrosive Atmosphere
3.2.5.1.2.11 Magnetic
3.2.5.1.2.12 Fog
3.2.5.1.2.13 Seismic
3.2.5.2 Induced Environments
3.2.5.2.1 Operating
3.2.5.2.1.1 Pressure
3.2.5.2.1.2 Temperature
3.2.5.2.1.3 Humidity
3.2.5.2.1.4 Shock and Vibration
3.2.5.2.1.5 Air Velocity
3.2.5.2.1.6 Acoustic
3.2.5.2.1.7 EMI/EMC
3.2.5.2.2 Non-Operating
3.2.5.2.2.1 Pressure
3.2.5.2.2.2 Temperature
3.2.5.2.2.3 Humidity
3.2.5.2.2.4 Shock and Vibration
3.2.5.2.2.5 Air Velocity
3.2.5.2.2.6 Acoustic
3.2.5.2.2.7 EMI/EMC
3.3 System Characteristics
3.3.1 Safety
3.3.2 Security
3.3.2.1 Physical Security
3.3.2.2 Information Security
3.3.2.3 Personnel Security
3.3.3 Interchangeability

---

[211] MIL-STD-490 consists of four phases: Exploration, Demonstration and Validation, Full Scale Development, and Production and Deployment.

[212] Systems Engineering Fundamentals, Supplementary Text, Defense Acquisition University Press, January 2001 consists of four phases: Concept and Technology Development, System Development and Demonstration, Production and Deployment, Sustainment and Disposal.

[213] Any deliverable designated by the contracting organization for configuration management such as hardware, software, etc.

3.2.5.1.1.8 Wind
3.2.5.1.1.9 Solar Radiation
3.2.5.1.1.10 Corrosive Atmosphere
3.2.5.1.1.11 Magnetic
3.2.5.1.1.12 Fog
3.2.5.1.1.13 Seismic
3.2.5.1.2 Non-Operating
3.2.5.1.2.1 Ambient Pressure
3.2.5.1.2.2 Ambient Air Temperature
3.2.5.1.2.3 Humidity
3.2.5.1.2.4 Fungus
3.2.5.1.2.5 Lightning
3.2.5.1.2.6 Precipitation
3.2.5.1.2.7 Sand and Dust
3.2.5.1.2.8 Wind
3.3.4 Human Factors
3.3.5 Miscellaneous

3.4 Logistics
3.5 Personnel and Training
3.6 Documentation
3.7 Major Component Characteristics
3.8 Precedence and Combined Characteristics

4.0 Verification Correlation
5.0 Delivery and Transition
6.0 Ethics
7.0 Sustainability
8.0 Notes

26.3.2 B Level Development Specifications

Type B Development Specifications state the requirements for the design or engineering development of a product. Each development specification effectively describes the performance characteristics that each item is to achieve when it evolves into a detail design for production. The development specification should be maintained current during production when it is desired to retain a complete statement of performance requirements. Since the breakdown of a system into its elements may have configuration items of various degrees of complexity which are subject to different engineering disciplines or specification content, there are sub-types: B1, B2, B3, B4, and B5.

26.3.2.1 Prime Item Development Specification

A Type B1 Prime Item Development Specification applies to a complex item such as an aircraft, training equipment, etc. A prime item development specification can be used as a functional baseline for a single configuration item development program or as part of the allocated baseline where the configuration item covered is part of a larger system development program. Normally configuration items needing a Type B1 specification meet the following criteria:

- The prime item will be received or formally accepted by the contracting organization.
- Provisioning action will be needed.
- Technical manuals or other instructional material covering operation and maintenance of the prime item will be required.
- Quality conformance inspection of each prime item, as opposed to sampling, will be needed.

1.0 SCOPE
2.0 Applicable documents
3.0 Requirements
3.1 Prime item definition
3.1.1 Prime item diagrams
3.1.2 Interface definition
3.1.3 Major component list
3.1.4 Government furnished property list
3.1.5 Government loaned property list
3.2 Characteristics
3.2.1 Performance
3.2.2 Physical characteristics
3.2.3 Reliability
3.2.4 Maintainability
3.2.5 Environmental conditions
3.2.6 Transportability
3.3 Design and construction
3.3.1 Materials processes and parts
3.3.2 Electromagnetic radiation
3.3.3 Nameplates and product marking
3.3.4 Workmanship
3.3.5 Interchangeability
3.3.6 Safety
3.3.7 Human performance / human engineering
3.4 Documentation
3.5 Logistics
3.5.1 Maintenance
3.5.2 Supply
3.5.3 Facilities and facility equipment
3.6 Personnel and training
3.6.1 Personnel
3.6.2 Training
3.7 Major component characteristics
3.8 Precedence
4.0 Quality Assurance provisions
4.1 General
4.1.1 Responsibility for tests
4.1.2 Special tests and examinations
4.2 Quality conformance inspections
5.0 Preparation for delivery
6.0 Ethics
7.0 Sustainability
8.0 Notes
Appendices

26.3.2.2 Critical Item Development Specification

Type B2 Critical Item Development Specifications apply to a configuration item, which is below the level of complexity of a prime item but which is engineering critical, or logistics critical.

Engineering critical occurs when the technical complexity warrants an individual specification. For example reliability of the critical item significantly affects the ability of the system or prime item to perform its overall function, or safety is a consideration. Also when the prime item cannot be adequately evaluated without separate evaluation and application suitability testing.

A critical item is logistics critical when repair parts will be provisioned for the item or the contracting organization has designated the item for multiple source re-procurement.

1.0 SCOPE
2.0 Applicable documents
3.0 Requirements.
3.1 Critical item definition
3.2 Characteristics
3.2.1 Performance
3.2.2 Physical characteristics
3.2.3 Reliability
3.3.6 Safety
3.3.7 Human performance / human engineering
3.4 Documentation
3.5 Logistics.
3.5.1 Maintenance
3.5.2 Supply
3.6 Precedence
4.0 Quality Assurance provisions

3.2.4 Maintainability
3.2.5 Environmental conditions
3.2.6 Transportability
3.3 Design and construction
3.3.1 Materials processes and parts
3.3.2 Electromagnetic radiation
3.3.3 Nameplates and product marking.
3.3.4 Workmanship
3.3.5 Interchangeability
4.1 General
4.1.1 Responsibility for tests
4.1.2 Special tests and examinations
4.2 Quality conformance inspections
5.0 Preparation for delivery
6.0 Ethics
7.0 Sustainability
8.0 Notes
Appendices

26.3.2.3 Non-Complex Item Development Specification

A Type B3 Non-Complex Item Development Specification applies to items of relatively simple design. During development, it can be shown the item is suitable for its intended application by inspection or demonstration. Acceptance testing to verify performance is not required. Acceptance can be based on verification that the item, as fabricated, conforms to the drawings. The end product is not software.

Example items are special tools, work stands, fixtures, dollies, and brackets. They can be adequately defined during the development by a sketch and during production by a drawing or set of drawings. If drawings will suffice to cover all requirements, unless a specification is required by the contracting organization, a specification for a particular non-complex item is not needed. However, when it is necessary to specify several performance requirements in a formal manner to ensure development of a satisfactory configuration item or when it is desirable to specify detailed verification procedures, a specification of this type is appropriate.

1.0 SCOPE
2.0 Applicable documents
3.0 Requirements.
3.1 Item Definition
3.2 Characteristics
3.2.1 Performance
3.2.2 Physical characteristics
4.0 Quality assurance provisions
5.0 Preparation for delivery
6.0 Ethics
7.0 Sustainability
8.0 Notes

26.3.2.4 Facility or Ship Development Specification

A Type B4 Facility or Ship Development Specification is applicable to each item, which is both a fixed (or floating) installation and an integral part of a system. Basic structural, architectural or operational features designed specifically to accommodate the requirements unique to the system and which must be developed in close coordination with the system. A facility or ship services which form complex interfaces with the system. A facility or ship hardening to decrease the total system's vulnerability; and ship speed, maneuverability, etc. A development specification for a facility or ship establishes the requirements and basic restraints / constraints imposed on the development of an architectural and engineering design for such facility or ship. The product specifications for the facility or ship are prepared by the architectural / engineering activity.

1.0 SCOPE
2.0 Applicable documents
3.0 Requirements
3.1 Facility or ship definition
3.1.1 Facility or ship drawings
3.1.2 Interface Definition
3.1.3 Major subsystems and component list
3.2 Characteristics
3.3 Documentation
4.0 Quality assurance provisions
5.0 Preparation for delivery
6.0 Ethics
7.0 Sustainability
8.0 Notes
Appendices

The main emphasis is on 3.2 Characteristics.

a. **Civil:** (1) Axle or wheel loads on roads, (2) Special lane width, (3) Turn and weight provisions for special vehicles, (4) Jack loads transfer requirements, (5) Parking (number of vehicles), (6) Grades on roads types pavement (flexible or rigid) type walks (flexible or rigid), (7) Special water and sewage requirements. Quantity and nature of water and sewage if special, (8) Special fire protection requirements (exterior), (9) Fencing and security, (10) Location and types of existing utilities if any (water gas sewer electrical storm drainage).
b. **Architectural:** (1) Personnel occupancy types hours per day, (2) Designation of use of areas within facility. Partition layout. Hazard areas. Special treatment areas, (3) Types of special doors required, (4) Floor level requirements. Floor drainage, (5) Window requirements if any, (6) Controlling dimension requirements, (7) Clear ceiling heights, (8) Exterior architectural treatment (concrete masonry brick etc.). Indicate whether treatment is to match existing if applicable, (9) Explosive safety requirements for construction.
c. **Structural:** (1) Crane and hoist location and loads. Control requirements. (2) Floor and roof loads. Special loads seismic loads wind loads, (3) Clear span and column-free areas, (4) Blast loads shielding requirements, (5) Personnel ladders elevators, (6) Transfer piers dock loads, (7) General configuration of building number of stories, (8) Barricades and shielding for explosive blast areas.
d. **Mechanical:** (1) Interior potable water. (2) Environment limits temperature humidity ventilation, (3) Compressed air, (4) Fire protection, (5) Vibration and acoustical requirements, (6) Equipment cooling requirements.
e. **Electrical:** (1) Power requirements - types and magnitude. (2) Light intensities, (3) Communications requirements, (4) Grounding.
f. **Equipment:** Provide layout and list each piece of equipment (1) Equipment name. (2) Units required (number), (3) Purpose of equipment, (4) Size of equipment (governing dimensions weight), (5) Power requirements - heat

gain BTU's per hour type cooling in-out temperatures relative humidity, (6) Minimum access requirements - front back sides. Ship characteristics shall include the consideration of the following as necessary.

g. **General:** (1) Limiting dimensions, (2) Weight control, (3) Reliability and maintainability, (4) Environmental conditions, (5) Standardization and interchangeability, (6) Shock noise and vibration, (7) Navy or commercial marine standards including certification of the latter.

h. **Hull structure:** (1) Structural loading and configuration, (2) Basic structural materials, (3) Welding riveting and fastenings, (4) Access features.

i. **Propulsion plant:** (1) Type and number of propulsion units, (2) Type and number of propellers, (3) Propulsion control equipment.

j. **Electric plant:** (1) Type and number of generator units \, (2) Power distribution system, (3) Lighting system.

k. **Communications and control:** (1) Navigation equipment, (2) Interior communication systems and equipment, (3) Electronics systems, (4) Weapon control systems.

l. **Auxiliary system:** (1) Air conditioning system, (2) Fuel systems, (3) Fresh and sea water systems, (4) Steering system, (5) Aircraft handling system, (6) Underway replenishment system, (7) Cargo handling system.

m. **Outfit and furnishings:** (1) Hull fittings boat storage and rigging, (2) Painting deck covering and insulation, (3) Special stowage, (4) Workshops and utility spaces, (5) Living spaces and habitability.

n. **Armament:** (1) Guns and ammunition stowage and handling, (2) Ship-launched weapon systems, (3) Cargo munitions handling and stowage.

26.3.2.5 Software Development Specification

Type B5 Software Development Specifications include both software and interface requirement specifications.

Software Requirements Specifications describe in detail the functional, interface, quality factor, special, and qualification requirements necessary to design, develop, test, evaluate and deliver the required software item.

Interface Requirements Specifications describe in detail the requirements for one or more software item interfaces in the system, segment, or prime item. The requirements are those necessary to design, develop, test, evaluate, and deliver the required software. The interface requirements may be included in the associated Software Requirements Specifications under the following conditions: (1) there are few interfaces, (2) few development groups are involved in implementing the interface requirements, (3) the interfaces are simple, or (4) there is one contractor developing the software.

1.0 Scope

3.8 Security requirements

1.1 Identification
1.2 CSCI overview
1.3 Document overview
2.0 Applicable documents
2.1 Government documents
2.2 Non-Government documents
3.0 Engineering requirements
3.1 CSCI external interface requirements
3.2 CSCI capability requirements
3.2.x Capability x
3.3 CSCI internal interfaces
3.4 CSCI data element requirements
3.5 Adaptation requirements
3.5.1 Installation dependant data
3.5.2 Operational parameters
3.6 Sizing and timing requirements
3.7 Safety requirements
3.9 Design constraints
3.10 Software quality factors
3.11 Human performance/human engineering requirements
3.11.1 Human information processing
3.11.2 Foreseeable human errors
3.11.3 Total system implications (e.g. training, support, operational environment)
3.12 Requirements traceability
4 Qualification requirements
4.1 Methods
4.2 Special
5.0 Preparation for delivery
6.0 Ethics
7.0 Sustainability
8.0 Notes
Appendices

26.3.3 C Level Product Specifications

Type C Product Specifications apply to any configuration item below the system level, and may be oriented toward procurement of a product through specification of primarily functional (performance) requirements or primarily fabrication (detailed design) requirements. There are Sub-types of product specifications to cover equipment of various complexities or using different outlines of form.

A product function specification states the complete performance requirements of the product for the intended use, and necessary interface and interchangeability characteristics. It covers form, fit, and function. Complete performance requirements include all essential functional requirements under service environmental conditions or under conditions simulating the service environment. Quality assurance provisions for hardware include one or more of the following inspections: qualification evaluation, pre-production, periodic production, and quality conformance.

A product fabrication specification is normally prepared when both development and production of the hardware are procured. In those cases where a development specification (Type B) has been prepared, specific reference to the document containing the performance requirements for the hardware are made in the product fabrication specification. These specifications state a detailed description of the parts and assemblies of the product, usually by prescribing compliance with a set of drawings, and those performance requirements and corresponding tests and inspections necessary to assure proper fabrication, adjustment, and assembly techniques. Tests normally are limited to acceptance tests in the shop environment. Selected performance requirements in the normal shop or test area environment and verifying tests may be included. Pre-production or periodic tests to be performed on a sampling basis and needing service, or other, environment may reference the associated development specification. Product fabrication specifications may be prepared as Part II of a two-part specification when the contracting organization desires close relationships between the performance and fabrication requirements.

26.3.3.1 Prime Item Product Specifications

Type C1 Prime Item Product Specifications apply to items meeting the criteria for prime item development specifications (Type B1). They may be prepared as function or fabrication specifications as determined by the procurement conditions.

26.3.3.1.1 Prime Item Product Function Specification

A Type C1a Prime Item Product Function Specification applies to procurement of prime items when a "form, fit and function" description is acceptable. Normally, this type of specification is prepared only when a single procurement is anticipated and training and logistic considerations are unimportant.

1. SCOPE
1.1 Scope
1.2 Classification
2.0 Applicable documents
3.0 Requirements
3.1 Item definition
3.1.1 Prime item diagrams
3.1.2 Interface definition
3.1.3 Major component list
3.1.4 Government-furnished property list
3.2 Characteristics
3.2.1 Performance
3.2.2 Physical characteristics
3.2.3 Reliability
3.2.4 Maintainability
3.2.5 Environmental conditions
3.2.6 Transportability
3.3 Design and construction
3.3.1 Materials processes and parts
3.3.2 Electromagnetic radiation
3.3.3 Identification and marking
3.3.4 Workmanship
3.3.5 Interchangeability
3.3.6 Safety
3.3.7 Human performance/human engineering
3.3.8 Standards of manufacture
3.4 Major component characteristics
3.5 Qualification Preproduction Periodic production inspection
3.6 Standard sample
4.0 Quality Assurance Provisions
4.1 General
4.1.1 Responsibility for inspection
4.1.2 Special tests and examinations
4.2 Quality conformance inspections
5.0 Preparation for delivery
6.0 Ethics
7.0 Sustainability
8.0 Notes

26.3.3.1.2 Prime Item Product Fabrication Specification

Type C1b Prime Item Product Fabrication Specifications are normally prepared for procurement of prime items when a detailed design disclosure package needs to be made available, it is desired to control the interchangeability of lower level components and parts, and service maintenance and training are significant factors.

1. SCOPE
1.1. Scope
1.2 Classification
2.0 Applicable documents
3.0 Requirements
3.1 Prime item definition
3.1.1 Major component list
3.1.2 Government furnished property list
3.2 Characteristics
3.2.1 Performance
3.3 Design and construction
3.3.1 Production drawings
3.3.2 Standards of manufacture
3.3.3 Workmanship
3.4 Pre-production sample
4.0 Quality Assurance Provisions.
4.1 General
4.1.1 Responsibility for inspection
4.1.2 Special tests and examinations
4.2 Quality conformance inspections
5.0 Preparation for delivery
6.0 Ethics
7.0 Sustainability
8.0 Notes
8.1 Intended use
8.2 Ordering data

26.3.3.2 Critical Item Product Specification

Type C2 Critical Item Product Specifications apply to engineering or logistic critical items and may be prepared as function or fabrication specifications.

26.3.3.2.1 Critical Item Product Function Specification

A Type C2a Critical Item Product Function Specification applies to a critical item where the critical item performance characteristics are of greater concern than part interchangeability or control over the details of design, and a "form, fit and function" description is adequate.

1. SCOPE
3.0 Requirements
3.1 Critical item definition
3.2 Characteristics
3.2.1 Performance
3.2.2 Physical characteristics
3.2.3 Reliability
3.2.4 Maintainability
3.2.5 Environmental conditions
3.2.6 Transportability
3.3 Design and construction
3.3.1 Materials processes and parts
3.3.2 Electromagnetic radiation
3.3.3 Identification and marking
3.3.4 Workmanship
3.3.5 Interchangeability
3.3.6 Safety
3.3.7 Human performance/human engineering 3.3.8 Standards of manufacture
3.4 Qualification (Pre-production) (Periodic production) inspection
3.5 Standard sample
4.0 Quality Assurance Provisions
4.1 General.
4.1.1 Responsibility for inspection
4.1.2 Special tests and examinations
4.2 Quality conformance inspections
5.0 Preparation for delivery
6.0 Ethics
7.0 Sustainability
8.0 Notes

26.3.3.2.2 Critical Item Product Fabrication Specification

A Type C2b Critical Item Product Fabrication Specification applies to a critical item when a detailed design disclosure needs to be made available or where it is considered that adequate performance can be achieved by adherence to a set of detail drawings and required processes.

1. SCOPE	4.1 General
3.0 Requirements	4.1.1 Responsibility for inspection
3.1 Critical item definition	4.1.2 Special tests and examinations
3.1.1 Government furnished property list	4.2 Quality conformance inspections
3.2 Characteristics	5.0 Preparation for delivery
3.2.1 Performance	**6.0 Ethics**
3.3 Design and construction	**7.0 Sustainability**
3.3.1 Production drawings	8.0 Notes
3.3.2 Standards of manufacture	8.1 Intended use
3.3.3 Workmanship	8.2 Ordering data
3.4 Pre-production sample	Appendices
4.0 Quality Assurance Provisions	

26.3.3.3 Non-Complex Item Product Fabrication Specification

A Type C3 Non-Complex Item Product Fabrication Specification applies to non-complex items. Where acquisition of a non-complex item is desired against a detailed design, a set of detail drawings may be prepared in lieu of a specification.

1.0 SCOPE
2.0 Applicable documents
3.0 Requirements
3.1 Non-complex item definition
3.2 Characteristics
3.2.1 Performance
3.2.2 Physical characteristics
3.3 Workmanship
3.4 Qualification inspection and samples
4.0 Quality Assurance Provisions
4.1 General
4.1.1 Responsibility for inspection
4.1.2 Special tests and examinations
4.2 Quality conformance inspections
5.0 Preparation for delivery
6.0 Ethics
7.0 Sustainability
8.0 Notes

26.3.3.4 Inventory Item Specification

A Type C4 Inventory Item Specification identifies applicable inventory items including their pertinent characteristics that exist in the inventory and which can be incorporated in a prime item or in a system being developed. The purpose of the inventory specification is to stabilize the configuration of inventory items on the basis of both current capabilities of each inventory item and the requirements of the specific application, or to achieve equipment or component item standardization between or within a system or prime item. This puts the organization on notice for the performance and interface characteristics that are required, so that when changes for an inventory item are evaluated the needs of the various applications may be kept in mind. If this is not done, design changes may make an inventory item unsuitable for the system. A separate inventory item specification should be prepared, as required, for each system, subsystem, prime item or critical item in which inventory items are installed or which require the support of inventory items.

1.0 SCOPE

2.0 Applicable documents.

3.0 Requirements. This section includes a paragraph for each inventory item covered by the specification. Each paragraph references an appendix for characteristics of the inventory item. Each appendix includes all of the functional and physical requirements of the inventory item that must be satisfied to assure compatibility with the system/configuration item.

4.0 Quality Assurance Provisions. This section invokes the quality assurance provisions contained in the appendix applicable to each inventory item.

5.0 Preparation for Delivery. This section invokes the requirements of the applicable appendix.

6.0 Ethics

7.0 Sustainability

8.0 Notes

Appendixes. These sections include a function specification for each inventory item. A separate appendix is prepared for each required inventory item and requirements and quality assurance provisions specified are limited to those necessary to ensure the form, fit, and function required to achieve its intended purpose in the system/configuration item. The function specification is prepared in accordance with the applicable appendix of this standard.

26.3.3.5 Software Product Specification

A Type C5 Software Product Specification applies to the delivered software item and is sometimes referred to as the "as built" software specification. This specification consists of the final updated version of the Software Top-Level Design Document, the Software Detailed Design Document, the Database Design Document(s), Interface Design Document(s), and the source and object listings of the software.

The Software Top Level Design Document describes how the top-level computer software components (TLCSCs) implement the Software Requirements Specification and the Interface Requirements Specification. The Software Detailed Design Document describes the detailed decomposition of TLCSCs to lower level computer software components and software units.

The Database Design Document describes one or more databases used by the software. If there is more than one database, each database may be described in a separate Document.

The Interface Design Document provides the detailed design of one or more software interfaces. When Interface Requirements Specifications have been prepared, associated Interface Design Documents are prepared as well.

These specifications consist of the final up-dated versions of the Software Top

Level Design Document, the Software Detailed Design Document(s), the Database Design Document(s), the Interface Design Document(s) and source and object code listings of the software that has successfully undergone formal testing.

26.3.4 D Level Process Specification

A Type D Process Specification applies to a service, which is performed on a product or material. Examples of processes are heat treatment, welding, plating, packing, microfilming, and marking. Process specifications cover manufacturing techniques, which require a specific procedure to achieve satisfactory results. Where specific processes are essential to fabrication or procurement of a product or material, a process specification defines the specific processes. Normally, a process specification applies to production but may be prepared to control the development of a process.

1.0 SCOPE
1.1 Scope
1.2 Classification
2.0 Applicable documents
3.0 Requirements
3.1 Equipment
3.2 Materials
3.3 Required procedures and operations
3.4 Recommended procedures and operations
3.5 Certification
4.0 Quality assurance provisions
4.1 Responsibility for inspection
4.2 Monitoring procedures for equipment used in process
4.3 Monitoring procedures for materials
4.4 Certification
4.5 Test methods
5.0 Preparation for delivery
6.0 Ethics
7.0 Sustainability
8.0 Notes
8.1 Intended use
8.2 Definitions
Appendices

26.3.5 E Level Material Specification

A Type E Material Specification is applies to a raw material (chemical compound), mixtures (cleaning agents, paints), or semi-fabricated material (electrical cable, copper tubing) which are used in the fabrication of a product. Normally, a material specification applies to production but may be prepared to control the development of a material.

1.0 SCOPE
1.1 Scope
1.2 Classification
2.0 Applicable documents
3.0 Requirements
3.1 General material requirements
3.1.1 Character or quality
3.1.2 Formulation
3.1.3 Product characteristics
3.1.4 Chemical, electrical and mechanical properties
3.1.8 Identification and marking
3.1.9 Workmanship
3.2 Qualification (Pre-production)
4.0 Quality assurance provisions.
4.1 Responsibility for inspection
4.2 Special tests and examinations
4.3 Quality conformance inspection
4.4 Test methods
5.0 Preparation for delivery
6.0 Ethics

3.1.5 Environment conditions.
3.1.6 Stability, Shelf life, aging, etc
3.1.7 Toxic products and safety

7.0 Sustainability
8.0 Notes
Appendices

26.3.6 Two Part Specification

A Two Part Specification combines both development (performance) and product fabrication (detail design) specifications under a single specification number as Part I and Part II respectively. This practice requires both parts for a complete definition of both performance requirements and detailed design requirements governing fabrication. Under this practice, the development specification remains alive during the life of the hardware as the complete statement of performance requirements. Proposed design changes are evaluated against both the product fabrication and the development parts of the specification. To emphasize that two parts exist, the same specification number identifies both parts and each part is further identified as Part I or Part II. Two-part specifications are not applicable when the product specification is a product function specification or when it is a computer software specification.

26.4 Example Specification Trees

Many systems are not massive infrastructure systems. What would a specification tree look like for a smaller project? In many instances someone has done a significant amount of work and captured requirements that need to be understood and converted to a working system.

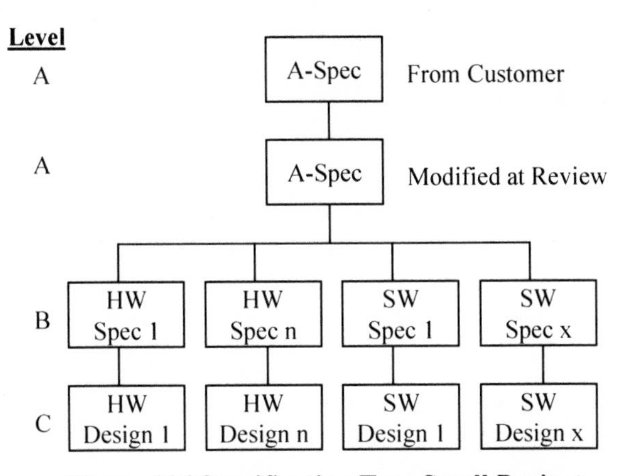

Figure 116 Specification Tree Small Project

A project starts when vendor receives a specification. The vendor works with the customer to develop a modified A-level specification. The discussions include feasibility, practical considerations, cost, and schedule. Eventually a modified A-level specification emerges and becomes the baseline. The vendor proceeds to decompose, design, and implement the system using various B and C level specifications.

C-level specifications provide design detail. Notice this tree does not contain D or E level specifications. Also the B and C levels can be various instances of their respective levels (e.g. B1, C3, etc).

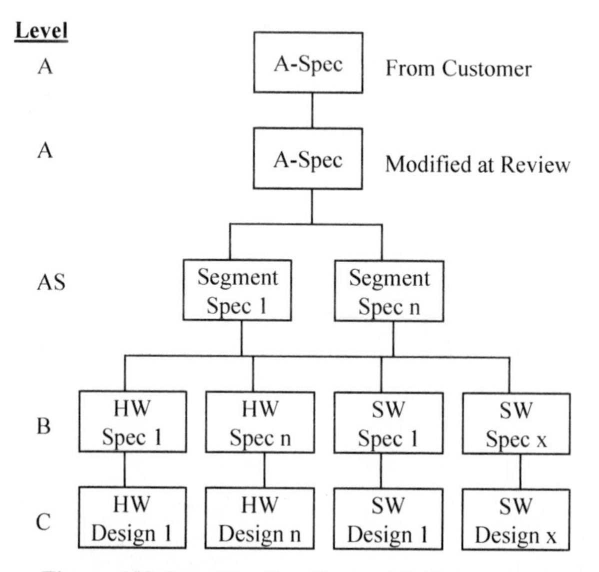

Figure 117 Specification Tree with Segments

Occasionally a segment specification is introduced. For example there may be the fielded system and then a duplicate system used for simulation, training, and support. The segment specification always presupposes architecture because it includes requirement allocations to its segment and allocates requirements to different subsystems in the segment.

Customers do not want to presuppose an architecture, which might be flawed without input from industry. So segment specifications are A-level specification but are labeled as “AS” for the special segment characteristic of the architecture being locked down. Who did that allocation how do you know it is correct?

There is some art in trying to determine if a design specification is needed. If you are buying a Commercial-Off-The-Shelf (COTS) item then perhaps all that is needed is a marketing glossy to maintain configuration control. However COTS configuration control is an oxymoron - why bother - in 6 months you will not be able to find the same COTS thing. So this is more of an art.

So you use A, B, and C to "slice and dice" your system in a way that makes sense for your needs. During the days of MIL-STD-490 the world was very different. There were:

- Engineering notebooks - housed raw information

- Typing pools - created papers and draft documents
- Drafting sections / departments - created blue prints
- Publications departments - created typeset documents / books
- Few computers - no complex models
- Prototypes - no computers to model the details just build a prototype

During that time there was a strong desire to not lose anything. This seemed to apply everywhere. Probably because it was so hard to create everything, things had great value and were not viewed as commodities.

Today with the computer we generate large amounts of paper because we can. So when an important document does need to be created it typically has no meat - it is filled with obvious fluff. This can lead to disaster in the case of specifications. As the years have clicked by, the names have changed and we dropped the requirements for MIL-STD-490 but the fundamentals still need to be present. A rose by any other name is still a rose.

26.5 Traceability

There are two aspects to traceability. The first is allocation to subsystems and links to lower level requirements. The second is the allocation to a test method and links to standalone test procedures, which can use the any of the test methods. As already stated, requirements should always appear in specifications. Studies surface requirements, including the extremely important study of surfacing the stakeholder needs and requirements.

There are different levels of requirement traceability. The levels of traceability are associated with different time frames and different goals. The levels are:

- Very Course Grain
- Course Grain
- Fine Grain
- Very Fine Grain

The terminology was taken from the processing domain where functions and their processing can be broken down into different size chunks. The finer the grain the smaller the chunk. As the numbers of chunks increase, the overhead or amount of effort increases. For traceability the differences between each chunk is very large. For example very course grain traceability might be between ten items. For the same system course grain traceability might jump to hundreds of items, fine grain traceability might jump to thousands of items and very fine grain traceability might jump to tens of thousands of items.

26.5.1 Very Course Grain Traceability

Very course grain traceability is at the document level. This is the simplest form of traceability to implement. It is a connection from a parent specification document to a child specification document. It is a specification tree. The tree can be shown as a block diagram or as an indentured list.

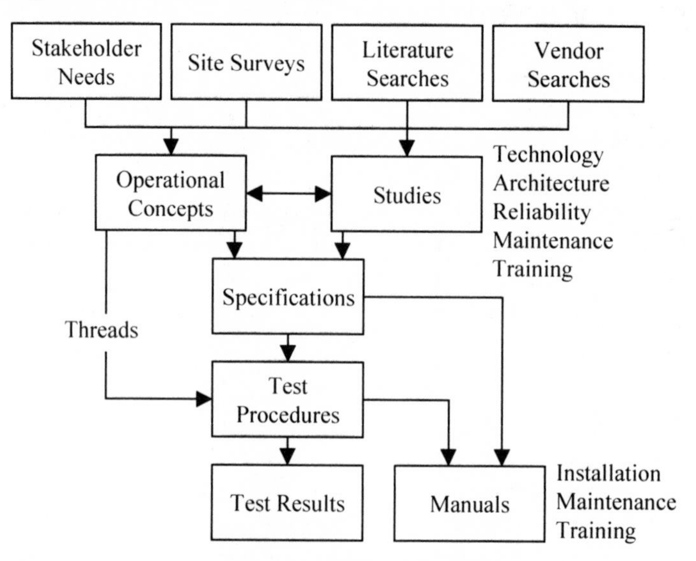

Figure 118 Document Tree

Many times traceability is also needed to other information products. Since the traceability is very course grain, this is not unreasonable. In fact showing how information products tie together is a good way to start to layout a program and understand how different activities interrelate.

Very course grain traceability is typically used in proposals and plans. It is also used as introductory material to show the context of an information product.

An indentured list specification tree is a decomposition of a specification block in the document tree. Each of the blocks can be decomposed. For example the studies consist of technology, architecture, reliability, maintenance, and training. These studies interrelate and can be shown in their own document tree.

Table 45 Indentured List Specification Tree
1.0 System/Segment Specification
1.1 Prime Item Development Specification
1.1.1 Prime Item Product Function Specification
1.1.2 Prime Item Product Fabrication Specification
1.1.2.1 Process Specification - production
1.1.2.2 Material Specification - production
1.1.3 Inventory Item Specification

Table 45 Indentured List Specification Tree

1.2 Critical Item Development Specification
- 1.2.1 Critical Item Product Function Specification
- 1.2.2 Critical Item Product Fabrication Specification
  - 1.2.2.1 Process Specification - production
  - 1.2.2.2 Material Specification - production
- 1.2.3 Inventory Item Specification

1.3 Non-Complex Item Development Specification
- 1.3.1 Non-Complex Item Product Fabrication Specification

1.4 Facility or Ship Development Specification
- 1.4.1 Inventory Item Specification

1.5 Software Development Specification
- 1.5.1 Software Product Specification

1.6 Inventory Item Specification - system
1.7 Process Specification - development
1.8 Material Specification - system

26.5.2 Course Grain Traceability

Course grain traceability is at the paragraph level. It is typically used to show requirement traceability when there are no automated tools such as a System Requirements Database (SRDB). The child document contains a Requirement Traceability Matrix (RTM) as an appendix. The RTM lists the current document paragraph numbers and headings in the left most column of a table. Each paragraph is a separate row. An additional column lists the paragraph number and optionally the heading of the parent document.

Table 46 Requirement Paragraph Level Traceability Matrix

Subsystem Specification	System Specification
3.2.1 Function 10	3.7
3.2.2 Function 11	3.7
3.2.3 Function 12	3.3, 3.4, 3.5
3.2.4 Function 13	3.1
3.2.5 Function 14	3.2

26.5.3 Fine Grain Traceability

Fine grain traceability is at the requirement statement level. It is used to show requirement traceability when there is an automated tool such as a System

Requirements Database (SRDB). Projects and programs that use an SRDB have one or more of the following system characteristics:

- Mission and or safety critical[214]
- Must be certified[215] before they can go operational
- Very large with multiple specifications at the same levels
- Anticipate significant growth and change
- Involve goods and services with significant financial risk
- Susceptible to failure due to poor or missing requirements[216]
- Software intensive

Typically requirements are initially captured in a document using office automation tools such as word processing or spreadsheets. Then the document is imported into the SRDB. As new requirements are developed and other requirements are changed the SRDB is the primary mode of authoring. When a customer delivery is needed or a baseline established the SRDB content is exported into a document format. Various techniques are used to ensure when an SRDB export is performed, a reasonable document is produced with minimal changes in word processing. At no time should a document and SRDB version be manually maintained. This will lead to inconsistencies, which are unacceptable for a project that demands the accuracy of an SRDB in the first place.

Occasionally the requirements are first entered into the SRDB. In these instances office automation tools may be used to create figures and tables, but rather than being imported they are copy and pasted into the SRDB.

Writing requirements destined for the SRDB has a slightly different style. The document headings stay the same, however the content at the paragraph level is parsed into stand-alone lines rather than clumped as one or more paragraphs clustered below a heading. When the document is imported into the SRDB then each line becomes a separate row in the SRDB with its own project unique identified (PUI) number. The PUI items then can be marked for traceability to other PUI items in upper or lower level specifications.

If the paragraphs are not parsed then the entire paragraph becomes a row in the database with one PUI number. The traceability then only is to the paragraph level rather than the requirement level, course grain rather than fine grain traceability. The advantage of the SRDB is then lost[217].

[214] Mission critical systems are systems where there can be loss of life or people can be harmed if the system fails.

[215] Certification usually involves an external authority outside the sponsor and user community that must approve of the system.

[216] Almost all software based systems suffer from this characteristic

[217] There is a separate section dedicated to the SRDB in this text that will provide more information.

Specifications using an SRDB include a parent child report in the appendix in place of the Requirement Traceability Matrix (RTM). While the RTM is at the paragraphs level showing the headings of the parent, the parent child report shows the PUI and text of both the parent and child requirements. Just like in the RTM the parent is the left column and the child is the right column. This obviously significantly increases the size of the specification and so the parent child report is typically delivered under a separate cover. The specification in this case just houses a matrix with just the PUI numbers of both the parent and child or the PUI and text of the parent and just the PUI for the child.

Table 47 Parent Child Report[218]

PUI	System Text	Subsystem PUI and Text
SSS-14	Requirement A	SRS-77 Requirement SRS-81 Requirement SRS-22 Requirement SRS-23 Requirement
SSS-15	Requirement B	SRS-77 Requirement SRS-78 Requirement
SSS-16	Requirement C	SRS-79 Requirement
SSS-17	Requirement D	Missing child
SSS-18	Requirement E	SRS-24 Requirement

Unlike course grain traceability, fine grain traceability includes a report showing that all the parent requirements trace to one or more child requirements. That means the parent document remains alive until the child document is complete. This process ensures that requirements are complete and consistent between levels. This is a critical step that significantly improves the quality of the requirements and attempts to surface all the requirements. A missing child link is a sign of a problem that must be corrected.

Table 48 Child Parent Report

PUI	Subsystem Text	System PUI and Text
SRS-77	Requirement	SSS-14 Requirement SSS-15 Requirement
SRS-78	Requirement	SSS-15 Requirement
SRS-79	Requirement	SSS-16 Requirement
SRS-80	Requirement	Missing parent
SRS-81	Requirement	SSS-14 Requirement

[218] Notice this table is associated with the parent document rather than just the child document. In fine grain traceability a table is produced for both the parent and child documents.

26.5.4 Very Fine Grain Traceability

Very Fine grain traceability starts with fine grain traceability and continues to the implementation level. For example software source code can be compiled and linked into an executable, the final implementation. In these cases the allocation continues to the implementation information products.

A-Spec Formal Module View: Export Test w/allocations

ID	Text	IsReq	Alloc	SRSPUI	Test	Rel
SYS -135	3.2.1 The system shall allow users to create a user account.	Yes	Software	SRS-98	12. Acct Mgmt	1.0

SRS Formal Module View: Peer Review

SRS	Text	IsReq	SYSPUI	CPC	File	Rel
SRS -98	3.7.3 Upon receiving an email acknowledgement from the user, account processing shall activate the users account.	Yes	SYS-135	ACCT	reg.cgi	1.0

reg.cgi file

Source code line 1

SRS-98 comment line

Source code line n

Figure 119 Very Fine Grain Traceability Approach

The approach is to take the requirement PUI and allocate it into the implementation products as comments. The requirement is placed as close to the end of where the requirement is satisfied as possible. Placing the requirement comment at a higher abstraction level will only obfuscate and defeat the purpose of the allocation to the implementation detail. Also bunching requirements at the end of a long process also defeats the purpose of the allocation.

26.6 Traceability Reviews

Normally you cannot inspect-in engineering work and quality. However in the case of traceability, verifying the traceability links through inspection is critical. This is a tedious demanding task. A single individual cannot do it. This is a team effort, however the leads of each area must take responsibility. They must know how their products trace from and trace to other products on the project or program.

26.6.1 Specification Parent Child Reports

Parent child reports are produced for each specification. If there are holes, this is an indication of missing requirements or requirements not needed for the system. There are two reports produced, parent and child. This cannot be done with a single parent or child report. Both perspectives and views are needed for closure.

The first report lists the parent requirements in the left-hand column and the child requirements in the right hand column. If a right hand column is blank, this is an indication of missing children. This represents forward traceability.

The second report lists the child requirements in the left-hand column and the parent requirements in the right hand column. If a right hand column is blank, this is an indication of a missing parent. This represents backward traceability.

26.6.2 Test Parent Child Reports

Just as in the case of specifications, parent child reports are produced to show the forward and backward traceability between a specification and its tests. If there are holes, they must be addressed. Requirements without test links are an indication of insufficient test coverage. Multiple tests with large numbers of common requirement links are a possible indication of repetitive tests that do not add to the test coverage. A test with large a large number of requirement links suggests that the test may be eligible for further decomposition.

26.7 Requirements Management

Requirements change as the system is developed and used. These changes need to be understood and tracked. Prior to the release of a specification, requirement changes are easier. The other team members are mostly working with the key requirements and anxiously awaiting the details that are in a specification that they will consume when it is released.

Once a specification is released the consuming team is impacted. Many times it is the consuming team that suggests the changes, as the decomposition is more fully understood. After the specification is released the requirement changes are tracked.

When the design and implementation starts there are also requirement changes. These changes are against the baseline specification so a more formal mechanism needs to be established to propose and either accept or reject the changes.

When the system goes operational there are always suggestions for improvement. These suggestions translate to requirements that need to be managed. Many times a successful system will evolve into a product line. The product line releases need to be managed.

The various requirement changes are best managed with a System Requirements Database Management (SRDB) tool that will capture the history of the changes. One or more attributes and a history of the changes are used to determine how various requirements started and evolved. The SRDB is described in the System Requirements Database Management section of this text.

At the highest level requirement management is established with a specification document, which has a title page that contains the title, sponsor or customer, date, and a version number. This establishes a baseline. As requirements change the specification is released with a new date and version number.

26.8 Exercises

1. Where did MIL-STD-490 come from, who actually wrote the original and why?
2. Where did the concept of Data Item Descriptions (DIDs) come from, who actually wrote the first one for a contract and why?
3. What existed before MIL-STD-490?
4. What existed before DIDs?
5. Develop a specification tree for your automobile and computer.
6. Develop a specification tree for wind, solar, and geothermal power stations.
7. Develop a specification tree for your house.

26.9 Additional Reading

1. Defense Acquisition Guidebook, Defense Acquisition University, August 2010.
2. Specification Practices, Military Standard, MIL-STD-490 30 October 1968, MIL-STD-490A 4 June 1985.
3. Systems Engineering Fundamentals, Supplementary Text, Defense Acquisition University Press, January 2001.
4. System / Segment Specification, DOD Data Item Description DI-CMAN-80008A, June 1986.

27 Architecture Concepts and Synthesis

The system architecture is your single most important information product. All the systems practices to this point were to get you to system architecture. So what is architecture and how is it depicted or represented?

27.1 Background and Definitions

When most people hear architecture they think in terms of buildings such as cathedrals, opera houses, museums, etc. Some think in terms of bridges, skyscrapers, and roads. Here are some traditional definitions of architecture and architecture related terms[219]:

- **Architecture** 1. The science, art, or profession of designing and constructing buildings, etc. 2. a building, or buildings collectively 3. a style of construction [modern architecture] 4. Design and construction 5. Any framework, system, etc.
- **System** 1. A group of things or parts working together or connected in some way so as to form a whole [a solar system, school system, system of highways] 2. A set of principles, rules, etc linked in an ordinary way to show a logical plan [economic system] 3. A method or plan of classification or arrangement 4. a) An established way of doing something; method; procedure b) orderliness or methodical planning in one's way of proceeding 5. a) the body considered as a functioning organism b) a number of organs acting together to perform one of the main bodily functions [the nervous system] 6. A related series of objects or elements, as rivers 7. A major division of stratified rocks comprising the rocks laid down during a period.
- **Framework** 1. a structure to hold together or to support something built or stretched over or around it [the framework of a house] 2. a basic structure, arrangement, or system 3. same as frame of reference
- **Frame of reference** 1. the fixed points, lines, or planes from which coordinates are measured 2. The set of ideas, facts, or circumstances within which something exists

The oldest text on architecture is the *De architectura* written by Roman architect

[219] Webster's New World Dictionary, 1982, Simon & Schuster, ISBN 0-671-41816-5 (edged) or ISBN 0-671-41816-3 (indexed)

Marcus Vitruvius Pollio[220]. It is a collection of "Ten Books on Architecture" and is a guide for building projects[221]. The ten books are:

1. **First Principles and the Layout of Cities** education of the architect, fundamental principles of architecture, departments of architecture, site of a city, city walls, directions of the streets, remarks on the winds, sites for public buildings
2. **Building Materials** the origin of the dwelling house, on the primordial substance according to the physicists, brick, sand, lime, pozzolana, stone, methods of building walls, timber, highland and lowland fir
3. **Temples** on symmetry in temples and in the human body, classification of temples, proportions of intercolumniations and of columns, foundations and substructures of temples, proportions of the base, capitals, entablature in the ionic order
4. **Temples** the origins of the three orders, proportions of the Corinthian capital, ornaments of the orders, proportions of Doric temples, the cella and pronaos, how the temple should face, doorways of temples, Tucson temples, circular temples and other varieties, altars
5. **Public Buildings** the forum and basilica, treasury, prison, and senate house, the theatre: its site, foundations, and acoustics, harmonics, sounding vessels in the theatre, plan of the theatre, Greek theatres, acoustics of the site of a theatre, colonnades and walks, baths, the palaestra, harbors, breakwaters, and shipyards
6. **Private Buildings** on climate as determining the style of the house, symmetry, and modifications in it to suit the site, proportions of the principal rooms, the proper exposures of the different rooms, how the rooms should be suited to the station of the owner, the farmhouse, the Greek house, on foundations and substructures
7. **Finishing** floors, slaking of lime for stucco, vaulting and stucco work, stucco work in damp places, decoration of dining rooms, decadence of fresco painting, marble for use in stucco, natural colors, cinnabar and quicksilver, cinnabar, artificial colors, black, blue, burnt ochre, white lead, verdigris, and artificial sandarach, purple, substitutes for purple, yellow ochre, malachite green, and indigo
8. **Water** how to find water, rainwater, various properties of different waters, tests of good water, leveling and leveling instruments, aqueducts, wells, and cisterns
9. **Sundials and Clocks** zodiac and the planets, phases of the moon, course of

[220] Vitruvius: Ten Books on Architecture, Cambridge University Press, Cambridge 1999, Editors D. Rowland, T.N. Howe, 2001 ISBN 0521002923.

[221] Vitruvius The Ten Books on Architecture, Herbert Langford Warren (Illustrator), Morris Hickey Morgan (Translator) Courier Dover Publications, 1960, ISBN 0486206459.

the sun through the twelve signs, northern constellations, southern constellations, astrology and weather prognostics, analemma and its applications, sundials and water clocks

10. **Machines** and implements, hoisting machines, elements of motion, engines for raising water, water wheels and water mills, water screw, the pump of ctesibius, water organ, hodometer, catapults or scorpiones, ballistae, stringing and tuning of catapults, siege machines, tortoise, hegetor's tortoise, measures of defense, note on scamilli impares

Today we naturally recognize that architectures are everywhere and range from physical entities like our houses to abstractions like our forms of government. In 15 BC when it is speculated that Marcus Vitruvius Pollio wrote De architectura, that may not have been the case. The following are some excerpts from this 2000 plus year old body of work[222].

From Book 1 Chapter III

THE DEPARTMENTS OF ARCHITECTURE

1. There are three departments of architecture: the art of building, the making of timepieces, and the construction of machinery. Building is, in its turn, divided into two parts, of which the first is the construction of fortified towns and of works for general use in public places, and the second is the putting up of structures for private individuals. There are three classes of public buildings: the first for defensive, the second for religious, and the third for utilitarian purposes. Under defense comes the planning of walls, towers, and gates, permanent devices for resistance against hostile attacks; under religion, the erection of fanes and temples to the immortal gods; under utility, the provision of meeting places for public use, such as harbors, markets, colonnades, baths, theatres, promenades, and all other similar arrangements in public places.

*2. All these must be built with due reference to **durability**, **convenience**, and **beauty**. Durability will be assured when foundations are carried down to the solid ground and materials wisely and liberally selected; convenience, when the arrangement of the apartments is faultless and presents no hindrance to use, and when each class of building is assigned to its suitable and appropriate exposure; and beauty, when the appearance of the work is pleasing and in good taste, and when its members are in due proportion according to correct principles of*

---

[222] Vitruvius, The Ten Books On Architecture. Translated By, Morris Hicky Morgan, Ph.D. Ll.D. Professor Of Classical Philology, Harvard University. Illustrations And Original Designs, Prepared Under The Direction Of, Herbert Langford Warren, A.M., Nelson Robinson Jr. Professor Of Architecture, Harvard University. Cambridge, Harvard University Press, London: Humphrey Milford, Oxford University Press, 1914.

symmetry.

From Book 1 Chapter II

THE FUNDAMENTAL PRINCIPLES OF ARCHITECTURE

1. **Architecture** *depends on Order, Eurythmy, Symmetry, Propriety, and Economy.*

2. **Order** *gives due measure to the members of a work considered separately, and symmetrical agreement to the proportions of the whole. It is an* adjustment according to quantit. *By this I mean the selection of modules from the members of the work itself and, starting from these individual parts of members, constructing the whole work to correspond. Arrangement includes the putting of things in their proper places and the elegance of effect which is due to adjustments appropriate to the character of the* work. Its forms of expression *are these: groundplan, elevation, and perspective. A ground plan is made by the proper successive use of compasses and rule, through which we get outlines for the plane surfaces of buildings. An elevation is a picture of the front of a building, set upright and properly drawn in the proportions of the contemplated work. Perspective is the method of sketching a front with the sides withdrawing into the background, the lines all meeting in the center of a circle. All three come of reflection and invention. Reflection is careful and laborious thought, and watchful attention directed to the agreeable effect of one's plan. Invention, on the other hand, is the solving of intricate problems and the discovery of new principles by means of brilliancy and versatility. These are the departments belonging under Arrangement.*

3. **Eurythmy** *is beauty and fitness in the adjustments of the members. This is found when the members of a work are of a height suited to their breadth, of a breadth suited to their length, and, in a word, when they all correspond symmetrically.*

4. **Symmetry** *is a proper agreement between the members of the work itself, and relation between the different parts and the whole general scheme, in accordance with a certain part selected as standard. Thus in the human body there is a kind of symmetrical harmony between forearm, foot, palm, finger, and other small parts; and so it is with perfect buildings. In the case of temples, symmetry may be calculated from the thickness of a column, from a triglyph, or even from a module; in the ballista, from the hol; in a ship, from the space between the tholepin; and in other things, from various members.*

5. **Propriety** *is that perfection of style, which comes when a work is authoritatively constructed on approved principles. It arises from pres*cription, *from usage, or from nature. From prescription, in the case of hypaethral edifices, open to the sky, in honor of Jupiter Lightning, the Heaven, the Sun, or the Moon: for these are gods whose semblances and manifestations we behold before our very eyes in the sky when it is cloudless and bright. The temples of Minerva, Mars, and*

Hercules, will be Doric, since the virile strength of these gods makes daintiness entirely inappropriate to their houses. In temples to Venus, Flora, Proserpine, Spring- Water, and the Nymphs, the Corinthian order will be found to have peculiar significance, because these are delicate divinities and so its rather slender outlines, its flowers, leaves, and ornamental volutes will lend propriety where it is due. The construction of temples of the Ionic order to Juno, Diana, Father Bacchus, and the other gods of that kind, will be in keeping with the middle position which they hold; for the building of such will be an appropriate combination of the severity of the Doric and the delicacy of the Corinthian.

6. ***Propriety*** *arises from usage when buildings having magnificent interiors are provided with elegant entrance-courts to correspond; for there will be no propriety in the spectacle of an elegant interior approached by a low, mean entrance. Or, if dentils be carved in the cornice of the Doric entablature or triglyphs represented in the Ionic entablature over the cushion shaped capitals of the columns, the effect will be spoilt by the transfer of the peculiarities of the one order of building to the other, the usage in each class having been fixed long ago.*

7. ***Finally, propriety*** *will be due to natural causes if, for example, in the case of all sacred precincts we select very healthy neighborhoods with suitable springs of water in the places where the fanes are to be built, particularly in the case of those to Aesculapius and to Health, gods by whose healing powers great numbers of the sick are apparently cured. For when their diseased bodies are transferred from an unhealthy to a healthy spot, and treated with waters from health-giving springs, they will the more speedily grow well. The result will be that the divinity will stand in higher esteem and find his dignity increased, all owing to the nature of his site. There will also be natural propriety in using an eastern light for bedrooms and libraries, a western light in winter for baths and winter apartments, and a northern light for picture galleries and other places in which a steady light is needed; for that quarter of the sky grows neither light nor dark with the course of the sun, but remains steady and unshifting all day long.*

8. ***Economy*** *denotes the proper management of materials and of site, as well as a thrifty balancing of cost and common sense in the construction of works. This will be observed if, in the first place, the architect does not demand things which cannot be found or made ready without great expense. For example: it is not everywhere that there is plenty of pitsand, rubble, fir, clear fir, and marble, since they are produced in different places and to assemble them is difficult and costly. Where there is no pitsand, we must use the kinds washed up by rivers or by the sea; the lack of fir and clear fir may be evaded by using cypress, poplar, elm, or pine; and other problems we must solve in similar ways.*

9. ***A second stage in Economy*** *is reached when we have to plan the different kinds of dwellings suitable for ordinary householders, for great wealth, or for the high position of the statesman. A house in town obviously calls for one form of construction; that into which stream the products of country estates requires* *another; this will not be the same in the case of money-lenders and still different for the opulent and luxurious; for the powers under whose deliberations the commonwealth is guided dwellings are to be provided according to their special needs: and, in a word, the proper form of economy must be observed in building houses for each and every class.*

27.2 Architecture Philosophy

In the 19th century, Louis Sullivan[223] suggested that form follows function. The concept is that the shape of a building or object should be based on its intended function or purpose.

With the introduction of computer automation, architecture took on a whole new meaning. Not only would the computer system itself embody an architecture that someone would need to define, but also its introduction into an organization would transform the architecture of the organization. Systems analysts surfaced which would attempt to understand current operations and how the computer might transform those operations. The transformation might include increased quality, greater capacity, or reduced staffing needs. Other examples include early satellite systems and spaceships that took people into Earth orbit and eventually to the moon. Everywhere questions were being asked: what should the architecture be for our: satellite system, lunar lander, computer based automation system, etc. The system boundaries were broad. These solutions included a mixture of technical solutions and policy solutions.

Probably one of the strongest architecture houses at the time was Hughes Aircraft. Known for building many first versions of systems that are now part of our infrastructure, they were driven by the concept of architecture, architecture analysis, and architecture selection. The idea, that "once the proper architecture is selected anyone could then go and implement it with relatively little risks"[224] was fundamental and in everyday technical conversation. Unlike many other entities who would conceive of architectures then walk away and let someone else try to implement the architecture, Hughes was known for making sure the system as driven by the architecture worked and worked well. Sometimes that meant they implemented the first version of the system. So implementation was also part of the recipe.

Architectures should be beautiful. They should raise goose pimples. They should be elegant, simple, and symmetric. Architectures should be well understood by everyone. Architectures should be characterized by science and engineering in

---

[223] Louis Henri Sullivan (September 3, 1856 to April 14, 1924) American architect called "father of modernism." Creator of modern skyscraper, mentored Frank Lloyd Wright, inspiration to Prairie School.

[224] This was the culture and the words used while I worked at Hughes in the early 1980's. The quote was consistent from the senior staff.

studies. These studies should use any and all techniques from 5000 plus years of civilization. New architectures require new scientific and engineering processes, tools, and techniques. Architectures should be born from the fire of tradeoff and alternatives from people seeking truth.

Patterns are fleeting and change with time and the challenges. Operational people tend to think in terms of patterns. As subject matter experts they tend to bring the current system into the new architecture. They naturally resist new patterns because they feel they are unproven.

By definition state-of-the-art has few patterns and true innovation or beyond state-of-the-art has no patterns. Now a very important observation, humans find patterns. Humans find patterns where no patterns exist. It is fundamental to the way in which humans solve problems. So patterns are fundamental to solutions and all its studies, but the patterns should be new otherwise only previous state-of-the-art is being duplicated, which may be ok.

Architectures are everywhere. A leaf is part of a family of leaves. It has architecture. The tire on your automobile is an architecture, one of many possibilities that someone selected. Architectures exist at all abstraction levels. Your home is an architecture, the kitchen is an architecture, the screws that hold your appliances together are an architecture. The point is that architecture is not something special that exists at some very high level. Choices need to be made at every level and care should be taken at all levels to ensure the choices are the best fit for the solution. For example, torx, phillips, and flat head screws are different architectures that address the same requirement of fastening.

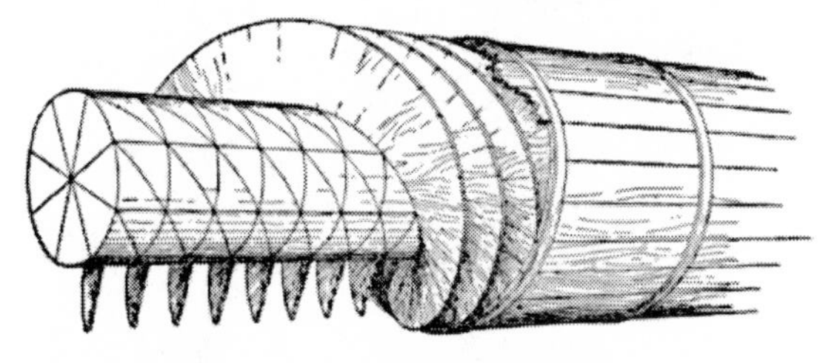

Figure 120 Construction of the Water Screw

Construction of the Water Screw is from the edition of "Vitruvius by Fra Giocondo, Venice, 1511".

27.3 Architecture Types

When we think of buildings and architecture images of Greek, Roman, Gothic, Victorian, and other building architectures appear in our minds. There are images and words that we use to describe these building architectures.

When we think of music we may not immediately equate different music genres to architecture types but we are able to detect differences in the music. We recognize classical, jazz, folk, gospel, rock, country, pop, disco, easy listening, etc. In the middle of the 1960's there were many musical groups experimenting with music. They were looking for new sounds or architectures as electronics had transformed musical instruments. Keith Relf and Jim McCarty offered a new musical "architecture" when they merged classical, jazz, folk, and rock styles to solidify and establish progressive rock as new genera of music[225]. This is an example of synthesis on a grand scale where alternative architectures are merged to form a new architecture.

Are there architecture types and do we consider them when we start to conceive our architectures? The short answer is yes. We cannot avoid it; our background always comes into play even if we are not fully conscious of the influences.

Architectures have structure, behavior, and internal connectivity. In the last century those engaged in computer architectures were consciously introduced to different architecture types and their behavior. These architectures were borrowed from observations of nature and human organization. We started to study and try to understand these systems using various techniques. So understanding these computer architectures provides a framework for future system architects irrespective of the system being developed.

27.3.1 Centralized Architecture

At one extreme is the concept of a centralized architecture. This is characterized by a very large central element responsible for most if not all the functions. It attaches to lower level entities with little or no processing power that usually are given the function of interfacing to the outside world, the peripherals. So a mainframe computer would do all the processing while printers, displays, keyboards, trackballs, switches, indicators, etc would interface to the outside world. Disk drives, tape drives, paper tape readers / punchers, card readers / punchers, etc would read and store all the data.

There are reasonable non-computer analogies. For example, in an organizational setting many people would be given a function with little or no decision authority while one individual or a very small group would have all the decision authority. In a structural setting one central element might be given the load bearing weight such as in a single beam in the basement of house. In a spatial setting such as a restaurant, one room is used to house all the customers and serve as

---

[225] In January 1969, former Yardbirds members Keith Relf, and Jim McCarty formed a new group devoted to experimentation between rock, folk, classical, and other forms. Relf guitar & vocals, McCarty drums, Louis Cennamo bassist, John Hawken pianist, and Relf's sister Jane Relf as an additional vocalist released albums on Elektra (US) and Island (UK-ILPS 9112), titled "Renaissance", produced by fellow ex-Yardbird Paul Samwell Smith. Renaissance reformed in 1971 with Annie Haslam vocals, Michael Dunford guitar, John Tout keyboards, Jon Camp bass and vocals, Terence Sullivan drums and percussion, Betty Thatcher the poet who worked with Michael Dunford to create the lyrics.

the area for all eating activities.

The advantages of a centralized architecture are:

- Little overhead to coordinate, everything is done in one place
- Consistent processing of all functions
- Least amount of parts to break
- Simple one direction connection control communications[226]

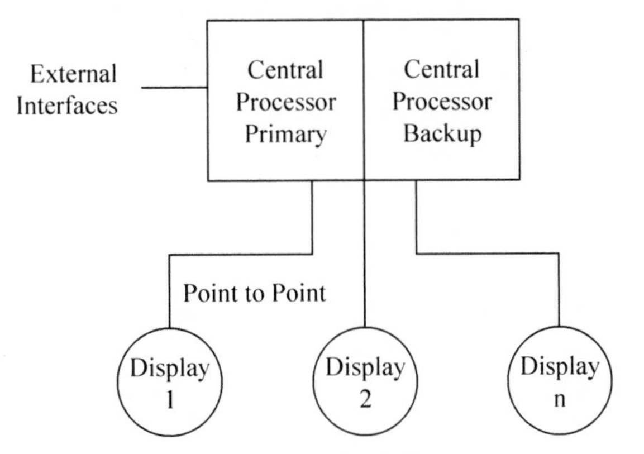

Figure 121 Centralized Architecture

The disadvantages of a centralized architecture are:

- Everything stops and fails when the central element breaks
- Everything stops or fails near maximum capacity limits
- Impact of Failure (IOF) is greatest
- Priorities are needed for running functions on the one element
- Slower response time as load increases
- Requests need to be queued and some are requests are always last

---

[226] Master/slave is a communication model where one device or process has one way directional control over one or more other devices. Examples are:

- A master database is the authoritative source; the slave databases synchronize to the master.
- Peripherals connected to a computer. Some peripherals can assume the role of a master such as when initiating a scanning session from a scanner front panel button.
- Multiple railway locomotives with the operation of all locomotives in the train slaved to the controls of the first locomotive.
- Operating controls on a master cassette tape or compact disc machine triggers the same commands on the slave units, so that duplication recording is done in parallel.

27.3.2 Fully Distributed Architecture

At the other extreme is the fully distributed architecture. This is characterized by many small-processing elements responsible for all the functions. The functions or data can be split across small processing elements. There can be combinations of split functions and data. An example of this is the personal computer. We can each sit down at our fully distributed element and perform any function using our data. Anyone else can sit down and perform the same function using their data. We also can log on to our bank computer and perform financial transactions from what we view as a centralized resource[227]. In the fully distributed architecture there is a maximum level of functional and data distribution. It is an idealized concept with compromise limiting the level of distribution as the system is considered.

The advantages of a fully distributed architecture are:

- No single point of failure
- Failure is isolated to the failed element
- Impact of Failure (IOF) is completely minimized
- No need to queue requests
- Processing performed on the spot with very fast response time

The disadvantages of a fully distributed architecture are:

- Significant overhead if processors need to coordinate
- Complex startup depending on coordination
- Greatest amount of parts
- Difficult to manage and assess status

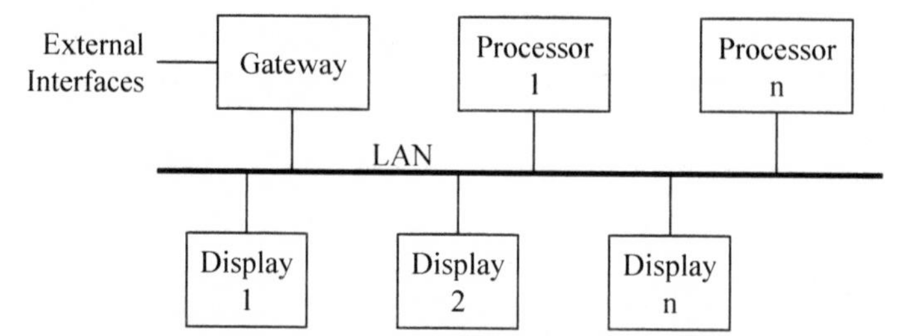

Figure 122 Fully Distributed Architecture

27.3.3 Partially Distributed Architecture

The partially distributed architecture attempts to merge the advantages of the

---

[227] A banking computer system is actually highly distributed but we all connect to it via the Internet in client server architecture. We perceive it as a central resource.

fully distributed and centralized architecture. Functions that can be performed at the lowest distribution level because they are standalone atomic functions with no interaction with other functions are allocated to the distributed elements. This may be because there is a clean split in the data or little functional coupling. Functions that span across the lowest level of the distributed elements are allocated to the central processor.

The advantages of a partially distributed architecture are:

- There is no single point of failure. If the centralized resource fails some functionality is offered by distributed elements.
- Impact of Failure (IOF) is lowered. If the centralized resource fails some functionality is offered by distributed elements.
- The overhead associated with functions that need to intercommunicate is minimized. They are allocated to the centralized resource.

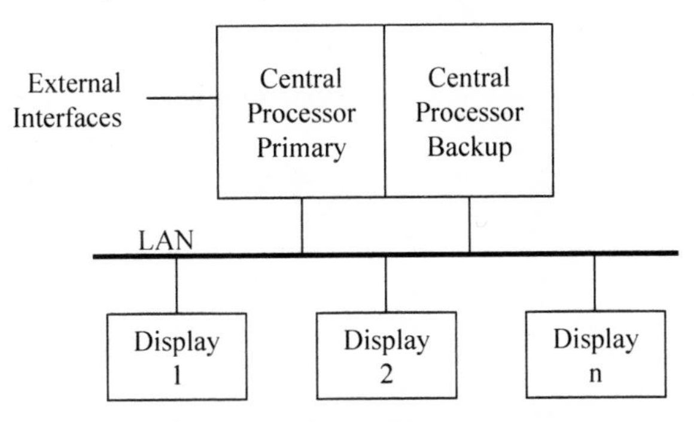

Figure 123 Partially Distributed Architecture

The disadvantages of a partially distributed architecture are similar to the fully distributed architecture but not at the same level:

- Some overhead if processors need to coordinate
- Startup depending on coordination
- Greater amount of parts
- Need to manage and assess status of distributed elements

In the ideal situation the bare-bones must-have functions are allocated to the distributed elements. This will allow the system to provide some level of service in the presence of a central resource failure, even if it is only for a limited time.

27.3.4 Hierarchical Architecture

The Hierarchical Architecture is born from the realization that some functions need to interact but not all functions. So the distributed elements at the bottom are conceptually grouped by these functions. This is a three-tier architecture where the centralized resource handles functions that span all the lowest level functions[228]. The lowest level functions are allocated to the fully distributed elements, and the functions in between are allocated to the middle resource.

The advantage of this architecture is that the lowest level distributed resource elements are not pushed to their capacity limits and intercommunication is minimized.

The disadvantage of this architecture is that processing of a single request may need to span all three levels rather that two or just one level. Each level translates to overhead and time delays as a resource needs to stop what it is doing to pass the information through itself to another level.

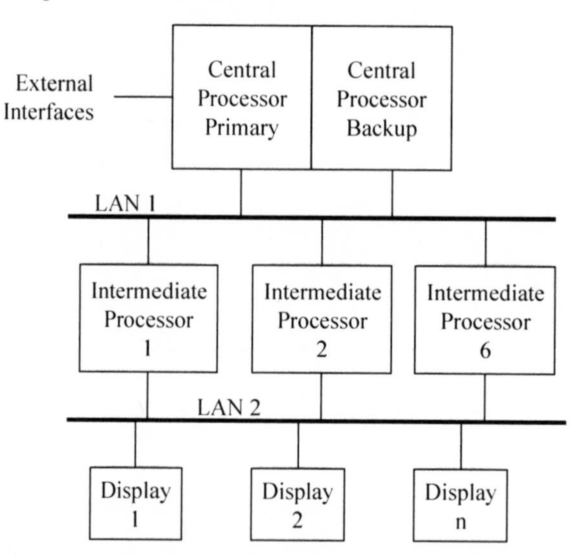

Figure 124 Hierarchical Architecture

27.3.5 Centralized Architecture Backups and Capacity

The centralized architecture has a long history of trying to deal with the single point of failure issue. There are various approaches that surfaced to build a centralized resource that would not fail. The two broad approaches are:

- Cold Standby Processor
- Hot Standby Processor

The single point of failure issue is also coupled with the fundamental issue that

---

[228] All of the sudden functional analysis and decomposition using formal DFD structured analysis becomes very important as the functions are identified and characterized.

the central resource is always capacity limited and compromises are needed. So approaches are coupled with building capacity and dealing with the single point of failure issue. These approaches are:

- Very Tightly Coupled Multiprocessing
- Tightly Coupled Multiprocessing
- Loosely Coupled Multiprocessing
- Very Loosely Coupled Multiprocessing

27.3.5.1 Cold Standby Processor

The simplest approach to deal with the central processor single point of failure is to have a standby processor. If the primary string fails then the standby string[229] is activated. There is no attempt to recover the data nor is there an attempt to perform a user transparent switchover. The user detects interruption.

The advantages of a cold standby are:

- There is little chance of a failure to propagate from the failed string to the new clean string coming up and eventually providing services

The disadvantages of a cold standby are:

- Loss of data resident in the failed string
- Backup string may be broken
- Time interval to bring up the new string and obvious disruption in service
- Large duplication of equipment, which uses floor space and maintenance resources
- Almost doubles the cost of the architecture without providing any additional services
- Since the switch over mechanism is rarely used it may be broken and not work when needed

27.3.5.2 Hot Standby Processor

The hot standby attempts to address the disruption of service issues associated with the cold standby. The approach is for the hot standby to parallel or shadow process the online string. There are different levels of the shadow processing but the

[229] String suggests backup is more than the central resource. It includes elements needed to duplicate ongoing operations. From a computer point of view it may include the central computer, switches, cables, but not the workstations with keyboard, trackball, display, switches and indicators.

closer the primary and secondary string match in processing, the less disruption to the user should a switch to the secondary string be initiated.

Since there is an attempt to preserve the online state and data, a latent data dependent failure that causes the primary string to fail may cause the secondary string to fail. Fortunately data dependent failures are significantly less probable than hardware failures in very mature systems. In immature systems there is insufficient time for the reliability growth program[230] to capture and minimize the latent data dependent failures.

The advantages of a hot standby are:

- Minimal loss of data
- Backup string is known to be working

The disadvantages of a hot standby are:

- Complexity in developing an effective shadow mechanism
- Susceptibility to latent[231] data dependent failures[232]
- Fault propagation from failed string to good string
- Large duplication of equipment, which uses floor space and maintenance resources
- Almost doubles the cost of the architecture without providing any additional services

27.3.5.3 Very Tightly Coupled Multiprocessing

Very Tightly Coupled Multiprocessing (VTCM) couples two or more central processing units (CPUs) via memory. The idea is to use an operating system to schedule functions for processing in the next available processor. Because of scheduling overhead, the processors are typically limited to six CPUs.

The advantage of this approach is that there is no special switch over mechanism. Since the multiprocessing mechanism is constantly in use it is always being tested and known to operate. In both the cold and hot standby architectures, the switchover mechanisms are rarely used and may actually be broken when

---

[230] A reliability growth program is part of a sophisticated maintenance strategy to increase the reliability of the system. It is usually established on mission critical software intensive systems.

[231] A latent failure is hidden in the system because testing did not surface the failure. This can be because of poor testing or data sequences found only in live operations over a long period of time.

[232] A data dependent failure surfaces with a certain sequence of data not found during testing. It is unique and surfaces after a long period of live operations.

needed. If a processing unit fails in a multiprocessing configuration there is no disruption in service other than increased response time. If the relative response time[233] is insignificant then there is no impact to the user.

27.3.5.4 Tightly Coupled Multiprocessing

Tightly Coupled Multiprocessing (TCM) is similar VTCM except the CPUs are coupled via disk drive. The disk drive interface is slower than the memory interface so inter processor communications will take longer. In a VTCM it may be possible to split a single algorithm across six CPUs and get 5X improvement in the response time but in a TCM architecture it might actually hurt the response time by adding CPUs. However TCM may work well for functions that are less tightly coupled and it might yield a 5X improvement in those cases.

27.3.5.5 Loosely Coupled Multiprocessing

Loosely Coupled Multiprocessing (LCM) is similar to TCM except the CPUs are coupled via high-speed computer input / output mechanisms. In the past this typically required close placement because distance introduces signal delay time and distance translates to bit error rates. Today the bit error rates are minimal with the use of fiber media. The distance still translates to signal delay time[234].

27.3.5.6 Very Loosely Coupled Multiprocessing

Very Loosely Coupled Multiprocessing (VLCM) is similar to LCM except that the computers are coupled via a network interface. The interface might be within a room, building, campus, or across the planet. The same issues associated with function coupling and response time surfaces as with all other coupling approaches. As the interface coupling becomes less (a slower interface) the associated function coupling also should be less or performance will degrade. So the coupling between the functions needs to be clearly understood so that they can be allocated to the appropriate multiprocessing approach.

27.3.6 Grid Computing

As functional coupling decreases the number of processors can increase. The communications overhead becomes less of an issue. Grid computing[235] is an

---

[233] Early human computer interface studies suggested links between operator stress and response times. The typing preview area should be sub second. An action should be less than 5 seconds, if longer a busy indicator provided. Progress offered if it exceeds 60 seconds. All actions should provide a positive indication that the action was accepted.

[234] This is about 1 nanosecond per foot.

[235] A program executes on the computer while it is not in use, such as when the screen saver is active.

architecture that uses the Internet to tie hundreds of thousands of computers together to work on a single problem or function. Obviously the data in this case is extremely loosely coupled.

27.3.7 Distributed Processing Potentials

The degree to which a multiprocessing or distributed architecture can improve response time is based on the level of parallelism in the system. If there is a high degree of parallelism then performance improvement is significant with the addition of processors and partitioning of either the functions and or data.

Amdahl's law is used to predict the maximum speedup possible using multiple processors. It simply states that the speed up of a program using multiple processors is limited by the time needed for the part that cannot be converted to a parallel form. This is the sequential fraction of the program.

If a program takes 20 hours using a single processor and a portion that takes 1 hour cannot be made parallel, while the other 19 hour portion (95%) can be made parallel then the maximum execution time cannot be less than 1 hour. So the maximum theoretical case scenario is a speed up of 20 times.

The equation is: $S = 1/[\ (1-P) + P/N\ ]$ where P is the proportion of the program that can be made parallel, N is the number of processors. If N is set to infinity then that is a maximum theoretical case scenario. This also assumes there is no overhead with creating, dispatching, and reassembling the parallel artifacts.

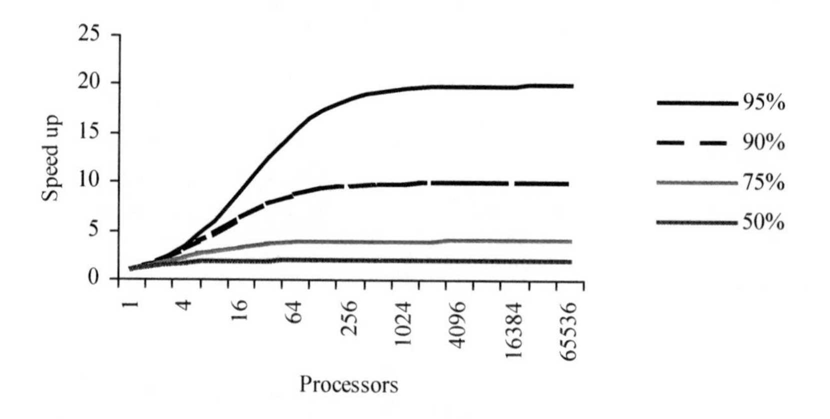

Figure 125 Amdahl's Law

27.3.8 Centralized Versus Distributed Architectures

The analogies of centralized versus distributed architectures are everywhere. For example, there can be a common dwelling where everyone shares living, dining, cooking, sleeping, and bathroom services. This is a typical college dormitory; centralized community architecture. The other approach is to have separate dwellings in the form of houses; distributed community architecture. Within a house the architecture can be centralized, fully distributed, or something in the middle. Most houses have common living, cooking, eating facilities and separate sleeping and partially shared bathrooms.

The centralized and fully distributed architectures have structure as suggested by the functional and data allocation. They also have connectivity. The connectivity or topology is inherent in the structure; they cannot be separated. The topology of an architecture becomes evident as soon as it is drawn in a block diagram. There is a geometric shape that is evident and suggests certain characteristics. There are relationships suggested by the sizes and shapes of the parts. For example a big box in relation to smaller boxes may represent a bigger machine capable of more processing or services. Circles may represent dumb peripherals with no processing abilities other than their specialized limited function usually viewed as fixed.

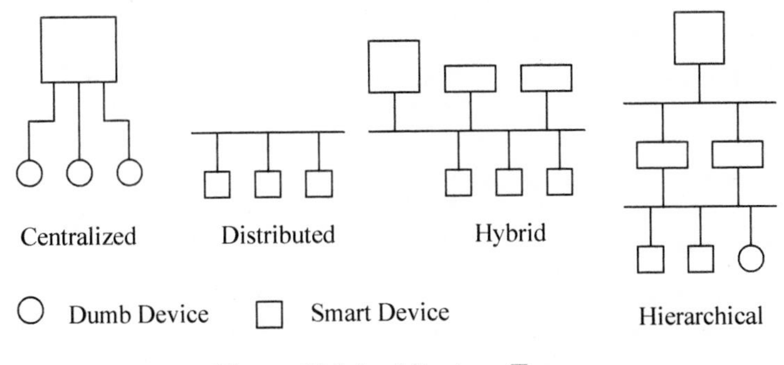

Figure 126 Architecture Types

27.3.9 Distributed Versus Multiprocessing

If there are six processors in a distributed architecture and six processors in a multiprocessing architecture, what is the difference between the architectures?

In a multiprocessing architecture there is a single queue that houses the requests and either each processor accesses the queue when not busy for the next task or there is a central controlling element that dispatches the requests to the least busy processor. In a distributed architecture there is a queue dedicated to each processor. If a processor gets bogged down the requests in its queue need to wait.

The analogy is obvious with a bank teller example. If separate lines form at each teller and there is no jumping between tellers then some people will be waiting while some tellers are not busy. The solution is to create a single line and pop the next customer from the single line queue when the next bank teller becomes available.

As hard as it is to believe many operations used distributed processing with dedicated queues as late as the 1990's. Examples include department stores and airline ticket counters. The banks were the first to realize the advantage of a single queue to form a multiprocessing service. Back then the department stores and airline ticket counters probably felt that they were proper in offering multiple customer service representatives rather than one central service provider. They could have reduced the number of customer service providers and increased service by using a single line queue.

The disadvantages of multiprocessing are associated with the single point of failure in the queue. If the queue is lost, then all processing stops. From a computer point of view the queue can be in shared memory, shared disk, or distributed in each processor memory as a common dynamically updating image. If the single queue is distributed in each processor memory then there is complexity associated with keeping the queues consistent (i.e. dynamically updating image).

27.3.10 Homogenous Heterogeneous Architecture

Distributed architectures can be made up of multiple elements of the same type or different types. If it is made up of the same type of elements it is a homogenous distributed architecture. If it is made up of different types of elements it is a heterogeneous architecture. An example of a homogenous distributed architecture is one that uses personal computers connected together via a local area network. Alternatively it also can use multiple mainframe computers connected together via a local area network. A heterogeneous architecture would mix mainframe and personal computers together in a network.

The advantages of a homogenous architecture are associated with maintenance and system management. The disadvantages are limitations associated with the basic element that makes up the homogenous architecture. The homogenous architecture limitations can be common failure modes, common maintenance issues, and other common problematic characteristics. The homogenous architecture is also less flexible, which can negatively impact growth and technology insertion.

The advantages of a heterogeneous architecture are inherent in it being more distributed than a homogenous architecture, even though both are distributed. The common problematic characteristics are removed from the system. However, the diversity may increase maintenance and system management challenges. So there are no common failure modes, common maintenance issues, or other common problematic characteristics. The heterogeneous architecture is also more flexible because all the diverse elements are made to work together as part of the system solution. This integration allows for more growth and easier technology insertion.

Example 44 Heterogeneous Architecture - Wind Farm

A wind power farm architecture can be based on 20 large category turbines. It can be implemented using the exact same type and manufacture. The model numbers for all 20 units are the same. This is a homogenous distributed architecture.

Alternatively the wind power farm architecture can be 10 large category turbines, 20 medium category turbines, and 40 small category turbines. The small category turbines might be further divided between propeller and cylindrical wind capture mechanisms. Also different manufacturers can be used to implement each category of the system architecture. A simple rule can be used, such as use the top three manufactures for each category. This is truly a heterogeneous architecture.

27.3.11 Federated Architecture

Architectures can be functionally distributed or data distributed. They also can be part of a single organization or multiple organizations. A federated architecture is an interoperable very loosely coupled highly distributed architecture that is typically managed and maintained by multiple organizations. However it does not need to be managed and maintained by multiple organizations. It tends to be associated with very large infrastructure oriented systems.

Example 45 Federated Architecture

In many ways a federated architecture is an architecture of architectures. For example air traffic control consists of voice communications, navigational aids, command and control centers, avionics, airports, air routes, backup systems, etc that are all stand alone architectures but seamlessly inter-operate.

27.3.12 Other Architecture Types

As we look out into our world we see different architecture types. They each have their advantages and disadvantages. Many times with just a word or word phrase we are able to visualize an architecture type and have some preconceived notions about its behavior, structures, and advantages / disadvantages relative to each architecture. In other cases we have no exposure to the domain and the word or phrase is meaningless[236].

In all cases, in a serious system architecture analysis effort the current architecture that might apply to the problem should be extensively researched.

With a broad and deep understanding of the functions, key requirements, and similar architectures we then are able to break new ground and conceive of new architecture alternatives for our unique system. The following is a list of words and phrases that may trigger visions of various architectures.

Table 49 Architecture Types

Town House Twin House Colonial Bi-Level	Chaotic Organization Functional Organization Vertical Organization Horizontal Organization	Cantilever Bridge Coal Fired Electric Plant Hydroelectric Plant Solar Electric Farm

[236] You are either in the business or not is an effective way to communicate with non-technical people disengaged from reality. You can enter the business area but that requires work.

Table 49 Architecture Types

Ranch	Clinic	Wind Electric Farm
Split Level	Doctors Office	Geothermal Plant
Row House	National Health Insurance	Nuclear Power Plant
AM Radio	Beveridge Health Care	Skyscraper
FM Radio	Bismarck Health Care	Office Building
Internet Radio	Private Health Insurance	Industrial Park
Analog TV	Out-of-Pocket Health Care	Downtown Business
Digital TV	Electric Motor	Betamax Video Recorder
Analog Cable TV	Synchronous Electric Motor	VHS Video Recorder
Digital Cable TV	DC Electric Motor	VHS Player
Internet TV	Pancake Electric Motor	DVD Player
Department Store	Trellis Bridge	DVD Video Recorder
Strip Shopping Center	Suspension Bridge	Record Player
Mall	Moveable Bridge	Reel to Reel Tape Recorder
Downtown Retail	Double Deck Bridge	Reel to Reel Studio Recorder
Intersection	Draw Bridge	Cassette Tape Recorder
Rotary or Circle	Lift Bride	8-Track Tape Player
Overpass	Power Boat	MP3 Player
Gas Automobile	Sail Boat	Steam Engine
Electric Automobile	Catamaran	Internal Combustion Engine
Hybrid Automobile	Trawler	Piston Engine
Hydrogen Automobile	Speed Boat	Turbine Engine
Compressed Air Automobile	Runabout	Wankel Engine
Diesel Automobile	Row Boat	Ion Engine
Propane Automobile	Planes	Rocket Engine
Methanol Automobile	Trains	Jet Engine
Matrix Organization	Automobiles	
Ecological Organization	Beam Bridge	
Hospital	Arch Bridge	

Notice that many times context is important. For example Ranch can refer to salad dressing or a house. In our case since it is in the group of words or phrases related to house, it is fair to assume the reference is to a house.

In many ways architecture types is the corner stone of all architecture efforts. In a technical paper the authors:

- Identify the problem
- State the current approaches
- State what is wrong with the current approaches
- Offer a proposed solution
- State why the proposed solution is the best solution

Notice that there is significant effort on the current state of affairs. That is also the heart of the architecture selection process and the key to understanding current and similar approaches before synthesizing a better approach or architecture.

27.4 Architecture Topology

Architecture Topology is the interconnection of various elements providing functional services. The topology can be physical or functional. For example someone can physically work in New Jersey and provide services to someone in New York. This can be via a permanent organization where this person organizationally reports to the New York office. Or it can be temporary as part of a contract on a project. Logical topology represents how information is processed and transferred. Physical topology represents the physical layout of the service providers. There are broad processing topology categories:

- Point to Point or Line or Point to Multi Point
- Star or Centralized Topology
- Bus
- Ring
- Mesh
- Fully Connected
- Tree

Do not confuse these architecture or processing topologies with network topologies. The network topologies were born from these processing topologies and their study. Also, these processing topologies do not apply to just computers. They can apply to machines, people, nature or any combination in our systems. They are organizational in nature.

27.4.1 Point to Point or Line or Point to Multi Point

The Point to Point Topology is the simplest topology. It is a permanent link between two endpoints. If multiple point to point nodes are linked in a chain, that is a line topology. There is no limit to the size of the chain. If one node If a central node links to multiple nodes where the other nodes do not interact, that is a point to multi point topology. It is a classic example of a central architecture where one large node interacts with other less capable nodes. The central node has a majority of the functionality.

Its advantage is the simplicity of the connection, which minimizes possible loss or corruption of data. It is also easiest to secure the link in this topology. There is only one simple element to consider. There is minimal overhead because there is no need to consider the needs of other nodes on the same connection. So channel conflict is non existent.

Its disadvantage is that if data or information needs to move from one node to a far node it needs to pass through other nodes that perform a data transport. In some cases the data needs to be stored before it is forwarded. This is a store and forward function.

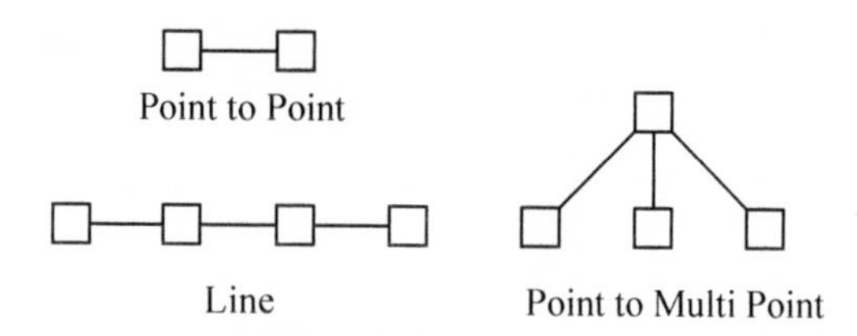

Figure 127 Point to Point / Line / Multi Point Topology

27.4.2 Star or Centralized Topology

The Star Topology uses a central resource to link to all other resources in a hub or spoke arrangement. All communications is through a hub processor. The hub is allocated functions such as processor scheduling, data fusion from the nodes, status monitoring management and other functions associated with two or more nodes. The nodes are allocated functions that minimize inter node communications. This is an example of a smart centralized architecture where the nodes have a significant amount of processing capability.

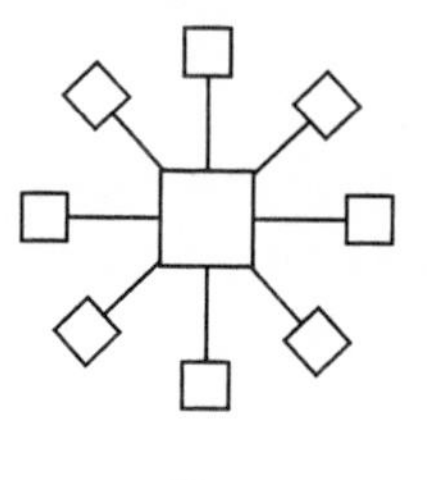

Figure 128 Star / Centralized Topology

The disadvantage of this topology is that the hub is a single point of failure. It is also a potential processing bottleneck. Growth is also limited if the hub cannot be expanded to accommodate more data or more centralized functions.

The advantage of this topology is that failure propagation is minimized if a node should fail. The other nodes continue providing services in the presence of the failed nodes.

27.4.3 Bus Topology

A Bus Topology is a distributed topology. It links all the processing nodes via a common communications mechanism. This allows any processing node to communicate with any other processing node. It also allows for broadcast so that one node can make its findings simultaneously available to all other nodes. However a bus needs rules so that the participants do not simultaneous access it for their needs. A bus is basically a backbone.

The rules start with some form of control where something identifies that the bus is about to be used by a node. There are multiple approaches to implementing a bus control mechanism but they fall into the following broad categories:

- **Interrupt:** In the interrupt approach a node raises an interrupt signal and this tells the other nodes to be silent until the interrupt is released. Since not all resources on the bus have the same urgency to deliver or receive data, a priority can be used on the interrupt mechanism. So if a higher priority interrupt is raised, the lower priority interrupt and transfer can stop, and release the bus to the higher priority interrupt. This allows for a significantly faster response time for high priority needs.
- **Master:** In the master approach there is a node on the bus that randomly solicits the other nodes for data. There can be variants where multiple masters using an interrupt mechanism allow one master to take control of the bus and solicit its selected nodes or resources.
- **Polling:** The polling approach is similar to the master approach except the solicitation for data is periodic rather than random.
- **Round Robin:** The round robin is an approach where each node is given a time slot to take control of the bus. The data is passed during that time slot or until the transfer is complete, depending on the round robin approach.
- **Collision Detection:** In the collision detection[237] approach each node listens to the bus. If it is silent and a node needs to access the bus, it just accesses the bus. It then listens to determine if the data is damaged. If two or more nodes access the bus at the same time the listening nodes detect their own respective damaged data. They then wait a random amount of time and try to access the bus with an new transmit of data. The collisions are rare when few nodes are present and there is little traffic.

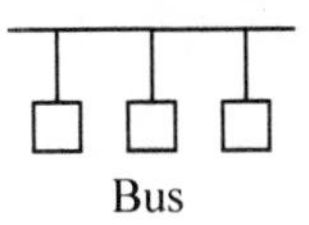

Figure 129 Bus Topology

Even though these ideas were used on internal computer buses, the interrupt, master, polling, and round robin techniques apply to people and organizations. Even Collision Detection a conceptual form of interrupt used in external computer communications applies and its analog is a shouting match among a group of people

[237] Ethernet became a very popular personal computer communications mechanism in the late 1990's. Its performance was managed with the use of network switches that limited the number of nodes and thus collisions.

in a heated meeting.

The function allocations to the nodes could be anything but one of the goals should be to minimize traffic on the bus. Usually there is a tradeoff between a heavily used node (while the others are under used) and high bus traffic. The trick is to balance the architecture such that the nodes and bus are effectively used from a timing-and-sizing point of view. This is difficult because there is a temptation to size all the nodes to be the same size. This has advantages from a maintainability and operations point of view. Making all the nodes the same is easier if the data can be sliced and allocated to the processors that then use all the same functions to process the different data sets. For example computer nodes process data[238] while healthcare facilities process different patients. The patients are analogous to data.

The advantage of a bus is that there is no single point of failure. Some might argue that the bus could be physically damaged or a node could fail in some way to prevent the bus from properly working. In most instances there are solutions for these failure cases including using redundant busses. The bus is excellent for broadcast messages. If the bus is implemented with a round robin or polling mechanism it exhibits many of the advantages of a token ring topology.

The disadvantage of this approach is that a standard must be developed and followed by all the nodes. If a node does not follow the standard, there is the potential for the node to seriously degrade the performance of the bus. As more nodes are added the bus traffic increases. At some point no more nodes can be added without the system failing to function properly. A bus is also limited by distance, depending on its implementation. As the distance increases the signal is susceptible to noise and timing delays which may lead to data errors. The bus is worst for point to point messages between nodes. If the bus is implemented with a collision detection mechanism, the collisions become excessive after approximately 30 nodes and performance significantly degrades. If the bus is implemented using an interrupt mechanism there is the risk that a node will hog the bus and prevent other nodes from accessing the bus. If the bus is implemented with a master mechanism, it then has a single point of failure.

27.4.4 Ring Topology

A Ring Topology is a distributed topology. It connects each node to exactly two other nodes. Data travels from node to node; each node is a repeater station where the signal is corrected before being sent to an adjacent node. The failure of a single node will stop the flow of information unless the ring can send data in both directions (clockwise and counter clockwise). From a processing point of view the ring brings order to the management of the individual processors. A token arrives at a node and the node gets the opportunity to provide data on the ring. If there is no data to be offered by the node, then the token is immediately passed to the next

---

[238] Computer processors transport, translate, and transform the data.

node. This is a very deterministic mechanism.

There are many advantages to the token ring. Its roots can be found in the round robin concept of internal computer buses. Since control is not random but based on the deterministic passing of the token, data integrity is very high. It cannot be lost because of random events. Its performance is also very predictable and stable. Every node on the ring is always given the opportunity to offer its data; other nodes cannot lock out nodes. Its structure allows for very high utilization rate. Basically the ring utilization can approach 100% without loss of data. No other topology offers that performance characteristic. There is no central controlling authority that can fail. The control is fully distributed to each node. When exceeding 30 nodes the token ring exhibits superior performance relative to collision detection buses.

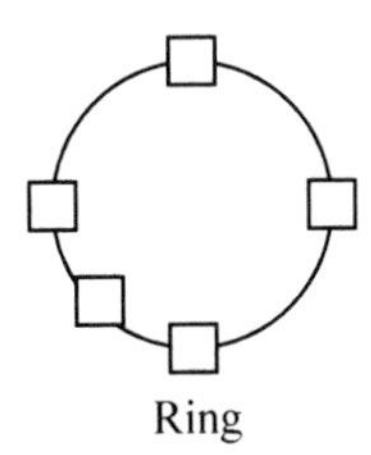

Figure 130 Ring Topology

The disadvantages are associated with the stopping of the token and the relative position of the nodes. If the token stops, there is the possibility the topology can fragment into two rings unless there is a backup mechanism such as reversing tokens and or redundant rings. Moving and adding nodes can affect response time relative to other nodes. For example if two cooperating nodes are adjacent the response time is minimal. If the cooperating nodes are at opposite ends of the ring the response time is maximized.

27.4.5 Mesh Topology - Fully or Partially Connected

A Mesh Topology is a distributed topology. It connects every node to every other node. In a mesh all nodes can simultaneous transmit to one or more other nodes. Most view this as point to point connections from every node to every node or multi point to multi point connections. However, this is just like a bus except that a bus has only one node transmitting at any given instant. From an implementation point of view this is difficult to imagine without time division or frequency division multiplexing and moving the time to infinitely small or the number of frequencies to infinity. From a practical point of view a bus can behave like a mesh if the bus transfer rates and delays are insignificant in the system timing budgets.

If the mesh is viewed as a multi point to multi point connection scheme then the number of physical connections is:

Connections = N*(N-1)/2

A mesh topology is self-healing. If one node breaks down or a connection goes bad the network can still operate. The network is very reliable because there are multiple paths to a node. Some can relate to the mesh topology in a wireless network setting.

In a partially connected mesh only some of the nodes are connected to other nodes. This allows for redundant paths in the event of failure while minimizing the connection complexity.

The advantage of the mesh is ability to continue to operate in the presence of multiple failures. It is the most survivable topology. There is no routing, so there is no place where data is only being transported and subjected to delay, possible corruption, or loss. The connection is direct from the source node to the destination node.

The disadvantages of the mesh are complexity associated with the connections. This translates to very complex physical connectivity, which becomes impossible with a large number of nodes. It needs a mechanism to route traffic to other nodes. Although every node is connected to every node and there is no forwarding as in the case of the Internet, something needs to decide which node should receive the data.

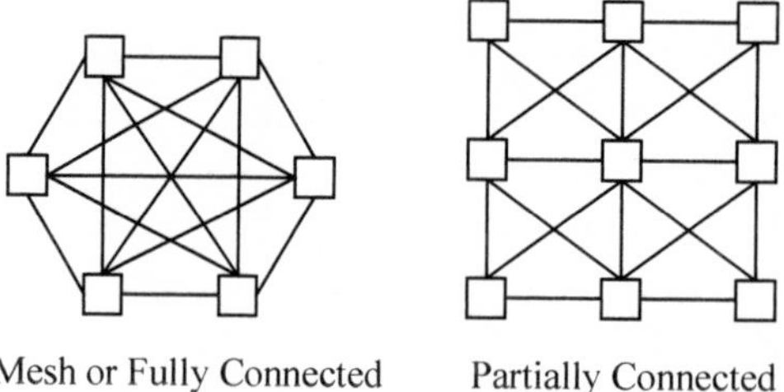

Figure 131 Mesh Topology - Fully Partially Connected

27.4.6 Tree or Hierarchical Topology

A Tree or Hierarchical Topology is a distributed topology. It must have a minimum of three levels since two levels is a Star or Centralized topology. The hierarchy can be implemented with point to point connections or a bus. Two levels of buses represent a two level hierarchy.

The advantage of this topology is isolation of traffic and processing to selected levels. This isolation allows for performance optimization, security isolation, and less fault propagation. It is also advantageous from a maintenance point of view. If online maintenance is not possible, then only selected levels or tree branches are affected during the down time.

The disadvantage of this topology is if data needs to traverse across different levels. The topology might start out as being ideal and fully optimized but as the system evolves, functions might be introduced that need to traverse different levels. So from a growth point of view the hierarchical topology might start to break down and not allow for future capabilities.

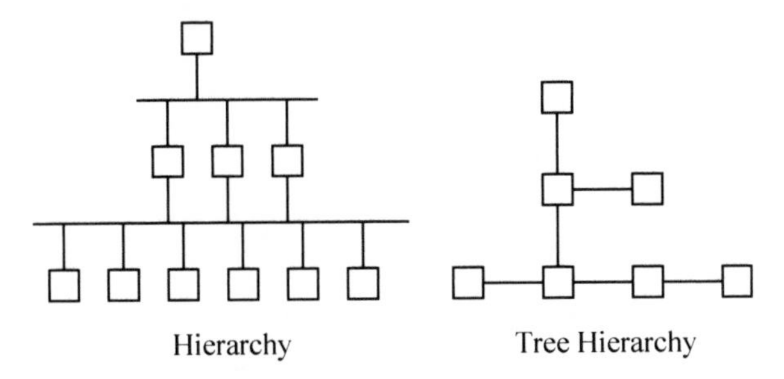

Figure 132 Tree or Hierarchical Topology

27.4.7 Hybrid Topology and Topology Control

A hybrid topology is a combination of two or more topologies. Most infrastructure systems are a hybrid topology. This is not because of a conscious effort to structure the system in a certain way. Instead the systems evolve and through trial and error the systems are optimized to their levels of capability. The dilemma is what to do when an existing system that evolved over 50 years reaches its limit. So all of the sudden system practitioners today may be faced with creating systems that use multiple topologies.

Buried within all the topologies is the idea of connectivity and control. The issue of control is significant. Being able to statically assign functions and data to different nodes in a topology is a challenging task. The modeling effort is significant.

Creating a mechanism that dynamically assigns functions and data to available resources is difficult if that mechanism is centralized. It is extremely difficult if the assignment mechanism itself is distributed and tries to take full advantage of a distributed topology. The concept of a fully distributed operating system that has full view of the whole system and simultaneously lives in all nodes so that one or more node failures will not lead to a failure of the distributed operating system is a significant challenge.

27.5 Architecture Analogies

Although the descriptions of architecture types and topologies is strongly related to computing and communications they strongly apply to systems where organizations of people and their activities are the heart of the system. Understanding these organizational structures will help when considering architecture alternatives for:

- Personal Transportation
- Electric Power
- Health Care
- Education
- Food Production
- Waste Management

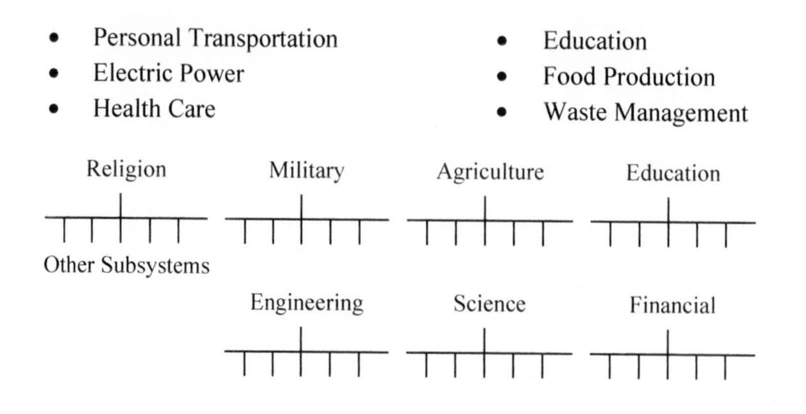

Figure 133 Centralized Architectures

The first questions that should surface are centralized or distributed. If distributed how distributed. What happens in the event of failure? What happens when the system needs to grow? How can technology be inserted? What are the maintenance ramifications? What system has the greatest longevity?

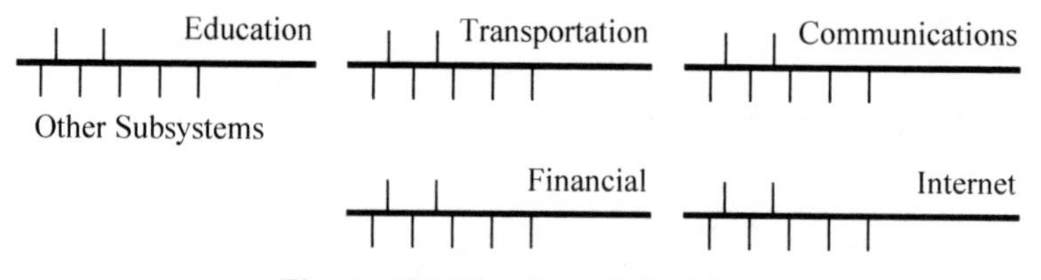

Figure 134 Distributed Architectures

Different subsystems have functionality that can be broadly characterized as processing, storage, distribution, or any combination. These subsystems then fall into some topology. For example a distribution subsystem can be the hub of a star topology or the central node of a top down centralized topology. A distribution subsystem also can take on the role of storage.

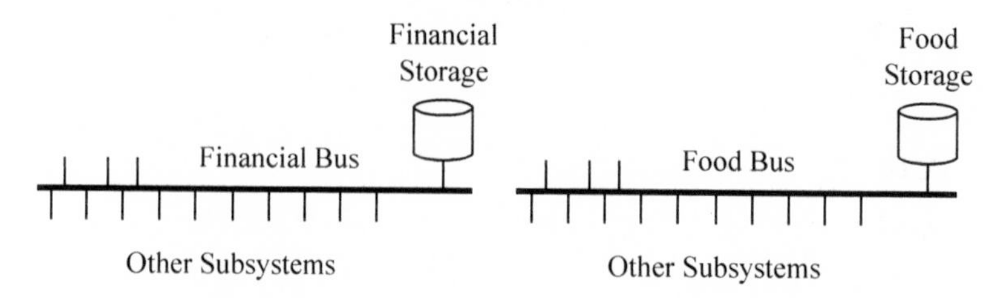

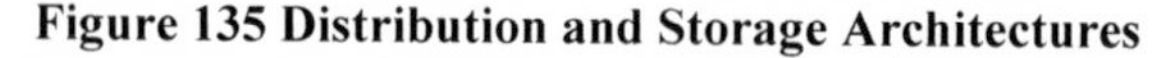

Figure 135 Distribution and Storage Architectures

Example 46 Schools Architecture Topology

Many towns in New Jersey, such as Cherry Hill, that empathize education have a distinct school architecture concept. Elementary schools are at the neighborhood level. This allows the children to walk to school or be bussed for very short distances in their neighborhoods. So there are large numbers of elementary schools. Middle schools are fewer in number and children are bussed form multiple neighborhoods, but walking is still a possibility. The high schools are usually one or two. Also the neighborhoods tend to be concentrated around the schools. Commercial and industrial areas do not break the neighborhoods apart, again allowing children easier access to the schools. So the school architecture interacts with the community architecture and is driven by zoning laws and master plans. In the last two decades of the last century many communities were created with no master plans or architecture concepts. Children in these communities travel long distances across commercial and industrial areas to regional schools and the concept of community is lost.

27.6 Function Types

As we go through the functional decomposition process patterns emerge. Certain functions have certain characteristics[239]. These characteristics are important because they will form the basis of allocation to the architecture. The following is a list of words and word phrases to consider when characterizing functions:

Table 50 Function Types

Safety Critical	Highly Parallel	Offline
Security Critical	Extremely Important	Manually Intensive
Processing Intensive	Less Important	Cognitively Intensive
Data Intensive	System Level	Rarely Executed
Event Driven	Communications	Continuously Executed
Cyclic	Display Intensive	Aesthetically Sensitive
Random	Human Interaction	Aesthetically Neutral
Storage Intensive	Backbone	Correlation Intensive
Communications Intensive	Function Overhead	Fusion Intensive
Data Transport	Mission	Signal Processing
Data Translation	Maintenance	Vision Intensive
Data Transformation	Support	Sound Intensive
Transcendental (Trigonometric)	Delivery and Installation	Motion Intensive
Floating Point	Training	Decision Function
Algorithmically Intensive	Diagnostic Self Test	Cognitive Processing
Highly Linear	Online	

[239] MIL-STD-490 is an excellent source for function types. It is partially included in modified form as an appendix to this text.

This list is not exhaustive but it should help in identifying the function types in a system. The function types need to be clearly understood and grouped, allocated, and applied in the architecture. For example highly parallel functions are ideal candidates for distributed processing. Aesthetically sensitive functions may trigger the need to add artists or industrial designers to the team who can impact the architecture.

27.7 Depiction

Architectures are fundamental collections of ideas and concepts that form a solution set. By definition all considered architectures should satisfy the requirements. However some architectures are better at addressing some requirements than other architectures. This is either because of a fundamental idea or concept or the resulting implementation detail of a particular architecture idea or concept.

So how do you capture architecture? How is it depicted? In many ways architecture depiction is a discovery process in much the same way the architecture ideas and concepts are a discovery process. What works for communicating architecture in one setting may not work in another setting.

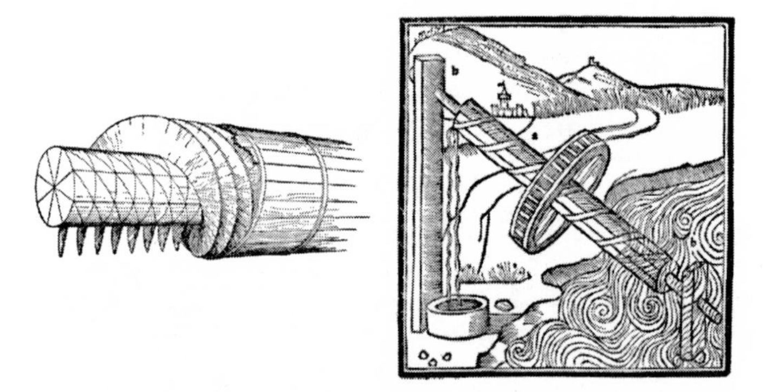

Figure 136 The Water Screw[240]

Mechanical drawing is probably the oldest form of architecture depiction. There are pictures from Vitruvious to Leonardo Davinchi. As our understanding of nature has unfolded through the centuries we have created other forms of art for representing architectures. The schematic of an electronic or electrical based device shows the architecture at some level. A chemical equation shows the architecture of a new synthetic material. An artist rendering of ranch house versus a colonial house shows architecture alternatives. A drawing of city streets and building types represents architecture for a new city in a master plan for a developing community.

Storyboards as previously described in this text are extremely effective in

[240] The Water Screw is from the edition of "Vitruvius by Fra Giocondo, Venice, 1511".

capturing and communicating architectures. A clear title, thesis statement, picture, and supporting phrases can capture the architecture. Similarly a storyboard can be used to show all the alternatives in one concise place for visualization and comparison. A stack of storyboards can capture all the architecture alternatives, the key requirements, key issues, and other important elements associated with the architecture analysis and selection.

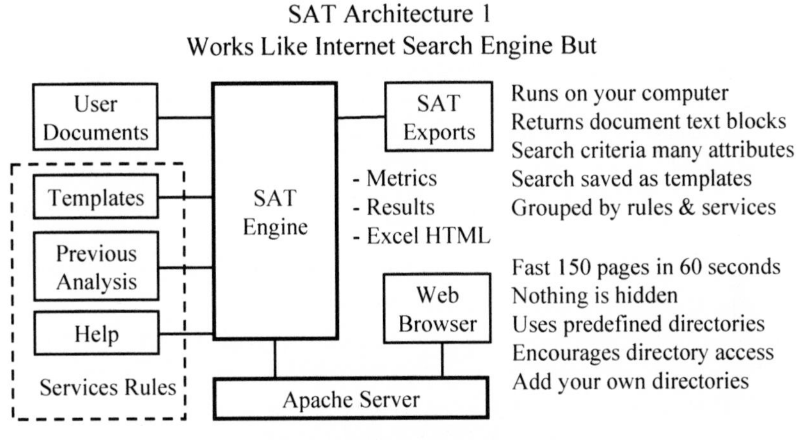

Figure 137 Architecture Depiction Storyboard

As with all storyboards, these storyboards should never be vague or general such that anyone can create any story from the depiction[241]. They are used to communicate in such a way so that everyone understands the message with little or no confusion.

27.8 Morphing Architecture or Transition

Many times especially in the case of infrastructure systems architectures must change over time. This change might include intermediate architectures before a final architecture can be put in place. For example, the current architecture might be highly centralized and the final architecture might be highly distributed. A transition strategy might be to move to a smart centralized architecture, then a hierarchy, and finally a fully distributed system or architecture. This transition or morphing architecture might take years or just a few weeks.

Example 47 Architecture Morphing - Business

As time moves forward we see that many industries move from decentralized architectures to centralized architectures. At one time farming was a very distributed

[241] Modern presentation practices encourage vague presentation slides. Storyboards should never be confused with presentation slides.

system. Today there are extremely large farming enterprises. Today the production of eggs is done in huge factory farm settings. A salmonella outbreak can affect hundreds of thousands of eggs[242]. At the start of this new century we saw two of the three US automobile companies (General Motors and Chrysler) and many financial companies being bailed out by the United States government[243]. They were deemed to be too big to fail. They were essentially highly centralized architectures.

Example 48 Architecture Morphing - Automation

Air traffic control systems have moved from more centralized towards more distributed architectures. The introduction of the first computers into the control centers were based on a centralized architecture concept where the workstation functionality was limited to what could be provide using analog hardware and human interfaces. As computer hardware became smaller and less costly, small computers migrated into the workstations to perform display-processing functions.

Transition is closely related to technology assessment and maturity. A transition strategy should include an understanding of the TRL at each stage. It may be possible that a transition strategy may be formed that everyone agrees is reasonable, but once viewed from a technology assessment and maturity level, new insights may surface and cause the transition strategy to change.

Another key element to the transition strategy is stabilization period. There are two aspects to the stabilization period. The first is how long should a current system capability exist before the next transition phase. The second is how long should the previous capability exist before the new capability is trusted and the old capability is removed. Sometimes the old capability may never be removed and it may act as either a backup mechanism or a load reduction mechanism (e.g. A new road that parallels an existing road).

The humans that use the system will probably drive the stabilization period. So the time period is probably measured in months between introductions of new capabilities. No new capabilities should be introduced until the humans master the current capabilities. The mastering of each new capability can be the trigger used to introduce a new capability.

Example 49 Infrastructure System Transition Challenges

Transition can be challenging for systems with long history and entrenched interests. This is a serious issue facing the U.S., which enjoyed the introduction of spectacular new systems after World War II, when everything was a clean slate.

---

[242] A half billion eggs produced by just two farms were recalled in August 2010 because of a salmonella outbreak.

[243] The United States government Troubled Asset Relief Program (TARP) program Signed into law by President George W. Bush October 3, 2008 purchased assets and equity from collapsing financial institutions and automobile companies GM and Chrysler.

However it became obvious that these systems with their entrenched interests have led to massive problems in the U.S. during the last 30 years. Third world countries are quickly rising and adopting new technologies with no internal resistance. It remains to be seen how they handle the various transition issues.

What is really buried in the transition discussion is the layout of a program or products line, which consists of multiple projects or products. The way the program is partitioned into projects, which includes the scheduled rollout of each capability and the sequence of the capabilities is the heart of transition planning and the layout of an effective program. A modified PERT[244] chart or product fishbone[245] diagram can be used to show how the various projects come together to form the program or product line.

27.9 Synthesis

There is usually a great deal written about architecture synthesis. In this case we begin with a definition, a brief description of the process, then an example.

Definition of Synthesis: 1. The putting together of parts or elements so as to form a whole 2. A whole formed in this way 3. The formation of a complex compound by the combining of two or simpler compounds, elements, or radicals[246].

Synthesis is about aggregation, putting things together, the opposite of decomposition. It is a reverse decomposition[247]. However it is done in an environment of analysis. The analysis includes models, alternative architectures for comparison, technology assessment, failure analysis, etc - many of the practices offered in this text. This analysis is embodied in tradeoff criteria and quality attributes as described in this text. So synthesis is tightly coupled to the architecture selection process.

Synthesis starts with aggregating a conceptual architecture, then the system architecture, then the subsystem architectures, then the assembly architectures, and then the component architectures[248]. As there is movement down the decomposition tree the lower level architectures may impact the upper level architectures just like

---

[244] A modified Program or Project Evaluation and Review Technique (PERT) chart can be used to show the phased introduction of capabilities, subsystems, and or systems.

[245] All products evolve and that evolution can be represented with a product fishbone where new features are added left to right with each new bone connected to the main backbone.

[246] Webster's New World Dictionary, 1982, Simon & Schuster, ISBN 0-671-41816-5 (edged) or ISBN 0-671-41816-3 (indexed)

[247] Reverse decomposition: take the known and generalize until the implementation details are removed.

[248] Do not get fixated on the definition of terms. The concept is to aggregate at different levels from top to bottom.

in functional decomposition. As there is movement down the architecture tree there is more reliance on existing components that are commercially available. Depending on the solution some of the components, assemblies, subsystems, and systems may not offer sufficient performance even at the state-of-the-art level and so new technologies may need to be developed.

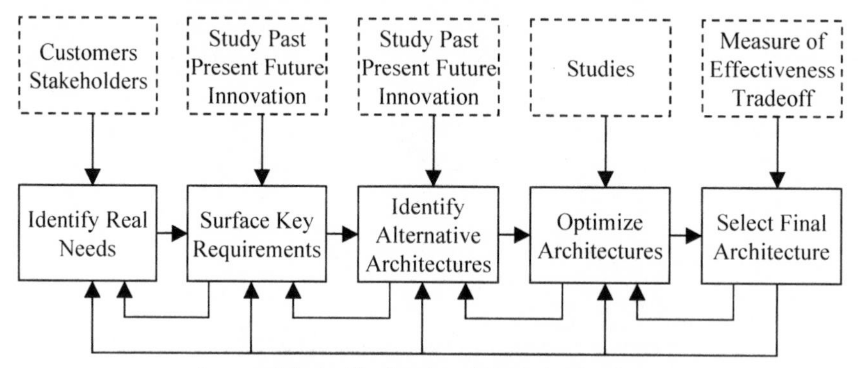

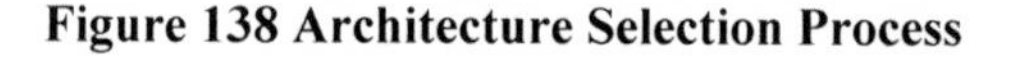

Figure 138 Architecture Selection Process

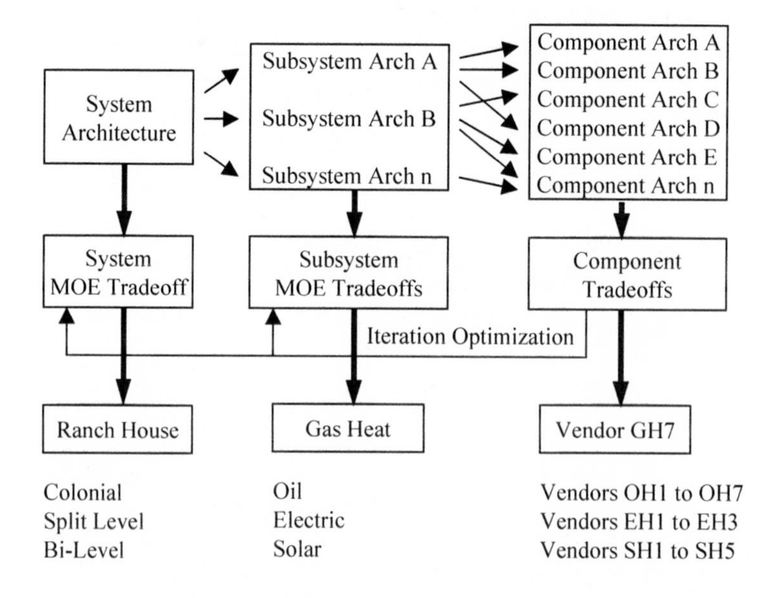

Figure 139 Architecture Decomposition - House

Architecture synthesis needs to balance all the child architectures in the system architecture. That balance includes making sure stress points are found and mitigated. This means understanding the function, performance, and technologies so that they can be effectively allocated into the balanced system architecture and balanced child architectures. This means analysis and models at various levels so that the correct tradeoff choices can be made.

Something as simple as a house includes multiple layers of architectures. All these architectures need to be selected based on reasonable tradeoffs at each architecture level. As each architecture level alternative is considered and optimized then selected, it may impact the previous architecture level. Subsystems and components may be purchased as commercial products, modified commercial products, new designs, or state-of-the art designs pushing one or more technologies. These architecture decisions are made as part of individual tradeoffs at each level.

Example 50 Synthesis - Air Traffic Control Center

Air traffic control is a classic example of a system that has evolved through the decades. As technology has surfaced it was added to the system.

In the beginning the system was nothing more than a pilot telling someone they were going someplace, maybe. There would be random sightings communicated via telegraph or radio.

At some point radio communications was introduced into the airplane cockpit in what was called a manual system. Using ground reference points and radio communications, the path of a flight was tracked in a manual command center.

With the introduction of RADAR, the actual position of an aircraft was tracked when RADAR coverage was available. In between RADAR coverage areas, flight data was used along with radio communications to manually track the progress of a flight. Since the early RADAR was analog it was not practical to run the RADAR video lines long distance. So the command centers were collocated with the broadband RADAR.

With the introduction of the computer, the analog RADAR was digitized and fed into the computer to perform a tracking function. This offered a position symbol and various alphanumeric data near the symbol associated with the airplane.

Prior to the computer, because of the limitations of broadband RADAR communications, multiple command centers were needed. After the computer, it was possible to consolidate all the command centers into one large center.

The United States has many command and control centers divided between Tower, Terminal, and Enroute air traffic. The terminal traffic is handled primarily by Terminal RADAR Approach Control (TRACON) facilities. Enroute Facilities or Area Control Facilities (ACF) handle Enroute traffic. In 1980 there were approximately: 23 Enroute Centers, 234 TRACONS, 400 Control Towers. The numbers change as facilities are consolidated and new facilities added.

Typically, outside the United States, air traffic control was merged with air defense in a common command center. Hughes Aircraft would analyze the needs of a nation state and usually offer 2 Command and Control Centers. In the event of disaster one command center would backup the other command center. This was possible because of relatively low air traffic and small landmass.

The primary functions in an Air Traffic Control center are:

- RADAR Processing (RDP)
  - Conflict Alert (CA)
  - Minimum Safe Altitude Warning (MSAW)
  - Restricted Air Space Detection
- Flight Data Processing (FDP)
- Workstation Processing
- Voice Communications
- Weather Processing (Wx)
- Notices To Airmen (NOTAMS)

Notice the generic term "processing" trails some functions. The repetition is irritating but sometimes needed to distinguish the noun use of a word from the verb use of a word.

The voice communications function is allocated to a different system and appears only as the peripherals on the air traffic controller workstation, specifically the speaker, microphone, headset, and control panel. The question is why?

Functionality can be provided by special purpose machines or by a general-purpose computer capable of many tasks as driven by internal configuration or programming. Before the combined efforts of converting analog signals to digital formats and using a program inside a general-purpose computer, many functions were automated with custom machines. The early dream of building a general-purpose machine that could perform many functions became a reality with SAGE as computers were moved from calculators to offering situational awareness. Today, because of digital voice it is possible to allocate the voice switching and control to the same processing complex that provides the air traffic controller with position data of aircraft in his or her airspace. So all of the sudden the functions can be reshuffled into:

- Situational Awareness: RDP, FDP, Wx, NOATMS
- Workstation Processing
- Voice Communications

The first architecture to consider is the fully centralized architecture. All the functions are allocated to the central processor. The workstation processing is also centralized in a display processor. The workstation is just the keyboard switches, trackball position sensors, panel switches, panel indicators, and Cathode Ray Tube (CRT) display.

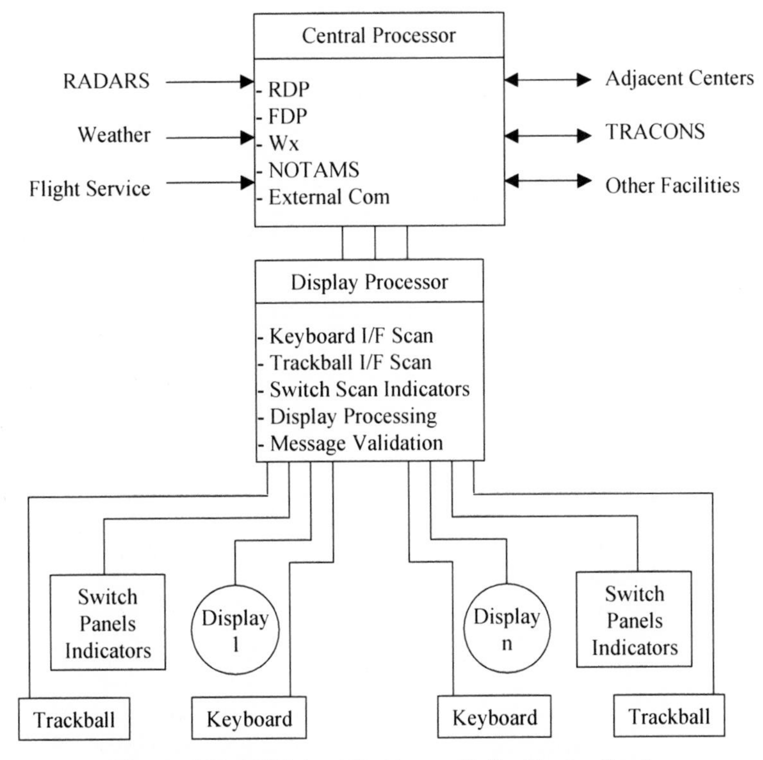

Figure 140 ATC Architecture - Fully Centralized

This was considered a practical solution in the late 1960's early 1970's. Since there was no keyboard, trackball, or switch local scan capability, the workstation cables were massive. The scanning was performed in the workstation processor as a separate custom hardware box. Terms such a RADAR Keyboard Multiplexer (RKM) or Non RADAR Keyboard Multiplexer (NRKM) were used to describe these devices. A general-purpose computer that was linked to the central computer via high-speed parallel interfaces performed message validation of the keyboard entries. Both the workstation processor and central processor were redundant to avoid single points of failure. The central processor itself was partitioned into an input output processors and central processing units that were tightly coupled via memory.

The second architecture to consider is the smart centralized architecture. In this architecture the workstation processor is distributed down to the physical workstation level. This is accomplished because of higher density circuits needing less space and lower cost. So it is practical to physically locate a dedicated workstation processor to each workstation. This significantly reduces the workstation cable plant size from the fully centralized architecture. The less cables and connectors in the system the more reliable the system, especially as the cable plant and connectors are periodically disturbed because of maintenance.

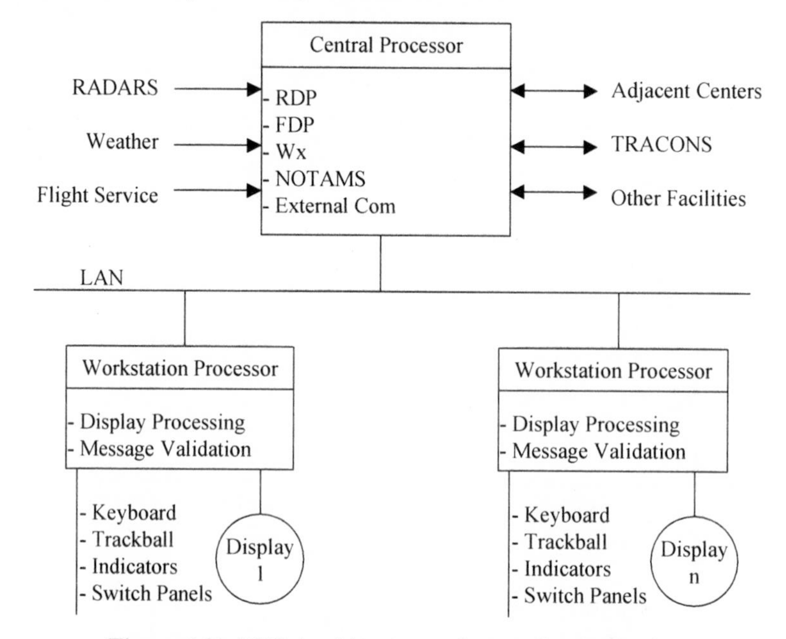

Figure 141 ATC Architecture - Smart Centralized

The interface between the smart workstations and the central processor can take many forms. In the early stages the interface was point to point using RS232 / 422. They were quickly replaced with deterministic buses like the Hughes Token ring. Hughes was also a strong proponent of Fiber Distributed Data Interface (FDDI), a redundant deterministic token bus. This architecture had a long life spanning from the late 1970's to the late 1980's.

By 1980 as computer costs were dropping and computer capacity was increasing, stakeholders were asking more from the system. Could it be made more available? Could it take on new functionality like cognitive processing[249]? Could air traffic control be re-thought to improve overall system operations? The following abstract captures the goal in the early 1980's:

"The Federal Aviation Administration's Advanced Automation Program (AAP) is described, and emphasis is placed on the program element referred to as Automated En Route Air Traffic Control (AERA). It touches on a few possible areas for future research in artificial intelligence. The AAP system replaces the National Airspace System Stage A with enhancements such as sector suite, functional enhancements and AERA. AERA has the objectives of allowing users to fly direct,

[249] Automated En Route Air Traffic Control (AERA) was a program that captured many of these ideas in the 1980's.

fuel-efficient routes, increasing controller productivity, and reducing operational errors. The AERA 2 functional capabilities are also discussed. The final phase AERA 3 has the goal of automatically generating conflict free, fuel-efficient clearances to pilots without controller intervention. There is a need to examine the proposed ATC system enhancements relative to the real and anticipated AI technologies to help FAA plan for their incorporation into future systems."[250]

Some of these ideas started to surface new and challenging requirements on the computer system. These ideas translated to:

- The system should never fail
- The system should never make a mistake

The cost of system failure would be switching the system to a backup mode with less functionality that might not be able to handle the increased load on the air traffic controller. All of the sudden the air traffic controller responsible for 100 planes would only have functionality that would allow the safe handling of only 20 planes. Within this environment, where the machine provides massive automation and the human cannot handle the load processed by the machine is the possibility that the machine might start making mistakes. How would the human be able to detect these errors and what would the human do in the presence of these errors if detected?

One of the architectures that surfaced at Hughes Aircraft to address these new challenging goals was referred to as Architecture A. It was during a time when the fully centralized architecture existed for U.S. Enroute air traffic control and the smart centralized architecture existed for air defense. The smart centralized architecture was just introduced at the Federal Aviation Administration (FAA) National Aviation Facilities Experimental Center (NAFEC)[251] as part of the Air Traffic Control Simulation Facility (ATCF) simulation test bed[252].

The fundamental idea behind Architecture A was that miniaturization would continue allowing hardware costs to drop as capacity continued to increase. With that simple assumption and the precedent set for display processing being comfortably performed at the console level, RADAR track processing was allocated to the workstation with a new dedicated single board computer. Tracking would

---

[250] FAA Air Traffic Control Directions, Report of the Workshop on Artificial Intelligence held at the National Academy of Sciences, Washington, D.C., October 23, 1985. Neumann, P J, Transportation Research Board, ISSN 0097-8515

[251] NAFEC was later renamed to FAA Technical Center.

[252] Sanders Graphic 7 with a single board computer (PDP-11/04) was used for generic ATC workstations and SEL 3277 super-mini computers with an early version of UNIX were the host processors.

now be distributed from a single point of failure central computer complex to 100 workstations representing 100 pieces of airspace[253].

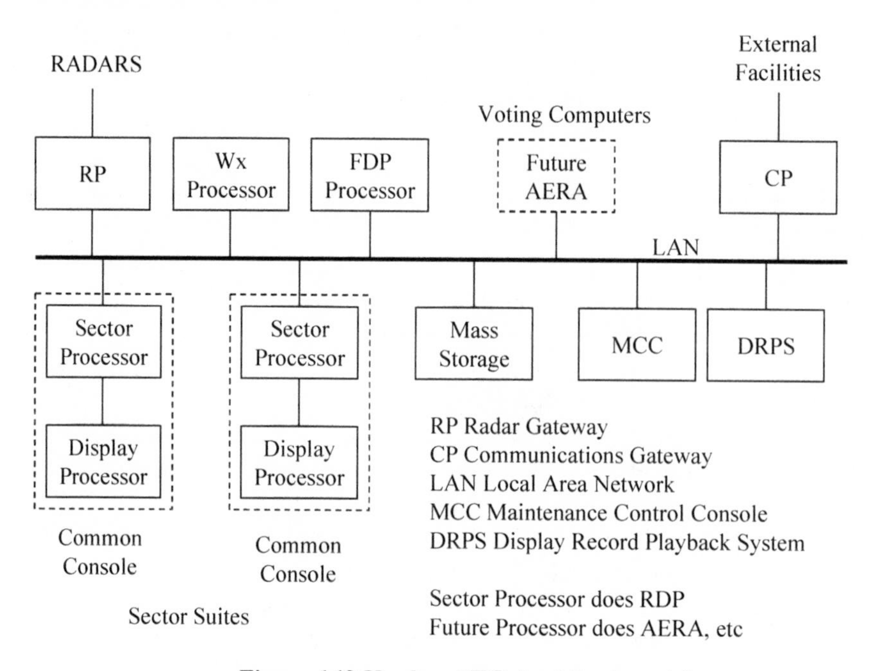

Figure 142 Hughes ATC Architecture A1

A radical idea but its advantage was obvious. All of the sudden the Impact of Failure (IOF) [254] could be at a single workstation level, 1/100 of the total facility responsibility. The allocation was taken all the way down to the lowest common denominator, the user. The implications were enormous. Not only could an emergency RADAR processing mode be guaranteed but also other functions might be eligible for the same allocation and advantage.

The driving force behind this allocation was a new phrase "Impact of Failure". With this allocation the impact of failure was fully minimized for the RDP function. Either each workstation would process all the RADAR returns and produce tracks for the entire facility airspace, or the workstations would produce tracks just for their piece of airspace and make the track data available to other workstations via the workstation token ring bus.

The distributed track database works because there is little coupling between the

[253] Although many at Hughes believed this was a valid assumption, Joe L at Hughes was brave enough to formally make that statement and force it with the simple analogy of from tubes to transistors to integrated circuits.

[254] Walt Sobkiw coined the term Impact of Failure at Hughes Aircraft and used his experience on the RDR and ATCSF upgrade to show how tracking might be distributed using filtering at the interface level and processing at the console level.

raw data representing position returns from the RADAR. They can be filtered from a range azimuth point of view[255] and each console could establish its own tracks. Boundary conditions between consoles could be handled with overlapping coverage. Once a track is established it is broadcast on the local area network (LAN) implemented as a redundant fiber optic token bus as other consoles bring in the broadcast track data. Any overlapping coverage redundant track data issues could be handled using mosaic approaches where a preferred track is selected not unlike RADAR mosaic processing in the ATC Tracker.

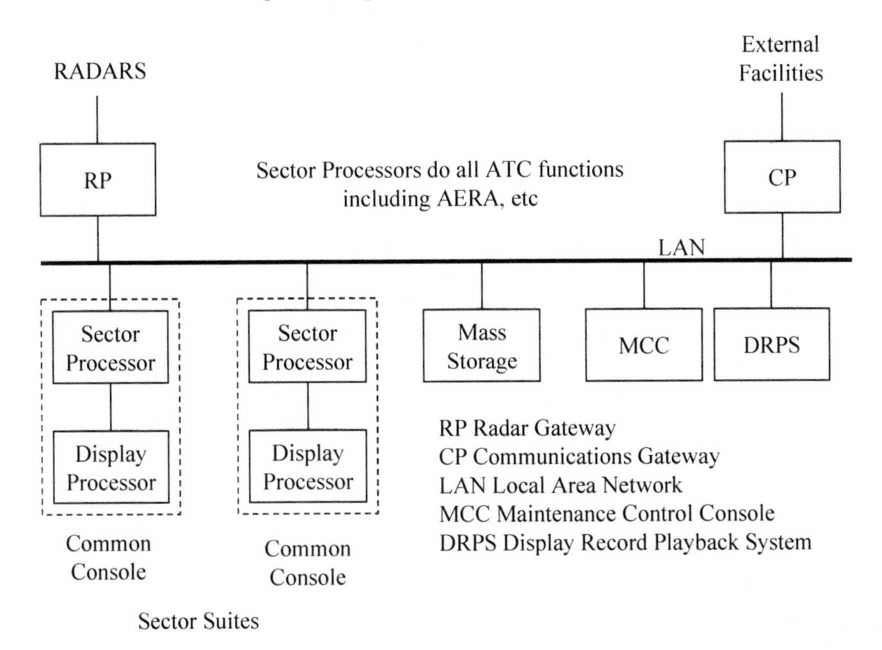

Figure 143 Hughes ATC Architecture A2

Once the distributed track database concept was analyzed and shown to be feasible and practical then eventually accepted by the team the architecture was accepted. The end result was two versions of Architecture A. The first was lower risk and acknowledged the need for centralized resources. The second was viewed as very high risk politically. Although the internal Hughes team was convinced that this was the best architecture, other stakeholders were involved. Today we see the second version of Architecture A everywhere.

---

[255] In the late 1970's NAFEC produced a RADAR Data Recorder (RDR) that filtered RADAR data from the Common Digitizer based on range and azimuth, allowing only selected data to be recorded.

27.10 Architecture Verification and Validation

Once architectures are identified, studied, selected, then optimized for the final solution how are they verified and validated? The process of selecting the architecture verifies and validates the architecture. This process includes formal and informal studies, analysis, and models.

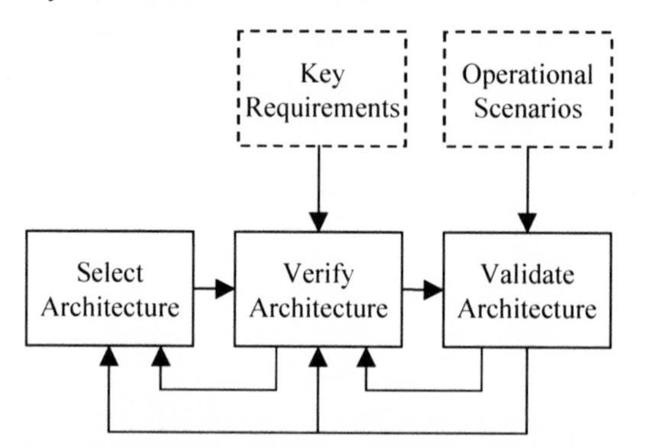

Figure 144 Architecture Verification and Validation

So the team points to the information products supporting the final decision and architecture approach. At the highest level the key requirements are used to verify the architecture and the operational concepts are used to validate the architecture. Many times the essence of the operational concepts may be captured in one or more models with various scenarios.

The next sections on Architecture Value Systems and Architecture Selection and Tradeoffs provide the framework for methodically capturing the verification and validation information products.

27.11 Exercises

1. Draw an architecture block diagram of our transportation system. What kind of architecture does the depiction represent? Do you think it is accurate?
2. Draw an architecture block diagram of our water system. What kind of architecture does the depiction represent? Do you think it is accurate?
3. Draw an architecture block diagram of our electric power system. What kind of architecture does the depiction represent? Draw an architecture block diagram of a sustainable electric power system. Is it different from the current architecture?
4. Draw an architecture block diagram of our health care system. What kind of architecture does the depiction represent? Draw an architecture block diagram of a sustainable health care system. Is it different from the current architecture?
5. Draw an architecture block diagram of our educational system. What kind of architecture does the depiction represent? Draw an architecture block

diagram of a sustainable educational system. Is it different from the current architecture?

27.12 Additional Reading

1. FAA Air Traffic Control Directions, Report of the Workshop on Artificial Intelligence held at the National Academy of Sciences, Washington, D.C., October 23, 1985. Neumann, P J, Transportation Research Board, ISSN 0097-8515.
2. Vitruvius The Ten Books on Architecture, Herbert Langford Warren (Illustrator), Morris Hickey Morgan (Translator) Courier Dover Publications, 1960, ISBN 0486206459.
3. Vitruvius, Fra Giocondo, Venice, 1511.
4. Vitruvius, The Ten Books on Architecture. Translated By, Morris Hicky Morgan, Illustrations And Original Designs, Prepared Under The Direction Of, Herbert Langford Warren, A.M., Nelson Robinson Jr, Cambridge, Harvard University Press, London: Humphrey Milford, Oxford University Press, 1914.
5. Vitruvius: Ten Books on Architecture, Cambridge University Press, Cambridge 1999, Editors D. Rowland, T.N. Howe, 2001 ISBN 0521002923.

28 Architecture and Value Systems

At some point you will need to select your architecture. Architecture selection is the most important activity on any effort. Pick the wrong architecture and people may spend decades implementing it only to find it has emergent properties that are catastrophic. Pick the wrong architecture and no one may be able to successfully implement it. Pick the wrong architecture and billions of dollars may be spent yet quickly displaced by another better architecture solution at a fraction of the cost. Pick the wrong architecture and everyone may get sick or die. For example some believe the fall of Roman Empire can be traced to the use of lead[256]. Pick the wrong architecture and your product is a failure.

Part of the answer to selecting the right architecture is to engage in the front-end systems practices suggested in this book. The other part is to make an impartial selection immune to any external influences. The selection needs to be based on evidence traceable to all the formal and informal studies. Buried in this is the concept of value. Which architecture has the greatest value?

28.1 Financial Metrics and Value

Financial metrics falls under the broad area of economics. They are based on two important economic principles: (1) Fixed and Variable Costs, and (2) Supply Versus Demand. Everyone is familiar with financial metrics and many erroneously believe all decisions should be based exclusively on one or more of these financial metrics. Examples of financial metrics are:

- Lowest Cost
- Highest Profit
- Greatest Return on Investment (ROI)
- Total Cost of Ownership (TCO)
- Internal Rate of Return (IRR)
- Economic Value Added (EVA)
- Real Option Value (ROV)
- Return on Assets (ROA)
- Return on Infrastructure Employed (ROIE)

However, value is not only financial but also non-financial. For example some value a day at the beach more than a day at the mountains. How do you capture the

[256] Ancient Rome lined their pipes and water tanks with lead, used lead plates and eating utensils, wine makers sweetened sour wine by adding lead powdered syrup. Some modern historians have suggested the fall of the Roman Empire may have been partially due to lead.

value of a bridge that connects two cities? How do you capture the value of a road that heads into a wilderness? These are interesting questions and they were addressed in the last century.

There is a technique that is based on a simple concept - value is more than apparent financial results. Apparent because it is all relative to your context or view. For example it might make sense to do something because it can translate into thousands of jobs or it can cause people to view a physical land area as attractive so they will move there and buy houses and or open businesses. These can be reduced to dollar numbers but it is more complex[257]. It uses the following elements:

- Life Cycle Costs (LCC)
- Measure of Effectiveness (MOE)

28.2 Fixed and Variable Costs

Fixed and variable costs are usually addressed in a manufacturing setting. However the concept of fixed and variable costs has broad application and some interesting relationships between fixed and variable costs can be applied in understanding architectures.

The fixed costs are described as costs that do not change over time. In a business setting they are costs that are not dependent on the production level. Time is relative and fixed to perhaps a quarter of a year. Examples of these costs are salaries, rent, and insurance.

Variable costs change in the time frame considered when other costs are fixed. These costs are volume related. Examples include the raw materials or components needed to produce an item.

What is missing from this production volume perspective of fixed and variable cost is the concept of unexpected costs. Unexpected costs are by their very nature variable and not readily predictable. You might be able to use history to predict unexpected costs and apply some risk assessment methods to the prediction. The result is that unexpected costs have the same effect as variable costs. They add to the total costs and change with time. In architecture analysis unexpected costs always should be considered. It can be an element of variable costs.

The total cost is the sum of the fixed and variable costs as a function of time. If we remove the time element and use this concept for a hypothetical business then we can surface the best, worst, and nominal case costs for the hypothetical business. If we view the business as a back box, then we are basically viewing the business as architecture. We can have several different architectures or businesses.

So this concept of fixed and variable costs can be applied to any architecture,

---

[257] This is a Formal Architecture Tradeoff with Measure of Effectiveness (MOE). Sustainable Development Possible with Creative System Engineering, Walter Sobkiw, 2008,

not just a business. Architectures have fixed and variable costs. The variable costs are heavily weighted with unknown costs and associated risk assessments. If we look at the relationship between fixed and variable costs for various architecture alternatives then there are three possible scenarios.

1. **Scenario 1 is where the fixed costs far surpass the variable costs.** In this situation there are no surprises. A reasonable example of this scenario is 99% fixed or known costs and 1% variable or unknown costs.

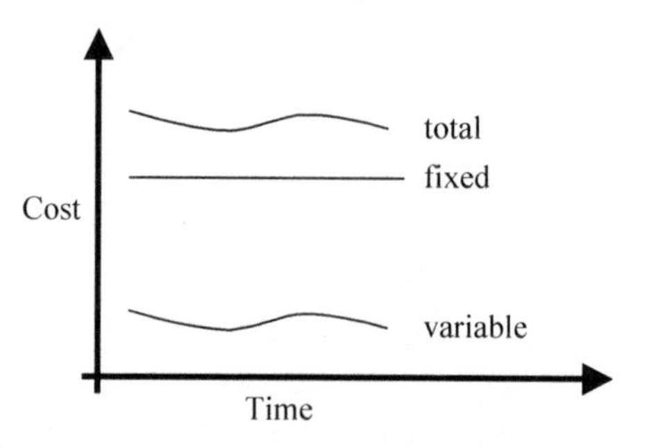

Figure 145 Low Risk Cost Scenario

2. **Scenario 2 is where the variable costs start to impact the total costs** in a serious manner such that the solution might be at jeopardy. . A reasonable example of this scenario is 80% fixed or known costs and 20% variable or unknown costs.
3. **Scenario 3 is where the variable costs impact the total costs in an unacceptable manner** such that the solution cannot move forward. An example of this scenario is 20% fixed or known costs and 80% variable or unknown costs.

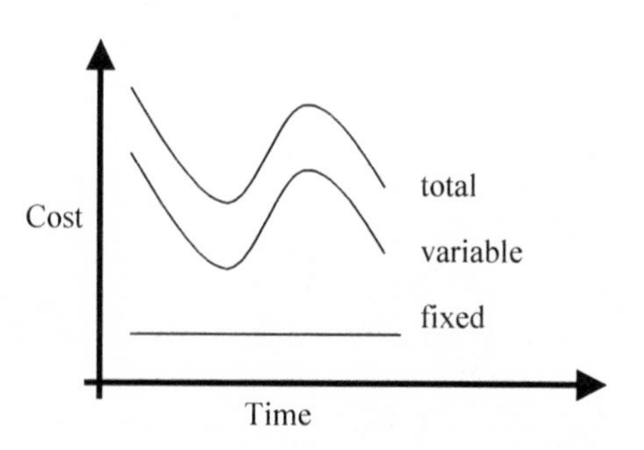

Figure 146 High Risk Cost Scenario

Obviously there are various shades of gray in the three scenarios. The three

ISBN 0615216307.

scenarios represent cost risk when architectures are being considered. So the cost risks can be as follows:

- Low Cost Risk is scenario 1
- Medium Cost Risk is scenario 2
- High Cost Risk is scenario 3

When architectures are considered they may have different cost scenarios. Architectures based on current approaches typically will have Low Cost Risk while architectures with little basis in current approaches will have High Cost Risk.

28.3 Supply and Demand

Supply and demand is a multidimensional concept that relates price, quantity, consumers, and producers. It offers several financial metrics and is a fundamental element of modern economics. Supply and demand is based on the following assumptions:

- Consumers behave rationally
- Producers behave rationally
- There are many independent consumers
- There are many independent producers

In this ideal environment of rational behavior and a level playing field of many independent consumers and producers the following relationships logically surface:

- Consumers as a group will buy more product items or service units (quantity) as price decreases
- Producers as a group will increase the number of product items or service units (quantity) as price increases

This relationship is shown in a supply and demand graph. A demand curve represents consumers and shows what price they are willing to pay for an item. As price decreases more consumers enter the market and so more items are sold. The supply curve represents the producers. As prices rise they are motivated as a group to find ways to offer more items. This may be with the introduction of new producers or new processes and or technologies that allow for more items to be made available. In both cases the quantity increases or decreased with price. The intersection of the supply and demand curves is the ideal price where there is no excess quantity or insufficient quantity. It is the point of equilibrium.

The lines in the supply and demand graph can be various shapes and slopes. Obviously infinite quantity is not possible. Eventually the supply costs start to increase rapidly with each increment in quantity. Eventually as the demand cost becomes very high, sales decrease rapidly with each increment in cost. So the lines are curved as they approach the limits of the market size and the technology.

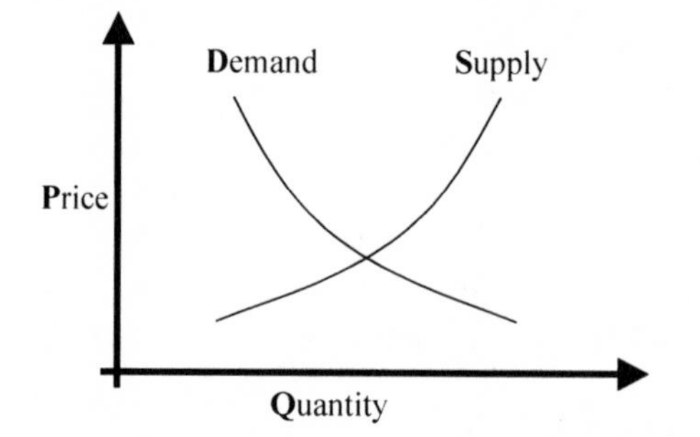

Figure 147 Supply and Demand Graphs

Elasticity is used to describe the slope levels of the supply and demand curves. The more vertical the supply slope the more inelastic the supply curve. The more vertical the demand slope the more inelastic the demand curve. If the supply curve is perfectly vertical it is perfectly inelastic. This means the quantity is fixed no matter what the price. Many practitioners use land as an example of this scenario. No matter how much someone is willing to pay for additional land more parcels cannot be created.

However the supply and demand model has flaws. As a teaching example the land analogy illustrates the concept, but it is wrong in practice. The reality is that there is always more land if one is willing to look outside the current context or boundary. There are other neighborhoods, towns, cities, regions, etc. There are even alternatives such as artificial islands or land fills and vertical locations such as skyscrapers.

Additionally consumers and producers do not behave rationally. The tulip mania of the 1630's is an example of a speculative bubble[258]. At the peak of the tulip mania in 1637 some tulip bulbs sold for more than the annual income of a skilled craftsman. There have been many economic bubbles in the ensuing centuries. Time and time again consumers convinced themselves that they should pay a price for an item that far exceeds its rational price point as they rode a speculative bubble and hoped not to be the holders of the items once the bubble burst.

Producers when acting in their own interest always try to increase market share. This eventually leads to buying out the last of the remaining weaker competitors. At

[258] From Mackay, Charles (1841), Memoirs of Extraordinary Popular Delusions and the Madness of Crowds: "In 1634, the rage among the Dutch to possess them was so great that the ordinary industry of the country was neglected, and the population, even to its lowest dregs, embarked in the tulip trade. As the mania increased, prices augmented, until, in the year 1635, many persons were known to invest a fortune of 100,000 florins in the purchase of forty roots."

some point barriers of entry such as capital, branding, regulations or any combination become so significant that essentially oligopoly or monopoly surfaces. The empirical evidence is everywhere in the USA, from Detroit being dominated by GM, Ford, and Chrysler for decades to Microsoft and its Windows operating system.

So consumers and producers do not behave rationally[259]. Although there are many consumers in this highly automated age, there are not many producers. Also in this age of mass media and mass influence the risk of irrational consumer behavior has risen beyond any expectation for consumers to behave rationally. This is not to say that they are stupid. They are just acting on the information they are offered and if that information is overwhelming and tainted, then there is no other expectation other than irrational behavior. The cottage industry model with consumers that have limited information access fits more appropriately with the supply and demand model; this model disappeared a hundred years ago.

Understanding the supply and demand facts of life was key to introducing new products and technologies in the last century. When electricity was first offered the general attitude was why. When the telephone was first offered the same issues surfaced. However realizing that consumers do not act rationally, strategies surfaced that stimulated interest in new products and services. The concept of Veblen[260] goods was applied to many new products and services in hopes of creating new markets.

Veblen goods and services are associated with high social status and exclusivity. Veblen goods and services have a different demand curve from normal goods. In a Veblen good or service, decreasing the price reduces the consumers desire to buy because it is no longer perceived as exclusive or high status. A price increase will increase the consumers' perception of status and exclusivity making the good or service more desirable. There are limits to the price increases and at some point demand will drop even in the exclusive high status market.

Veblen combined sociology with economics in the book "The Theory of the Leisure Class" (1899)[261]. The book suggested that there was a difference between industry, run by engineers, which manufactures goods for the general population, and business, which exists only to make profits for a leisure class. Unlike industry, which is productive, business allows the leisure class to engage in "conspicuous consumption", where the economic contribution is "waste" that contributes nothing

---

[259] Irrational human behavior happens in all systems. It needs to be acknowledged and respected. Users will find ways to misuse the system.

[260] Thorstein Bunde Veblen, (July 30, 1857 – August 3, 1929) an American sociologist and economist was the first to introduce the term "conspicuous consumption" and concept of status seeking.

[261] Thorstein Bunde Veblen (1899) The Theory of the Leisure Class, An Economic Study of Institutions, London: Macmillan Publishers.

to productivity.

Example 51 Value Systems - Modified Supply and Demand

So how does this relate to architecture value system? Architectures in the early stages are groups of ideas and concepts. These ideas and concepts are circulating in what is essentially a small test market with the participating stakeholders. In many ways these stakeholders are an early indication of how the consumers or users will receive the product or architecture. In this setting many questions can be asked. For example what is the incremental cost of adding a certain feature that separates one architecture or product from another and what is the impact on stakeholder acceptability? So conceptually each architecture or product approach can have a set of supply and demand curves.

If we replace the x-axis Quantity on the supply and demand chart with Approaches where approaches can be different architectures, products, or clumps of features as viewed by consumers then some interesting ideas surface. For example features typically characterize architectures and products. Further there is a continuum from least costly to produce to most costly to produce based on the feature groups. The project stakeholders can be extrapolated to the broader market consumers and or users. They have a price they are willing to pay for each architecture or product increment feature block. This is the internal stakeholder demand curve. This demand curve could be augmented with a market study.

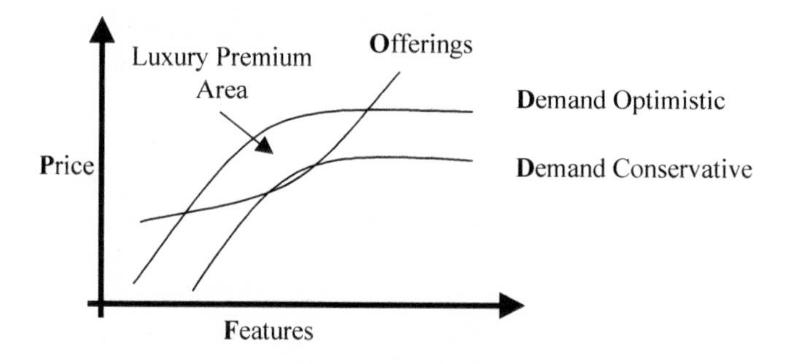

Figure 148 Offerings and Demand

As we examine the demand-offering chart there are several observations. The Offerings curve shows that the cost to produce each architecture or product increases with increasing features. It is not linear or smooth. The Demand curve shows that consumers will only pay a certain price for the architecture or product regardless of the level of features. It also shows that if the demand intersects the offering at some point, there is an area where a premium price can be charged for some increment of features. These features fall into the luxury category and represent opportunity to increase margins via the offering of luxury features.

As we will see the modified supply and demand model is the closest financial metric to the system engineering Measure of Effectiveness (MOE). This is also your first step into creating a new method based on an academically accepted method. Supply and Demand as modified to yield something new, Offerings and Demand.

Example 52 Value Systems - Early Suburban Housing

Imagine buying and building a house in 1960's suburbia USA. Lets start with the consumer. Here you are living in the city. You have a car and there are new roads to take you into a different world. As you travel these roads you see various farms being sold to developers who are building houses. You stop in, wonder through a few sample homes, pick up some brochures, and drive through the new streets. You leave the small development of 50 - 300 houses and drive around the surrounding community. What value system can you use to make your final decision? Are you considering the price with the various features? One developer offers a larger lot at the expense of a smaller house. Another developer offers a large house and large lot but at a different location up the road.

Imagine being a builder. The war is over and you partnered with a brilliant architect you met in the military. You have dreams of building a new world. What features do you offer to separate you from the competition. How do you attempt to speculate what people might want when they themselves are not sure. The Veblen effect does not work. These are people who just lived through depression and war, they are looking for a different way. What do you do?

Fortunately in this scenario everyone got what they wanted except for the people left behind in the cities. It took decades for the Veblen effect to take hold of the new cities being born after the World War II. Housing bubbles that led to whole communities being bulldozed in some insane frenzy of Veblen effect and massive speculation did not surface until the new century.

28.4 Lowest Cost

Cost is the total cost to purchase an item if you are buying it or the total cost to build an item if you are building an item. Most people focus on lowest cost no matter the perspective. However if you are buying an existing item you run the risk of a latent quality issue that does not surface until the product is in use. If you are buying an item yet to be built you run the risk of cost over runs.

The government, when it buys a system based on lowest costs examines multiple bids and looks for cost reasonableness. If a particular proposal comes in at a significantly lower price than the competition, the bid may be tossed out of the competition because of cost reasonableness based on the bids of other vendors and internal cost estimates. The same may be true of a high bid. It should be closely examined to determine if it is the reasonable bid and the others are not reasonable as they try to buy into a contract.

Most people who work on fixed price contracts, based on lowest cost tend to fixate on the "project triangle". The idea is the corners of the triangle are labeled good, fast, and cheap[262] (quality, schedule, and cost). The area of the triangle is project scope. So if you want something at low cost then quality and or schedule must decrease. The alternative is to create a smaller triangle (less area) which is a project with less scope.

However, this is a very myopic view of the problem. Cost is driven by quality, up to a point. There comes a point where no amount of money will increase quality and yet the quality may be too low to field the solution. Cost is driven by schedule, but there comes a point where increasing the cost will not reduce the schedule. This is like trying to give birth to a 9-month baby in 3 days. It is a hilarious but sad situation in an organizational setting to see these ideas taken to unreasonable extremes.

When these ideas surface in an organization most people will start to look for all the hidden costs. Itemized lists are then developed. They try to find the collection of items to cut and yet have the project, organization, system, etc still work. This mindset eventually leads to cost shifting strategies where the burden is moved to another area.

For example if an organization is big enough it may choose to move operations to another country with lower labor and environmental costs. However new costs surface that far exceed the savings to the entity. The entity does not pay for these cost shifted elements. For example a military to protect the assets.

All these views are superficial and unfortunately do not consider traditional economies of scale lessons from the previous century. This typically is a sign of panic or irresponsible behavior. Sometimes it genuinely can be because of a myopic view, which leads to a lack of understanding of the lowest cost. A broader view of cost is to understand what actually drives the cost. Lowest cost is defined as:

Cost ~ Quality + Features / (Process + Automated Tools)

So cutting features or reducing quality can reduce cost. Typically that is not possible in a healthy competitive environment. However introducing process and automated tools can reduce the cost. Further tuning the process and automated tools is what will separate a vendor from its competition. The process and automated tools may have such an impact that quality increases and the solution now becomes viable not only from a cost point of view but also from a quality point for view (e.g. chip yields from a foundry). In many ways this simple relationship is the story of the industrial revolution.

Process and automated tools have a cost. However if their cost is less than the cost savings they offer, then they are effective and the indirect costs of the process and automated tools are worth the investment. If the leaders focus on the short term

---

[262] Hughes Aircraft defined cheap as a low quality solution while low cost was defined as a solution with no quality loss.

and strip the assets by cutting investment in process and automated tools then the organizations projects will eventually start to fail.

28.5 Highest Profit

Profit is all the money that comes in minus all the costs. There are typically three types of profit to capture the different categories of costs. They are:

Gross Profit = Sales Revenue - Cost of Goods Sold
Where: Cost of Goods Sold = Production Cost

Operating Profit = Gross Profit - All Operating Expenses
EBIT = Operating Profit
(Earnings before Interest and Taxes)
Operating Expenses = General overhead like Research and Development (R&D), General and Administrative G&A.

Net Profit = Operating Profit - Tax - Extraordinary Expenses
Net Income = Net Profit

So there are different slices of "profit" that stakeholders use to try to assess and apply value to an organization, project, program, etc. The bottom line is that anyone can show great short term profits while deferring costs such as R&D, G&A, and even taxes. Selling off assets or reducing staff also can increase or maintain profits.

This is relevant in a system architecture discussion about a company, industry, or organization. An architecture that leads to maximum profits at the expense of no future is not sustainable and is a fatally flawed architecture.

28.6 Return on Investment

Return on Investment (ROI) is the ratio of the net revenue generated by a business that is divided by the cost to establish and maintain the business. It is also the ratio of cost savings generated by the introduction of a tool, process, method, etc divided by its cost. From an investors point of view it is the rate of return, typically annualized on an investment such as a bank certificate of deposit. The goal is to maximize the ROI.

From a business point of view the ROI is typically associated with a period of time. This is the investment period and the goal is to have the full investment plus additional returns provided back to the investors by the end of the investment period. Depending on the length of the investment period and the inflation rate on the local currency, the cost of money must be factored into the total costs. Typically an ROI plan has the goals of an investment period of 1-3 years with an ROI of 20%.

For complex high stake systems the context of the ROI becomes extremely important. For example, expenditures on one project may not result in planned ROI numbers for that project, but its off shoots are so significant that an entire company, industry, country, or the world is transformed. For example when computers were introduced into air traffic control it was for the purpose of reducing staff. What actually happened is the staff increased as people thought of new ways to use the new system with its new levels of automation. This allowed air transportation to expand beyond anyone's dreams at the time. It gave birth to our modern society of air travel that everyone now takes for granted. How do you calculate the ROI for that investment?

28.7 Total Cost of Ownership

Total Cost of Ownership (TCO) is a financial metric that attempts to capture all costs from the perspective of the owner. It typically includes purchase, operation and maintenance. The problem with TCO is that it is from the perspective of the owner. It does not take into account cost shifting or indirect costs, which must be paid by other stakeholders or even non-stakeholders.

For example airline companies are commercial companies and certainly have costs, however the indirect costs of the air traffic control system which includes airports, air traffic control centers, navigational aids, voice communications, satellites, etc are paid for by the tax payer. In addition the indirect costs associated with aircraft include research, development, and technology transfer from military aircraft to the commercial sector. Building and creating maintenance and support mechanisms for something that does not fall from the sky has huge indirect costs.

There are other indirect costs such as pollution that are borne by other elements of society. Noise pollution results in large tracts of land being reserved for special use around airports. Carbon pollution recently has started to enter the dialog and it is a cost that someone must bear.

28.8 Internal Rate of Return

Internal Rate of Return (IRR) is similar to ROI. It is the internal rate of return on an investment. It is the annualized effective compounded return rate that can be earned on the invested capital. It does not include the cost of money or inflation.

28.9 Economic Value Added

Economic Value Added (EVA) compares multiple approaches using their respective ROI numbers, however the comparison is with leaving the money in an external investment instrument, such as a bank versus making the investment within the organization. It is essentially adding the cost of money to the investment. So you are asking if it better to make the investment in the idea, business, process, method, tool, etc or should you just leave the money in the bank.

28.10 Real Option Value

Real Option Value (ROV) is a modified ROI where the value is not associated with the current project, but with future projects. Essentially the costs are amortized across several projects, programs, business units, etc that stand to benefit from the investment but each alone could not pay for the investment. This is a form of infrastructure investment. The infrastructure investment could have huge benefits for the organization and its current and future projects and programs. This is also a form of investing in the future. The benefit may not surface for several years.

A classic example is the introduction of new tools into an organization. The first project will most likely suffer from a cost and schedule point of view but future projects stand to significantly benefit once the new tools are understood and internalized by the organization.

Another example is the introduction of a capital-intensive machine such as bulldozer into a third world country. The first project cannot pay for itself and using human labor with picks and shovels will yield the most cost-effective approach for the project. However, once the bulldozer is amortized over several projects and the laborers are freed to perform more productive tasks the third world country will transform itself into a first world super power. Buying the first bulldozer is the hard part in any organization without vision.

Investing in the future is exemplified by traditional research and development or deciding to go to a university rather than enter the workforce. In both cases an investment is made with the intent of a large payoff in the future.

28.11 Return on Assets

Return on Assets (ROA) is the net income divided by the value of the assets being used to generate the income. If the income is less than could be derived by the selling off of the assets, then a case can be made to sell off the assets. Coupled in this analysis is time. If the assets can perform for 10 years versus 5 years then the decision to sell at a particular factor of the value of the assets will change. Capital intensive industries have small returns on assets. They also have a large barrier to entry into the market. It is difficult to arrange for the purchase of the "expensive" assets.

Fir example, it was only a few decades ago that computers were extremely capital intensive. Only governments and extremely large companies were able to purchase computers. The ROA for these machines were very small.

28.12 Return on Infrastructure Employed

Return on Infrastructure Employed (ROIE) is similar to ROA except it is applied to infrastructure services. For example a computer support department provides services at some cost to various projects, programs, business units, etc. The ROIE may be different for in house versus outsourced computer support services.

28.13 Life Cycle Costs

Unlike traditional financial metrics which are based on artificial system boundaries that attempt to maximize the benefit of one stakeholder at the expense of another stakeholder[263], Life Cycle Costs (LCC) is a broad metric used to capture all the costs from all the stakeholders.

Each generation since World War II has attempted to identify the elements of LCC. Today with the realization of environmental impacts the LCC has become even broader than in the past.

At the highest level of LCC there are direct costs attributed to a project or program and then there are indirect costs that are attributed to infrastructure[264] or other costs not typically allocated to the project or program. Typically it is the indirect costs that significantly impact the selection of the architecture.

The following is a starting point for your LCC analysis:

LCC = R+D+P+O+M+S+I+E or PROMISED

Where:

R = Research	S = Shut Down & Disposal
D = Development	
P = Production	I = Infrastructure
O = Operation	E = Environment & Waste
M = Maintenance	

In the USA the direct costs have typically been borne by commercial companies and indirect costs have been borne by the taxpayer. For example automobiles will not sell unless there is a road infrastructure. Airplanes cannot fly unless there are airports and an air traffic control system. Further airplanes would not exist without defense projects that invested in aircraft. Spaceships carrying satellites that offer commercial services cannot fly without NASA launch and tracking systems.

It is always an interesting moral situation when companies forget that if the true

---

[263] Essentially the total costs in traditional financial metrics are shifted and the true cost burden is hidden or obfuscated.

[264] Executive Order 12803 - Infrastructure Privatization April 30, 1992, (b) "Infrastructure asset" means any asset financed in whole or in part by the Federal Government and needed for the functioning of the economy. Examples of such assets include, but are not limited to: roads, tunnels, bridges, electricity supply facilities, mass transit, rail transportation, airports, ports, waterways, water supply facilities, recycling and wastewater treatment facilities, solid waste disposal facilities, housing, schools, prisons, and hospitals.

costs of the products and services were passed on to the customer, assuming they could actually implement the infrastructure, they would quickly go out of business. This is a modern political discussion that has resurfaced approximately 40 years after the end of World War II and has not been resolved in the USA at this time.

28.14 Measure of Effectiveness

Minimizing life cycle costs, although extremely beneficial when compared to other financial metrics, still does not offer a true indicator of the value of one architecture over another architecture.

Value is not only financial but also non-financial. For example some value a day at the beach more than a day at the mountains.

The Measure of Effectiveness (MOE) is based on the concept that the tradeoff criteria in an architecture tradeoff matrix should not include cost or requirements. It is a given that all solutions will satisfy the known requirements at some cost.

Cost is used at the bottom of the tradeoff matrix where each approach total rating (sum of all the tradeoff criteria) is divided by cost. This essentially identifies the ***goodness of each approach per unit of cost***. This is called the Measure of Effectiveness or MOE. This is expressed as the following equation:

MOE = Sum of tradeoff criteria/cost to produce approach

The MOE equation is the heart of systems engineering. It is fundamental and drives everything. The tradeoff criteria ratings are the result of analysis, which may take years and involve hundreds of stakeholders.

A word of caution, most institutions do not base their decisions on the MOE or measure of goodness of an approach. So there is confusion on what the MOE actually represents. For example some refer to the MOE in terms of what is actually the tradeoff criteria. They state that the MOE is based on operational needs and then are silent on the relationship between the sum of the tradeoff items and cost. So the idea of MOE is obscured and lost. The following is a description of MOEs from NASA:

"MOEs are the "operational" measures of success that are closely related to the achievement of mission or operational objectives in the intended operational environment. MOEs are intended to focus on how well mission or operational objectives are achieved, not on how they are achieved, i.e., MOEs should be independent of any particular solution. As such, MOEs are the standards against which the "goodness" of each proposed solution may be assessed in trade studies and decision analyses. Measuring or calculating MOEs not only makes it possible to compare alternative solutions quantitatively, but sensitivities to key assumptions regarding operational environments…[265]"

Notice the term MOE has become plural, the plural MOEs usage and the description clearly represent tradeoff criteria. However in this ontology the criteria are lost and MOE is obfuscated. Some practitioners are familiar with the subtle distinction between the singular and plural MOE term and use them appropriately in the process[266]. However for clarity, especially in highly controversial settings the MOE term should not be used with the tradeoff criteria for the ontology[267]. Do not use MOE and MOEs together in any process description. Use MOE and tradeoff criteria terminology.

Another potential trap where the MOE could be lost or inappropriately used is if people attempt to equate it to cost benefit analysis. Cost benefit analysis is primarily performed in the financial domain while the MOE with its use of non-financial tradeoff criteria works outside the financial domain. Also cost benefit analysis is typically used to justify an intervention as opposed to picking an architecture approach. The intervention is meant to disrupt the status quo with the intent of reducing future costs.

The MOE is the only value system that uses non-financial variables in the equation. In many ways it is like shifting from the time domain to the frequency domain to perform circuit analysis in electrical engineering. The view that the MOE offers is not possible when using strictly financial variables. Examples of other financial analysis that might be equated with the MOE are:

- Cost-Effectiveness Analysis
- Economic Impact Analysis
- Fiscal Impact Analysis
- Social Return On Investment (SROI) Analysis

Example 53 Value Systems - US Interstate Highway System

Today we marvel at the US interstate highway system. However 45 years ago people drove their automobiles on these new roads for miles and there was nothing. Today cities exist around these roads. What possessed the previous generation to build roads into the middle of nowhere? What did they use as a value method? Were these roads built exclusively for survival from a war or was something else going on in the minds of the people[268]? It is interesting to see that the goal of building a

---

[265] NASA Systems Engineering Handbook, NASA/SP-2007-6105 Rev1, National Aeronautics and Space Administration NASA Headquarters Washington, D.C. 20546, December 2007.

[266] At Hughes Aircraft I / we used the single MOE to select architecture approaches. This can be referred to as the architecture MOE.

[267] These controversial settings are described in Sustainable Development Possible with Creative System Engineering, Walter Sobkiw, 2008, ISBN 0615216307.

[268] Interstate Highway System authorized by Federal-Aid Highway Act of 1956 also

national network of roads was decades in the making. It is also interesting to see that one of the primary arguments was funding sources and national defense. The Federal-Aid Highway Act of 1956 also known as National Interstate and Defense Highways Act of 1956 is partially included as an appendix in this text.

There is a bridge connecting NJ with Philadelphia that was built in 1929, the Benjamin Franklin bridge. It is a huge bridge able to support traffic capacity in this new century. Not much was going on in 1929 in southern New Jersey. What possessed the people to build this incredible bridge that still works? What did they use as a value method?

Figure 149 Needs of the Highway Systems 1955-1984

The argument of the role of government and how projects would be funded surfaced time and time again after World War II as the society was moving into a modern post nuclear world. Eventually the concept of Dual Use Technology surfaced. It refers to technology that can be used for both peaceful and military purpose[269].

Example 54 Value Systems - Space Exploration

Today we take satellites for granted. They give us television, position

---

known as National Interstate and Defense Highways Act of 1956. In addition to private and commercial transportation, it would offer transport for military supplies and troop deployments in emergencies or foreign invasion.

[269] Dual Use Technology, Dual Use Processes, Dual Use Products all refer to Dual Use where peaceful and military use is possible. It became part of the national dialog after World War II. Early references are not easily found. A Survey of Dual-Use Issues, IDA Paper P-3176 Prepared for Defense Advanced Research Projects Agency (DARPA).

information, personal communications across massive distances, weather information, geological surveys, etc. How did we get here? What were the arguments between some futuristic vision, defense, and business? The National Aeronautics and Space Act of 1958 is included in this text as an appendix.

Figure 150 Hughes NASA Surveyor Lunar Lander

Hughes Aircraft designed and built the Surveyor spacecraft[270]. Surveyor 1 was the first lunar lander to explore the Moon. The NASA Jet Propulsion Laboratory managed the program. It was launched May 30, 1966 and landed June 2, 1966. Its landing weight was 596 lbs. (270 kg). There were 11,237 images transmitted to Earth.

Syncom stands for synchronous communication satellite. Syncom, a NASA program started in 1961 to build geosynchronous communication satellites. All the

[270] A spacecraft system consists of various subsystems, depending on the mission: Life support, attitude control, guidance navigation and control, communications, command and data handling, power, thermal control, propulsion, structures, payload, ground segment, launch vehicle, and a great stereo.

Syncom satellites were developed and manufactured by Hughes Space and Communications. Syncom 2, launched in 1963 and was the world's first geosynchronous communications satellite. In the 1980s, the Syncom continued with Syncom IV which were much larger satellites, also manufactured by Hughes. They were leased to the United States military under the Leasat program.

Figure 151 Hughes Syncom 2 First Geosynchronous Satellite

Hughes Aircraft built the Galileo spacecraft at its El Segundo, California plant. The Jet Propulsion Laboratory (JPL)[271] managed the Galileo mission for NASA. Germany supplied the propulsion module. NASA's Ames Research Center managed the probe. It was controlled by six RCA 1802 Cosmac[272] microprocessor CPUs. The 1802 CPU had previously been used onboard the Voyager and Viking spacecraft.

Galileo was launched on October 18, 1989 from the Space Shuttle Atlantis on the STS-34 mission. It arrived at Jupiter on December 7, 1995. The spacecraft was the first to perform an asteroid flyby, discover an asteroid moon, orbit Jupiter, and launch a probe into Jupiter's atmosphere. Because of Galileo scientists think that the Europa moon may have a salt-water ocean beneath its frozen surface. So on

---

[271] JPL is a federally funded research and development center.

[272] Each microprocessor clocked at about 1.6 MHz was fabricated on sapphire (silicon on sapphire) a radiation static hardened material ideal for spacecraft. It was the first low power CMOS processor. It was previously used onboard the Voyager and Viking spacecraft.

September 21, 2003, after 14 years in space and 8 years of service in the Jovian system, Galileo's mission ended by sending it into Jupiter's atmosphere. The goal was to avoid any chance of contaminating local moons with bacteria from Earth.

Figure 152 Hughes Galileo Spacecraft

The Galileo spacecraft and its attached Inertial Upper Stage booster are released from the payload bay of Atlantis on October 18, 1989.

Figure 153 JPL Viking Lander and Mars Landscape

Viking 1 and 2 space probes were sent to Mars. Each probe had an orbiter designed to photograph the surface of Mars and a lander designed to study the planet from the surface. The orbiter also provided communication for the lander once it touched down. The landers conducted biological experiments designed to detect life in the Martian soil. Three separate teams designed the experiments. One experiment turned positive for detection of metabolism. However another test failed to reveal any organic molecules in the soil. Most scientists became convinced that the positive results were likely caused by non-biological chemical reactions from highly oxidizing soil conditions. The total cost of the Viking project was approximately $1 billion US.

Pioneer 10 and Pioneer 11 explored the outer planets and left the solar system. They carry a golden plaque, showing a man and a woman and information about us on Earth. The Pioneer plaque image is engraved into a gold-anodized aluminum plate, 152 by 229 millimeters (6 by 9 inches), attached to the spacecraft's antenna support struts to help shield it from erosion by interstellar dust.

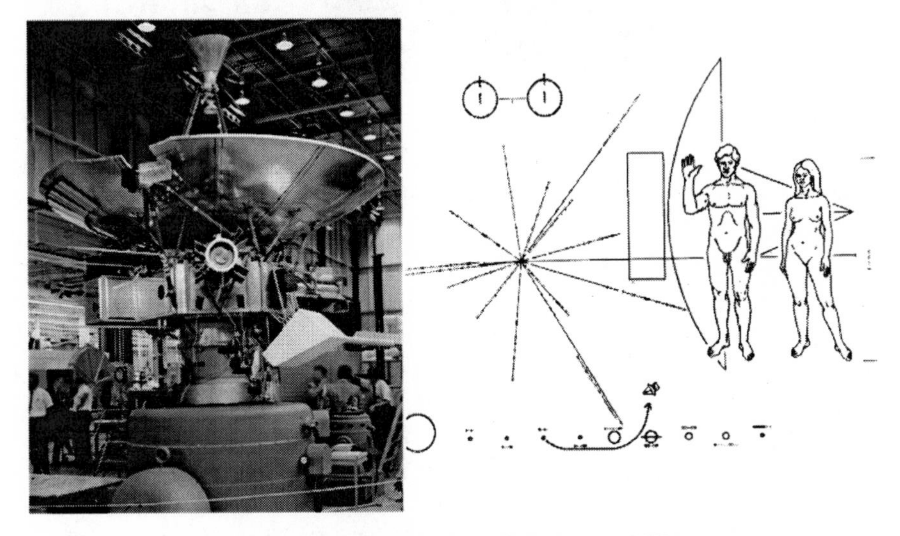

Figure 154 TRW Pioneer 10 Probe and Plaque

Pioneer 10 in the final stage of construction at the TRW[273] plant in Southern California. 20 December 1971 NASA Ames Research Center.

Voyagers 1 and 2 space probes were launched in 1977. The original mission was to study Jupiter and Saturn, however the probes were able to expand their mission and continue into the outer solar system. They are currently on course to exit the

[273] Simon Ramo and Dean Wooldridge left Hughes Aircraft and formed the Ramo-Wooldridge Corporation with financial support from Thompson Products. The company became TRW Inc in July of 1965. TRW was active in development of missile systems and spacecraft. They were involved in the development of the U.S. ICBM program via the Teapot Committee led by John von Neumann. They pioneered systems engineering and served as the primary source of systems engineering for the U.S. Air Force ballistic missile programs. In 1959 the US Congress issued a report recommending that the companies Space Technology Labs (STL) be converted to a non-profit organization. With nearly half of STL's employees the Aerospace Corporation was formed in June of 1960.

solar system. These probes were built at JPL. Voyager 1 has crossed into the heliosheath, the region where interstellar gas and solar wind start to mix. Voyager 1 is the farthest human made object from Earth.

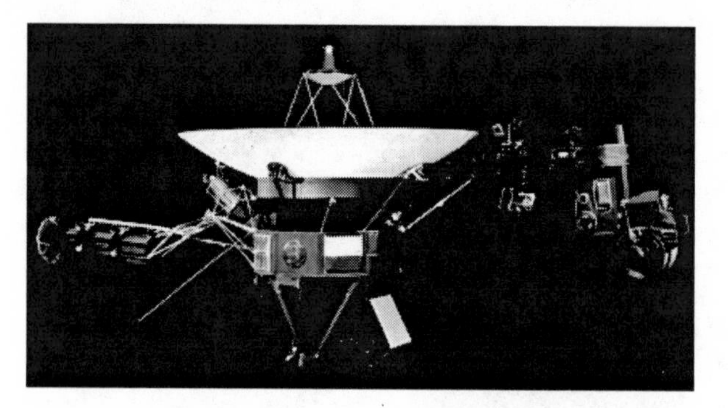

Figure 155 Voyager Space Probe

Many point to technological achievements of NASA. Few point to the approach of phased missions, where each is a stepping stone to the next more challenging mission. Few also point to the simulations used to not only train astronauts and mission personnel but also understand the systems on Earth. This relentless dedication to stepping stones and simulation is an important lesson for all major system practitioners. The irony is that even in that non-reactive systems driven environment Apollo 1[274] was struck with tragedy. At the senate inquiry following the Apollo 1 fire Senator Anderson questioned Astronaut Frank Borman, “What caused the fire?” Frank Bormans’ response was:

"**A failure of imagination.** We've always known there was the possibility of fire in a spacecraft. But the fear was that it would happen in space, when you're 180 miles from terra firma and the nearest fire station. That was the worry. No one ever imagined it could happen on the ground. If anyone had thought of it, the test would've been classified as hazardous. But it wasn't. **We just didn't think of it.** Now who’s fault is that? Well, it's North American's fault. It's NASA's fault. It's the fault of every person who ever worked on Apollo. It's my fault. I didn't think the test was hazardous. No one did. I wish to God we had."

Could you imagine going to the moon without these fundamental concepts in place? How do you measure the value of phased missions and simulations?

Author Comment: Many stakeholders view the interstate highway system as an example of what went wrong and an unsustainable future with massive suburban sprawl. However people forget that millions of people were trapped in dirty cities with bad air and no connection to nature. The cities were becoming increasingly

[274] Apollo 1 was to be the first manned mission to the moon. On January 27, 1967 a fatal fire killed all three crewmembers (Command Pilot Virgil "Gus" Grissom, Senior Pilot Edward H. White, and Pilot Roger B. Chaffee). This happened during a pre-launch test of the spacecraft on Launch Pad 34 at Cape Canaveral.

unlivable. Although the following message is a broad message about the environment it captured what many people were subjected to in the cities:

"...there are already more rats in New York than people. By 1980, you will need a gas mask and a flashlight to get to the office. There won't be any birds or bees or trees unless we stop what we are doing, we are going to kill ourselves.[275]"

The interstate highway system allowed millions of people to move to a place of clean air, cool nights, grass, trees, and the sound of birds and views of rabbits and squirrels (nice versions of rats). This would never have happened without the interstate highway system. Certainly trains, trolleys, and other public transportation systems existed before the rise of the interstates and automobiles and yet people by the millions were trapped in bad cities.

Were there other alternatives? Was there a failure of imagination? Or was this an example of massive imagination with abandoned emergence that needed to be addressed by the next generation? What if the automobile were to be powered with a perfectly clean energy source, would the view of interstate highways change?

What happens to people who are unable to drive because of health or old age? Do they move back to the cities that are now showing signs of relieved stress and reintroduce the stresses? Many roads in more enlightened communities were built with large grass medians to accommodate future public transportation needs. However rather than consider light rail or other alternatives the medians were used to add more lanes to the existing road. The argument was always cost as the current generation decided to base the economy on services rather than build infrastructure for the next generation. What would a system of shared ubiquitous public transportation and ubiquitous roads look like and would it be internally and externally sustainable[276]?

28.15 Exercises

1. What value systems did you use when you purchased a garment of clothing, food item, house, car, cell phone, computer, etc? List them separately, what are the differences if any?
2. What is the relationship between maximizing profits and providing services in an oligopoly or monopoly?
3. Why is the concept of the same price for the same goods and services important? Do you think this concept separates the modern world from the third world? Do you think inconsistent pricing is an overhead that someone

[275] Earth Day Public Service Announcement (archival) from the transcript of the documentary, Earth Days, Documentary Film, December 2009, story of growing awareness and understanding of environmental crisis and emergence, during the 1960's and '70's.

[276] Internal and external sustainability is addressed in Sustainability section of this text.

must pay and does it use extra resources needlessly?

4. What value systems should be applied to a small family restaurant, large franchise restaurant, hospital, electric provider, cell phone provider, cable TV provider? If they are different, why and if they are the same why?
5. If you feel there is a relationship between maximizing profits and getting the most effective health care, generate a mathematical proof or the equation(s) of a model that shows this relationship.
6. Is charging for meals, toilet access, and / or baggage a valid business model for airlines? If so what is the overhead associated with tracking these cash inflows? How do these revenue streams compare with the true cost of flying which includes all the indirect cost of the air traffic control system, airports, and continuing research of aircraft?
7. What would happen to air travel if all the indirect costs were shifted to the price of a ticket?
8. Do the indirect costs of providing a road for a restaurant equate to the indirect costs of air travel?
9. Can a restaurant exist without access via a road? What are some indirect cost examples for other items in a modern society of cell phones, GPS, automobiles, electricity, water, sewage, etc.
10. Is converting flight tickets to a commodity such that one person pays more than another for the same flight and seating area a valid business model? If so what is the overhead associated with tracking these cash inflows? Is this possible without computers? Is this a misuse of computer technology and human resources?
11. How would you measure the value of the Interstate Freeway system?
12. How would you measure the value of NASA?

28.16 Additional Reading

1. NASA Systems Engineering Handbook, NASA/SP-2007-6105 Rev1, National Aeronautics and Space Administration NASA Headquarters Washington, D.C. 20546, December 2007.
2. Sustainable Development Possible with Creative System Engineering, Walter Sobkiw, 2008, ISBN 0615216307.

29 Architecture Selection and Tradeoffs

Architecture Selection and Tradeoffs is based on the simple concept that value is more than apparent financial results. Value is relative to your context or view. For example it might make sense to do something because it can translate into thousands of jobs or it can cause people to view an area as attractive so they will move there and buy houses and or open businesses. These can be reduced to dollar numbers but it is more complex and that complexity is captured as a Tradeoff with Measure of Effectiveness (MOE) for each architecture approach.

Architecture selection and tradeoff is a series of simple steps that anyone can execute at anytime. The complexity surfaces as science and engineering are applied to the various elements of the steps with all the stakeholders and their views. The steps are:

- Identify the architecture alternatives
- List the advantages and disadvantages of each approach
- Identify the tradeoff criteria using the list of advantages and disadvantages
- Create a tradeoff matrix listing the tradeoff criteria with the alternatives shown in columns
- Using science and engineering to fill in each cell of the matrix
- Determine the life cycle cost (LCC) of each approach
- Divide the total rating of each approach by the LCC to calculate the measure of effectiveness (MOE) of each approach

29.1 Architecture Alternatives

Architecture alternatives come from two primary areas. The first is from current and similar approaches. The second comes from the participants on the team using creativity, innovation, and invention.

There always should be more than one participant involved in identifying the alternatives. It does not matter how simple or complex the architecture.

The best way to illustrate this is through the concept of metrics. More or less one participant is packed with 100 units of creativity, innovation, and invention - an IQ (invention quotient). There is no way to measure IQ, but if you could measure IQ and compare different participants it would not differ by more than a factor of ten, pick a number, it does not matter. So you could have one participant identifying alternatives or 1000 participants identifying alternatives. At what point do the addition of more participants' result in no new identification of alternatives? That is an interesting question.

In an organizational setting you could start with 3 participants, let them identify alternatives, then start adding more participants over time. At some point diminishing returns start to surface. At some point gestalt surfaces and the IQ of the participants is no longer a simple sum but becomes non-linear. For example 10 participants with an individual IQ of 100 might result in a team IQ of 100,000 rather than 1000.

There always should be more than one alternative being considered. If there is only one alternative you are not engaged in a meaningful activity. Two is not the right answer either. That is normally a sign of someone stacking the solutions towards a vested interest. The universe is more complex than left right, up down, centralized distributed, etc. The issue surfaces when you are engaged in a true system engineering effort and you are really trying to apply science and engineering principles to each alternative. That costs money. At some point pruning needs to happen in the early studies so that when the serious studies begin not only are costs controlled but also the stakeholders can actually follow the science and engineering. The number seven plus or minus two is the answer. Eventually there are two competing alternatives.

Any of the latter stage architecture alternatives can be made to work. One architecture alternative might be strong in one area and weak in another area. Money can be applied to address the weak area. Each architecture alternative will display this result and each will cost about the same in the end.

This is an extremely powerful observation, and shakes true engineers and scientists engaged in these activities to the core. It usually surfaces in the heat of battle of selecting system architectures.

So why go through the process? Why not pick an alternative using one participant and just make it work? Well the alternative from the one participant may not have made the short list once understood by the team because of a fatal flaw. Also the weaknesses of the alternative will not surface because there is no debate. This process forces the team to understand each architecture alternative and surface their emergent properties.

29.2 List of Advantages and Disadvantages

In the beginning of the architecture trade study there is nothing, just a blank sheet of paper. The first step is to identify alternatives, no matter how bizarre. One of the alternatives should be the current approach. Another alternative should be the dream approach. They form two extremes. The other approaches should be everything in between. List the alternatives on a single sheet of paper and create two columns and label them advantages and disadvantages. This is basically the Benjamin Franklin method of decision-making.

Keep this on a single sheet of paper so you can visualize it and cut to the key issues. It is easy to get lost in the noise. This also quickly cuts down on the alternatives. If you only have two alternatives something is wrong. If you have ten alternatives something is wrong. The answer is in between and should include the impossible alternatives.

Table 51 List of Advantages Disadvantages

Architecture	Advantages	Disadvantages
Approach 1	Advantage 1 Advantage 2 Advantage 3	Disadvantage 1 Disadvantage 2
Approach 2	Advantage 1 Advantage 2	Disadvantage 1 Disadvantage 2
Approach n	Advantage 1	Disadvantage 1

The advantages-disadvantages or plusses-minuses tables are very important. It is at this time that the key tradeoff criteria start to surface. These tradeoff criteria will be used in the next phase of the architecture trade study. At this point we depart the Benjamin Franklin method and start moving into the heart of system engineering.

29.3 Tradeoff Criteria

Using the List of Advantages and Disadvantages and input from the stakeholders identify the tradeoff criteria. These criteria can be general and applicable to any architecture: resilience, growth, flexibility, technology insertion, maintainability, testability, user acceptance, performance, supportability, transition, fragility, brittleness. They also can include unique items associated with a domain like sustainability: carbon pollution, water pollution, land pollution, air pollution, noise pollution, visual pollution, fuel sustainability. The domain of electronic medical records might be flexibility, adaptability, interoperability, security, maintainability, transition, user acceptance, etc. There is overlap between tradeoff criteria and quality attributes discussed in the Quality Section.

29.3.1 Traditional Tradeoff Criteria

The following is a list of traditional tradeoff criteria and their descriptions. There is overlap in the list. The intent is to stimulate discussion on the tradeoff criteria that matter for the system being developed.

1. **Resilience**: Resilience to failure is the capability to withstand multiple failures and still offer services. Resilience to change is the ability to accept change and offer current and the new services.
2. **Robustness**: The ability to easily tolerate and recover from unexpected failures and abnormal conditions, strongly constructed, sturdy, durable.

3. **Ruggedness**: Ruggedness is a measure of how tolerant a solution is to its external natural environment. The natural environment includes wind, rain, snow, sleet, humidity, temperature extremes, barometric pressure extremes, dew point, salt fog, mold, mildew, fungus, insects, animals, and radiation (cosmic, etc). The more rugged a system the more survivable it is in a hostile environment.
4. **Survivability**: Survivability is a measure of how a system behaves once the environment exceeds its design requirements. A system is more survivable if it slowly loses capabilities, as the environment becomes more hostile. Sudden loss of all capability with just a small environmental stress level above the design requirement is not a very survivable system. Massive excess in environmental stresses with slow loss of capability is a very survivable system.
5. **Brittleness**: The degree to which something will shatter or completely break down when touched by failure or change. Something that is brittle can be very strong and withstand all assaults, which may otherwise lead to failure. Unfortunately when a failure does happen the system shatters, it is catastrophic.
6. **Flexibility**: The ability to adapt to new, different, or changing requirements or needs.
7. **Reconfigure-ability**: How easy is it to reconfigure the system in the presence of a failure, different load conditions (night time versus daytime operations), training needs, and upgrade needs
8. **Scalability**: This is the building block level. The ability to add more functionality and capacity in small increments. Can the system be sized to support very small operations and very large operations or will the system have over capacity at the small locations.
9. **Stability**: Does the system stop operating, need random human intervention, offer the same performance and response especially to humans interacting with the system.
10. **Control Ability**: The ability to easily control and reconfigure the system. Is it from a single management source or from multiple dispersed locations? The level of control offered or is the system just on / off with no ability to continue offering services such as during maintenance or upgrade.
11. **Elegance**: Suggests significant effort and refinement to reflect tasteful richness of design, precision, neatness, effectiveness, and simplicity.
12. **Symmetry**: Balanced proportions, harmonious or appropriate proportionality and balance. For example is the load equally distributed across the architecture or is one part of the architecture needlessly stressed.
13. **Beauty**: In balance and harmony, admired, perfection, very good, attractive, or impressive, useful, or satisfying feature, qualities that make something pleasing and impressive, exalts the mind or spirit, graceful.

14. **Simplicity**: Easy to understand or explain, few parts, few connections, believed to have no unexpected behavior, linear, untangled.
15. **Reasonableness**: Not extreme or excessive, moderate, fair, based on sound judgment, logical, sensible, sound, valid, well-founded, well-grounded, coherent, rational, good sense, plausible or acceptable. Some approaches on the surface are immediately viewed as reasonable while others are not. As the approaches are studied and the details surface the level of reasonableness will change.
16. **Transition**: This is a measure of how difficult it is to transition into the new system. For example some approaches may need special throwaway subsystems or elements. Other approach may use the transition supporting elements as part of backup mechanisms. Transition is passage from one state, stage, subject, place, style, concept, etc to another.
17. **Interoperability**: The ability to easily interface and seamlessly operate with existing and new systems.
18. **Usability**: The ability for any operator to easily use the system with little or no training under all load conditions and scenarios.
19. **User Acceptability**: Some approaches may be more acceptable to the users. Care needs to be exercised in this area because users will tend to resist change. They understand the status quo.
20. **Comfort**: Some approaches offer better human factors characteristics. The elements include ergonomics, field of view, reach, tactile feedback, cognitive processing loads, user exposure to temperature, noise, humidity, etc.
21. **Availability**: The percent of time a system is offering services in a system operating 24 hours a day, seven days a week, every day of the year. Availability has little meaning for a system that is permitted to go offline for periods of time.
22. **Fault Tolerance**: The ability to tolerate one or more faults and provide full or degraded services in the presence of the faults. Ideally all functionality should be preserved with response time being reduced in the presence of a fault. At some point when response time can no longer be reasonably increased then functionality is shed.
23. **Graceful Degradation**: This is a measure of how the system behaves as failures are introduced. Is the loss of function and or performance small or are large pieces of function and performance lost? How many failures does it take before the system completely collapses?
24. **Ability to Meet Requirements**: It is a given that all approaches should be able to satisfy all the requirements. However some approaches are able to meet the requirements with greater ease either because the technology is a better match or an approach is more mature in a particular area.
25. **Performance**: It is a given that all approaches should be able to meet the

performance requirements. However some approaches perform better in some areas than other either because the technology is a better match or an approach is more tuned to a particular area.

26. **Model results**: A comparison of various model results from various alternatives under consideration.
27. **Capacity**: The ability to accept more functionality and or more loads without adding new technology or new system elements.
28. **Excess Capacity**: Some approaches use large building block that result in excess capacity, which comes at a cost.
29. **Efficiency**: Minimal or no waste.
30. **Inherent Capacity**: Some approaches have inherent capacity that is provided for free.
31. **Growth**: This ability to accept new system elements of the same type to support new functions and performance.
32. **Technology Insertion**: The ability to accept new technology to support new functions or performance. This includes replacing old equipment with new technology as it ages and becomes failure prone or difficult to maintain.
33. **Produce Ability**: Some approaches may be ideal solutions in limited quantities. As quantities increase the ability to produce the system repeatedly and cost effectively surfaces.
34. **Testability**: Some approaches are inherently more testable than other approaches. This tends to surface when white box testing is needed which means there must be access to the system internals.
35. **Reliability**: This is a measure of how often something breaks in a system. If an approach uses more parts than another approach or if the parts are inherently less reliable in an approach a qualitative assessment can be made before the actual reliability numbers are produced.
36. **Maintainability**: Some approaches are more maintainable than other approaches. The issues are inherent reliability, access, lowest replaceable parts, skill level, tools, etc. Some approaches may offer better diagnostic tools, other approaches may need special diagnostics.
37. **Training**: Some approaches may have special training needs. Also the skill levels may be different. Even the forms of training may change - class room instruction versus classroom instruction plus simulator training plus on the job training.
38. **Supportability**: This is tied to maintainability but focuses on the logistics elements. Some approaches may need onsite maintenance staff and spares while other approaches will work with a maintenance hub.
39. **Sustainability or Internal Sustainability**: Most systems will stop operating in a very brief period of time without human intervention. Even with human intervention, the system may eventually collapse. This is a

measure of each approach to keep working year after year decade after decade, human generation after human generation. The Roman Aqueducts are an example of a very sustainable approach to water delivery. This is exclusive of maintenance. Internal sustainability can be compromised in many ways.

40. **Effectiveness**: This is a measure of how effective the system is once it goes operational. Notice the term system is used rather than approach. It is assumed all the approaches meet the same requirements so they will have the same effectiveness. This is a reality check to see how these approaches compare with the previous system or another possible system.
41. **Safety**: Some approaches are inherently more safe than other approaches. The safety considerations include possible harm to users, maintainers, stakeholders relying on the system, the community where the system is located. For example one approach may blow up and destroy an entire community while another approach may just gracefully shut down.
42. **Security**: Some approaches are inherently more secure than other approaches. Using the commercial open Internet even with sophisticated encryption techniques is less secure than using an inaccessible custom communications network.
43. **Vulnerability**: Some approaches are more vulnerable to external undesired influences that other approaches. A custom computer system with limited knowledge of its structure is less vulnerable than an open system computer system with large numbers of installations and practitioners.
44. **Deployment**: Some approaches are easier to deploy than other approaches. This can be because of the size and or weight of the largest elements, the number of elements and interconnections, physical needs during transition, installation skill levels, etc.
45. **Shutdown**: Some approaches are easier to shut down than others. For example a highly distributed approach may require coordination between many elements while a centralized approach needs no coordination. There are also issues associated with preparing for a graceful shutdown so that there are no issues. This translates to shutdown time and complexity.
46. **Disposal**: Some approaches may have disposal challenges. For example highly exotic materials in an approach may be unsafe and may require special handling and processing for safe disposal.
47. **Technology Maturity**: Some approaches use less mature technology than other approaches.
48. **Technology Stability**: Some approaches use new technology that is viewed as less stable or old technology that is known to be less stable.
49. **Time to Obsolescence**: Some approaches use older technology than other approaches, which leads to quicker obsolescence.
50. **Degree of Obsolescence**: Some approaches use a higher proportion of

obsolete or nearing obsolescence technology than other approaches.

Author Comment: This is a rather large list. How can you prioritize this list? All I can say is that back at Hughes, architecture discussions always eventually revolved around **Elegance, Simplicity, Robustness, Resilience, Fault Tolerance, Efficiency,** and **Effectiveness**. These are highly subjective terms yet people found phrases and analysis data to support or refute them for each architecture approach. The other criteria were also tracked but they were not the discriminators. They tended to be a washout across the architecture approaches in the tradeoff analysis.

29.3.2 Sustainability and Regeneration Tradeoff Criteria

The following is a list of Sustainability and Regeneration tradeoff criteria and their descriptions. The intent is to stimulate discussion on the tradeoff criteria that matter for the system being developed.

1. **Sustainability or External Sustainability**: Development that meets the needs of the present without compromising the ability of future generations to meet their own needs[277]. Meet current as well as future mission requirements worldwide, safeguard human health, improve quality of life, and enhance the natural environment[278]. We also see that sustainability was a traditional tradeoff criterion associated with the ability of a system to sustain itself through years of operations. So a qualifier is added to distinguish the difference between internal or system and external or community sustainability.
2. **Ethics:** What are the ethical ramifications of an approach. One approach may be more ethical than the other approach.
3. **Regeneration**: The ability to re-establish a damaged environment and then improve it beyond its original state before the introduction of a system. For example an approach that is better able to improve water quality than another possible approach.
4. **Integrity**: Does the approach actually work. This is interesting because the approach must meet the stated requirements. However, not all requirements may be captured and a loophole may be found where the system meets the requirements, appears to be verified and validated in some setting[279], but actually does not work. Some might argue this is a failure of systems

[277] World Commission on Environment and Development, Brundtland Commission, 1987, Our Common Future.

[278] The Army Strategy for the Environment, October 2004.

[279] An incandescent light bulb can be made with two supporting posts or three supporting posts. Both can meet the same lifetime requirements, however a two post light bulb used in a ceiling fixture of a two story wood frame house will fail very quickly from normal floor to roof vibration.

engineering, however that is irrelevant[280] when an approach is being offered.

5. **Diversity**: The level of diversity possible when offering an approach. A more diverse approach is less brittle and more resilient to stresses. For example can only one company offer the approach or can many companies offer the approach.
6. **Resilience**: Ability to withstand aging or stress and not regress from original levels of regeneration or sustainability.
7. **Acceptability**: The level of acceptance from the regenerative and sustainable community stakeholders.
8. **Growth**: Ability to grow the system in a regenerative or sustainable way.
9. **Technology Insertion**: Ability to accept new technology in a regenerative or sustainable way.
10. **Quality of Life**: Ability of an approach to increase the quality of life for people in a regenerative or sustainable way.
11. **Freedom Liberty**: Ability of an approach to increase freedom and liberty of a people in a regenerative or sustainable way.
12. **Population Growth**: Ability of an approach to allow human population to increase in a regenerative or sustainable way.
13. **Standard of Living**: Ability of an approach to increase the standard of living in a regenerative or sustainable way.
14. **Social Mobility**: Ability of an approach to increase the social mobility in a regenerative or sustainable way.
15. **Pursuit of Happiness**: Happiness is elusive and no one can force someone to be happy. However elements can be established so that people can reasonably pursue and find happiness. A reference to Maslow's pyramid[281] is appropriate.

29.3.3 Environmental Impact Tradeoff Criteria

The following is a list of Environmental Impact tradeoff criteria and their descriptions. The intent is to stimulate discussion on the tradeoff criteria that matter for the system being developed.

1. **Noise**: Cumulative analysis, single event analysis, noise sensitive area

---

[280] New high efficiency heaters exhaust from the side of a building. In a residential setting the neighbor is exposed to noise and exhaust pollution.

[281] Motivation and Personality, Abraham Harold Maslow, HarperCollins Publishers, 1954, 3d Sub edition January 1987, ISBN 0060419873. Maslow's hierarchy of needs usually shown as a pyramid with the most fundamental levels of needs at the bottom and self-actualization at the top. Usually lower level needs must be satisfied before upper level needs.

analysis

2. **Compatible Land Use**: Land uses surrounding the facility/ action, future land uses projected in proximity of facility/ action
3. **Social**: Relocation of residences, relocation of businesses, community disruption
4. **Socioeconomic**: Shift in population and growth, public service demands, change in business and economic activity
5. **Air Quality**: Area air quality status, National Ambient Air Quality Standards emissions inventory, conformity with Local and State authorities
6. **Water Quality**: Additional impervious area, requirements for additional water supplies or waste treatment capacity, aquifer or sensitive ecological areas, erosion and sediment control
7. **Laws**: Potential impact, coordination with officials having jurisdiction
8. **Historic**: Includes architectural archaeological and cultural resources, properties in or eligible for inclusion in the National Register of Historic Places, potential impact (e.g. noise, air pollution), coordination with State Historic Preservation and Advisory Councils on Historic Preservation
9. **Flora and Fauna**: Biotic communities, potential for loss of habitat, coordination with wildlife agencies and US Fish and Wildlife Service for streams or other water bodies to be controlled or modified by the project
10. **Endangered and Threatened Species**: Flora and fauna, listed or proposed species and / or designated or proposed critical habitat from US Fish and Wildlife Service and or National Marine Fisheries Service, biological assessment if listed endangered and threatened species are potentially in the area to be disturbed
11. **Wetlands**: Wetland delineation, potential disturbance limits
12. **Floodplain**: Existing floodplain, potential changes to the floodplain limits
13. **Water Table**: Water tables and wells within the area of potential effect, consult agencies for consistency
14. **Rivers and Lakes**: Rivers and lakes, within the area of potential effect, consult agencies for consistency
15. **Inter coastal**: Estuaries, inlets, bays, inter coastal water ways, beaches, ocean within the area of potential effect, consult agencies for consistency
16. **Farmlands**: Prime, unique, statewide local importance farmland in area of potential effect, farmland converted to non-agricultural uses, coordinate with local Soil Conservation Service or State Conservationist
17. **Energy Supply and Natural Resources**: Irreversible expenditures of natural resources, fuel, construction materials, cost of the project compared to benefits, energy efficiency
18. **Light Emissions**: Light requirement and location of necessary lights, compatibility of light emissions with existing and future land uses
19. **Solid Waste**: Amount and type of waste to be generated

20. **Construction**: Construction noise, all impacts of construction, sediment control, dust impacts
21. **Visual Impact**: Design art and architectural, areas of natural beauty or historic or architectural significance, aesthetics, design, art and architecture
22. **Environmental Justice**: Disproportionate impacts on selected populations, low income populations, or minority populations, noise, residential relocation
23. **Cumulative and Other Considerations**: Categories with adverse impact, past, present or reasonably foreseeable actions that add to adverse impacts, consistency with state and local plans

29.4 Tradeoff Matrix

List the tradeoff criteria in rows and place the alternatives in columns. Rate each criterion for each alternative. You can use 1-3, 1-4, 1-10, or rank each relative to the other. If you start with using high medium low just translate all your word ratings to numbers. In the beginning of the trade study, try to fill in each cell of the matrix with a rating. Literally play with each approach. Do this in one day.

Now comes the hard part. Look at each criterion and architecture alternative (A1, A2, AN). Look at each intersection or cell. Now start to identify studies techniques methods approaches from 5000 plus years of civilization to convert those initial gut-based ratings into ratings backed on sound scientific and engineering principles. Do this as a group and use hard science, soft science, and everything else in that order (your logic) to back up your numbers. Document that logic even if it is just bullets on a chart.

Table 52 Architecture Tradeoff Matrix

Criteria	**A1**	**A2**	**AN**	**Wt**	**A1**	**A2**	**AN**
Criterion 1	5	7	9	1	5	7	9
Criterion 2	5	5	8	5	25	25	40
Criterion 3	5	5	6	1	5	5	6
Criterion 4	5	6	6	2	10	12	12
Criterion 5	5	9	8	3	15	27	24
Criterion Y	5	9	8	1	5	9	8
Total	30	41	45	-	65	85	99

This process may take years. Take snapshots and change the cells of the tradeoff matrix. Add weights to the criteria based on your continued refined analysis of the problem. Some criteria may disappear and others may surface. Don't be afraid to call the team in and have everyone enter their view of the rating for each cell, no matter how detached they may be from the detailed studies. At some point some criteria will become a wash and go away while others become very different or new criterion are added.

Sum the total for each approach. This is the rating for each architecture alternative. Keep it simple at first and do not use weights until you get some initial results. As you gain more insight you can apply different weights to each criterion and change the total.

Develop initial costs and total life cycle costs (LCC) for each approach. Take each rating and divide it by the initial cost. That is your initial MOE. It is a measure of goodness of each approach for each dollar spent. Do the same thing for the life cycle cost and see if they are different. You pick the architecture that has the highest MOE when all costs are considered (the LCC). So you get the biggest advantage for each unit of cost.

Table 53 Architecture MOE

Criteria	A1	A2	AN	Wt	A1	A2	AN
Total	30	**41**	**45**		65	85	99
LCC	1	1.2	1.4		1	1.2	1.4
MOE	30	**34**	**32**		65	**71**	**71**

1. The LCC is normalized.
2. Adding cost shows that Arch 2 is more effective than Arch N.
3. Including weights shows that Arch 2 becomes less effective and matches Arch N.

What should your tradeoff criteria include? That is really your call. It is part of the discovery process. However a word of caution. The tradeoff criteria should not include cost or requirements. It is a given that all solutions will satisfy the known requirements at some cost. Cost should be used at the bottom of the tradeoff where each approach total rating is divided by cost. This essentially identifies the goodness of each approach per unit of cost. This is called the measure of effectiveness or MOE.

MOE = Sum of tradeoff criteria/cost to produce approach

Author Comment: This is a terrifying experience for those who want control and have hidden agendas. Vested interests both visible and hidden hate this approach. This is the heart of being Systems Engineering driven and there is management that understands this process and knows how to effectively manage it, but these managers are born only in system engineering driven organizations. This process is based on truth and it is fully transparent so that everyone understands the science and engineering, even the grandparents of the participants. That means that the studies and tradeoffs need to be communicated so that all stakeholders quickly grasp everything[282]. This requires real genius.

[282] STOP (Sequential Thematic Organization of Publications) invented at Hughes Aircraft Fullerton in 1963 was first applied to proposals, by the 1980's it was applied to important national and international studies.

Some claim the MOE in many areas but the MOE is actually a measure of goodness per unit cost[283]. That is a significant statement, because it removes cost from the decision and yet considers cost. The decision process is leveled and requested from the people using scientific and engineering principles of analysis when possible, then processed either formally or informally (early in the process) using the MOE.

System engineering is a process for solving problems using system engineers (always more than one in a true system engineering based effort). The system engineers "herd" all the stakeholders (hardware, software, mechanical, civil, chemical, maintenance, training, support, etc) so that all the criteria can be vetted and a reasonable decision can be made by reasonable people always using the MOE. This is like F=MA in physics. It is fundamental - it drives everything. The systems engineering grand unifying equation: MOE = Sum of tradeoff criteria / cost to produce approach[284].

29.5 Sensitivity Analysis

Don't be afraid to perform a sensitivity analysis. Look at the matrix and do sensitivity analysis by changing some values that were previous points of contention. You can then just change values in a big way as part of the sensitivity analysis. It may seem odd but the same architecture tends to surface as part of the answer. If it does not it is because you are missing a key criteria item. Sometimes if you re-list the advantages and disadvantages of each approach new criteria items surface that will help to close the trade study.

Table 54 Architecture Tradeoff Sensitivity Analysis

Criteria	**A1**	**A2**	**AN**	**Wt**	**A1**	**A2**	**AN**
Criterion 2	5	5	8	**4**	20	**20**	32
Total	30	41	45		65	85	99
LCC	1	1.2	1.4		1	1.2	1.4
MOE	30	34	32		60	**67**	**65**
Criterion 2	5	5	8	**6**	30	**30**	48
Total	30	41	45		65	85	99

[283] Some suggest that the criteria are MOE. However, there is only one MOE for the architecture and a decision needs to be made based on that MOE.

[284] This is how I was taught system engineering at a place that no longer exists - Hughes Aircraft.

Table 54 Architecture Tradeoff Sensitivity Analysis

Criteria	A1	A2	AN	Wt	A1	A2	AN
LCC	1	1.2	1.4		1	1.2	1.4
MOE	30	34	32		70	**75**	**76**

The sensitivity analysis can fold into the tradeoff matrix. At some point the studies start to reach a level of diminishing returns and an approach starts to surface.

29.6 Architecture Selection Big Picture

So architecture selection is a trade study. The most important aspect of the trade study is not the results but the journey. It is during the journey that the stakeholders learn things about the alternatives and ramifications of those alternatives that would normally never surface. Many people try to complicate the architecture tradeoff study because they attempt to document the journey without realizing they are documenting the journey. Once that simple realization sinks in then the documented journey is a pleasure to read and understand.

Author Comment: I would like to offer a personal example, which I believe, hits the nail right on the head relative to all value systems. I will never forget when my new wife and I went on our first trip to a challenged country in 1980. We stayed at a fabulous resort right out of a James Bond Movie. Eventually we took a pink jeep into town and I will never forget what I saw. As far as I could see there were people with picks and shovels digging a trench by the side of the road. The reality is if you just look at traditional financial numbers, you can never justify the cost of a bulldozer over the cost of these poor people digging with a pick and a shovel. This is the trap of third world thinking. Instead of buying a bulldozer and freeing these people to become bulldozer operators they were stuck in the mud and dirt. You need to punch through to another level of thinking.

Author Comment: When you investigate the MOE topic in other documents be aware that some confuse tradeoff criteria with MOE. The MOE is one thing and it was defined as the sum of all the tradeoff criteria numbers divided by the total life cycle cost. This is done for each architecture approach. In these documents they erroneously call the tradeoff criteria the MOE and then they mention it in the plural as MOEs. Then they never really say how you pick an approach. They sprinkle cost into the middle of the tradeoff items. I attribute this to the finance types running amok in the past 30 years.

Author Comment: There are also those that confuse the MOE with cost benefit analysis[285]. The problem with cost benefit analysis is that financial metrics drive the answer rather than function and performance. So when they address a problem they

---

[285] From the Macnamara wiz kids era.

usually get a very low quality solution or a non-working solution.

There should be typically 5 architectures tracked for a very long period of time. The LCC people should not be called until the system architectures are deeply understood from a functional and performance point of view. The architectures must survive the criteria ratings first. The analysts should not be tainted by early views of the costs until the architectures have been fully understood.

Example 55 Architecture Selection - Hospital Laptops

You need to decide which laptop to provide a staff of hospital physician assistants. They will primarily use them to do their charts. They can select any laptop however you would like to offer a suggestion based on a tradeoff analysis. These are the characteristics of the laptop alternatives:

Alternative 1 Cost: $1140

Processor, Memory, Motherboard, and Operating System

- Hardware Platform: Mac
- Processor: 2.4GHz Intel Core Duo
- Number of Processors: 1
- RAM: 4GB DDR3
- Memory Slots: 2
- Mac OS X v10.6 Snow Leopard Operating System

Hard Drive, DVD / CD

- Size: 250 GB
- Manufacturer: Serial ATA
- Type: Serial ATA-150
- Speed: 5400 rpm
- 8x DVD/CD SuperDrive

Graphics and Display

- 13.3 inch LED-backlit display
- LCD Native Resolution: 1280 x 800
- NVIDIA GeForce 320M graphics processor with 256 MB of shared memory

Cases and Expandability

- Weight: 4.5 pounds

Power

- Rated Charge (normal use): 10 hours
- Built-in lithium-polymer 63.5 watt-hour

Warranty and Support

- 1 Year

Alternative 2 Cost: $1719

Processor, Memory, Motherboard, and Operating System

- Hardware Platform: MAC
- Processor: 2.4 GHz Intel Core i5

- Number of Processors: 1
- RAM: 4 GB DDR3
- Memory Slots: 2
- Mac OS X v10.6 Snow Leopard Operating System

Hard Drive, DVD / CD

- Size: 320 GB
- Manufacturer: Serial ATA
- Type: Serial ATA
- Speed: 5400 rpm
- 8x DVD/CD SuperDrive

Graphics and Display

- 15.4 inch LED-backlit display
- LCD Native Resolution: 1440 x 900
- Graphics RAM: 256 MB
- Intel HD Graphics and NVIDIA GeForce GT 330M with automatic graphics switching and 256 MB dedicated graphics memory

Cases and Expandability

- Weight: 5.6 pounds

Power

- Rated Charge (normal use): 8 hours
- Built-in lithium-polymer 77.5 watt-hour

Warranty and Support

- 1 Year

<u>Alternative 3</u> Cost: $1749

Processor, Memory, Motherboard, Operating System

- Processor: 2.8 GHz Intel Core Duo
- System Bus Speed: 1066
- Number of Processors: 2
- RAM: 4 GB DDR3
- Mac OS X v10.6 Snow Leopard Operating System

Hard Drive, DVD / CD

- Size: 500 GB
- Manufacturer: Portable
- Type: Serial ATA
- DVD/CD SuperDrive

Graphics and Display

- 17 inch LED Display
- LCD Native Resolution: 1920 by 1200 pixels
- Graphics RAM: 512 MB
- NVIDIA Geforce 9400M + 9600M GT Graphics

Cases and Expandability

- Weight: 6.6 pounds

Power

- Rated Charge (normal use): 8 hours

- Built-in lithium-polymer 95 watt-hour

Warranty and Support

- 1 Year

What would be a reasonable set of tradeoff criteria? The first step is to identity the advantages and disadvantages of each approach.

Table 55 Advantages Disadvantages Charting Alternatives

Alternatives	Advantages	Disadvantages
Approach 1	Light weight Small package Long lasting battery	Small screen
Approach 2	Medium weight Larger screen	Less battery time More heat 78 watt
Approach 3	Great screen Great resolution	Heavy Lots of heat 95W

Since entering chart data is not processing or storage intensive, features related to the processor and storage are not relevant unless an alternative cannot meet the basic requirement. We do see some items that appear to be discriminators. These items start to form the tradeoff criteria for the next step of the analysis. The following is a list of tradeoff criteria that surfaced from the advantages and disadvantages list:

- Weight
- Physical Size
- Display Size
- Battery Time
- Heat

There are choices when creating the tradeoff table. Since actual specification numbers are available in some cases these numbers can be entered into the table. An alternative is to enter the normalized specification numbers for each item. Yet another alternative is to let the stakeholders enter values from 1-3 or 1-4 or 1-10 based on their view of the features list and their perception of value. Each of these alternatives is shown in the following tables.

Table 56 Architecture Tradeoff Actual Specification

Criteria	Arch 1	Arch 2	Arch 3
Weight (lbs)	4.5	5.6	6.6
Physical Size (in)	13	15	17

Table 56 Architecture Tradeoff Actual Specification

Criteria	Arch 1	Arch 2	Arch 3
Display Size (in)	13	15	17
Battery Time (hrs)	10	8	8
Heat (watts)	64	78	95

Table 57 Architecture Tradeoff Normalized Specification

Criteria	Arch 1	Arch 2	Arch 3
Weight (lbs)	100	80	68
Physical Size (in)	100	87	76
Display Size (in)	76	88	100
Battery Time (hrs)	100	80	80
Heat (watts)	100	82	67
Total	476	417	392
Cost ($)	1140	1719	1749
MOE	.417957	.242763	.22414
MOE (normalized)	186	108	100
Total (normalized)	122	106	100

The numbers can be normalized to 1 or 100. In this case the numbers were normalized to 100. This decision is based on the audience and participants. As always the goal is to clearly communicate to all the stakeholders. Once the numbers are normalized it is interesting to see the difference between the MOE and Total of the criteria ratings. This lets stakeholders see the impact of cost on the MOE. Some approaches, which appear to be relatively, close in total criteria rating separate significantly when the MOE is examined. This also can form the basis of negotiation when the cost is based on existing products where business decisions affect the vendor price or solution provider cost.

A reasonable question is when should the stakeholder ratings table be developed. Should it come before or after the specification based tradeoff tables? Will the stakeholders be influenced by the numbers especially the normalized numbers?

Table 58 Architecture Tradeoff Stakeholder Ratings

Criteria	Arch 1	Arch 2	Arch 3
Weight (lbs)	10	8	5
Physical Size (in)	10	7	5
Display Size (in)	8	9	10
Battery Time (hrs)	10	5	4

Table 58 Architecture Tradeoff Stakeholder Ratings

Criteria	**Arch 1**	**Arch 2**	**Arch 3**
Heat (watts)	10	5	1
Total	48	34	25
Cost ($)	1140	1719	1749
MOE	.042105	.019779	.014294
MOE (normalized)	295	138	100
Total (normalized)	192	136	100

This tradeoff is relatively easy because the normalized actual specifications match the stakeholder ratings. Also the cost tracks the results of the sum of the tradeoff criteria.

The MOE results for the normalized actual specifications show that approach 2 is 8% more effective than approach 3 and approach 1 is 86% more effective than approach 3. The MOE for the stakeholder ratings show that approach 2 is 38% more effective than approach 3 and approach 1 is almost 3 times more effective than approach 3.

The disparity between the normalized actual specifications and the stakeholder ratings is attributed to the value system of the stakeholders. They tend to value some items more than other items. Weights could be used in the normalized actual specification table however the non-linearity of each criterion rating would be lost that is captured by the basic stakeholder ratings. For example heat varies by only 33% from a specification point of view but from a stakeholder point of view the variation is 10 times. This is significant and demonstrates how value systems of the stakeholders affect the tradeoff results.

Notice the tables are sequenced suggesting a time line for each table and the concept of no table or artifact before its time. This is important because you do not want to influence the individual cell entries with a visible total result. There should be a mixture of surprise and no surprise at the results. So the sequence is:

- Advantages Disadvantages Table, preferably on 1 sheet of paper
- List of Tradeoff Criteria, some of which come from the advantages disadvantages table
- Tradeoff Ratings Table, containing only the ratings
- Tradeoff Totals Table, with the original entries
- Tradeoff MOE Table, with the totals and original entries

Once the results are shown you can never go back. At this point the team should now focus on sensitivity analysis and refining the individual entries as they question or agree with the results.

If the tradeoff went well there should be dissension and much discussion. If everyone agrees, then the tradeoff missed identifying the key architecture discriminators or the analysis is not broad or deep suggesting an unacceptable level of ignorance. This usually happens in non-system engineering driven organizations where management drives the project with their only view of cost and schedule[286].

Conversely the team may fail to converge and fall into a "paralyses by analysis". This is where strong technical leadership is needed to bring the team to a reasonable conclusion. What does the technical leadership do? They ask questions and add team members when needed. For example, if there is a lack of intellectual disagreement other creative people can be added into the team. At what point do the technical leaders know they have converged? When the addition of new creative team members does not result in any new significant questions or insights. If the technical leaders can imagine more than the team, then there is an issue.

29.7 Architecture Maturity Levels

As architecture alternatives surface and they are subjected to tradeoff analysis each alternative matures. In the early stages all the architectures have fatal flaws. They are easy to invalidate. As the arguments are presented and the alternatives modified the architectures become stronger - harder to invalidate or knock down. These architectures can be broadly categorized as:

- Straw Man: Easy to invalidate.
- Iron Man: Serious and significant analysis needed to knock down the architecture.
- Stone Man: Unable to knock down the architecture with the current analysis and everyone agrees the analysis is complete.

Just because an architecture achieves Stone Man status, it does not mean it has no flaws or disadvantages. It is just the best representation of that architecture concept. Further changes / improvements would mean shifting to another architecture concept which might be a new architecture or movement into an existing architecture alternative.

---

[286] As suggested in Sustainable Development Possible with Creative System Engineering: The role of project management is cost and schedule, but they are only one stakeholder in the project and must not be permitted to takeover the effort or there will be catastrophic failure. All projects must have a financial box big enough to successfully execute the project. If the financial box is tool small project management must address this issue by getting more funding, schedule, and or resources or stopping the project.

29.8 Processing Functions and Allocation Alternatives

There are various alternatives for allocating processing functions. The decisions are difficult and should be made with formal architecture alternative tradeoffs. However it is never to early to start to identify the possible approaches, technologies, and alternatives for implementing the functions that perform processing. These alternatives are broadly split into people and machines. The machine alternatives include mechanical, electrical, software, chemical. The electrical machines can be further broken down into analog and digital.

Author Comment: Much of the world has shifted to digital machines. There are technology misuse issues that have surfaced from this trend that is discussed in the technology misuse section of this text. There are characteristics associated with digital approaches that are also important to understand. Unlike analog alternatives, digital is very brittle. Analog systems can be subjected to extensive noise and degradation and the function is still provided. For example an analog radio signal can have a great deal of static but the human is still be able to extract the signal. Digital systems stop in the presence of anomalous conditions. So there is a compete loss of signal. This is evidenced by complete loss of television signals from digital media and digital transmissions. With the introduction of digital television, people who were able to get "snowy analog" pictures found themselves with no digital signal and complete loss of all content. The failure in the digital case is catastrophic, not graceful. The idea of failures, graceful degradation, resilience, brittleness, etc will be found in later sections of this text and are an important theme and element of systems practices.

29.9 Exercises

1. Identify 3 or more alternative power generation systems. Develop an advantages and disadvantages list. Identify tradeoff criteria. Develop a tradeoff matrix. Populate the tradeoff matrix. Estimate the LCC of each approach. Find the MOE for each alternative.
2. Perform a sensitivity analysis of the above tradeoff. What insights have you gained that were not known prior to the tradeoff study?

29.10 Additional Reading

1. NASA Systems Engineering Handbook, NASA/SP-2007-6105 Rev1, December 2007.
2. Sequential Thematic Organization of Publications (STOP): How to Achieve Coherence in Proposals and Reports, Hughes Aircraft Company Ground Systems Group, Fullerton, Calif., J. R. Tracey, D. E. Rugh, W. S. Starkey, Information Media Department, ID 65-10-10 52092, January 1965.
3. Sustainable Development Possible with Creative System Engineering,

Walter Sobkiw, 2008, ISBN 0615216307.

4. Systems Engineering Fundamentals, Supplementary Text, Defense Acquisition University Press, January 2001.
5. The Army Strategy for the Environment, October 2004.
6. The Theory of the Leisure Class, An Economic Study of Institutions, Thorstein Bunde Veblen, London: Macmillan Publishers, 1899.

30 Decision Making

If you started reading this text sequentially, you have just been taken from a blank sheet of paper to the point where an architecture selection is made. The architecture is part of a group of architectures that have been studied from a functional and performance point of view using several systems practices. This culminated in a decision making process referred to as architecture tradeoff. The architecture selection is the most important selection decision in the entire system effort. However there are other decisions that need to be made from selecting tools, processes, machines in the system, feature rollouts, proposal bid no bid decisions, etc. This section will review the previous decision making approaches and introduce Probability Based Decision (PBD) and Analytical Hierarchy Process (AHP).

30.1 Advantages Disadvan tages List

As previously described in the architecture selection section of this text, this list identifies the advantages and disadvantages of each approach. It is on a single sheet of paper so that the whole picture can be visualized. If multiple sheets are used they are spread on a physical table so that the full view can be provided[287]. The list is qualitative and identifies items associated with the mind, heart, and stomach[288]. In professional settings the list is usually oriented towards the mind, in anticipation of converting to quantitative measures in follow on analysis techniques.

The advantage of this approach is that it is fast and stimulates discussion. The discussion process starts to push the team into a knowledge discovery mode about the problem space.

30.2 Tradeoff Matrix

As previously described in the architecture selection section of this text, the tradeoff matrix lists the criteria, assigns ratings for each approach to the criteria, and then sums the criteria. The criteria come from different sources but the advantages disadvantages list is usually the starting point for identifying the criteria.

The assignment of numbers uses ad hoc methods decided by the team. This process evokes discussion of how much better one approach is over another

---

[287] This is one of the massive limitations of the computer. A full view cannot be visualized. Using a table or even an entire room with items posted on walls visually immerses the analysts in the problem space.

[288] The mind, heart, and stomach are elements in a phrase referring to logic, compassion, and basic needs. Basics needs are food clothing shelter usually equated to money.

approach for each criterion. Occasionally analysis yields quantitative data that is traceable to the values in a particular cell in the matrix. Eventually the values are normalized to some scale such as 1-10, 1-4, 1-3, etc.

To select the best approach a cost is determined for each approach and divided into the total rating of each approach. Arbitrarily setting values to zero or maximum ratings is used to perform sensitivity analysis. Weights are also added to different criteria based on the teams' ad hoc view of the analysis.

The advantage of this approach is the ability to visualize all the alternatives on a single sheet of paper. While assigning values to each cell the team is engaged in intense debate and forced to produce quantitative and qualitative analysis to justify the rating in each cell. Different team members perform analyses of the same areas as they try to justify their original guess and gut level reactions to a rating. Even though the analysis may be based on hard science and engineering it is always modulated by assumptions that originate from different perspectives of the problem.

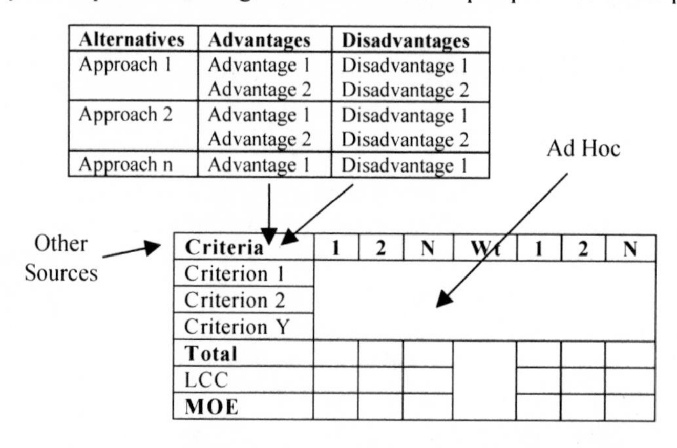

Figure 156 Alternative Lists and Tradeoff Decisions

Eventually the team converges and the matrix and resulting MOE value remain unchanged[289]. There is convergence in a setting where everyone has had a view of the problem, a chance to participate, and the ability to easily view the results captured in the tradeoff matrix. Inclusion and visualization of the decision is key and everyone is able to describe how they arrived at the selected approach by just talking to the matrix and referencing analysis that led to each cell in the matrix.

30.3 Probability Based Decisions

The best way to discuss Probability Based Decisions (PBD) is through the example of deciding to pursue a contract that requires a proposal submission in a

[289] It should never be turned off because of money or schedule constraints. The concern should be has the team converged too quickly, should other participants be added to offer new perspectives.

competitive space. Assume there are four possible competitors pursuing a request for proposal (RFP). Also assume there is no other knowledge other than interest in pursuing the contract. So they are all equal. So the probability of win for any one of the four vendors is 25% (1 in 4 or 1:4). This is represented as a probability tree. As marketing starts to investigate the opportunity more information is uncovered that changes the win probability distribution. The Win probability tree starts to decompose into lower levels of information.

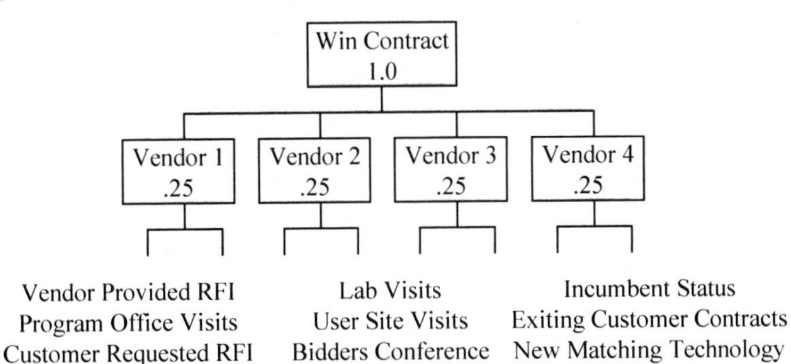

Figure 157 Win probability Tree

This can be converted to a table as an alternative view. The win probability can be calculated by just assigning the number 1 to each item a vendor satisfies (V1-V4), totaling all the columns for each vender, then dividing each vender total by the total for all the vendors. This approach treats all elements of marketing data equally. For example Incumbent status is valued in the same way as Lab Visits.

Table 59 Win Probability

Marketing Data	**V1**	**V2**	**V3**	**V4**
Customer requested RFI	1		1	
Provided RFI	1	1	1	
Program Office Visits	1	1	1	1
Lab Visits	1		1	
User Site Status Visits	1		1	
Bidders Conference	1		1	
Existing Contracts With Customer	1			
Incumbent	1			
Total	8	2	6	1
Win Probability (Total / 17)	.47 or 47%	.11 or 11%	.35 or 35%	.06 or 6%

30.4 Analytical Hierarchy Process

Those who have developed probability models, fault trees, reliability models, decomposition trees will recognize the basic approach of using a hierarchy to represent a decision using Analytical Hierarchy Process (AHP). The AHP process is summarized as follows[290]:

1. **Define the problem**. Determine the kind of knowledge sought.
2. **Structure a decision hierarchy**. At the top is the *decision or goal.* Below the decision are the broad *objectives.* Continue the decomposition of the intermediate levels. These are usually *criteria* on which subsequent elements depend. The lowest level is usually the set of *alternatives.*
3. **Develop pair-wise comparison matrices**. Each element in an upper level is used to compare the elements in the level immediately below in the tree.
4. **Use priorities obtained from comparisons to weigh priorities in level immediately below**. Do this for every element. Then for each element in the level below add its weighed values and obtain its overall or global priority. Continue this process of weighing and adding until the final priorities of the alternatives in the bottom most level are obtained.

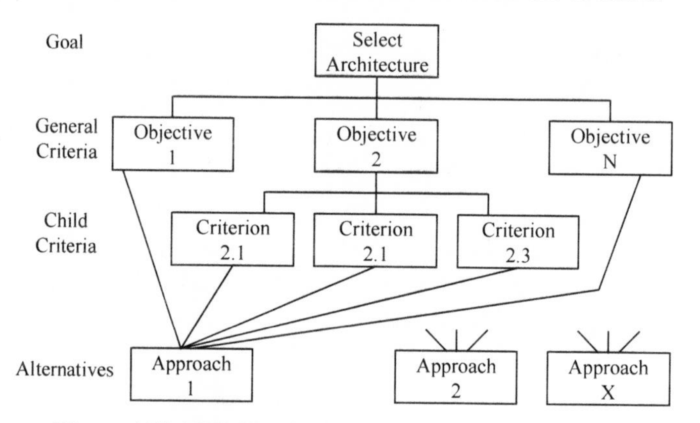

Figure 158 AHP Vertical Hierarchy or Tree Diagram

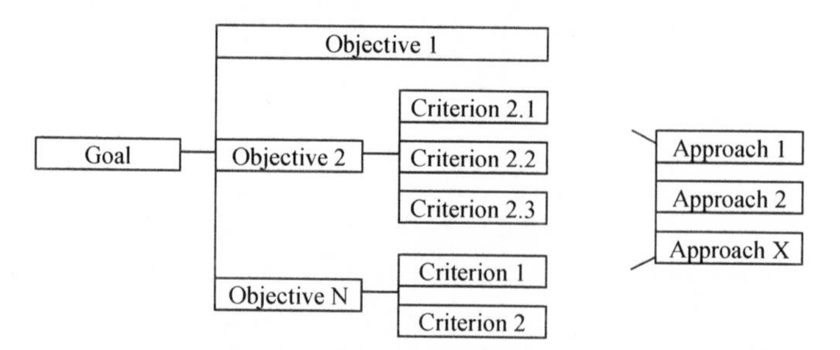

Figure 159 AHP Horizontal Hierarchy or Tree Diagram

290 Thomas L. Saaty, Decision Making with the Analytic Hierarchy Process, Int. J. Services Sciences, Vol. 1, No. 1, 2008.

The AHP decision hierarchy is a decomposition tree diagram. It can be oriented vertically or horizontally. The more levels in the hierarchy, the more voluminous the calculations because pair-wise matrices are developed for each relationship.

A pair wise matrix is titled using the parent name or designator and the child nodes are rows and columns in the matrix named using the child names or designators. The values assigned in the matrix are based on a scale. The matrix must be consistent so when a whole number is entered in a position its reciprocal is automatically entered in the transpose position.

Table 60 AHP Assignment Scale

Intensity of Importance	Definition	Explanation
1	Equal Importance	Two activities contribute equally to the objective
2	Weak or slight	
3	Moderate importance	Experience and judgement slightly favor one activity over another
4	Moderate plus	
5	Strong importance	Experience and judgement strongly favor one activity over another
6	Strong plus	
7	Very strong or demonstrated importance	An activity is favored very strongly over another; its dominance demonstrated in practice
8	Very, very strong	
9	Extreme importance	The evidence favoring one activity over another is of the highest possible order of affirmation
Reciprocals of above	If activity i has one of the above non-zero numbers assigned to it when compared with activity j, then j has the reciprocal value when compared with i	A reasonable assumption
1.1–1.9	If the activities are very close	May be difficult to assign the best value but when compared with other contrasting activities the size of the small numbers would not be too noticeable, yet they can still indicate the relative importance of the activities.

Example 56 AHP Matrix Calculations

AHP matrix calculations are best illustrated with an example. To keep the example simple only one level of criteria is considered. This actually represents the case of the Architecture Tradeoff Matrix previously discussed in this text. In this case the numbers entered into the tradeoff cells are based on AHP rather than some ad hoc method.

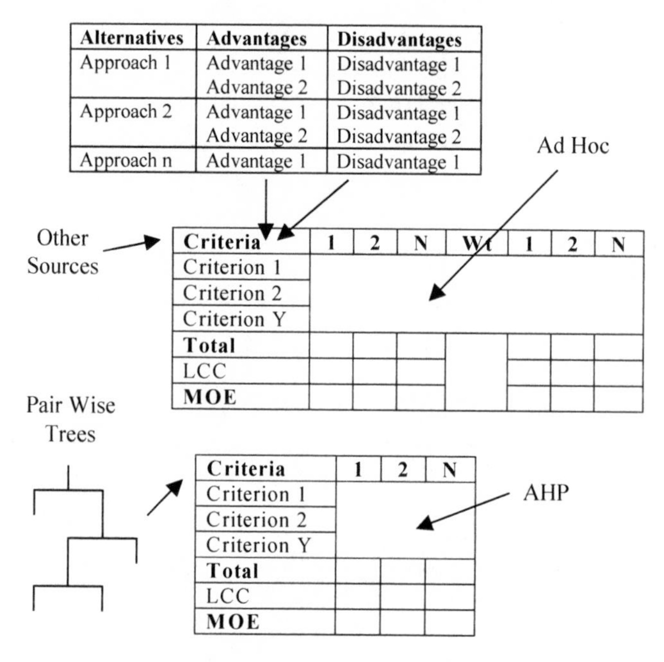

Figure 160 AHP Compared to Alternative List and Tradeoff Matrix

The specific steps for executing AHP are as follows:

- Decompose the goal into its parts, progressing from the general to the specific. At its simplest this consists of a goal, criteria, and alternatives. The more criteria included, the less important each individual criterion becomes as the mix is diluted.
- Create pair-wise matrices representing the decomposition. In this case there is a matrix for each criterion. The rows and columns are the alternatives.
- Populate the cells using the AHP assignment scale.
- Normalize the values in the columns, sum the rows of each matrix, and divide the sum of each row of each matrix with the number of alternatives.
- Sum the criteria for each alternative. This is the value of the alternative. Apply weight to each criterion using lower level AHP analysis if present.
- Find the cost of each alternative and normalized it.
- Divide the value of each approach by its cost. This results in an MOE. Normalize the MOE to determine level of differences between alternatives.

- Alternatively plot the value versus cost to visualize the degree of separation between the alternatives.

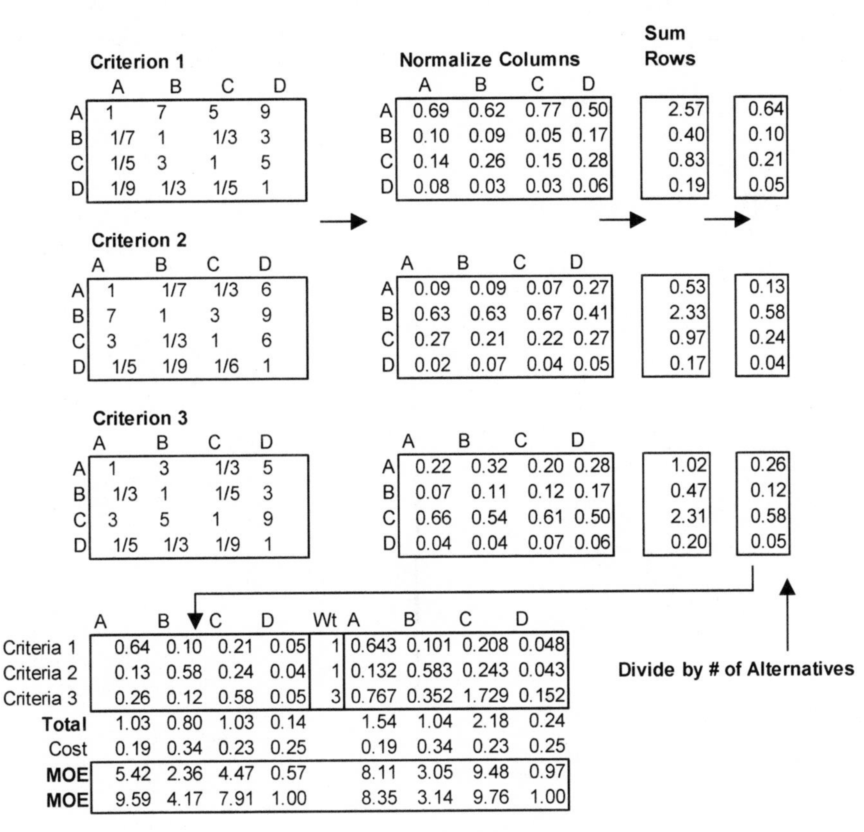

Figure 161 AHP Example

30.5 Exercises

1. Identify different system activities other than architecture selection where formal decision making is important. Why should formal decision making be used in these cases?
2. Use probability based decision to make a decision.
3. Use AHP exclusively to make a decision.
4. Identify 3 or more alternatives. Identify the advantages and disadvantages of each alternative. Identify tradeoff criteria. Populate a tradeoff matrix for each alternative using the ad hoc tradeoff method. Use AHP to populate the matrix. Are the conclusions different? If so why if not why not?

30.6 Additional Reading

1. NASA Systems Engineering Handbook, NASA/SP-2007-6105 Rev1, December 2007.
2. Systems Engineering Fundamentals, Supplementary Text, Defense Acquisition University Press, January 2001.
3. The Analytic Hierarchy Process: Planning, Priority Setting, Resource Allocation (Decision-Making Series), Thomas L. Saaty, Mcgraw-Hill, January 1980, ISBN 0070543712.
4. Decision Making with the Analytic Hierarchy Process, Thomas L. Saaty, Int. J. Services Sciences, Vol. 1, No. 1, 2008.

31 Other Practices

Other practices have surfaced in recent years that are mostly associated with software[291]. These practices are available to systems practitioners and are described from a systems practitioner's point of view in the following sections. As with all practices, there is no silver bullet that will satisfy every aspect of systems analysis. Instead there are practices in the plural that should be understood and applied if they are appropriate to the system under development.

31.1 Inheritance

Inheritance has been popularized by object oriented programming[292]. However inheritance is an old concept. Inheritance in biology uses a family tree and probability to see how certain attributes of one generation manifest in another generation. Set theory can be used to show these relationships but it becomes extremely complex when dozens, hundreds, or thousands of items need to be tracked. An inheritance tree can easily communicate this information.

As with all system practices inheritance has its proper place in communicating some aspect of the system. A good example is a system that uses a large number of display screens[293]. The main screen contains all the properties for the font size, type, color, use of scroll bars, titles, etc. All child screens can be set to inherit the properties of the parent screen. When a graphic user interface (GUI) builder is used the check boxes called "inherit" from the parent are checked and the fields are all pre-populated with the parent information. Selected changes can be made as needed.

So inheritance allows a system to be internally consistent. Without inheritance system inconsistencies proliferate. The implication is that the development process and system solution needs to allow and support inheritance.

There are many ways to represent inheritance in a system but a caution is needed. For example there may be thousands of attribute instances where the values are inherited. It may not make sense to model these details. Instead high level requirements and or diagrams showing the inheritance may be appropriate.

---

[291] There has been an attempt to create a software engineering discipline with emphasis on different modeling techniques and languages.

[292] Inheritance in software is an approach to facilitate code reuse. This is called implementation inheritance. This is easily extrapolated to a physical item such as a screw or a beam in a physical structure.

[293] Hundreds of screens is not out of the ordinary in many systems.

Author Comment: Inheritance should not be used in place of decomposition, which is based on set theory. Many show inheritance, where decomposition is the actual relationship because they do not want to prepare decomposition views. There are times when inheritance is less precise and offers a confusing picture. Inheritance should augment functional decomposition not replace it.

31.2 Polymorphism

Polymorphism has been popularized by object oriented programming. However, polymorphism is an old concept. Polymorphism in the broadest sense is related to final use. For example a screw can be used to fasten two pieces together in a home appliance such as a toaster. The same type of screw can be used in a radio on a ship. The behavior of the screw is determined at operator use time. This is the binding time for the screw[294].

Example 57 Polymorphism - Air Space Management

An interesting polymorphism example is an air defense system being used for air traffic control. In one instance the mission is to bring the airplanes together in another instance it is to keep the airplanes apart. A set theory view or functional decomposition would show that each system has unique functions, but the overlap in functionality is sufficiently large enough for air defense to support air traffic control. However, because of missing intercept and weapons assignment function the air traffic control system cannot be used for air defense.

31.3 Attributes and Properties

An attribute consists of a parameter and its setting. It is defined by the relationship: parameter = value. The parameter element is generically referred to as a property. For example FontColor=RED sets the font color to RED for the font properties. Fonts have multiple properties and attribute settings. The font properties can be color, style, and size.

Attributes and properties become extremely important when inheritance and polymorphism are addressed. It is the properties and attributes that allow the practitioners to describe the inheritance and polymorphism is a system.

31.4 Model Based Development

By 1988 the UNIX workstation was able to support extremely high performance graphics. In this environment X-Windows was born which included client server networking and a windows Graphic User Interface (GUI). When the UNIX

---

[294] In software operator overloading allowing various numerical types (integer, float, double, etc) is polymorphism. Also some operators such as "+" or "*" have multiple meanings depending in context is polymorphism.

platforms had progressed to the point where windows became possible, the fundamental issue was how to consistently build the unique windows[295] interfaces for different applications. These windows used thousands of lines of code.

The concept of drawing a static window with all the buttons, radio boxes, check boxes, text areas, etc was adopted and GUI builders were created that generated the code that would represent a complete window or family of interconnected windows. It was now possible to build a consistent GUI that followed an accepted style guide like MOTIF[296] with hundreds of windows by just drawing pictures of the actual GUI. Once the GUI was tested and accepted, code would be automatically generated. The code would then have callbacks that would connect to the application. The generated code would easily approach several hundred thousand lines and what would take 100 person years to complete was now done in 30 person days[297]. Air defense and air traffic control simulators were built in the early 1990's using this approach[298]. These simulators used hundreds of screens to support site and scenario authoring.

This approach had its beginnings in the 1960's with the start of higher order languages. In the beginning there was assembly code. With a High Order Language (HOL) the abstraction layer was raised such that one line of HOL might equal 10 or 15 lines of assembly. So the big question is can a graphical method of representing systems be created, in much the same way as GUI builders, so that dragging and connecting a few boxes in 30 days will translate to 100 million lines of code[299]. The dilemma is at what point does one move to a model that generates code. For example, does it make sense to spend 30 days populating a model that generates 100 lines of code?

This idea of drawing pictures in a highly abstract manner and pressing buttons is tempting and may be possible for software and automated manufacturing. However you can't press a button and get: water power sewage healthcare education systems, cities, office buildings, houses, roads, space systems, airplanes, trains, automobiles, radio broadcasting, telephone systems, television broadcasting, etc.

---

[295] The X Window System (commonly X or X11) a network protocol provides a graphical user interface (GUI) for networked computers.

[296] Motif: set of user interface guidelines developed by Open Software Foundation (OSF) circa 1989, specifies look and feel of X-Window system application.

[297] X-Designer a product used at Contraves in Tampa Florida for simulators and trainers.

[298] Approaches To Air Traffic Control / Air Defense Workstation Simulation And Training, Thirteenth Interservice / Industry Training Systems Conference (ITSC) Fall 1991, Walter Sobkiw.

[299] So far only the signal-processing domain has made progress in this area. Dragging and dropping signal processing elements will generate executable code.

31.5 SysML and UML

System Modeling Language (SysML) is a systems engineering modeling language that is an extension to the software Unified Modeling Language (UML). The Object Modeling Group (OMG) created the UML standard. OMG[300] is a consortium whose original goal was to set standards for distributed object-oriented systems but is now focused on modeling and model based standards. They are both extremely formal methods with extensive syntax and semantics. In many ways they were born in a time when computers had become ubiquitous. While the other practices were created in a time of paper and pencil, UML and SysML really do need a computer-based tool. UML has 14 diagrams.

31.5.1 UML Structure diagrams

Structure diagrams show the structure and elements in the software system being modeled. They are used in documenting the architecture of the software system.

1. **Class Diagram:** Shows the classes of the software, their interrelationships (including inheritance, aggregation, and association), and the operations and attributes of the classes. They are drawn as boxes with three sections, the top shows the name of the class, the middle lists the attributes of the class, and the bottom lists the methods.
2. **Component Diagram:** Shows the software components and the inter-component interfaces. The external interfaces are also shown.
3. **Composite Structure Diagram:** Describes a run time view of interconnected instances.
4. **Deployment Diagram:** Shows the hardware used in the system and the software allocated to the hardware.
5. **Object Diagram:** Shows a complete or partial view of the structure of a modeled system at a specific time. It focuses on some particular set of object instances and attributes, and the links between the instances to provide insight into how an arbitrary view of a system is expected to evolve over time. They are more concrete than class diagrams, and are often used to provide examples, or act as test cases for the class diagrams. Only items that are of current interest need be shown on an object diagram.
6. **Package Diagram:** Organizes model elements into groups, making UML diagrams simpler and easier to understand. They are shown as file folders and can be used on any of the UML diagrams. They are most common on use case diagrams and class diagrams because these diagrams have a tendency to grow.
7. **Profile Diagram:** Defines custom stereotypes, tag definitions, and constraints

---

[300] Formed in 1989 by several companies including Hewlett-Packard, IBM, Sun Microsystems, Apple Computer, American Airlines and Data General. OMG's modeling standards includes the Unified Modeling Language™.

in a dedicated UML diagram. Customizations are defined in a profile, which is then applied to a package.

31.5.2 UML Behavior diagrams

Behavior diagrams show what happens in the system being modeled by illustrating the behavior of a software system. They are used to describe the functionality of software systems.

1. **Activity Diagram:** Shows the step-by-step work flows of components in a system. It shows the overall flow of control.
2. **State Machine Diagram:** Shows the states and state transitions of the system.
3. **Use Case Diagram:** Shows what system functions are performed for an actor. This can include sequences of actions and roles of the actors. An actor is a person, organization, or external system. Actors are drawn as stick figures. Associations are shown as lines connecting use cases and actors to one another, with an optional arrowhead on one end of the line. The arrowhead is used to show the initial start of the relationship or to show the primary actor in the use case. The arrowhead is not data flow.

31.5.3 UML Interaction diagrams

Interaction diagrams are a subset of behavior diagrams. They show the flow of control and data among the elements in the software system being modeled.

1. **Communication Diagram:** Shows the interactions between objects or parts as sequenced messages. They are a combination of information taken from Class, Sequence, and Use Case Diagrams describing both the static structure and dynamic behavior of a system.
2. **Interaction Overview Diagram:** Provides an overview of the interaction diagrams. It is similar to the activity diagram except activity in the interaction overview diagram is pictured as frames that contain interaction or sequence diagrams.
3. **Sequence Diagram:** Shows how processes operate with one another and in what order. Vertical lines are different processes or objects that live simultaneously and horizontal arrows are the messages exchanged between processes, in the order in which they occur. It is representing a runtime scenario in a graphical manner.
4. **Timing Diagrams:** Used to explore the behavior of objects in a given period of time. It is a special form of a sequence diagram. The differences between timing diagram and sequence diagram are the axes are reversed so that the time is increased from left to right and the lifelines are shown in separate rows. Its roots are traced to hardware timing diagrams.

31.5.4 SysML Diagrams

SysML has 11 diagrams. Six diagrams are unchanged from UML, three diagrams are modified UML diagrams, and two diagrams are new for SysML.

Table 61 SysML Versus UML Diagrams

SysML	SysML Purpose	UML
Activity diagram	Describes the flow of control and flow of inputs and outputs among actions.	Modified Activity diagram
Block Definition diagram	Defines features of blocks and relationships between blocks such as associations, generalizations, and dependencies.	Modified Class diagram
Internal Block diagram	Captures the internal structure of a block in terms of properties and connectors between properties.	Modified Composite Structure diagram
Parametric diagram	Describes the constraints among the properties associated with blocks. This diagram is used to integrate behavior and structure models with engineering analysis models such as performance, reliability, and mass property models.	N/A
Package diagram	Organizes the model by partitioning model elements into package elements and establishes dependencies between the packages and/or model elements within the package.	Package diagram
Requirement diagram	Shows the requirements, packages, other classifiers, test cases, and rationale.	N/A
Sequence diagram	Describes the flow of control between actors and systems (blocks) or between parts of a system.	Sequence diagram
State Machine diagram	Shows the system states and the state transitions.	State Machine diagram
Use Case diagram	Shows behavior in terms of the high level functionality and uses of the system.	Use Case diagram
Allocation tables dynamically derived tables, not really a diagram type	Shows allocations (e.g. functional allocation, requirement allocation, structural allocation). Useful for verification and validation and gap analysis.	N/A
N/A	N/A	Component diagram
N/A	N/A	Communication diagram
N/A	N/A	Deployment diagram

Table 61 SysML Versus UML Diagrams

SysML	SysML Purpose	UML
N/A	N/A	Interaction Overview diagram
N/A	N/A	Object diagram
N/A	N/A	Timing diagram

31.5.5 Paper and Computer Based Practices

There are some significant differences between paper and the new computer based practices. They each have their advantages and disadvantages. Although a paper based practice can be automated with a computer, that is not necessary for the basic elements of the practice to be useful. On the other hand many would suggest that UML and SysML should be practiced exclusively with a computer-based tool.

The advantages of paper based practices are:

- **Visualization**: It is easy to place pieces of paper on a table and get the big picture. It is not unreasonable to use a 72inch X 30inch table to spread out 9 columns by 2 rows of paper. It is also not unreasonable to be surrounded by 2 or 3 tables that show 54 views that are all legible, easily modified, and can be sorted and or stacked.
- **Transportability**: It is easier to carry and use a stack of paper and add content when inspired than a computer. The computer requires a login, is subject to display washout in sunlight, where people get significant insight, and there is limited time because of power needs.
- **Three Dimensional Immersion**: The paper can be placed on walls in a room allowing multiple people to enter a three dimensional space of the analysis. This immersion is extremely powerful and leads to new insights not possible in personal two-dimensional limited views of the analysis. It allows everyone to easily see duplicates, poor organization, undeveloped or poorly developed areas, and internal inconsistencies.
- **Low Noise**: Because it is harder to start with a blank sheet of paper, there is a tendency to ignore irrelevant details. However the devil is in the details and making those arbitrary detail choices can lead to problems. Peer reviews are one way to try to surface important details that were originally missed or consciously ignored.

So what are the advantages of computer based practices? The advantages of computer based practices are:

- **Internet Networking**: The computer has the ability to quickly network with geographically disbursed team members. The analysis practice is easily augmented with modern Internet practices such as message boards and instant messaging.
- **Hardcopy**: The computer content can always be printed and capitalize on the advantages of table and wall reviews. After markups the information can be updated on the computer.
- **Executable Model**: The model can be executable. In the paper approach the model is just a collection of static pictures.
- **Reuse**: The computer allows for quick access to large databases of similar information. This does require organizational competence to establish a reuse database. However the danger is creativity and innovation may suffer as practitioners rely too heavily on previous work that is very easy to access and incorporate.
- **Fine Detail**: The level of detail can be extremely high with the computer. Between the ability to reuse existing information and populate screen after screen, the detail can actually translate to implementation as in the case of software and automated manufacturing.

31.6 System of Systems

System of systems is a controversial subject because all systems contain subsystems. However the subject may now be appropriate because we naturally perceive different systems that appear to interact in some greater system, either consciously or unconsciously. So what is a system of systems and why is it different from a system?

Systems of systems are the results of years, decades, or centuries of evolution. They are characterized as a system that uses what everyone acknowledges as multiple standalone systems. The standalone systems have their own operational organizations and function streams. Sometimes the different systems use radically different technologies. Yet all these systems are able to interact in what appears to be a single system.

Example 58 System of Systems Air Traffic Control

Air traffic control is a good example of a system of systems. The current system is the result of decades of evolution as different technologies were accessed when needed to offer some form of air traffic control (ATC) services. Voice radio was one of the early systems. Navigational aids based on radio technology were added to the system. This was followed by broad band RADAR. All of the sudden the ATC command center was using radio and RADAR. The RADAR was digitized and fed into a computer. Now fours systems were being used in the ATC system: radio, navigational aids, broad band RADAR, digitized computer tracking. At some point it may make sense to investigate how these various systems might be more tightly integrated to offer even greater services. An example of such an effort is the integration of air ground communications with the computer data available from the ATC command center now processing RADAR, weather, fight plans, and other ATC relevant data.

The challenge is a social and technical challenge. The social challenge is to get all the various systems stakeholders to coordinate, understand everyone's needs, capabilities, and technologies then postulate various system-of-systems approaches. An approach to capturing and managing a system of systems is to offer a long-term plan that captures each systems anticipated upgrades and integration[301]. The danger in this activity is to consolidate and merge various projects, programs, and even systems without appreciation for the complexity inherent in each project, program and system[302]. A long-term plan (ten-year plan) can be used to address the system of systems challenge and may contain the following key elements:

- **Section 1.0:** A description of the system of systems with a picture, block diagram, or graphic of each system and their high level interfaces. A statement of the challenges of each system and the future goals. Include a timeline for each system and identification of the new major services by year.
- **Section 2.0:** A two-page description of each system from a planned projects point of view. List all the projects with a brief description. Show the rolled up anticipated costs. Include a timeline for each project and identification of new major services by year. If there is an interaction or merge with another system show it in the timeline.
- **Section 3.0:** A two-page description of each project or program anticipated for each system grouped by system. Include anticipated costs. Show a graphic with the expected start and end dates for each project. Do not use a schedule Gantt chart. Use a timeline or modified PERT chart[303].

31.7 Value Engineering

Value engineering originated at General Electric during World War II to deal with shortages of skilled labor, raw materials, and parts. They looked for acceptable substitutes and noticed that the substitutions could reduce costs, improve the product, or both. The process is based on identifying functions using a two-word

---

[301] In 1980 the FAA published the National Airspace System (NAS) Plan known as the brown book.

[302] Some point to the FAA Advanced Automation System (AAS) as an example of over merging, such that AAS became too big to successfully complete.

[303] The modified PERT (Program Evaluation and Review Technique) is described in the Transition section of this text.

statement consisting of an active verb and measurable noun (what is being done - the verb - and what it is being done to - the noun). A cost is then estimated to provide the function. Its value is the ratio of the function to the cost: **Value = Function / Cost.**

In 1996 Congress passed Public Law 104-106, National Defense Authorization Act For Fiscal Year 1996 which requires value engineering to be applied within the government. It is viewed as a cost cutting measure.

Notice the equation does not include performance. Yet performance is significant in a system. The same functions can be offered but at significantly different performance levels. For example a portable radio can tune in AM and FM stations and play music. Three basic functions are available. However the performance of the receiver will determine how many stations can be located with minimal noise and drift and the performance of the audio subsystems will determine the quality of the sound in terms of frequency response, distortion, and other performance parameters.

This is an interesting philosophical issue because either systems are over engineered, thus costing more than they should, or system practices have failed and the wrong system with extraneous functionality and performance are produced. Alternatively value engineering could be leading to reduced performance that stakeholders may be unaware of until they have purchased and used the system.

As with all practices, care needs to be taken in what is actually happening during value engineering activities. This does not mean that value engineering cannot add value. It just means the holistic transparent view inherent is systems practices needs to be preserved when value engineering is initiated and major decisions made about the system. Everything needs to be disclosed to all stakeholders.

Author Comment: When GE took over RCA some old timers decades later suggested that GE removed components from a Television until it stopped working. They would then add the last offending component back into the Television and throw away the extra components sitting on the bench and declare success. No one would care about the quality of the picture, sound, or reduced life of the product. Whether it is true or not, this is offered as an example of a downward spiral in quality that can happen if this practice is misused.

31.8 Agile Iterative Incremental Development

In recent years there has been emphasis on what some think is a new form of software development[304]. The fundamental element that binds these ideas is that of going through a full development cycle on a small set of functionality so that the stakeholders can review the results on a weekly or monthly basis. There is also an

---

[304] Agile, iterative, incremental, extreme programming, lightweight processes are some of the terms used to reflect these ideas.

emphasis on a work setting that spurs innovation, creativity, and ability to quickly respond to new insights as the effort unfolds. What many have forgotten is that this is classic research and development (R&D) and prototyping. A traditional R&D prototyping environment is characterized by[305]:

- Senior staff with enthusiastic young people
- Small number of team members
- Stakeholders and developers work together daily
- Motivated individuals
- Technical excellence
- Face to face communications
- Self organizing team no artificial hierarchies
- Cross functional teams
- Culture that thrives on chaos
- Changing requirements based on findings throughout effort
- Solutions evolve through collaboration
- Early frequent (weeks to a few months) delivery of valuable solutions
- Simplicity minimizing work
- Environment and support provided instantly without question
- Hands off management trusts team to get the job done
- Working prototypes are primary measure of progress
- Balance and respect so everyone can maintain a indefinite constant pace
- Team reflects on findings then tunes and adjusts its behavior

For software, an evolved solution may be good enough to transition into an operational environment. In other cases where hardware and mechanical elements are involved, the body of work in the form of a prototype may need to be converted to a production version. The concept of taking something and making it production ready has been forgotten by many.

Even though software in less critical areas may be fielded from an agile iterative incremental development, other software may need to undergo a process where it is cleaned up and made production ready. This may include removing unsafe code, removing unused code, restructuring for maintainability and growth, optimizing for performance, etc.

Solutions from this environment are relatively small using small teams of less than approximately 20 people. However multiple solutions can be bought together to form very large systems. This presupposes that there is a larger system architecture view and some subsystems, usually of very high technology, are bought

---

[305] This was my early working career at the FAA and Hughes Aircraft circa 1979. Slowly R&D disappeared as everyone refused to pay for it and wanted only production solutions.

together from R&D or prototyping activities that may or may not have been made production hardened. So Agile Iterative Incremental Development may not be a standalone process but part of a much larger process depending on the system being developed.

There is a reason why software is called software and hardware is called hardware. Software is malleable and easy to change. Hardware instantiated as a building, bridge, automobile, etc is very difficult to modify. For example no one would build an office building with one floor, they proceed to add 5 more floors based on the findings of the first floor. Yet that may be reasonable for software. An interesting item to note is that after the first office building is produced; it is the prototype for the next office building. There are always lessons learned that move to the next challenge. After the third or fourth office building or house of the same type then the lessons learned will drop off dramatically and one can say they are truly in production mode.

Since software is so malleable, many will evolve into a solution. The problem is that this software eventually becomes a "rats nest" that is unable to be expanded or supported. This was the dilemma of early software intensive systems circa 1979, thus the emphasis on using hardware engineering principals and the conscious slowing of software R&D prototypes. As with everything there is a balance and it needs to be understood so that the right choices can be selected.

31.9 Exercises

1. Are there other practices that you can think of that are not discussed in this text that can be applied to systems development?
2. Discuss this section with your classmates. What are your observations?

31.10 Additional Reading

1. National Airspace System Capital Investment Plan for Fiscal Years 2009–2013, Federal Aviation Administration, 2011.
2. National Airspace System Plan; Facilities, Equipment, and Associated Development, Federal Aviation Administration, 1982, April 1987.
3. R&D Productivity Second Edition; Hughes Aircraft, June 1978, AD Number A075387, Ranftl, R.M., "R&D Productivity", Carver City, CA: Hughes Aircraft Co., Second Edition, 1978, OCLC Number: 4224641 or 16945892. ASIN: B000716B96.
4. System Engineering Management, Military Standard, MIL-STD-499 17 July 1969, MIL-STD-499A1 May 1974.
5. Systems Engineering Fundamentals, Supplementary Text, Defense Acquisition University Press, January 2001.
6. Systems Engineering Management Guide, Defense Systems Management College, January 1990.

32 System Requirements Database

A System Requirements Database (SRDB)[306] is a relational database used to enter clean requirements, search for requirement anomalies, and to manage them once they are surfaced. The SRDB can be used on existing projects or new clean slate projects.

If you are in the middle of a program that is yet another spin of an established industry, product, or product line, then the requirements exist and they come from the organization. Typically the requirements are part of the culture and locked in people's heads. If the organization is mature, then there are libraries containing information products such as previous studies and specifications. If the organization is very mature then the requirements are in a SRDB repository. In this setting you are duplicating what has been done in the past with a slight spin and perhaps a new element which requires a few new requirements. Creating new requirements in this established organization now becomes tricky because you may need to follow some of the principles of requirements development that are used for new "clean slate" programs.

If you are on a new program with a clean slate, then the requirements generation process is a sight to behold, if it's done "properly". There is a picture. You have to visualize it. Draw it on a napkin if needed as you read the next few sentences. The input shows personal inputs, brainstorm sessions, round table discussions. However, this does not do justice to the process that is actually used to create requirements from nothing for your clean slate program. What really happens is affectionately called studies and there are large numbers of studies and participants. The reason for the large number is to gain all the perspectives before the implementation machine is activated. For example, you might have a group of people examining a half dozen implementations of existing similar systems. There might be a group of people performing architecture tradeoff studies allocating equipment within buildings across the planet and its orbit. There might be a group of people studying different algorithms for accomplishing a very narrow task like tracking commercial aircraft in various stages of flight. There might be people creating fault trees looking for single points of failure that need to be removed from the system. People might even be creating prototypes of various aspects of the future system. There could be simulations using existing infrastructure of machines and people. The participants

[306] The concept was created and the term SRDB was coined at Hughes Aircraft circa 1983 as part of the FAA Advanced Automation System (AAS) pre-proposal phase. The SRDB was implemented as part of the Design Competition Phase (DCP) effort.

can range from subject matter experts who were or are operational personnel on similar systems to a lone Ph.D. implementing a thesis as a new technology is being developed. These studies no matter how small or how grand result in requirements.

The requirements need to be made consistent, complete, and testable. They also need to be organized into groups that represent similar in kind (functional and temporal cohesion) and similar in level (system, segment, subsystem, or configuration item like hardware or software). The requirements are captured in specifications. Requirements are never held within analysis / study information products. They always must be transferred to the specification tree.

Why the emphasis on organization of the requirements? It's driven by the fact that we are humans and that we can only handle about 6 things at any given instant. It's also driven by a process, which tries to identify missing requirements. This process is known as traceability and the goal is to see that all parent and child requirements have links. For example, a software designer might surface a requirement from the software design perspective, which was never seen from other related perspectives. That means a high level requirement might be missing that this lone designer surfaced as a result of that painful detail. That high level requirement might now impact a dozen other designers that need a similar detailed requirement in their part of the system. The same thing can happen at a higher level, where a subject matter expert (SME) walks in, reviews the system, and sees a fatal flaw that exists in other similar systems that surely will exist in this system. That one flaw might translate to one high level requirement which might translate into hundreds of lower level requirements.

So the purpose of traceability is to support the great hunt for missing and conflicting requirements.

32.1 SRDB History

The story begins over 30 years ago, circa 1977 - 1980. At the time people were excited about a process called structured system analysis and structured programming. The process acknowledged that people could only relate to about 6 things at any given moment. So the idea was to create a process that did just that, captured a system view at a particular level, which did not exceed 6 things, well ok, maybe 15 or 20. That view would then be decomposed with another view that would contain about 6 things. The decomposition would continue until the analysts could go no further, they would reach the leaves of a wondrous tree. At the same time people were claiming that most project failures could be traced to bad requirements. They were missing, incomplete, inconsistent, vague, conflicting, etc. Additionally, even though computers were an expensive piece of infrastructure at the time, there were people also starting to think of new computer based tools to organize information and the relational database was a topic of discussion.

So, you have these three elements flowing around in the community: structured analysis, bad requirements, and relational databases. Like all great moments in history, someone made a connection and the idea was born to place requirements into a relational database. Those requirements could then be grouped (allocated to functions), leveled (placed into the system, segment, subsystem, or component levels), and links could be created to look for missing requirements.

32.2 Traceability Using an SRDB

Visualize a table where each row contains a single requirement statement. Next add columns which are attributes of the requirement text. At this point there is a table. The table represents a particular abstraction level of the requirements. For example an initial A-Level specification document can be imported into the SRDB. The requirement text can be changed and attributes added and modified as they change. The requirements can then be exported from the SRDB into a document format.

Multiple tables representing multiple abstraction levels of requirements (specifications) populate the SRDB. Links can be formed between each row in each table to represent the traceability between the requirements.

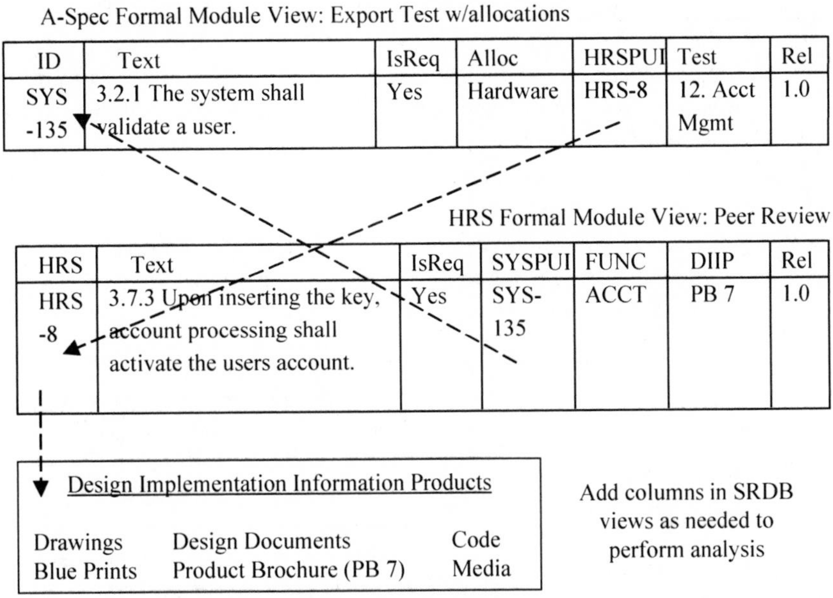

A-Spec Formal Module View: Export Test w/allocations

ID	Text	IsReq	Alloc	HRSPUI	Test	Rel
SYS -135	3.2.1 The system shall validate a user.	Yes	Hardware	HRS-8	12. Acct Mgmt	1.0

HRS Formal Module View: Peer Review

HRS	Text	IsReq	SYSPUI	FUNC	DIIP	Rel
HRS -8	3.7.3 Upon inserting the key, account processing shall activate the users account.	Yes	SYS-135	ACCT	PB 7	1.0

Design Implementation Information Products

Drawings	Design Documents	Code
Blue Prints	Product Brochure (PB 7)	Media

Add columns in SRDB views as needed to perform analysis

Figure 162 SRDB Traceability to Design and Implementation Information

The SRDB can be used to produce reports. The reports can show the forward traceability from the parent requirements to the child requirements as implemented by the links. They also can show the backward traceability from the children to the parents. The attributes also can be shown in the traceability reports. Parents without children and children without parents are an indication of a requirement hole that must be addressed. Inconsistent attributes are an indication of requirement, architecture, and or design issues depending on the level of the requirements being reported.

32.3 Establishing the Schema

The first step in establishing an SRDB is to define the database schema in a five to ten page document such as a technical memorandum. This needs to happen at the start of the project as part of the planning process when the information products are identified. The details of the SRDB may be unknown until the architecture solidifies. For example a team may not know a system will have software until the architecture analysis reaches a certain level of maturity. When it's realized that a system has software then one or more software requirement specifications (SRS) need to be produced and they need to be in the SRDB. The SRDB schema document then needs to be updated to reflect the new tables and the addition of new attributes unique to the software introduction. There are several elements that should be in the SRDB schema document:

Information Products List: Start the SRDB schema document by listing all the deliverable information products. Update the list to indicate what is placed into the SRDB. The SRDB tables should then be named. These names should correspond to the information product names; there should be no confusion. Typically all the specifications are placed in the database.

All requirements must be moved into the specifications. This includes requirements from studies and interface documents. Requirement Traceability is maintained by linking the specifications from one level to a lower level. Test traceability is maintained by linking test headings to one or more requirements.

Table 62 SRDB Information Product List

Information Product	SRDB
Statement of Work (SOW)	No[307]
Customer Requirements Specification (CSR)	Yes
System or System Segment Specification (SSS)	Yes
Studies	No
Subsystem n Specification (SnS)	Yes
Software Requirements Specification (SRS)	Yes
Interface Requirement Specification (IRS)	No
Hardware Requirements Specification (HRS)	Yes
Interface Design Document (ICD)	No

[307] All requirements must be in specifications. If there are non-work requirements in the SOW, they should be moved to the specifications. The SOW can be imported to the SRDB for project management but not for system requirement traceability.

Table 62 SRDB Information Product List

Information Product	**SRDB**
System or System Segment Tests (SST)	Yes
Subsystem n Tests (SnT)	Yes
Software Requirements Tests (SRT)	Yes
Hardware Requirements Tests (HRT)	Yes
Environmental Requirements Test (ERT)	Yes
Operational Requirements Tests (ORT)	Yes
Certification Requirements Test (CRT)	Yes

Traceability Diagram: There should be a diagram showing the traceability links between the database tables that represent the different specifications which will be developed.

It should be represented in a simple picture that shows each table as a box and lines only showing one input from a parent table to a child table. Parents can have multiple outputs to different child tables. If this is a top level table there is no input. The lines should all have arrows and they should all flow in the same direction. If there is a change in direction in the flow the traceability will be broken at that point. The direction is irrelevant as long as the flow is not broken and there is a clear indication of parent, child, grandchild, and other lower level relationships.

In addition to the specification traceability links, there also should be links for the test cases. Unlike the specifications, which have the full content in the SRDB, the test SRDB tables can be a list of test cases with the actual test steps captured in documents or an external test tool. This allows the test steps to change without needing to update the SRDB. The test traceability is not to the steps but to the atomic test case.

Attributes List: The attributes, their settings, and their default values are listed for each table. The values change with time as the architecture unfolds. Many attributes are used to capture broad allocations. The broad allocations are then used for the next level of decomposition. In many ways it is a form of course grain traceability. The detailed traceability is then captured in links where requirements are allocated to something that represents the next level of decomposition.

Table 63 SRDB Attributes

Attribute	**Settings**
System	Function, characteristic, segment, subsystem, test methods
Automation	Hardware, software, operations
Function	RADAR, weather, flight data processing, account processing, kitchen, bedroom,
Characteristic	Sustainability, safety, reliability, environment, maintenance, training,

Table 63 SRDB Attributes

Attribute	Settings
	Integrated Logistics Support (ILS)
Segment	Air, ground, space, sea, under water, or facility type 1, facility type 2 (e.g. Tower, terminal, en-route, field support center, training center, research center)
Subsystem	Subsystem 1, subsystem 2, (e.g. Communication processor, router, modem, workstation, display, peripherals, heating ventilation air conditioning (HVAC), lighting, engine, transmission, suspension
Test Methods	Inspection, analysis, demonstration, test, certification
Test Types	System, software, hardware, environment, certification,
Certification	underwriters labs (UL), Leadership in Energy & Environmental Design (LEED)
Software Allocation	Computer Software Configuration Item (CSCI) 1, CSCI n
Hardware Allocation	Hardware Configuration Item (CSCI) 1, CSCI n
Software Files	File 1, file 2, file n
Software Libraries	Library 1, library 2, library n

SRDB Links: The SRBD links, although shown in the traceability diagram should be restated in text and named. Many will use general links to capture traceability between many information products. For example traces to, specified by, tested by, etc are used. However using one link to house multiple traceability paths can lead to difficulty when generating reports.

Each link is uniquely used and named. For example SSS-SRS1 is used to capture the traceability between the System Specification and Software Requirements Specification 1, while SSS-SRS2 is used to capture the traceability between the System Specification and Software Requirements Specification 2.

Table 64 SRDB Links

Links	Settings
Requirements Traceability	
CSR-SSS	This link connects individual Customer System Requirements to individual contractor or developer SSS requirements.
SSS-S1S	This link connects individual SSS requirements to individual subsystem 1 requirements.
SSS-SnS	This link connects individual SSS requirements to individual subsystem n requirements.
SSS-SRS1	This link connects individual SSS requirements to individual SRS 1 requirements.
SSS-SRSn	This link connects individual SSS requirements to individual SRS n requirements.
SSS-HRS1	This link connects individual SSS requirements to individual HRS 2

Table 64 SRDB Links

Links	Settings
	requirements.
SSS-HRSn	This link connects individual SSS requirements to individual HRS n requirements.
Test Traceability	
SSS-SST	This link connects individual system SSS requirements to one or more system test headings.
SnS-SnT	This link connects individual subsystem SnS requirements to one or more subsystem test headings.
SRS1-SRT	This link connects individual SRS 1 requirements to one or more software test headings.
SRSn-SRT	This link connects individual SRS n requirements to one or more software test headings.
HRS1-HRT	This link connects individual HRS 1 requirements to one or more software test headings.
HRSn-HRT	This link connects individual HRS n requirements to one or more software test headings.
SSS-ERT	This link connects individual SSS requirements to one or more system environmental test headings.
SSS-ORT	This link connects individual SSS requirements to one or more system environmental test headings.
SSS-CRT1	This link connects individual SSS requirements to one or more system certification 1 test headings.
SSS-CRTn	This link connects individual SSS requirements to one or more system certification n test headings.

The linking process is the heart of traceability. Traceability is complete when all the parents have one or more children and all the children have one or more parents. The quality of the linking is paramount. If everything links to everything, there is no traceability. This tactic sabotages the fundamental premise of traceability. This is called over linking and the review process must negate this activity should it surface.

Intermediate Reports: There should be a list of intermediate reports and the reviews that they will support. As a minimum there should be a review of the parent child reports so that everyone agrees on the traceability. The attributes, which show the allocations to the subsystems and lower level elements, also should be reviewed and made consistent with the parent child report.

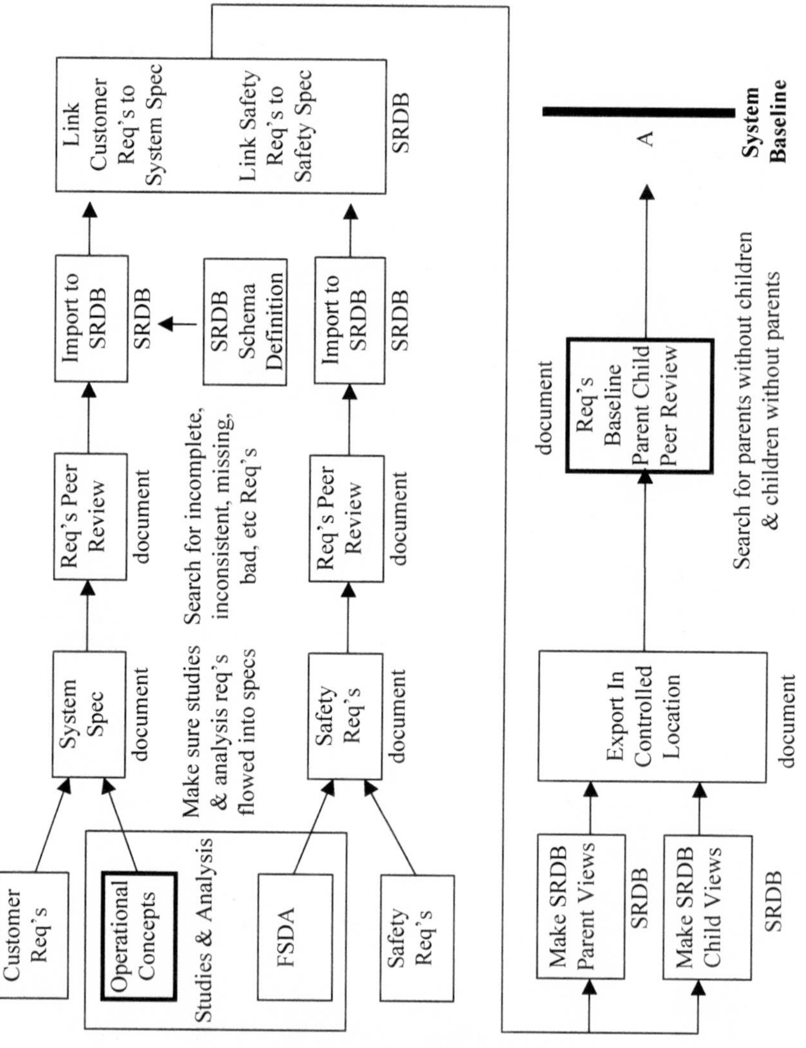

Figure 163 SRDB Process 1

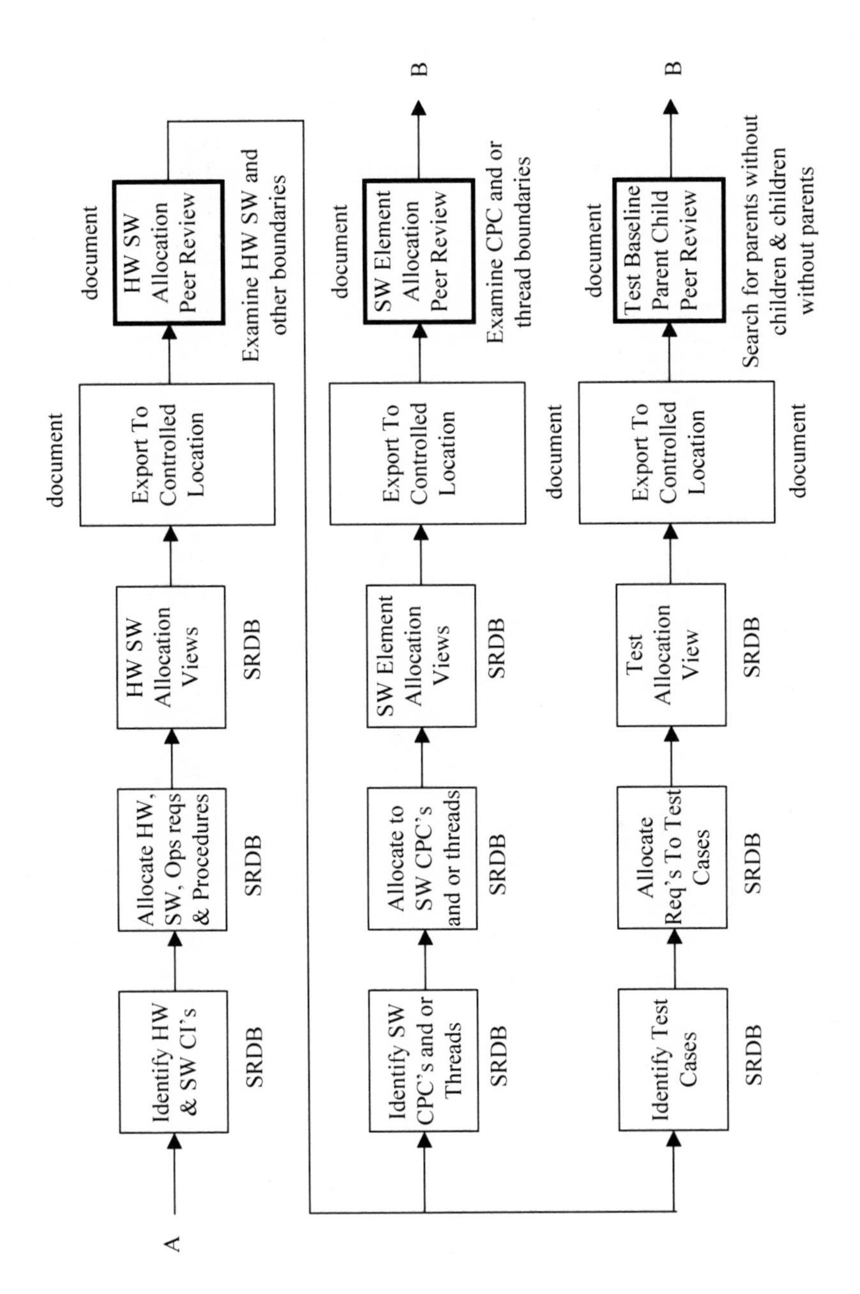

Figure 164 SRDB Process 2

Author Comment: There was an old saying when databases were first being considered for many applications: "Garbage in garbage out". It is critical that the SRDB not be based on garbage. One of the ways to avoid that trap is to carefully define and apply the attributes and links. Every time an attribute is set or a link is made, it needs to be reviewed, validated, and maintained. Maintenance is critical as a system evolves. For example someone could take the time to link requirements to individual test steps, but what would be the benefit in offering "very fine grain traceability" to a test step? Meanwhile as the system evolves the traceability to the test steps need to be maintained. There is a significant difference in maintaining links to 100 atomic tests versus 5000 test steps.

32.4 Exercises

1. Draw a tree like diagram representing an SRDB schema for a system with a system specification, hardware specification, software specification, and system test. Add hardware and unit level tests to the SRDB schema. How does this compare with your classmates?
2. Write a 2 to 3 page document describing an SRDB schema for a project.
3. How can you show traceability to products not in the SRDB?
4. Describe the dangers of garbage in garbage out.
5. What are the maintenance issues associated with an SRDB and how can you minimize maintenance and inconsistencies in the database?
6. Identify reviews that must be performed to ensure the database does not contain errors or inconsistencies.

32.5 Additional Reading

1. Specification Practices, Military Standard, MIL-STD-490 30 October 1968, MIL-STD-490A 4 June 1985.
2. System Software Development, Military Standard Defense, DOD-STD-2167 4 June 1985, DOD-STD-2167A 29 February 1988.

33 Integration Test Verification & Validation

You can't know a thing until you stress and break it. Test integration verification and validation is a broad subject much like architecture development and integrated logistics support. However they need to be addressed together in a holistic approach because of the impacts one area has on another area.

Test is a broad term that can mean any form of testing. For example verification is a form of testing and validation is a form of testing. It is an overloaded term and context is important when using the term test.

Integration is the assembly of the system from its parts. Integration is usually discussed in the same general area of test strategy because the integration approach or strategy is affected by the test strategy.

33.1 Integration Approaches

There are three broad approaches to integration. The first is to start integrating or assembling the system as soon as the parts are available. The second is to start integrating when collections of parts that are logically related in some way are available. The third approach is to wait until all the parts are available and then start assembling the system. These three approaches may differ when in development versus production. Also these three approaches impact the test strategy.

If the first integration approach is selected - integration as soon as the parts are available, the testing is unplanned and driven by parts arrival. This results in test holes until all the anticipated parts are integrated.

If the second integration approach is selected - integration when logically related parts are available, the testing can be planned and performed for an integrated unit. This allows full test coverage with no test holes for the integrated unit. Since the units are smaller, troubleshooting and debugging is easier than on a full system with many interactions. Also early testing allows for future tests to be modified and optimized based on knowledge gained from previous tests.

With the third approach of waiting until the very end to perform integration, if the system is new and complex, the system will not work the first time. Troubleshooting and debugging is significantly more difficult than with the peace meal integration approach. The impact is delayed testing. Also there is no benefit of learning from early tests and so the tests may be flawed, needing changes adding further delay.

Obviously a development integration environment is different from a production integration or assembly environment. The best integration strategy for a development environment might be the second approach. The best integration or assembly strategy for a proven production environment might be the third approach, assuming test points are fully accessible.

33.2 Test Approaches

Early in the system development, operational concepts are developed along with a functional decomposition. The tests can be organized using these groupings. So the tests can be based on operational threads, functions, or both threads and functions.

Most test strategies use a combination of thread and functional organization. It makes no sense to test something 50 times while ignoring something else, which may result if the dogma of just thread based testing is adopted. Also if the dogma of just functional testing is adopted the system will never even be tried in an operational scenario. In theory nothing should be ignored if the tests are traceable to the requirements. However different initial conditions, configuration, and ending conditions not captured in the requirements or operational concepts should result in tests and these tests may uncover significant issues.

33.3 Characteristics of Goo d Tests

Good tests have certain characteristics. These characteristics are:

- **Atomic**: This means the test procedures are standalone. The test procedures are not dependent on other test procedure or a test sequence.
- **Traceable**: The atomic test procedures are traceable to a reasonable number of requirements. This traceability comes into question if large numbers of requirements are linked to an atomic test case test procedure. Many times it is helpful to encode the requirement text or portions of the requirement text in the test step where the requirement satisfaction is claimed. Obviously this does not mean a copy and paste of the requirements into the test procedures.
- **Thread Oriented**: Test threads may differ from operational threads, although they may appear to be very similar. Test threads try to minimize the number of configuration changes between atomic test procedures. Although each test is atomic, they should take advantage of certain previous tests. For example power cycling and reinstalling software between each test may add several minutes to each test which can translate into hours or days. This wasted time can be used for additional test coverage.
- **Standalone**: Test procedures should be completely standalone. No additional information products should be needed to complete the test. This does not mean that a test procedure does not include other information products such as a product flyer, datasheet, etc. It just means the test procedure contains everything needed to complete the test.
- **Signature Block**: Each test procedure should end with a signature block for

the witnesses to mark that they have seen the test executed as written or as redlined and initialed. The signature block defines the boundary of the atomic test procedure.

Example 59 Test Procedure Document Organization

In a book of 150 test procedures there are 150 signature blocks and any test can be executed in any sequence. However the ideal sequence is organized from beginning to end. The ideal sequence minimizes the test setup and configuration time as each test builds upon the next test. Just because the ideal test sequence is embedded in the test procedure book, it does not mean that sequence must be followed. The sequence can change as needed with time.

There are different strategies that can be used to select the test execution sequence of any test procedure document or book. One approach is to start with the easy tests and progress to the more complex tests. Another approach is to organize the sequence to minimize the test setup time, the natural order of the test procedure document. Test setup time may be a big issue if it takes hours or even days to setup for a few 10-minute tests. Yet another approach is to start with the complex tests or tests that are suspected to have issues. This allows for time to address these issues while the simple tests are being executed in parallel.

33.4 Test V-Diagram

The Test V-Diagram[308] was created to show a new approach to test and integration for a new system that would start to automate the cognitive processes of its system operators. It was driven by the new challenge to surface latent defects in the software. It was recognized that software could be untested using existing test approaches. So to offer full test coverage of the software the concept of bottom up integration and test was suggested. It was later modified and adopted by others[309] to represent systems engineering at a high level[310].

It is unclear if it originated with the VEE Heuristic[311]. It would make sense

[308] Test V-Diagram first appeared at Hughes Aircraft circa 1982 on the Federal Aviation Administration (FAA) Advanced Automation System (AAS) program pre-proposal effort.

[309] The reason the V is such a powerful and enduring image comes from the Hughes culture of coupling all text to powerful multidimensional images. This was the foundation of Sequential Thematic Organization of Publications (STOP) invented at Hughes in 1963.

[310] The Relationship of System Engineering to the Project Cycle, Kevin Forsberg and Harold Mooz, National Council On Systems Engineering (NCOSE) and American Society for Engineering Management (ASEM), 21–23 October 1991.

[311] Educating, Gowin, D.B., Ithaca, N.Y., Cornell University Press. 1981. Gowin developed the VEE heuristic to help students understand knowledge structure (relational networks, hierarchies, and combinations) and understand the process of knowledge construction. It assumes knowledge is not absolute but depends on concepts, theories, and

because so many people from education, psychology, cognitive processing, knowledge management, etc were working with computer systems in the 1970's. A side oriented V appeared in DOD-STD-2167 circa 1982 to represent the software development cycle. In this case it showed parallelism between established hardware development and emerging software development. It suggested an approach for building the next generation software systems and showed the different abstraction levels of the design associated with the major milestones used to manage the effort.

Decomposition as a top down activity was fully accepted at that time. Test as a bottom up activity was obvious in most cases but not fully articulated. The thought process was that the bottoms up test approach would complement the top down decomposition and offered an alternative view of the system. Strong traceability would be used between the tests and the associated requirements. Strong bi-directional traceability also would be used between the various specifications.

The bottoms up activity blended well with the natural progression of integration. So tests would be performed at the lowest levels so that all conditions and cases could be evaluated and not hidden or prevented from stimulation by a higher level abstraction. Also as the integration continues the interfaces could be fully tested.

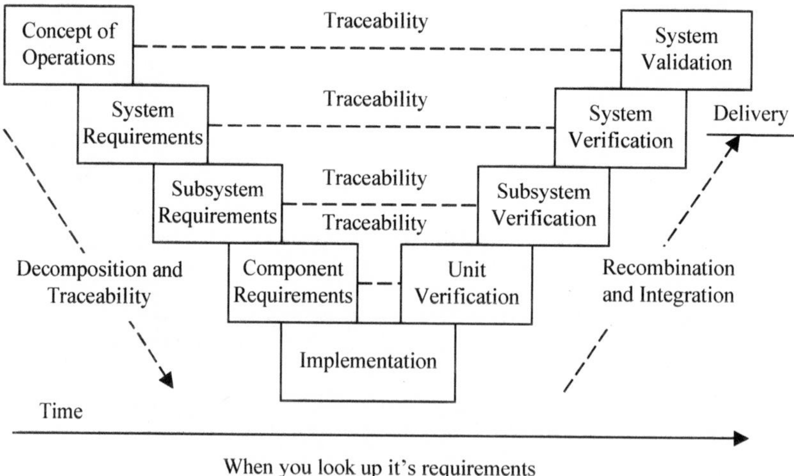

Figure 165 Test V-Diagram

33.5 Verification

Verification testing is used to determine if the system is designed and implemented according to the specifications. The requirements are allocated to

---

methodologies used to view the world.

different test levels and to different test procedures and cases within the procedures. That allocation represents the bi-directional traceable path between each requirement and each atomic test procedure.

The test procedures can be grouped as needed for the unique system but should respect the concept of unit, subsystem, and system level verification. There may be many unit test procedure books, many subsystem test procedure books, but only a few system test procedure books. Examples for splitting the system procedures into separate books are:

- System Environmental Test Procedures
- System Functional Test Procedures
- System Performance Test Procedures
- System Recovery Test Procedures
- System Interoperability Test Procedures
- System Certification Test Procedures

Specifications are imperfect and changes may be made to the specifications as a result of the tests. However these changes should be minimal if a good system engineering activity was performed.

As the tests are dry run in preparation for formal test witnessing, defects will surface. These defects and resulting changes need to be formally tracked using a defect-tracking tool. Whenever a defect is detected as much information about the defect should be gathered and documented. This should include an initial assessment of the defect[312].

Table 65 Priorities Used For Classifying Defects

Priority	Applies if a problem could
1 Grave	a. Prevent accomplishment of an operational or mission essential capability b. Jeopardize safety, security, or other requirement designated critical c. Result in loss of life or health
2 Catastrophic	a. Adversely affect the accomplishment of an operational or mission essential capability and no work-around solution is known b. Adversely affect technical, cost, or schedule risks to the project or to life cycle support of the system, and no work-around solution is known
3 Critical	a. Adversely affect the accomplishment of an operational or mission essential capability but a work-around solution is known b. Adversely affect technical, cost, or schedule risks to the project or to life cycle support of the system, but a work-around solution is known
4 Marginal	a. Result in user operator inconvenience or annoyance but does not affect a required operational or mission essential capability

[312] MIL-STD-498 Appendix C Category And Priority Classifications For Problem Reporting. There is an overlap with the FMEA categories found in MIL-STD-1629A.

Table 65 Priorities Used For Classifying Defects

Priority	Applies if a problem could
	b. Result in inconvenience or annoyance for development or support personnel, but does not prevent the accomplishment of those responsibilities
5 Negligible	Any other effect

33.5.1 Verification Plans

The purpose of the verification plan or plans is to capture all the elements of the test program so that personnel, equipment, and facilities can be prepared and made available when needed. The test plan also captures the test philosophy so that the stakeholders can agree on the approach. For example some test programs may be large needing multiple test plans addressing specific areas. Some test approaches are highly automated using state-of-the-art tools while others are manually intensive using large numbers of people. If there are multiple test plans a master test plan introduces the plans and their interrelationships. It also serves as the parent source of common practices and tools such as problem tracking.

Table 66 Test Plan Suggestions

• Master Test Plan (The Plan Of Plans)	• Stress Test Plan
• Hardware, Software, Integration Test Plans	• Interoperability Test Plan
• Safety Test Plan	• System Test Plan
• Security Test Plan	• Production Test Plan
• Environment Test Plan	• Final Acceptance Test Plan

Table 67 Test Plan Outline Suggestion

1.0 Introduction
1.1 Brief description of system under test
1.2 Relationships to other test plans, procedures, and reports
2.0 Reference documents
3.0 Test environment
3.1 Test configuration
3.2 Test tools, data recording, and data reduction approach tools
3.3 Problem tracking approach and tools
3.4 Requirement traceability
4.0 Test logistics
4.1 Test location and security considerations
4.2 Participating organizations, roles, responsibilities, and personnel (test director, testers, quality assurance, sponsors, customers, etc)
4.3 Orientation and training needs
4.4 Process used to validate tests were executed
4.5 Planned test schedule and relationship to other milestones

Table 67 Test Plan Outline Suggestion

5.0 List of planned tests
Appendix - Test Verification Cross Reference Matrix (VCRM) listing tests and requirements verified by each test.

33.5.2 Verification Procedures

The verification procedure takes the verification plan and updates the current plan content with the latest information and then introduces test procedures with test steps in a new section. The VCRM is also updated to reflect the changes surfaced while adding the test procedures and their associated test steps.

The test procedures are atomic or standalone blocks starting with the test purpose or objective and ending with a signature block where all the test witnesses sign off that the procedure as written or as marked up was executed.

Table 68 Test Procedure Outline Suggestion

1.0 - 5.0 Updated from plan
6.1 Test Procedure Name 1 (from test list in section 5.0)
Objective, Requirements verified, Test material, Test configuration settings, Initial conditions, Special test considerations
Test Steps
Test Step 1
Test Step 2
Test Step x
Witness Signatures
6.2 Test Procedure Name 2
6.n Test Procedure Name n
Appendix - Test VCRM

Ideally the tests should be structured so that they are independent and can run in any sequence. However a sequence should be developed that takes advantage of previous configuration setup sequences so that the total test run time for the entire procedure is minimized[313].

33.5.3 Dry Run Test Execution

Once the procedures are written in the office they need to move from the Ivory Tower to the real test environment. This is known as dry running the test procedures. As part of the dry runs the procedures are updated and defects are

[313] Many latent defects surface as configurations change and sequence changes. Even though the test procedure is optimized for the most efficient formal test run, during dry run the sequences should be naturally permitted to change. This will help to surface sequence and configuration dependent latent defects.

uncovered. At some point new defects stop surfacing, old defects have been addressed, and a formal test can be performed. The fully dry run procedures are complete and are submitted for review and acceptance.

As part of the review process requirement traceability is reviewed to ensure all requirements trace to one or more tests. Also, the tests are reviewed to ensure they really verify the requirements. This may be open to interpretation and more test cases may be added to ensure a requirement is verified in all conditions. The conditions might be configuration dependencies, range of values, etc.

33.5.4 Formal Test Execution

The dry runs are rehearsals to get all the bugs out of the show. The analogy is very appropriate. The formal run is the real show with the customer and internal Quality Assurance (QA) witnesses in attendance. The test director[314] typically hosts a Test Readiness Review (TRR) where the artifacts are examined and a determination is made if a test can be started. The items considered in the TRR are open defect reports, test configuration issues, known open requirement issues, etc.

As the tests are executed the steps are typically checked off. If there is a minor problem with a step it is redlined using hand written text and signed, initialed, stamped by QA and the witnesses. At the conclusion of a test all the witnesses sign it off in a witness area. If the test fails it is noted on the procedure with a note of the failed step or steps. This approach is used to validate that the tests were executed as written.

33.5.5 Test Report

The test report is a log of the formal test run. It includes the as run procedures with redlines, signatures, and new information reporting the test and its results. It is captured in a concise document with the following content.

- **Introduction**: The test performed and requirements verified. References test plan, procedures, and specifications.
- **Summary**: Tests performed, test participants, start and end dates, defects written, and a statement noting if the test was completed successfully.
- **Test Log**: Organized by date captures the daily events.
- **Detailed Results**: List of the tests and redlines, issues, and defects found for each test.
- **Defects**: List of the defects uncovered during the test.
- **Conclusions and Recommendations**: Statement of the next steps if there

---

[314] A test director is associated with a plan, its procedures and the report. They are in charge of the entire test effort for that body of work. There may be multiple test directors each with their unique set of plans, procedures, and reports.

are significant redlines, any failed tests, and significant defects.

- **Appendix**: As run test procedures with handwritten checkmarks, redlines, and signatures.

33.6 Validation

Validation is used to determine if the right or correct system was designed and implemented. This comes from the user perspective. At one level the operational concepts are used to test the system. At another level the community revisits the system mission and either the system successfully satisfies the mission or not.

Validation can start at the development facility where the system is staged before delivery. Operators can be bought in and the system can be put through various operational scenarios using simulated inputs. Validation can continue at the development facility using some combination of mixed simulation and live data. At some point the system is delivered.

The first site is a key site and is a significant gate. Unlike the operators that may have been used at the development facility, these operators must live with this system. What they say and do is significant. Just like at the development facility, the testing can begin with simulated data, then a combination of simulated and live data, then a session using only live data during a off peak time, such as midnight.

Once confidence is developed, the key site will decide to take the system live with the ability to go back to the previous system. There may be set backs at anytime during this validation sequence, however, eventually all the bugs are worked out and the site goes live with the new system and one or more other sites are recommended for the new system roll out.

The system is not fully validated until all the sites have been delivered and operating with the new system for a reasonable period of time. This time period may be as long as a year. At some point the old system is phased out as the new system has proven itself. It is validated.

33.7 Accreditation and Certification

Accreditation or certification is given by the organization best positioned to make the judgment that the system is acceptable. The organization may be the operational user or client, program office or sponsor, or a third party entity such as a government agency, depending upon the purposes intended.

Certification usually includes analysis of the process and the solution. It can involve proof like elements. The developers need to show that there was a consistent process that was followed to design and implement the system. Deviation from the process raises concern because it may represent compromises in system. The solution needs to be documented and show strong traceability, especially to the tests. Without the traceability there is concern because once again it may represent compromises in the system.

In many ways the ultimate systems approach needs to be strongly followed with very high levels of accuracy, precision, rigor, and methodical behavior.

33.8 Various Tests, Verifications and Validations

There is a large endless list of tests, verifications and validations. This is just a small sampling of the kinds of tests that may exist in a domain. It is offered to broaden and deepen the perspective on test, verification, and validation.

33.8.1 Highly Accelerated Life Test

Highly Accelerated Life Test (HALT) is used to identify and resolve design weaknesses. It is typically performed during engineering development. It can be performed at the unit, subsystem, or system level. The approach is to progressively apply more environmental stresses in a methodical measured manner until a failure is detected. Obviously this should be beyond the design requirements however surprises do surface when components are integrated into an assembly. There are commercial environmental test chambers available for this type of testing. The environmental stresses are:

- Cold step and Hot step
- Rapid temperature cycles
- Stepped vibration and shock
- Power switch and fluctuation
- Combined stress tests

33.8.2 Safety Tests

Safety tests are used to find unsafe elements in a system. Many times standards are invoked for safety requirements. These standards are highly specialized and typically include many tests with a certificate that is issued when the system or product is shown to meet the standard. A good example is a certificate from Underwriters Labs (UL)[315].

33.8.3 Capacity Stress Testing

Capacity Stress Testing takes selected performance requirements and subjects the system to progressively more stringent performance requirements until a failure is detected. This testing can be isolated to a single performance item or a group of performance requirements that represent a scenario of ever-increasing challenge. Examples include:

- Continuously add more load to a physical structure
- Continuously increase the number of transaction requests into a computer

---

[315] Found in 1894 Underwriters Labs ® (UL) has been testing products and writing standards for safety for more than a century.

based system until thrashing[316] occurs

- Continuously increase the speed of a motor until it flies apart
- Continuously increase the number of simulated aircraft in an air traffic control system until tracks are dropped
- Continuously increase the number of simulated aircraft in an air traffic control system until the operators are overloaded and near misses increase and more aircraft are placed into holding patterns

33.8.4 Destructive Testing

Tests can be destructive such that the item under test is compromised or destroyed. However fully verifying a requirement may call for a destructive test. This is best described with an example of an electrical fuse. A non-destructive test for a fuse is to check its continuity. A destructive test for a fuse is to continuously increase the current and measure when and how quickly the fuse blows (i.e. works). However once the fuse is blown it is of no value. An approach to destructive testing is to sample a production run to verify that an entire batch is working as expected. The implication being that the production run is consistent and the process is not compromised if a selected sample set behaves as expected.

33.8.5 Characterization Tests

Characterization Tests are performed to characterize an existing device, subsystem, system, product, etc. It is used to compare against alternative choices and compare against possible future system upgrades or outright replacement. Informed consumers are aware of tests and associated resulting specifications for audio products and automobiles. Some times when characterization tests are performed the original design specifications are exceeded.

In the past it was common engineering practice to de-rate components and include extra capacity and room for meeting performance requirements in the design. As design has become more precise with miniaturization, automation, and computers many solutions are offered at the specification limits. So a bridge built in the 1960's may catastrophically fail[317] while a bridge in the same location built at

---

[316] A resource stops doing useful work because there is contention. Once started, thrashing is typically self-sustaining until something occurs to remove the contention. For example, multiple processes accessing a shared resource repeatedly and neither willing to relinquish control so the resource can complete one of the processes while the other patiently waits.

[317] I-35W Mississippi River bridge (Bridge 9340) eight-lane, steel truss arch bridge opened in 1967 collapsed during evening rush hour August 1, 2007. Some have suggested modern precise engineering practices were a factor as the bridge reached its original design life. Older bridges with the same design life included less precise over engineering practices, yielding long life.

the start of the previous century is still standing and offering service.

33.8.6 Architecture Verification and Validation

This text places great importance on architecture. However can the selected architecture be tested, verified, and validated before it is designed and implemented? Is the architecture tradeoff analysis sufficient evidence to show that the architecture was tested, verified, and validated?

The case for the selected architecture is embodied in the architecture studies and models. The typical studies and models that test and verify the architecture are against the system key requirements and include:

- Timing, sizing, load studies and models
- Reliability, maintainability, availability studies and models
- Functional decomposition and allocation
- Requirements decomposition allocation and allocation

The studies and models that validate the architecture include:

- Operational sequence diagrams
- Operational simulations
- Operational simulation prototypes, and demonstrations

33.8.7 Software Intensive System Verification

Software intensive systems have a challenge because so many logical and arithmetic elements are within the total software set. All of these elements are suspect until verified. The problem is many of these elements are rarely executed at the system level and if they are executed, their limit conditions and internal paths may be unexercised without inserting computer failures. So software is tested at different abstraction levels and on different environments outside the normal target execution environment.

- **Unit Testing**: This is the smallest atomic test performed. The test might be something as simple as to perform a compile and link. Depending on the application, the testing may include developing unit stubs and drivers to stimulate the inputs and observe the outputs. It may go further and exceed the input limits and or stimulate all possible values in all possible conditions. These exhaustive unit level tests are associated with extremely critical software units.
- **Configuration Item Testing**: Once multiple units are grouped into a configuration item such as a standalone application, it can be tested. The goal is to surface the unexpected emergent behavior associated with

multiple unit interactions and their interfaces. This testing may or may not be in the target environment. It also may or may not include a debugger, which can provide white box testing services such as inserting faults and examining internal intermediate values.

- **Integration Testing**: Once all the software configuration items are available they are typically loaded on the final target environment and subjected to integration testing. This may be as simple as verifying that the software executes on the hardware to verifying all the software requirements as specified in the software requirement specification(s).

33.8.8 Interoperability Tests

Interoperability tests are a difficult subject because initially everyone assumes some gold standard test fixture is developed to support interoperability tests of devices and systems from various vendors. However, no one wants the responsibility for validating the gold standard or even paying for its development. So there tends to be a mad rush to be the first out of the gate so that they become the gold standard that everyone else needs to meet. Other will need to make the changes, if needed, not the one who first made it to the gate.

33.9 Exercises

1. Develop a test plan for a wind turbine farm.
2. Develop a test plan for a consumer product.
3. How does testing differ between an infrastructure program and a consumer product?

33.10 Additional Reading

1. CMS Testing Framework Overview, Department of Health and Human Services, Centers for Medicare & Medicaid Services, Office of Information Services, Version: 1.0, January 2009.
2. Test And Evaluation Handbook, Federal Aviation Administration, Version 1.0, August 21, 2008.
3. Test Inspection Reports, DI-NDTI-90909A March 1991, DI-NDTI-90909B January 1997.
4. Test Plan, Data Item Description, DI-NDTI-80566, April 1988.
5. Test Procedure, DI-NDTI-80603, June 1988.

34 Configuration Management

Configuration Management is about being able to find and if needed duplicate something tomorrow, next week, next month, next year, ten years from now, a generation from now. This is accomplished by following certain practices. These practices are extremely important and if they are ignored or bypassed, configuration management will be compromised and works will be lost to history.

34.1 Nomenclature

Nomenclature uniquely names something so that it can be identified, tracked, and characterized. This is a configuration item. An item is internally designated with a nomenclature by the originating organization and externally designated by either a customer or a third party. For example this book has an ISBN[318] that identifies it uniquely. The ISBN is associated with a title, author, and publisher[319].

34.2 Configuration Item

An item that is destined for formal configuration control is a configuration item. Configuration items tend to be associated with external delivery rather than internal items not leaving the organization. An item may start within the organization, undergo several internal changes, be designated as a configuration item, but not placed into formal configuration control until the first delivery. Prior to the first delivery items are under informal configuration control.

34.3 Libraries

If this book is placed in a library it is given a unique library number based on the Dewey Decimal System[320]. The Library of Congress Classification[321] is an

---

[318] An International Standard Book Number (ISBN) is assigned to each edition and variation of a book. An ISBN authority issues the ISBN.

[319] In 1870 Frederick Leypoldt published the first edition of the Annual American Catalogue, a precursor to Books in Print. In 1872 he published the first issue of Publishers Weekly, and in 1876 the first issue of Library Journal. In 1878 Richard Rogers Bowker acquired Leypoldt's company. Leypoldt and Bowker founded Literary Marketplace and Ulrich's Periodicals Directory. Bowker is the official U.S. ISBN Agency.

[320] The Dewey Decimal System is a system of library classification developed by Melvil Dewey in 1876.

[321] The Library of Congress Classification is a system of library classification developed by the Library of Congress in 1897.

alternative to the Dewey Decimal System. So this book can be tracked by two separate systems from a library perspective. This is in addition to the ISBN.

Libraries are very important to an organization. The library is the mechanism used to house and track the body of work of an entire organization. Without a library it is not possible to maintain organizational knowledge and bring new staff into the group. Libraries are the global memory of the organizational gestalt.

34.4 Internal Configuration Management

Internal configuration management is less formal however rules should be followed by the organization. The goals are the same as for formal configuration management: to be able to find items and duplicate them regardless of time and organizational changes.

Computer files should be consistently named using descriptive text. When there are changes the old files should be archived in an archive folder in the same location as the current working file. As part of the archive, the file should be made read only and the file name should be appended with an encoded date (e.g. filename-120410.doc). When someone makes comments the file name should be augmented with their initials (e.g. filename-ws.doc).

When hand written materials are produced they should be marked in a consistent location, the lower right hand corner. The mark should show the date and name of the originator or their initials. Informal electronic drawings should be marked in the same way.

All the authors of a document and their dates of contribution based on official releases should be listed within the first few pages of the document. It should acknowledge people who substantially contributed to the effort. Too many times people take authorship of an information product that they did not originate and they cannot speak to its core.

34.5 Formal Configuration Management

Formal configuration management is applied to all material released from an organization. Devices are marked with labels that are not easily removed or defaced. The labels contain key configuration information such as nomenclature, model number, serial number, and date. Information products are given cover pages with a date and release number. This information is found in a configuration management database. The physical products are placed in a vault and the virtual products digitally stored are archived and the archives are placed in a vault. These items and the database exist within the vault for the life of the organization.

The same devices and information products may be under formal configuration control by many organizations. These organizations include system developers, subcontractors, vendors, and customers.

34.6 Configuration Freeze

During development there are multiple items in various stages of maturity. There are times when one or more items need to be frozen where a line is drawn in the sand changes are temporarily stopped and all the items are placed into a configuration. The configuration is frozen and released.

For example in preparation for a review a snapshot is taken of all the individual configuration items and placed into a single configuration with its own nomenclature. The review might include documents, video, sound recordings, devices, software, data, etc. The goal is to make sure all the items in the configuration are consistent, so there is always the desire to make one more final change in different items. This can lead to frenetic activity and loss of configuration control as members try to slip in one more change. This should be avoided. Inconsistencies should be expected and everyone should know an intermediate configuration is a snapshot in time.

At the end of an activity, in preparation for a delivery, the system configuration is frozen. The configuration is given a nomenclature and a tracking number that represents its release or revision.

34.7 Functional and Physical Configuration Audit

As part of delivery the frozen formal configuration of the system is reviewed. This is a physical inspection to verify that all the individual configuration items used in the system exist, behave as expected, and are under configuration control. This is typically a review of the information products and a physical inspection of the physical devices. Every item including the physical devices has a marking tag. Today the tag is typically a universal bar code with additional information to allow a human to immediately identify the configuration item by model, release, revision, serial number and or other designations. In the past it tended to be a marking plate.

34.8 Change Tracking and Control

There is an old saying that you cannot inspect quality into a product or system. There is a corollary to that which states that you cannot fix your way into quality. Quality needs to be inherent in the solution. The process, architecture, design, implementation, need to use practices that maximize quality. However defects surface and changes are requested. When these changes are made they need to be fully addressed in a methodical, rigorous, accurate, traceable way. This is change tracking and control.

Change tracking does not necessarily suggest there is a problem. Change can be the result of new needs and insights. It also can be the result of extreme success and the desire to add new capabilities. Change also can be the result of detecting an issue or a problem.

Change tracking is applied to all elements in the system including the process. It can be implemented with a massive database or a federated collection of autonomous databases. Most organizations use a federated change tracking and control architecture.

The change tracking and control process must be documented. This includes identifying the tools, their key attributes, how the stakeholders will interact with the tools, and the approval authorities.

34.9 Requirement Change Requests

Requirement change requests track changes to the requirements. The proposed requirement changes can come from any source. They typically come from the team working the next level of system development.

A periodic review is scheduled. If there are no proposed changes the stakeholders ignore the meeting notice. The periodic nature of the meeting ensures that the process is not ignored or bypassed by the team. Quality representatives track the meeting and produce findings if the process is compromised.

Every requirement change request has a unique tracking number that cannot be deleted. This number is appended with a prefix to form a unique project identifier of the requirement change request.

The requirement change request attributes include: requirement change number (RCN), requirement project unit identifier (PUI), origination data, old requirement text, proposed requirement text, state (proposed, accepted, SRDB[322] updated), old requirement attributes, new requirement attributes, old requirement link[323], new requirement link, impacted requirements by PUI, impacted documents, close date. This information can be captured in a separate relational database tool, the SRDB, or in a document using a table format.

34.10 System Change Requests

Most organizations do not use system change requests. Instead they track changes based on using separate databases such as software, hardware, or engineering change notices. System change requests track all changes for all information products, deliverables, and configuration items in one location using a single relational database.

It typically starts immediately with the start of a project. The requests are used to track peer reviews, walk through sessions, releases, and change requests. The system change requests can be enhancements or the result of detecting an issue or problem. They can surface at anytime from any source but typically originate from a

---

[322] SRDB – The System Requirements Database is an automated tool for managing requirements and maintaining tractability.

[323] Links are used to show up down traceability to other requirements and test cases in a parent child relationship.

peer review, walk through, or test.

Every system change request includes a unique tracking number that cannot be deleted. This number is appended with a prefix to form a unique project identifier. The system change request attributes include: unique change request number (CRN), title, description, state (initiated, accepted, started, completed, tested), originator, origination date, severity (grave, catastrophic, critical, marginal, negligible), planning priority, system element, resolution, closing rational, risk level, final risk, risk mitigation if not closed.

Other attributes include: configuration item, component, release, type (defect, improvement, release), source (internal, customer, marketing, etc), phase detected, responsible contact, planned release, due date, related CRNs, information product (specification, design, drawing, manual, code, etc), segment (air, ground, space, facility type 1, facility type n, etc), subsystem (computer, display, voice communication, RADAR, engine transmission, body, electrical, water, sewage, etc).

34.11 Software Change Requests

Software change requests track changes to the software. It typically starts when software implementation starts. The requests are used to track peer reviews, code walk through sessions, releases, and change requests. The change requests can be enhancements or the result of detecting an issue or problem. The requests can surface at anytime from any source but typically originate from a software peer review, software code walk through, or test.

Some organizations do not invoke formal software change tracking until test has started. In this case the peer review and code walk through findings are captured in action item tables or databases. It also suggests that the peer review does not include unit level test findings.

Every software change request includes a unique tracking number that cannot be deleted. This number is appended with a prefix to form a unique project identifier. The software change request attributes include: unique software change request number (SRN), title, description, state (initiated, accepted, started, completed, tested), originator, origination date, severity (grave, catastrophic, critical, marginal, negligible), planning priority, system element, resolution, closing rational, risk level, final risk, risk mitigation if not closed.

Other attributes include: software configuration item, software component, release, type, source, phase detected, responsible contact, planned release, due date, related SRN, information product, segment, subsystem.

34.12 Hardware Change Requests

Hardware change requests include all electronic, mechanical, and structural elements of a system. Typically hardware has less change than requirements and software. However with the advent of VHDL code and programmable gate arrays, electronic hardware has started to experience the same level of changes as software and requirements[324]. In the past hardware change requests were tracked with drawing releases. Today change requests use a relational database with attributes similar to software change request attributes.

34.13 Engineering Change Notice and Proposals

An Engineering Change Notice (ECN) typically originates from the customer. It includes a budget and schedule. The ECN might request new features and or fixes to deferred problems found during verification and validation and captured in the change requests.

The ECN is addressed in an Engineering Change Proposal (ECP). The key ECN and ECP elements are: identification number, title, customer, deadline, reason for the change, indication if the change is required or optional, description, priority, related documents and activities, anticipated introduction.

The ECP should be treated like any other proposal. The proposal needs to include the process, analysis, alternatives, selection, verification, validation, cost, and schedule. Depending on the ECN, the ECP can be small, a few pages, or extremely large with a technical, management, and cost volume submission.

34.14 Exercises

1. Look at your computer information and try to find its nomenclature and other configuration management data.
2. Identify the configuration items in an automobile, computer, wind power plant, solar power plant, geo thermal plant.
3. Does your project maintain a library? If not why not?
4. What internal configuration management approaches do you use?
5. At what point in a project should formal configuration management be established? How should informal configuration management products be transferred to formal configuration management?
6. Identify the attributes for change tracking and control, requirement change requests, system change requests, software change requests, hardware change requests, and engineering change notices and proposals. Are there common attributes? Should these elements be part of one database or separate databases? If so why, if not why not?

---

[324] VHSIC hardware description language (VHDL) is used to design digital and mixed-signal systems on programmable gate arrays and integrated circuits. In 1980 the United States started the VHSIC joint tri-service Army Navy Air Force project. This led to VHDL and advances in integrated circuit materials, lithography, packaging, testing, algorithms, and computer aided design tools.

34.15 Additional Reading

1. Baseline Description Document, DOD Date Item Description, DI-CMAN-81121, February 1991.
2. Configuration Management Guidance, Military Handbook, MIL-HDBK-61 30 September 1997, MIL-HDBK-61A (SE) 7 February 2001.
3. Engineering Change Proposal (ECP), Date Item Description, DI-CMAN-80639C, September 2000.
4. Notice of Revision (NOR), DOD Date Item Description, DI-CMAN-80642C, September 2000.
5. Request for Deviation (RFD), DOD Date Item Description, DI-CMAN-80640C, September 2000.
6. Specification Change Notice (SCN), DOD Date Item Description, DI-CMAN-80643C, September 2000.

35 Maintainability

Maintainability is the practice of optimizing the ease and effectiveness of time and resources needed to keep a system in its original functional and performance condition. This suggests that personnel perform preventative and corrective maintenance with established skill levels using approved procedures and resources at each level of maintenance and repair.

There are quantitative and qualitative practices associated with maintainability. The quantitative practices are captured in the Reliability Maintainability Availability (RMA) section of this text and the qualitative practices are captured in this section. The qualitative practices are more important because they drive the architecture, which drives the quantitative RMA results. The numbers are the numbers, they cannot be denied.

The maintainability practice provides input into architecture selection, a maintenance philosophy, installation manuals, maintenance manuals, a Failure Reporting and Control Action System (FRACAS)[325], and a field support system. In addition the maintainability approach is just like developing the system. It is a system within the system. This means that many systems practices apply to developing the maintainability approach.

35.1 Maintainability Interactions

Maintainability is a function of design and impacts the design. It includes various elements of a system such as production or manufacturing, safety, human engineering, safety, diagnostics[326] and testability, and logistics support. Each domain has unique elements that need to be identified and considered. The following is used as starting point to stimulate thoughts on maintainability.

Production (or manufacturing) processes need to capture the maintainability approach established during development. It is important that manufacturing not compromise the maintainability concept. This suggests manufacturing stakeholders are involved at the start of system development. Without their contributions, maintainability design features or approaches may make the product difficult or too expensive to manufacture. For example, access panels that are included for maintainability might be placed in an area that has compound curves. Moving the panel to an area that is flat or has a simpler surface might still allow good access and improve producibility.

---

[325] FRACAS is addressed in the RMA and Quality sections of this text.

[326] Diagnostics is addressed in the Fault Tolerant Computing sections of this text.

Human Factors addresses the safety, effectiveness, role, and integration of people in the operation, use, and maintenance of a system. As part of the total system design process, human factors examines how the design of the system affects human and how people interact with the system. These people include users, operators, and maintainers of the product. The physical structure and mechanical operation of the human body and functioning of human senses determine how people can interact with a system. This interaction is usually referred to as the human-machine interface. The level of ease and economy to perform maintenance is impacted by human limitations and abilities related to strength, perception, reach, dexterity, and biology.

Maintainability is directly related to the anthropometric and psychological characteristics of the people who maintain the system. Maintainability stakeholders collaborate with human factors, and consider human factors during design efforts, to ensure the required range of expected human maintainers can accomplish the tasks. Anthropometric characteristics determine access openings size, the need for stands, how far replaceable units may be placed inside a compartment and still be reachable, etc. Psychological factors determine types of warnings, which way a calibration knob should turn, whether a continuously variable or detented knob should be used, etc.

Safety becomes very important when a system is being subjected to maintenance. This is not a normal system state and so there is increased possibility of safety issues surfacing during maintenance. Safety includes designing the system and maintenance procedures to minimize the possibility of damage to the system during maintenance and to minimize the possibility of harm to maintenance and operating personnel. From the safety practice come warning labels, precautionary information for maintenance and operating manuals, and the procedures for disposing of hazardous materials and the system elements.

Diagnostics and testability allow the maintainers to troubleshoot a system and verify that once parts have been replaced the system is operating as expected. Diagnostics consist of manual, automatic, and semi-automatic maintenance hardware, software, and procedures used to determine status, detect faults, and isolate faults. The hardware, software, and procedures depend on the maintainability concept. A highly maintainable system needs the least amount of support equipment and the fewest and the simplest procedures. This usually translates to highly automated diagnostics.

Logistics support is significantly affected by maintainability decisions. The results of maintainability analyses are used by logistics in planning for personnel and staff levels, support and test equipment, facilities needs, training, sparing, and technical manuals.

The actual logistics support provided during system operations affects the maintainability. Even if the inherent maintainability meets or exceeds the design requirement, the observed operational maintainability is as expected only if the required logistics support is provided. The support concept and any customer constraints or requirements related to technical data, support equipment, training, personnel turnover, field support, spares procurement, contractor depot support, mobility, and support personnel needs to be understood and considered during tradeoffs and analyses.

A critical aspect of logistics is obsolescence of internal and piece parts. Sometimes parts vanish because the underlying manufacturing processes are eliminated for ecological or economic reasons. Sometimes parts are displaced by ones that use new technology but are not identical in form, fit, and function. Parts and process obsolescence is a critical issue and needs to be addressed. Life buys are one way of coping with obsolescence.

35.2 Self Healing

We would like to think that our systems are infallible. They never break, make mistakes, harm anything, and never deteriorate. The reality is that all systems eventually break. The question then becomes how often and what are the impacts.

We are first introduced to breaking systems as children when we fall and skin our knees. Eventually we learn that our knees heal and once again we challenge ourselves to skin our knees. This is our first exposure to self-healing systems. Systems that we build include maintainers who are the elements to implement self-healing. If we exclude human maintainers then we would need to build systems that never fail or systems that use non human elements to accomplish self-healing.

35.3 Maintainability and Architecture

When developing a system, maintainability is always addressed. Entropy is the natural state of the universe and it requires energy to counter entropy. So we can expect any system to eventually require maintenance.

Maintainability tends to be bound by the architecture approach. Some architecture approaches are inherently more maintainable because of easier access, fewer parts, less exotic parts, less complex parts, fewer internal interfaces, less diverse internal interfaces, higher reliability, etc.

Some architecture approaches may drive the overall maintenance philosophy. For example one architecture approach may work well with a regional maintenance hub while another architecture must have onsite staff with onsite spares.

Maintainability needs to be articulated with each architecture approach in the front end of the systems activity and addressed in detail after the architecture is selected and implemented. The types of maintainability issues to consider during architecture selection are:

- Meant Time Between Failure (MTBF) numbers
- Mean Time to Repair (MTBF) numbers

- Reliability numbers
- Availability numbers
- Lowest replaceable unit
- Diagnostic, fault isolation, and troubleshooting tools
- Personnel skill levels, training, and certification needs

After the architecture is selected, during design and implementation, the original maintenance issues addressed during the architecture selection are fully addressed with more detail and final information. In addition more issues are addressed:

- Automation to track field failures
- Approach to preventative maintenance and its impact on the system
- Preventative maintenance schedules
- Maintenance depot approach
- Support staff location
- Spares location
- Centralized filed support automation systems
- Daily system certification
- Fault trees
- Automated expert system support including Internet message boards

35.4 Maintainability Design Considerations

In many ways maintainability and support is based on previous experience. To deny that body of knowledge is to run the risk of developing an ineffective maintainability approach. This knowledge tends to be bound to the application domain area, but there are generic items that can be considered and offered to stimulate thoughts for a more maintainable system[327].

- Optimize the size and weight of replaceable parts so that they can be easily handled by a single individual
- Avoid or minimize the use of exotic or custom parts and tools
- Avoid tight corners and difficult to access areas
- Smooth all edges and corners to prevent cuts and bruises
- Avoid or minimize the amount of disassembly to get to replaceable parts
- Minimize the number of different types of maintainers and total number of maintainers needed to properly maintain the system
- Minimize the needed skills and skill levels to properly maintain the system
- Use keyed connectors to prevent wrong connections and design interfaces

[327] MIL-STD-1472 has large body of knowledge associated with design for maintainability or maintainability design considerations.

so that there is no risk of damage if items are improperly connected or interfaced together

- Minimize the number of different parts and tools, try to use commonality as much as practical
- Maximize the modularity of the system so that individual modules are logically related and offer minimal disruption during maintenance activities

35.5 Maintainability Models

Models and diagrams are used to augment tradeoff studies. They are developed for alternative system concepts, configurations, proposed design changes, allocation, and prediction. Maintainability models are developed and used in the same way. Maintainability models consider the operational maintenance concept, safety, applicable levels of the system hierarchy, list of Lowest Replaceable Units (LRUs), and organizational policies.

A maintenance activities block diagram is a graphical representation of maintenance tasks and is used to assess compatibility with maintenance needs. It is typically augmented with a description that contains elements such as task description, number of maintenance personnel, equipment parts, and activities.

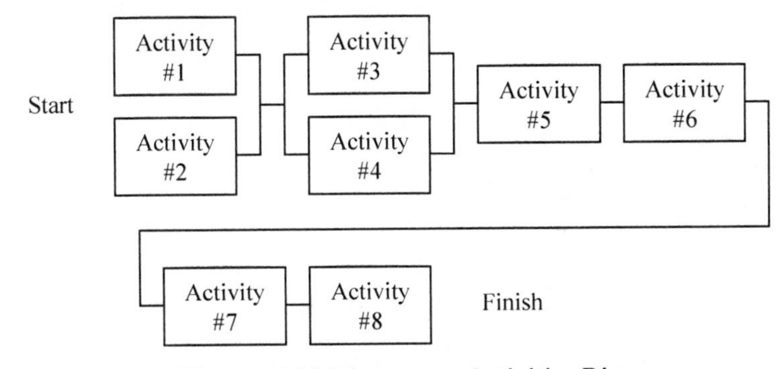

Figure 166 Maintenance Activities Diagram

Activities that can be done simultaneously (in parallel) are identified. Although it may be possible for one person to perform a task by doing each activity serially, two people make the job easier and safer. Notes should be offered when both people are not needed during the entire maintenance action. The individual not performing parallel activities can perform other work in the vicinity. The activities are listed in step format using action phrases that start with action such as get, move, open, close, disconnect, reconnect, unscrew, install, start, stop, power off, apply power, etc. They clearly trace to a Maintenance Activities Diagram.

35.6 Maintainability Allocation

Maintainability allocation assigns portions of the system level maintainability requirements to lower levels of assembly. They are apportioned to each subsystem, components and equipment within subsystems, and finally modules and parts. Maintainability allocation analyzes the system architecture and uses knowledge of the characteristics of systems, subsystems, components and parts. Allocations are, made primarily for corrective maintenance requirements. System level requirements are difficult to fully assess without history, a prototype, or first production version of the system. So allocations are used to assess the progress being made toward achieving the system maintainability requirements. An allocation diagram coupled with a table is an effective way to visualize and track the maintainability allocation.

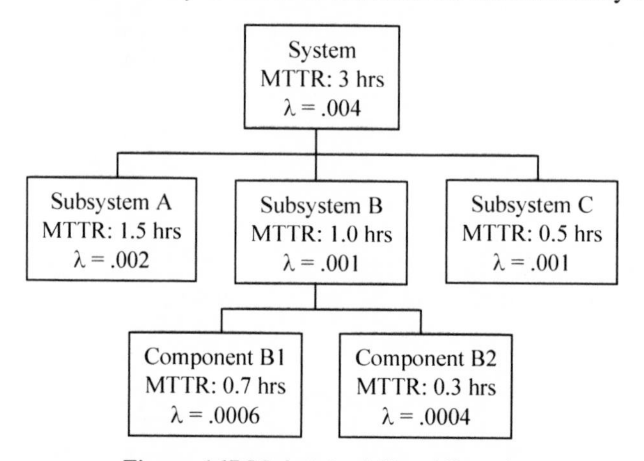

Figure 167 Maintainability Allocation

As with all allocations the maintainability allocation is an iterative process. The feasibility of meeting the initial set of allocated values is evaluated and if the allocated values are not reasonable, the allocation is revised.

35.7 Troubleshooting or Diagnosis

Troubleshooting is a procedure for locating and diagnosing malfunctions or breakdowns in equipment using systematic checking or analysis[328]. System maintainers are tasked with troubleshooting or diagnosing problems with systems and then providing corrective actions. Diagnosis is strictly just identifying the problem while troubleshooting is both identifying and correcting the problem.

Many people develop this skill in an ad hoc fashion and are able to apply troubleshooting or diagnosis to any system or device without any training or knowledge of the system or device. It is a problem solving skill where the troubleshooter or diagnostician:

---

[328] Electronic Reliability Design Handbook, Military Handbook, MIL-HDBK-338B, October 1998

- Identifies the high level subsystems
- Deduces functionality and performance of subsystems
- Detects aberrant behavior
- Correlates aberrant behavior to a function
- Deduces subsystem from aberrant functional or performance behavior
- Decomposes the subsystem and isolates the fault to the lowest level using the same techniques that lead to the defective subsystem

The process of observation and deduction requires significant time even with seasoned troubleshooters or maintenance staff. Knowledge of similar systems helps to shorten the time. It is possible to offer information to help the maintenance personnel troubleshoot a system. One of the most powerful tools is a graphical troubleshooting tree or table.

A troubleshooting tree as a hierarchical graphical representation quickly communicates the decomposition of a system. It also immediately suggests functional allocation and performance of the subsystems and components. A troubleshooting tree follows the rules of decomposition and is usually a family of trees spread across several logically grouped pages using numbering schemes similar to functional decomposition.

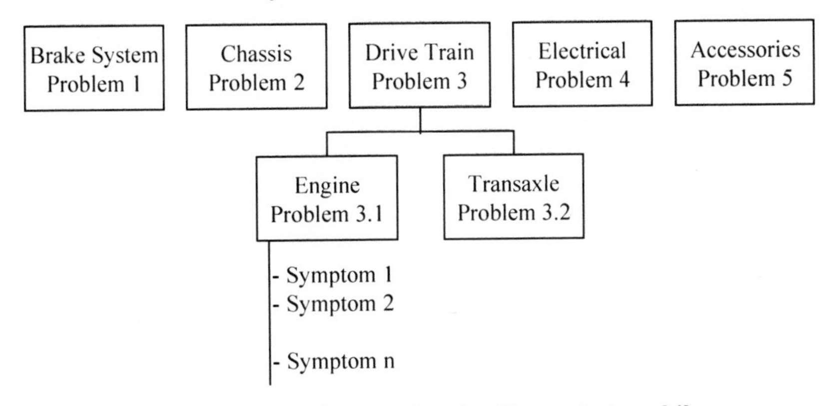

Figure 168 Troubleshooting Tree - Automobile

Both for a troubleshooting tree or table, a pointer can be offered to a sheet containing more information. The information fully describes the symptom and all possible elements to deal with the symptom. The following is a guideline:

- **Symptom**: Description in title
- **Reasons**: Reasons why the problem exists
- **Confirm**: Ways to confirm the problem including tests and diagnostics
- **Solutions**: Suggested solutions or fixes
- **Avoidance**: Suggested ways to avoid the problem from reoccurring

- **Metrics**: Data to collect for tracking the problem

A troubleshooting tree can be augmented with a troubleshooting table that contains additional information. In some cases only a table is offered. A troubleshooting table tends to be a list of symptoms and possible corrective actions. It does not focus on communicating decomposition. A graphical representation of the system and specifically the problem area is more effective at communicating the hierarchy and relationships than a table.

Table 69 Troubleshooting Table - Automobile Engine

Symptoms	Corrective Actions
Engine will not crank starter spins	See page x
Engine will not crank	See page z
Engine cranks will not start	Try action 1, action 2
Engine cranks starts hard when cold	
Engine cranks starts hard when hot	
Engine idle speeds varies	
Engine misses at high speed	
Engine misses at low speed	
Engine stalls at idle engine cold	
Engine stalls at idle engine hot	
Engine stalls during deceleration or on quick stop	
Engine hesitates stalls during acceleration	
Engine has poor acceleration	
Engine has less than normal power	
Exhaust system backfires	
Engine surges	
Excessive fuel consumption	
Excessive oil consumption	
Engine noise	
Spark knocks or ping	
Engine vibration	
Engine fails to reach normal temperature	
Loss of coolant or engine overheating	
Smoke from exhaust system	
High floor pan temperatures	

35.8 Exercises

1. What are the maintainability elements affecting ocean wind farm alternative architectures?
2. What are the maintainability elements affecting urban wind farm alternative architectures (3 or more architectures)?
3. What are the maintainability elements affecting land based wind farm

alternative architectures (3 or more architectures)?

4. What are the maintainability differences between ocean, urban, and land based wind farms (3 or more architectures)?
5. Identify the worst maintainability experience you have had to deal with and what was your approach?
6. Identify the best maintainability experience you have had and why?
7. Draw a maintainability allocation diagram for your automobile. Do you have any historical data to back up your allocation to the subsystems?
8. What are the maintainability issues associated with onsite versus depot maintenance?
9. Draw a troubleshooting tree for your computer. Draw a troubleshooting tree for your cell phone. Draw a troubleshooting tree for your automobile. How are they similar, how are they different?
10. Provide an example of your last troubleshooting adventures. How is this similar to problem solving? How is it different from problem solving? Can you draw a troubleshooting tree of the adventure?
11. What is the most sustainable LRU level? Compare it with your classmates and identify the differences. After the discussions what is the most sustainable LRU level?

35.9 Additional Reading

1. Designing And Developing Maintainable Products And Systems, DOD Handbook, MIL-HDBK-470A August 1997, MIL-HDBK-470 June 1995, MIL-HDBK-471 June 1995.
2. Electronic Reliability Design Handbook, Military Handbook, MIL-HDBK-338B, October 1998.
3. Human Engineering Design Criteria, DOD, MIL-STD-1472C May 1981, MIL-STD-1472D March 1989, MIL-STD-1472E October 1996, MIL-STD-1472F August 1999.

36 Training and Education

Training is the process of transferring knowledge and skills to individuals so that they can effectively perform a job. This is distinctly different from education, which is the process of transferring knowledge, skills, and values, to positively shape the mind and character of the next generation. There is an overlap between education and training. For example reading, writing, and basic mathematics fundamental to education also applies in training. The ability to perform research, think critically, and problem solve are also critical skills that overlap both education and training. Training can be used to:

- Prepare staff to go overseas and interact with foreign cultures
- Install, maintain, operate, and manage systems and subsystems
- Support various mission and organizational needs
- etc

System engineering companies in the past century employed great educators to establish curriculum to train people to operate and manage complex systems. Some of these companies had the additional challenge of doing this at the international level with different cultural and educational backgrounds of the native people[329].

These training programs include a combination of classroom instruction, Computer Based Instruction (CBI), simulation, entry level On-the-Job-Training (OJT) and progressive movement into more complex areas of the systems with more OJT at each system complexity level. A training program can include one or all of the training approaches. It was not unusual to spend an entire lifetime in a training career, which would peak at the pinnacle of developing training programs for the most sensitive aspects of systems.

36.1 Classroom Instruction

Classroom Instruction is a traditional training approach. This approach uses approved or certified instructors and course material that follow an appropriate pedagogy[330]. The course materials follow a syllabus with clearly stated goals for each course and its elements. A curriculum is structured as a set of courses used to

[329] Hughes Aircraft delivered air defense and air traffic control systems and trained people in Europe, Middle East, Pacific Rim, Asia, etc.

[330] Pedagogy is the art, principles, practice, and profession of teaching. It is also the study of teaching methods.

build the necessary knowledge skills and abilities for a particular area such as RADAR maintenance. The materials can include books, handouts, presentation slides, audio, videos, papers, physical materials, computer-stored information, etc. During the course, homework is assigned, tests are given, and lab time is provided with appropriate lab facilities. A final test is usually given that can be used to perform a final assessment of the student against the class goals and the grading rubric[331].

Classroom instruction can be live or virtual. The virtual classroom can use Internet technologies to network students together in geographically distributed locations. The lecture can be archived for access by students who may be unable to interact live with the instructor, but can view the lecture, hear other student questions, and email questions if needed to the instructor. The virtual classroom also can be implemented with pre-recorded video so that it can be played on site using local video players. Prior to the Internet, this was a very popular approach for a virtual classroom approach.

36.2 Computer Based Instruction

Computer Based Instruction (CBI) is sometimes referred to as interactive CBI because of its ability to interact with the student. This is instead of classroom instruction, which many times can be more of a lecture rather than a student instructor interaction. CBI captures the course content, presents it to the student using audio video, and engages the student by asking for responses at selected points in the training sequence. The response can be used to determine the next content to offer the student. The ability to modify the training as the student is engaged in the training session and stay in an area until it is mastered is a significant advantage of CBI. The CBI is also self paced. The student decides the pace of the training while working with the CBI progression mechanisms.

36.3 Simulator Training

Simulator training is used to imprint automatic behavior on a student. Simulators are used to train airline pilots, air traffic controllers, plant operators, astronauts, etc. The most valuable aspect of a simulator is its ability to introduce the student to catastrophic situations and have them develop automatic responses to recover from the situations.

This imprinting of safe automatic responses is accomplished with repetition of correct responses. This is accomplished using a multi step process where the

---

[331] A rubric is used to consistently score subjective assessments such a papers and projects. It lists criteria and standards that are traceable to learning objectives. The rubric is usually provided ahead of time so that a student can respond to the assignment and understand the grade or assessment provided by the teacher.

simulation can be halted to allow for instructor debriefing so that the student can correct behavior. The simulation can then proceed, rewind to some known point, or restart as determined by the instructor. All performance during a simulation can be recorded. The recorded information can then be used to debrief the student. It can be coupled with a replay of the recorded simulation exercise so that the student can see their behavior[332].

36.4 On the Job Training

On the Job Training (OJT) can be accomplished during live normal operations or during low level activities. It is not prudent system management to support significant OJT during peak loads. If there is an onsite simulation capability, simulation data can be mixed with live data before someone proceeds to full operational status. Once in full operational status an individual apprentice is engaged in OJT under the direction of a journeyman or a master. At some point the individual progresses from apprentice via OJT to journeyman level. At this point the individual can be trusted to work without the close watch of the designated journeyman or a master teacher.

36.5 Training and Architec ture Selection

Many training programs tend to have a progression of training methods. An example is the student begins with classroom instruction, progresses to CBI, moves on to simulation, then is assigned to a job where OJT completes the training. The training may bypass classroom instruction and go directly to CBI. Or it might just be exclusively CBI then OJT. The system-training specialists develop the curriculum and the training approach. This translates into people, skills, tools, and facilities to support the training approach.

The training approach varies based on the architecture approach and system needs. It is a key element in the early architecture tradeoff studies. For the various architecture approaches and system needs, the trainers identify the training approaches and needs. For example some architecture alternatives may need very sophisticated maintenance training while others can use very simple maintenance training approaches[333]. This translates into cost and complexity of the training program. Based on the system, the training approach may need new state-of-the-art tools and techniques. So the architecture of the training program is like the architecture of the system.

---

[332] Although these features are very challenging they became available circa 1992. Source: this author while developing air traffic control, air defense, and flight simulators.

[333] For example mainframe computer specialists versus personal computer specialists

36.6 Exercises

1. Identify the training needed for an automobile mechanic before they can maintain an entire automobile. Think in term of subsystems.
2. Identify the training time needed for each of the training courses for an automobile mechanic.
3. Identify the training methods for each course for an automobile mechanic.
4. How do you treat an individual who grew up maintaining cars since the age of 12 when they ask for a job in your garage when they are in their early 20's?

36.7 Additional Reading

1. Integrated Logistics Support Guide, Defense Systems Management College, May 1986.
2. Systems Engineering Fundamentals, Supplementary Text, Defense Acquisition University Press, January 2001.
3. Systems Engineering Handbook A Guide For System Life Cycle Processes And Activities; INCOSE-TP-2003-002-03.1; version 3.1; August 2007.

37 Integrated Logistics and Support

Integrated Logistics Support (ILS) groups certain practices under a common organizational umbrella. These practices are interrelated and experience suggests a better system is established if these disciplines share their work in an iterative way from the start to the end of the project or program.

For example the maintainability approach impacts the training and logistics needs. If the maintainability philosophy is to replace an entire subsystem then the training, tools, and parts are different than if the philosophy is to replace a block within the subsystem or a component on the block of a subsystem. Conversely the logistics may force a certain maintainability approach. Maintaining a system in a difficult location with limited space may force a component based maintenance approach.

As the maintenance approach moves from subsystem to component level the skill level and training of the personnel increases. This tends to increase the cost of the maintenance personnel. So there is a tradeoff between training and maintenance philosophy.

The maintenance philosophy does not just include level of repair. It also addresses issues associated with time to repair, which impacts logistics. Should the parts be collocated with the operational system? Should the maintenance personnel be collocated with the operational system? These interactions continue, as the system is understood.

The life cycle of a system, extending from cradle to grave, can be divided into exploration, development, production, deployment, operations, and disposal including storage and reclamation. ILS starts at the beginning of a project and follows through the full life cycle of the system. It addresses and acknowledges the interaction between:

- Supply Parts Spares
- Training and Support
- Facilities and Installation
- Personnel Staffing Levels and Skills
- Reliability, Availability, Maintainability, System Safety
- Maintenance Support and Test Equipment
- Packaging Storage Shipping and Handling
- Information Products Documents

37.1 Maintenance

Maintenance analysis identifies, plans, and develops maintenance concepts and requirements with the best possible equipment and capability at the lowest possible life cycle cost (LCC). It establishes the maintenance concepts and requirements for the life of the system. This includes, levels of repair, repair times, testability requirements, support equipment, training, staffing levels, skills, facilities, mix of repair responsibility, site activation, preventive maintenance, sustainment, software support, etc. Maintenance planning significantly impacts other logistics elements.

37.2 Personnel

Personnel analysis identifies, plans, and hires internal and external personnel to the organization, with the skills needed to operate, maintain, and support the system over its lifetime. This includes identification and recruitment of personnel

37.3 Supply support

Supply support analysis identifies, plans, and buys the best repair parts, spares, equipment, capability, and all supplies at the lowest possible LCC. It establishes management actions, procedures, and techniques needed to buy, catalog, receive, store, transfer, issue and dispose of spares, repair parts, and supplies. This means having the right spares, repair parts, and all classes of supplies available, in the right quantities, at the right place, at the right time, at the right price. The process includes provisioning for initial support, as well as acquiring, distributing, and replenishing inventories.

37.4 Support Equipment

Support equipment analysis identifies, plans, and implements' actions needed to purchase and support equipment needed to support the operation and maintenance of the system. This includes ground handling and maintenance equipment, trucks, air conditioners, generators, tools, calibration equipment, and test equipment.

37.5 Information Products

Information product analysis identifies, plans, and implements' actions to develop and buy information to operate, maintain, and train personnel on the equipment. This also may include the original engineering data to allow for future system growth. This is information, regardless of form or character such as equipment technical manuals and engineering drawings, engineering data, specifications, standards, reports, etc. Technical manuals and engineering drawings are the most expensive and probably the most important data acquisitions made in support of a system. The information also should include training and diagnostic fault isolation procedures.

37.6 Training and Training Support

Training and training support analysis identifies, plans, and implements' a cohesive integrated strategy to train operational, maintenance, and management personnel throughout the life cycle of the system. This includes identifying, developing, and purchasing Training Aids, Devices, and Simulators. It includes policy, processes, procedures, techniques, and provisioning for the training facilities including equipment used to train personnel to acquire, operate, maintain, and support a system. This includes new equipment, institutional, sustainment, crew, unit, collective, and maintenance training using, formal, informal, on the job training (OJT), and sustainment proficiency training methods.

37.7 Computer Resources Support

Computer resource support analysis identifies, plans, and purchases facilities, hardware, software, documentation, and personnel needed to support computer hardware and software in the system. The technical data or information products come in many forms such as, specifications, design artifacts, Computer Software Configuration Item (CSCI) definitions, test descriptions, operating environments, user and maintenance manuals, and computer code. The approach maybe to use the developer to initially maintain the system. As the system ages the support may be offered to a support company or internal resources may be used to maintain the system. These strategies should be described in the computer resource support analysis with transition time frames and issues clearly described.

37.8 Facilities

Facilities analysis identifies, plans, and purchases the best facilities for operations, training, maintenance and storage of the operational system and the logistic support system at the lowest LCC. It consists of analysis and recommendations for permanent and semi-permanent facilities, types of facilities, facility improvements, location, space needs, environmental and security requirements, and equipment. Security, physical access, power and communications are some of the issues that should be addressed.

37.9 Delivery

Delivery analysis identifies, plans, and purchases packaging, preservation, handling, storage and transportation needs for the system. This includes the resources, processes, procedures, design, considerations, and methods to ensure that all system, equipment, and support items are preserved, packaged, handled, and transported properly, including environmental considerations, equipment preservation for short and long term storage, and transportability. Some items may need special environmentally controlled containers or shock isolated containers for transport to and from repair and storage facilities via various modes of transportation (land, rail, air, and sea).

37.10 Design Influence

ILS starts at the beginning of the project and has influence on the system throughout its development. ILS participates in the systems engineering process and facilitates the ILS elements. ILS related design influence includes the following:

- System safety
- Human factors
- Corrosion
- Transportability
- Energy management
- Nondestructive inspection
- Survivability and vulnerability
- Environmental quality factors such as assessment of air, water, and noise pollution.
- Reliability, availability, maintainability
- Standardization and interoperability
- Human machine interface usability
- Hazardous material management

37.11 Exercises

1. Identify the key ILS elements of a consumer product like a computer and an automobile. Are there differences? How are the different ILS elements affected?
2. Identify the key ILS elements of a wind power farm, a solar power plant, a geothermal plant, and a nuclear power station. Are there differences? How are the different ILS elements affected?

37.12 Additional Reading

1. Integrated Logistics Support Guide, Defense Systems Management College, May 1986.
2. Logistic Support Analysis, DOD, MIL-STD-1388-1 October 1973, MIL-STD-1388-1A April 1983, MIL-STD-1388-2A July 1984, MIL-STD-1388-12B March 1991.
3. System Engineering Management, Military Standard, MIL-STD-499 17 July 1969, MIL-STD-499A1 May 1974.
4. System Software Development, Military Standard Defense, DOD-STD-2167 4 June 1985, DOD-STD-2167A 29 February 1988.
5. Systems Engineering Management Guide; Defense Systems Management College, January 1990.

38 Installation Deco mmissioning Disposal

Installation is a very good time in the system life cycle. Many challenges were overcome and the stakeholders decided to press the button and establish a new system or system element.

Decommissioning is a sad time because the stakeholders have decided the current system has outlived its useful life. It also can be a day of celebration if the old system has been of great service and or if it is being replaced by a new system.

38.1 Installation

Installation and validation is like all systems practices. It is methodical, rigorous, and based on logic and reason. A team does not just wander into a site and say, "we're here where de we set up our system".

Installation and validation begins with site surveys to gather information about the physical location. It continues with training to introduce the users, maintainers, administrators, and managers to the system. It then can proceed with installation and a series of steps that take the system into full-scale operation. This can be captured in a site installation and test procedures plan. The plan elements include:

- Identification of physical locations and roll out sequence
- Site surveys of each location
- Training needs and training rollout sequence

The test procedure elements include:

- **Pre-shakedown Tests**: This is the first step and these tests ensure full installation, check all system elements, ensure proper placement, and make sure the installation meets all safety needs. This can include mechanical and electrical measurements and alignment and adjustments.
- **Shakedown Tests**: This is the next step and these tests ensure that the system meets the performance requirements in the installed environment. Marginal parts and material a replaced.
- **Limited Operational Tests**: These tests are started after everyone agrees the system is correctly installed. These tests are operationally oriented but are performed in a way so that there is minimal disruption to the existing operations. They demonstrate that the system is performing as expected.
- **Onsite Interoperability Tests**: These tests demonstrate that the system

interfaces and operates with live external systems as expected. Lab oriented interoperability tests were performed prior to the onsite interoperability tests and all issues were addressed. There is nothing worse than being surprised while onsite.

- **Performance Tests**: These tests demonstrate that the system can handle the full peak operational load once commissioned. This test may not be possible until the system is permitted to be fully used and provide its services. In this instance the previous system remains in place as a backup just incase problems are detected.
- **Full Scale Operations**: The system is allowed to support full-scale operations. If there is a previous system that is being replaced, the previous system is kept in a backup mode just in case it needs to be activated. Once the new system has proven itself over extended operations including several peak load operations, the old system can be removed.

Obviously for a physical structure like a building or a bridge the movement from the Pre-Shakedown tests where everything is physically inspected to full-scale operations is very fast. For a computer-based system the movement is very slow and can take weeks or months.

38.2 Validation Operational Suitability Commissioning

Validation, operational suitability, and certification is accomplished by following a process that builds confidence in the users and other stakeholders that the system does offer proper services and is operationally acceptable. Examples of this process are found in this text in the Test Validation section and the previous discussion on installation. The elements of the process are:

- Minimize risk by taking small steps by picking one or more key sites, then at the key sites take small steps
- Progressively add more responsibility to the new system in a controlled manner with a fall back position
- Once the system is operational let it work for a period of time so it can prove itself, if available keep the previous system in place as backup
- Commissioning or certification is a stamp of approval given by a body charged with ensuring the system behaves as expected, they provide the approval based on their level of confidence

Obviously the stakeholders will decide the best approach to validation, operational suitability, and commissioning or certification.

38.3 Shutdown Decommissioning Disposal

Shutdown, decommissioning, and disposal is the end of a system. One would think that all systems could grow, be injected with new technology, and keep on getting better. However there are times when systems are allowed to die either through neglect or conscious thought and justification - and that is as it should be. Some systems are just bad and need to go. There are three possible shutdown levels and one must be selected.

- Shutdown with transfer to another system owner at another location[334]
- Shutdown with the remote possibility of resurrection in a distant future
- Shutdown with no prospect of transfer or future resurrection

38.3.1 Shutdown Criteria

Shutdown is the process of stopping system operations. In a company setting, shutdown might be associated with a product, product line, division, group, or the entire company. The criteria for shutting down a company are relatively easy to determine since they are based on financials.

Revenue (R) and Total Cost (TC) are the primary financial criteria used to determine if there should be a shutdown. If the R < TC then there is a strong argument for shutdown. The TC can be decomposed into Fixed Cost (FC) and Variable Cost (VC). If R > VC then the company (system) is covering its VC and since FC will exist regardless of shutdown status, it just makes sense to keep the operation up and running until either the FC is minimized or the R can cover both the VC + FC. So shutdown is driven by the impact of shutdown costs, which includes the FC.

For a system that is not based on free markets, shutdown is more problematic. This is especially true for monopolies and government systems. There is an old saying that it is impossible to remove an existing entrenched bureaucracy or system. There are a few approaches to force the shutdown:

- Reorganize it out of existence such that duplicate operations are merged and optimized[335].
- Legislate it out of existence. For example enforcing existing laws such as antitrust, restraint of free trade and other laws can enable the breakup of an

---

[334] Punching Out: One Year in a Closing Auto Plant, Paul Clemens, January 2011, ISBN 0385521154.

[335] The U.S. Department of Transportation (DOT) established in 1967 with responsibilities for air and surface transport subsumed several existing organizations such as, Federal Aviation Administration, Federal Highway Administration, Federal Railroad Administration, Coast Guard, and Saint Lawrence Seaway Commission.

extremely strong monopoly (system)[336].

- Allow it to be consumed by its own extreme philosophies. This is pushing an organization to its limits so that it gets whatever it wants until it self-destructs in its ignorance. This is extremely difficult to watch as significant damage is usually unleashed and the financial or other collapse is catastrophic with many innocents harmed[337].
- Cut off the source of funding. This is impossible in a monopoly situation and extremely difficult in a government situation.

38.3.2 Shutdown Preparation

Once a shutdown is finally accepted the preparations need to be made. The preparation is different for each level of shutdown. The items include:

- Identify key personnel
- Inventory all physical items
- Gather and organize all existing policies and procedures
- Gather and organize all information products such as manuals, plans, specifications, design information
- Archive all computer system files using disk imaging techniques
- Pack all physical items and mark for shipment and storage
- Close all contract commitments
- Terminate staff

If the system is being transferred to another system owner at another location there are additional considerations:

- Identify key personnel that will go to the new location either on a full time or part time assignment
- Capture all intellectual material, this may include video recording existing operations
- Invite future personnel to the new location to get some on-the-job-training (OJT) and experience

Many times especially in the case of government systems there is a desire to shutdown and store the system. It is not uncommon for the government to resurrect

[336] In the case of Hughes Aircraft tax laws forced its eventual breakup and demise.

[337] Enron is offered as a possible example. This is a classic theme through out literature. Little or no oversight and unaccountability that leads to obscene, revolting, uncontrolled behavior that eventually results in complete and total collapse.

a system several years after it has been mothballed[338]. In this setting, quality assurance, configuration management, training, maintenance, and logistics play a significant role in making sure the entire system is captured, packed, shipped, and properly stored.

38.3.3 Decommissioning

A system is decommissioned prior to its final shutdown. Its services are stopped in a controlled manner so that there are no serious unintended consequences. Decommissioning is the term used to represent the removal of a license or approval authority from a system so that it must be shutdown. It cannot provide services. This can be a planned event as part of a controlled shutdown mothball retirement disposal process or an unplanned event such as a safety finding. In a planned decommissioning, there is usually a ceremony acknowledging the system and its service.

38.3.4 Disposal

Disposal is the final phase of a decommissioning shutdown. The process includes:

- Identifying hazardous material and its proper disposal
- Placing the physical site in a condition where it can be safely used by others
- Transferring items to museums, educational institutions, and libraries
- Marking items for sale or auction and facilitating the transfer of ownership
- Identifying items for recycling and properly offering them to fully vetted and approved recycling centers
- Identifying remaining items for disposal and properly disposing of them at fully vetted and approved disposal facilities
- Closing out final financial records and transferring funds to final recipients

38.4 Exercises

1. Using your move to campus housing identify what elements of installation, validation, and commissioning did you encounter?
2. Using your move away from campus housing identify what elements of shutdown, decommissioning, and disposal did you encounter?
3. You have landed your first job and one-year into the adventure you are charged with installing a new system that you have been developing. You have moved across the country a few times and you have a personal feel for installation. You develop an installation plan following the principles in this section and your personal experience. Your manager yells at you and says

[338] Poplar slang term used to represent safe long-term storage.

there is no time or money for any of this silliness. What do you do?

38.5 Additional Reading

1. Installation and Acceptance Test Plan (IATP), DOD Data Item Description, DI-QCIC-80154A, January 1994.
2. Installation Test Procedures, DOD Data Item Description, DI-QCIC-80511, January 1988.
3. Site Preparation Requirements and Installation Plan, FAA Data Item Description, FAA-SI-004.

39 Quality

There are four broad areas to consider when addressing quality from a systems perspective. They are quality perceptions, quality assurance, quality control, and quality improvement. What may not be apparent is that making the fundamental decision to be systems driven will force many of the elements associated with quality to be practiced. So there is a direct relationship between using systems practices and movement into high quality.

39.1 Quality Perceptions and Attributes

From a systems practices point of view quality is about doing everything well. The implication of that statement is that function and performance is pushed to the limits in a balanced reasonable solution.

Quality perception is best illustrated by examples. In these examples it becomes apparent that there is some connection between quality perceptions and the Measure of Effectiveness (MOE) described in 28.14.

Example 60 Automobile Quality

An automobile can accelerate form 0-60 miles per hour in 6 seconds. However it may idle roughly, not start in all temperatures, poorly handle turns and road bumps, consume large amounts of oil and gas, and have the paint fade in 1 year. Alternatively another automobile also may go from 0-60 miles per hour in 6 seconds, run smoothly, start all the time, handle really well, consume very small amounts of gas, have a paint job that allows you to see your eyelashes 10 years later, and have a great sound system. Obviously the second automobile is higher quality[339].

As we look deeper into this comparison we clearly see the second automobile has a higher MOE. So there is a relationship between MOE and quality. However quality includes more attributes[340] than the MOE criteria in an architecture study.

---

[339] This was the difference between a muscle car and a high-end luxury car circa 1969. Many luxury cars were very fast with standard positraction rear ends or front wheel drive and large block engines with massive carburetors. Rather than sit and burn rubber they would chirp and just fly like bats out of hell (e.g. Riviera, Toronado, Eldorado, Imperial, etc).

[340] The 1960's were noted for publications that always compared meaningful attributes of various products. In many cases the consumer was educated on the engineering specifications or the science prior to the comparison. In recent years critical assessment as been overcome by hype and uneducated consumers are unable to easily make informed choices.

These attributes become apparent as different solutions are compared. These comparisons should include all the stakeholders including the consumers of the systems[341].

Example 61 Audio Sound System Quality

It is easier to compare the quality of two approaches that use the same basic architectures (e.g. comparing gas engine automobiles) than two approaches with different architectures. (e.g. gas versus electric automobiles). What if we tried to compare a reel-to-reel tape recorder with a cassette tape recorder, CD player, or MP3 recorder player? The problem becomes even more interesting if the stakeholders are different such as an audio aficionado and a casual music listener.

The reel-to-reel tape recorder may have spectacular sound quality with frequency response, harmonic distortion, and signal to noise ratios that bring tears of joy to an audio aficionado. It also uses storage media that has been shown to last a lifetime.

Meanwhile an MP3 player is not even measured for frequency response, harmonic distortion, or a new distortion associated with digitization called quantization distortion because the numbers are so poor. Music is lost as files disappear and music sound quality changes as files of different compression and sampling ratios are recorded.

The casual music listener will prefer the MP3 player to the reel-to-reel because of portability. For the music aficionado the reel-to-reel is higher quality compared to the MP3 player. For the casual listener the MP3 player is higher quality than the reel-to-reel until of course they hear a reel-to-reel sound system play their favorite songs. Then they start their quest for higher quality sound. They become educated.

This is an interesting scenario because this can result in a downward trend of quality for an entire class of systems. The MP3 player audio quality was compromised for cost and portability. However as time marches on, the reel-to-reel systems disappear from the market because there is more money to be made with MP3 players. A new generation is never exposed to what once existed; they do not know any different. So there is no desire to improve the audio quality of the MP3 player. It is good enough; not realizing what is being missed and what was lost. The end result is devolution in quality.

Quality attributes are sometimes referred to as the "ilities". They are non-functional and are very similar to tradeoff criteria when selecting architecture alternatives. They are:

- Accessibility, Autonomy, Accountability, Credibility
- Accuracy, Precision, Correctness, Integrity

---

[341] Sample publications are / were Consumer Reports, Stereo Review first published in 1958 by Ziff-Davis, Popular Mechanics first published 1902.

- Adaptability, Customizability, Composability, Configurability, Tailorability
- Availability, Dependability, Reliability, Durability
- Deployability, Distributability, Discoverability
- Efficiency, Effectiveness, Affordability
- Elegance, Simplicity, Demonstrability
- Failure Transparency, Fault Tolerance, Resilience, Recoverability, Orthogonality, Degradability
- Fidelity, Responsiveness, Seamlessness
- Growth, Flexibility, Modifiability, Scalability, Upgradability, Evolvability, Extensibility
- Interchangeability, Compatibility, Reusability, Interoperability, Ubiquity
- Maintainability, Modularity, Installability, Serviceability, Debugability
- Manageability, Operability, Learnability, Administrability, Auditability, Understandability, Usability
- Mobility, Portability
- Predictability, Repeatability, Reproducibility, Determinability
- Relevance, Timeliness
- Robustness, Stability, Survivability, Safety, Securability
- Self Sufficiency, Self Sustainability, Sustainability
- Standards Compliance, Testability, Provability, Traceability

39.2 Quality Assurance - Design in Quality

Quality Assurance is the prevention of defects. Quality cannot be inspected or tested into a product or system. Quality needs to be designed and built into the product to minimize defects. The ideal goal is to have zero defects at the end of design and implementation. Evidence of high quality includes:

- Small tolerances and extreme accuracy such as narrow compartment lid seams, perfect spacing between adjoining areas, straight lines, no burs or surfaces that can cut
- Sensitivity to maintainability such as color coded or stripped wires, hoses, pipes, modularity, ease of access
- HMI that is easy to use, complete, consistent, properly uses colors, text size, field of view, controls, alerts, indicators, tactile feedback
- Ergonomics that uses proper sizes, weights, reach, and adjustments
- Ability to not only handle design peak load but above design peak load with no loss of function, performance, or damage to system
- Excellent finishes and coatings with uniform application, no drips, thin areas, blotches, changes in reflection, immune to scratching and rub off
- Ability to do everything well without degradation, attention to ALL the details no matter how far hidden from users

39.2.1 Quality Assurance - Design

The first step to take when designing in quality is to realize that a quality stakeholder should be involved at the earliest stages of a project. They should be part of the architecture team and follow the project as the system unfolds. That quality stakeholder needs to pull in other quality and quality related stakeholders as needed. The quality information, which will influence the architecture and system evolution, is as follows:

- Identify disasters from similar systems
- Identify catastrophic failures from similar systems
- Echo the concerns of the stakeholders
- Identify classes of problems that may be encountered
- List the top 10 problems from similar systems
- Identify the potential showstoppers in this system

Every system can be killed. These are the system showstoppers. Use the history of previous systems to identify the current system showstoppers. The system approach should mitigate these quality issues. There should be documented evidence of this mitigation, especially the showstoppers. At some point systems practices such as Failure Mode Equipment Analysis (FMEA) are used as part of an integrated quality management system[342] for quality assurance.

Some designs are inherently more prone towards higher quality. For example some approaches have fewer parts, interfaces, and process steps. Elegantly simple designs and low complexity lead to inherently higher quality approaches. When a tradeoff is performed, use the quality-criterion as a tradeoff criterion and consider the past history and the design complexity when rating the approach.

In many ways adopting the systems approach for the project will significantly enhance the quality. Many elements associated with quality are normal systems practices:

- Understand past quality problems - history is critical
- Do FMEA early and regularly update - start simple then expand
- Quality function deployment (QFD)[343] - design the right system
- Use multi-functional teamwork - basic to systems engineering
- Begin with the concept or architecture - start as early as possible
- Simplify the design - less interface, parts, complexity

---

[342] ISO 9000 an example of standards for a quality management system.

[343] System engineering is about capturing the stakeholder needs so that the correct system can be developed. QFD and capturing the voice of the customer is a restatement of basic System Engineering.

- Use high quality parts - pay me now pay me later
- Design with fewest parts - less things to break
- Use highest quality automated tools - intuitively obvious
- Raise and resolve issues early - so others learn from you
- Identify critical dimensions, tolerances, and precision parts
- Reuse proven designs - learn from others
- Document thoroughly and completely - no finger pointing or blame games
- When designing always address quality - make statements

39.2.2 Quality Assurance - Implementation

Quality assurance practices during implementation overlap the quality assurance practices during design. The concepts are the same but the environment might be the manufacturing floor, the computer software development environment, or job sites for buildings, roads, bridges, etc. They are:

- Understand past quality problems
- Continue the FMEA
- Use multi-functional teamwork
- Simplify the process
- Use high quality parts
- Use highest quality automated tools
- Raise and resolve issues early
- Screen critical parts
- Reuse proven processes
- Document thoroughly and completely

39.3 Quality Control

Quality control emphasizes testing of products to uncover defects. In a system intensive setting quality personnel do not produce test procedures and execute tests. Instead they witness the testing and stamp off the test documentation to indicate that the procedure was executed as written or as modified by relines developed and documented on the physical test procedure.

In many small commercial organizations quality is given the task of testing a product. There may or may not have been systems analysts involved in conceiving, documenting, and specifying the product requirements. So the quality representative may actually need to develop tests without any requirement documentation. In this setting there is no traceability. This is a challenge and it is not representative of a systems practice driven organization. However it may make perfect sense and the products may be wonderful but not necessarily important. For example no ones life is in danger, no industry is dependent on the products performance, no significant financial investment is involved.

In more significant commercial and industrial companies, and government settings, test is a complex subject. It requires specialists able to understand the requirements and develop tests that fully verify that the system behaves as specified in the requirement specifications. Quality representatives are used to ensure that the tests are executed as written with no deviations unless approved by the test director and clearly documented on the spot on the actual test procedure information product.

As tests are executed, defects may surface. The defects are recorded in a defect tracking mechanism usually embedded within the change request mechanism for the project. Quality attends the meetings associated with defect tracking and eventually witnesses the tests associated with the corrective actions.

The concept of monitoring and control is broadened even further in the significant organization. All critical processes are monitored and controlled by quality. For example they monitor and signoff on all deliverable documents. In this case they review documents and other information products for defects such as inconsistencies, omissions, and content. This is typically performed via spot checking which challenges the other reviewers (authors of the information product) who are not part of quality. If there is a trend suggesting reviews are not being adequately performed, quality can withhold approval of the document and prevent delivery.

Whenever a defect is detected as much information about the defect should be gathered and documented. This should include an initial assessment of the defect[344].

Table 70 Priorities Used For Classifying Defects

Priority	Applies if a problem could
1 Grave	a. Prevent accomplishment of an operational or mission essential capability b. Jeopardize safety, security, or other requirement designated critical c. Result in loss of life or health
2 Catastrophic	a. Adversely affect the accomplishment of an operational or mission essential capability and no work-around solution is known b. Adversely affect technical, cost, or schedule risks to the project or to life cycle support of the system, and no work-around solution is known
3 Critical	a. Adversely affect the accomplishment of an operational or mission essential capability but a work-around solution is known b. Adversely affect technical, cost, or schedule risks to the project or to life cycle support of the system, but a work-around solution is known
4 Marginal	a. Result in user operator inconvenience or annoyance but does not affect a required operational or mission essential capability b. Result in inconvenience or annoyance for development or support personnel, but does not prevent the accomplishment of those responsibilities
5 Negligible	Any other effect

Author Comment: There is a severe quality downward spiral that easily surfaces with defect tracking. The danger is that the team finds defects and fixes

344 MIL-STD-498 Appendix C Category And Priority Classifications For Problem Reporting. There is an overlap with the FMEA categories found in MIL-STD-1629A.

them in isolation. There is no desire to spot trends or worse no desire to fix a collection of defects even if the defects are located in the same area. In these settings it is not uncommon to see teams touch a single element such as a software file dozens of times rather than touch it once and fix all the defects at once. The excuses are usually cost and schedule. Meanwhile most of the cost is in the touching of the element not in the individual fixes. An approach to recover from this downward quality spiral is to create a single defect whose sole purpose is to improve quality and let the personnel work in a level of effort mode to increase the quality. There are dozens, hundreds, or even thousands of elements that can not be micro-managed, but a focused team could address. This is easy for software and problematic when physical elements are involved.

39.4 Organizational Alternatives

There are different organizational alternatives for Quality. Depending on the organizational structure there are different checks and balances and organization efficiencies. Ideally Quality would never be compromised regardless of organization, however as systems practitioners we know organization behavior is a function of organizational structure. This is not any different than any system we might conceive. The organization alternatives are:

- Quality reports directly to the CEO
- Quality reports to Engineering
- Embed within the Test Organization and reports to CEO
- Embed within the Test Organization and reports to Engineering

39.5 Quality Improvement

There is an axiom that states you cannot inspect in quality. Its corollary is that you cannot fix in quality. However, there is an approach to systematically improve the quality of a system. The approach is to build upon and expand the Failure Reporting, Analysis and Corrective Action System (FRACAS) that is part of the reliability program. So the FRACAS not only tracks failure, incidents such as safety, but also elements related to poor quality. The sources of the poor quality reports include:

- Field feedback from user, customers, maintainers, managers
- Internal development and support staff
- Marketing and sales staff
- Independent analysis of FRACAS reliability and incident reports

Quality staff uses this database and engages in the same process as described in the FRACAS reliability section of this text. Their goals are to improve quality rather than reliability. Reliability may be related to quality, but it is not the whole set that represents quality.

Typically high quality products, systems, services do everything well. Poor quality is usually perceived when one aspect is done well while the other aspects are ignored, usually because of cost, and so they are of poor quality. For example an automobile can have great acceleration, but corner poorly, stop poorly, be uncomfortable, etc. However the acceleration is great. This is opposed to another automobile, which executes everything very well and matches or exceeds the acceleration of the fast but poor quality automobile. So high quality encompasses functionality, performance, and perception.

Poor quality also can manifest from disrespect of the system users or customers. This is usually present in a monopoly, oligopoly, or unregulated utility situation. Money is not even the issue; it is just a poor attitude on the part of the managers of the system or product developers. Improving quality in this setting is not possible without using anti trust laws or severe regulation and penalties.

39.6 Root Cause Analysis

Root cause analysis is based on the concept of causality or cause and effect. A result is the final detected consequence of a chain of events. Root cause analysis attempts to find the first cause in the cause and effect chain. The reality is that it is usually difficult to pin down one cause that leads to a problem. Usually multiple root causes surface.

The best way to view root cause analysis is as a backward chain of events analysis. For the observable event a cause is postulated. That cause is then treated as an event so that its cause can be postulated.

Obviously finding the root cause is secondary to preventing the same problem from resurfacing. The following steps are suggested:

- Define the problem
- Gather data and evidence
- Create a cause and effect chain
- Identify solutions that address the root cause
- Implement the recommendations
- Observe the fix for effectiveness
- Investigate to see if this root cause applies elsewhere
- Share the findings (e.g. Internet web page)

Example 62 Root Cause Analysis - Emissions Failure

Assume an automobile fails its annual emissions inspection. The cause and effect chain can be represented as: A <= B <= C <= D

A: Car fails emission inspection
B: Electric air pump has failed
C: No airflow clogs the air injection valves
D: Car not maintained

In this case the root cause analysis points to a car not being properly maintained. However if we dig deeper into the analysis of why the electronic air pump failed we see: A <= B <= C <= D

A: Car fails emission inspection
B: Electric air pump has failed
C: Water sucked into air pump
D: Poor air pump intake design

In this case we see that the design is at fault. The air pump is sucking road water during normal driving conditions. This is almost impossible to detect because water has evaporated long before the pump is removed and examined. It is only after the pump is disassembled that it is detected that the pump is frozen because the bearing is covered with calcium deposits. When the pump installation is physically inspected it is clear the air intake can easily suck water in heavy rain and minor flood conditions.

So in this case the root cause analysis points to two separate root causes. The first is poor design and the second is poor maintenance, which leads to air injection valve failure[345]. When the new air pump and air injection valves are installed the air pump intake should be moved to prevent future water contamination.

There are different broad lists of root causes depending on industry. The following are some common root causes.

- **The 8 Ms used in manufacturing:** Machine or technology, Method or process, Material, Manpower, Measurement, Milieu or Environment, Management, Maintenance
- **The 8 Ps used in service industry:** Product or Service, Price, Place, Promotion, People, Process, Physical Evidence, Productivity and Quality
- **The 4 Ss used in service industry:** Surroundings, Suppliers, Systems, Skills
- **The Ishikawa Diagram:** is a fishbone diagram[346] that suggests the root causes can be categorized into: Equipment and Materials, Process and Environment, People and Management

Author Comment: All problems are traceable to Bad Process. However all processes can be subverted. This is the dilemma of humanity and soft elements surface such as personal integrity, confidence, and motivation that influence quality.

---

[345] This is an actual root cause analysis of the authors' daughters 1997 automobile.
[346] Kaoru Ishikawa proposed Ishikawa diagrams in the 1960s.

In the absence of the soft elements the second root cause is Management. It is management's responsibility to make sure the processes work, the technology and tools are current, and the financial box defined by cost and schedule is sufficient for a successful high quality project.

39.7 Parts Tracking

When part tracking is mentioned most people think in terms of initial manufacturing of a device. The parts tracking mechanism is established to meet some level of quality in the manufactured item. A system may start out as a high quality system, however as time progresses its quality may start to drop. This is usually because of increasing failures due to age, wear, or abuse. There is also the possibility that functionality may disappear because of neglected maintenance and decreased performance because of inferior parts being used in maintenance. A parts tracking mechanism is an approach to minimize quality degradation from the introduction of poor quality parts.

39.8 Exercises

1. Identify a high quality product. Why is it a high quality product?
2. Identify a low quality product. Why is it a low quality product?
3. Is there any difference between quality attributes and tradeoff criteria? If yes state why. If no state why? Use examples in your rational.
4. What can you do if an organization is caught in a quality downward spiral?
5. What are some of the root causes for a quality downward spiral in an organization?
6. What is the quality impact when an organizations market is disappearing?
7. Is there a relationship between quality and technology maturity?

39.9 Additional Reading

1. Quality Assurance Terms And Definitions, DOD, MIL-STD–109C 2 September 1994, MIL-STD-109B 4 April 1969.
2. Quality Program Requirements, DOD, MIL–Q-9858A 16 December 1963, MIL–Q-9858 April 1959.
3. System Software Quality Program, Military Standard Defense, DOD-STD-2168, 29 April 1988.
4. Systems Engineering Fundamentals, Supplementary Text, Defense Acquisition University Press, January 2001.
5. Systems Engineering Management Guide, Defense Systems Management College, January 1990.

40 Planning and Tracking Progress

There are different levels of planning and tracking progress. In all cases the planning begins with the simplest level of establishing goals and progresses to more rigorous levels of planning and tracking process. Many times people will attempt to immediately jump to the most rigorous complicated planning. This will always lead to problems. In many ways this is like establishing different views of an architecture or a system. Unless these views are developed unexpected emergence will surface. The suggested planning path is from simplest to most complex is:

- Establish the goals
- Describe the process and methodology
- Develop a schedule
- Develop a PERT chart[347]
- Develop a time line
- Establish a WBS[348]
- Establish EVMS[349]

Typically there is a natural evolution of a project or program from inception to full commitment. As that evolution unfolds the various planning and tracking mechanisms naturally unfold.

40.1 Goals Basic Concepts

List seven goals plus or minus two with expected due dates. This allows everyone to understand the effort. How the team gets there is a different question. The goal driven plan is captured on one sheet of paper or presentation slide. A goal driven plan is extremely effective and should never be bypassed.

- Goal 1 month 03
- Goal 2 month 09
- Goal n month 15

---

[347] PERT - Project / Program Evaluation Review Technique

[348] WBS - Work Breakdown Structure

[349] EVMS - Earned Value Management System

40.2 Process and Methodology Overview

The process and methodology identify the steps that will be followed to execute the activity. It also lists the tools to be used for each step. A block diagram is developed showing each step as a block. The blocks are connected together in sequence from left to right top to bottom with key feedback loops shown using a different line type. Text is then offered describing each block and its expected tools.

The process and methodology should be less than five pages, something that is easily read and understood in a single sitting. The process and methodology should trace to the goals. This eventually is used to help develop the WBS.

40.3 Schedule Gantt Chart

A schedule takes the blocks in the process and methodology overview, decomposes the blocks into lower level tasks, adds the start and end dates, and allocates resources to each task. When automated tools are used skill levels with costs are also assigned that can be used to track progress as part of EVMS. The schedule also shows major milestones[350]. The major milestones are usually listed at the top of the schedule. A schedule also can be developed using the following steps:

- Develop a list of tasks
- Sequence the list of tasks
- Determine the dependencies between the tasks
- Identify the duration of each task
- Note the resulting project start and complete dates (duration)

Tasks	Resp	Jan	Feb	Mar	Apr	May	Jun	Jul
Milestones			M	M			M	
Gather Data	Sys	SSSSSSSSSSNNNNNNNNNNNNNNNN						
- site surveys	walt	CCCCCCC						
- literature search	claudia		SSSSNNN					
- industry survey	cass		NNNNNNNNNNNNNN					
- vendor search	lizzy			NNNNNNNNNNNNNN				
		S start, C complete, O not started, M milestone						

Figure 169 Schedule Gantt Chart

A schedule should never be considered complete until the goals are identified and a process and methodology is developed. Many feel that the schedule should be developed from the WBS. However developing a WBS before understanding the goals, process, and methodology will lead to missing elements in the WBS. So the

---

[350] A milestone is definable achievement in a project or program.

suggestion is to develop the schedule after the process and methodology overview. It should trace to process and methodology, and the goals. The schedule should then be modified based on the Project Program Evaluation and Review Technique (PERT) chart analysis and the WBS.

A schedule can be layered and decomposed just like a functional decomposition. So it can be shown at different levels depending on the communications and analysis needs. It is captured with a stack of paper using a traceable numbering scheme.

40.4 PERT Chart

The PERT[351] chart is used to show the sequence of tasks to complete a project or program. Most activities have dependencies. The PERT allows the planner to identify the dependencies and link them using a graphical representation. When the ideal PERT is developed it represents the minimum amount of time to complete a project, even though the start and end times of the individual elements may be unknown. It is an event driven view of the activity.

Initially the start and end dates may be unknown. However as the dates become available they are added to the PERT chart. This usually happens after the initial schedule is developed. The PERT chart is used to verify and update the schedule dependencies identified when the Gantt chart is developed. The task dependencies can be shown in a network diagram called a PERT chart or represented in a modified Gantt chart that shows dependencies between the tasks. Usually both visualizations are developed.

The ideal PERT may not be possible. Usually access to resources such as equipment, people, tools, etc is limited and things that in an ideal world can be planned in parallel actually must be planned sequentially (in series) based on the real world dependencies.

Some planers prefer to first build a PERT chart before the Gantt chart. This choice is a function of the complexity of the project or program. For large efforts with many complex dependencies and many possible parallel activities, it may make sense to start with the PERT then move to the Gantt chart. The PERT process includes several common terms:

- **PERT Event**: Marks the start or completion of one or more activities. It consumes no time and uses no resources. All the activities leading to that event must complete.

---

[351] Volume 4 Chapter 4, Contract Pricing Reference Guides, a set of five reference volumes developed jointly by the U.S. Federal Acquisition Institute (FAI) and the U.S. Air Force Institute of Technology (AFIT). Maintained by the Office of the U.S. Deputy Director of Defense Procurement and Acquisition Policy for Cost, Pricing, and Finance (DP/CPF) they provide instruction and guidance for contracting personnel.

- **Predecessor Event**: Immediately precedes some other event without any other events intervening. An event can have multiple predecessor events and can be the predecessor of multiple events.
- **Successor Event**: Immediately follows another event without any other intervening events. An event can have multiple successor events and can be the successor of multiple events.
- **PERT Activity**: A task that consumes time and requires resources such as labor, materials, space, machinery, tools, etc. It is the time effort and resources needed to move from one event to another event. A PERT activity cannot be performed until the predecessor event has completed.
- **Optimistic Time (O)**: Minimum possible time to accomplish a task, everything proceeds better than normally expected.
- **Pessimistic Time (P)**: Maximum time to accomplish a task that, everything goes wrong.
- **Most Likely Time (M)**: Best estimate of time to accomplish a task, everything proceeds as normal.
- **Activity Time (AT)**: Best estimate of time to accomplish a task using the following equation: $AT = (O + 4M + P) \div 6$
- **Float or Slack**: Maximum time that a task can be delayed without causing a delay. Free float is a delay in subsequent tasks. Total float is a delay in project completion. See slack.
- **Critical Path**: Longest continuous time path from an initial to a terminal event that will delay the project or program. Each project or program may have multiple critical paths.
- **Critical Activity**: An activity that cannot be delayed without delaying other activities. It has no float or slack time. A zero float activity may or may not be on the critical path.
- **Lead Time**: Time when a predecessor event must complete so that there is sufficient time for a specific successor event to complete.
- **Lag Time**: the earliest time a successor event can follow a specific predecessor event.
- **Earliest Start Time (TE)**: This is the earliest starting time for an event. Beginning with the starting event, work across the PERT network and determine how long it will take to complete the project. The times developed by working from the beginning to end are the Early Start Times or TE.
- **Latest Start Time (TL)**: This is the latest starting time an event can start and still complete the project on time. The TL is calculated the same way as TE except the calculation is done from the end of the project back to the beginning.
- **Slack or Float**: Amount of time a task can be delayed without causing a

delay in the project or program. Slack = TE-TL. Fast tracking is performing more critical activities in parallel. See float.

- **Crashing Critical Path**: Shortening duration of critical activities by adding more resources to shorten the duration or schedule. There are reasons the duration was originally set, (available personnel, skill levels, learning curves, natural time, etc) and those reasons must be properly addressed. **Author Comment**: You cannot give birth to a nine-month baby in one month by adding more people.
- **Fast Tracking**: Taking previously serial tasks and converting them to parallel tasks. There is a reason the original tasks were serial (such as equipment, people, tools, etc) and those reasons must be properly addressed.

There are two graphic approaches used to represent a PERT chart: (1) Arrow Diagramming Method (ADM) or Activity-on-Arrow (AOA) and (2) The Precedence Diagram Method (PDM).

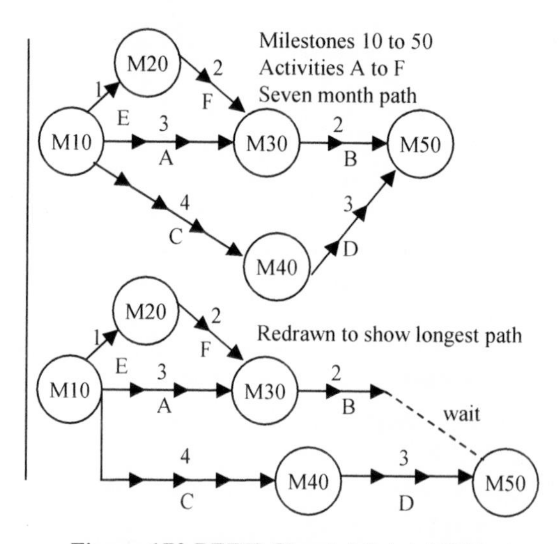

Figure 170 PERT Chart AOA / ADM

The AOA (also known as ADM) shows activities as lines with arrowheads. The length of the line represents the duration. The lines terminate at the major milestones of the project.

The PDM uses nodes to represent activities and connects them with lines in a graphic that shows sequence. The nodes include the following information: early start date, late start date, slack time, early finish date, late finish date, duration, critical non-critical path identification and WBS reference. A critical path is the sequence of project network activities, which add up to the longest overall duration[352].

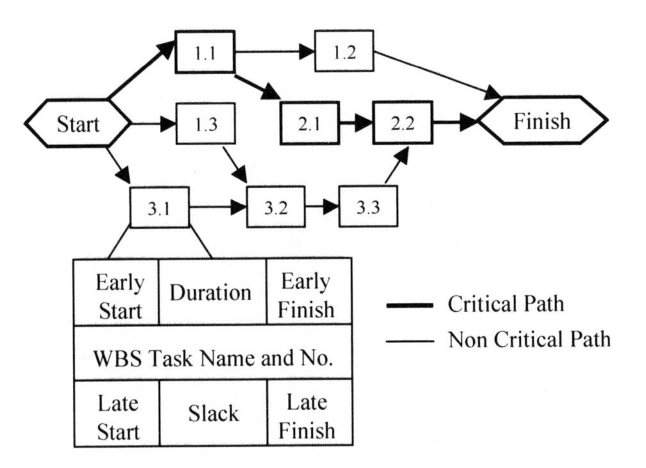

Figure 171 PERT Chart PDM

A PERT analysis can be layered and decomposed just like a functional decomposition. So it can be shown at different levels depending on the communications and analysis needs. It is captured with a stack of paper using a traceable numbering scheme.

40.5 Work Breakdown Structure

A Work Breakdown Structure (WBS) groups project or program work elements to organize and define the total work. Many people erroneously assume that the WBS is another view of the architecture or its functionality. Instead the WBS should be viewed as buckets of money. When a particular activity completes the associated bucket of money is empty. If there is surplus, it gets redistributed to other WBS elements. If there is a shortfall and the bucket will empty before the activity successfully completes, money must be found.

Author Comment: If a WBS element runs out of money or hours and the activity stops, then there is a high probability the work was arbitrarily stopped rather than reaching its natural conclusion. In a proper activity there is always additional money or hours left in a WBS element that is then redistributed when the activity properly completes.

---

[352] Critical Path is the longest path of planned activities from the start to the end of a project or program. Critical Path Method (CPM) surfaced with the Manhattan Project. Management of the Hanford Engineer Works in World War II, How the Corps, DuPont and the Metallurgical Laboratory fast tracked the original plutonium works, Harry Thayer, ASCE Press, pp. 66-67, ISBN 0784401608.

The WBS is not static and can change as the project unfolds. Some changes are trivial with little impact across the WBS. Other changes are the result of project problems and corrective actions must be taken thus significantly impacting the WBS.

A WBS is a tree where each descending[353] level of the WBS is an increased level of detail. The leaves of the tree represent specific work items while the aggregation box (parent) represents the sum of the lower level work items (children). Time and cost is tracked by using a unique charge number for each WBS lowest level (leaf) element. The top level WBS element represents the total sum of all the work.

Once a WBS is developed it does have some look and feel that is similar to an organizational chart. However just as in the case of architecture and functionality it is distinct and separate from the organization chart. Keeping the concept of collections of money in mind and when the money should be released will yield the most effective WBS.

Author Comment: Some suggest that some projects do not have a WBS. However that is not the case. If money and or work products are being transferred then someone has either a graphical view or an indentured list of the work being performed. It just may not be visible to the participants.

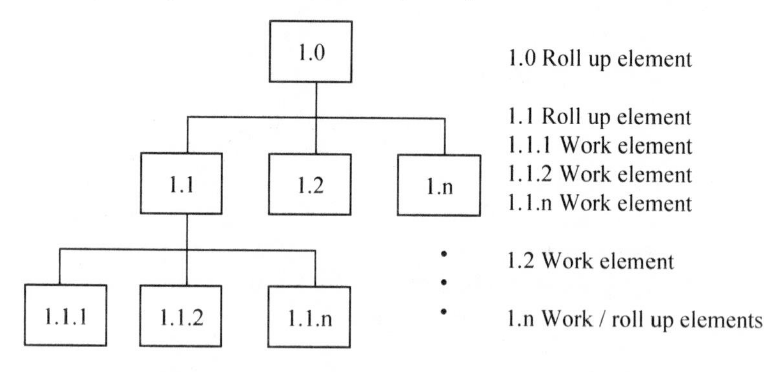

Figure 172 Generic Work Breakdown Structure

The steps for developing the WBS are broadly associated with deliverables, progress payment, level of detail, and change. The steps and mechanics of developing the WBS are as follows:

- Identify what must be delivered for a successful project.
- Identify the major deliverables. As major deliverables are competed, typically payment is made. These form the progress payments to keep the project funded and move to the next level of work.

[353] Decomposition once again is a key approach to a practice.

- Add additional levels to a WBS to satisfy the customers and projects management and control needs. Each project is different. A WBS can be started with a template but it must be updated to satisfy the customer and project staff.
- Review and refine the WBS until all stakeholders agree. There are planning and reporting elements that surface with a WBS. Just as in the case of micro management a WBS can become too detailed and impede progress.
- If new lower levels are needed for a WBS element, the entire WBS probably will be impacted. The same lower level elements probably will be needed across the WBS, not just the selected elements. This is a restructuring of the WBS and a re-plan where cost tracking and reporting need to be modified.
- If the WBS level of detail is too low, the lower levels can be rolled up to the higher level. In this case there is little impact to the WBS. No restructuring of the WBS or re-planning is needed.

Example 63 Work Breakdown Structures

Some projects have work that is performed in parallel. Other projects have work that is performed in serial. One work activity cannot start without the successful completion of another work activity.

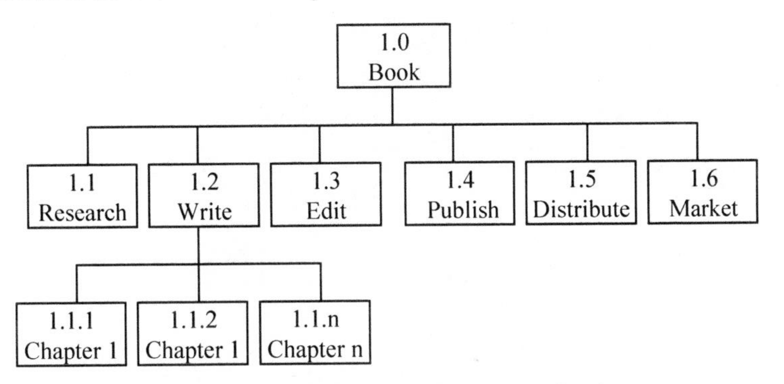

Figure 173 WBS - Writing a Book

Notice that in both cases it is not possible to show the full WBS in graphical format. The approach is to use the same technique that is used to decompose data flow diagrams.

Take an element that needs further decomposition, mark it on the parent diagram as being shown on a child diagram, and decompose it on a separate view. Continue the decomposition until the WBS is fully developed. So the WBS is not a single sheet of paper but a stack of papers.

A WBS dictionary describes the effort needed to accomplish the work and the expected work products. It is written at the leaf level (lowest level) of the WBS. The WBS dictionary is typically used to start the development of a statement of work (SOW).

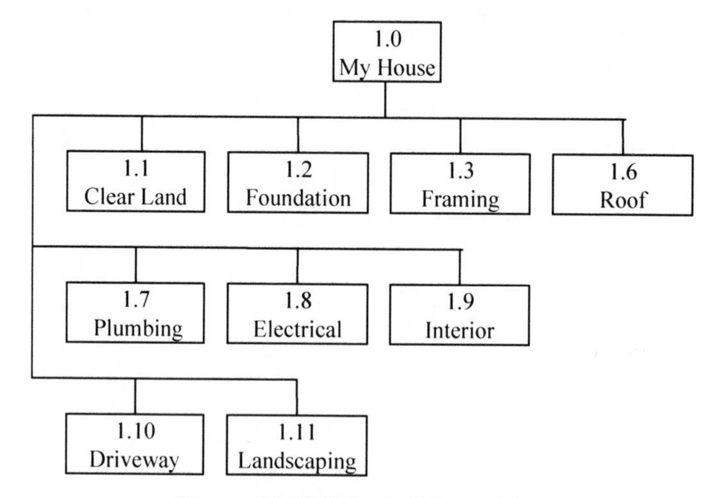

Figure 174 WBS - Building a House

The WBS is executed by an organization. The organization houses the authority, responsibility[354], and resources needed to complete the WBS project. An organization chart is developed to show the structure of the organization that will execute the WBS. It is a tree structure and can be represented as a graphical figure or an indentured list.

Table 71 Responsibility Assignment Matrix

Organization	**Resp**	**1.1 Test Plans**			**1.2 Test Procedures**			
		1.1.1	**1.1.2**	**1.1.3**	**1.2.1**	**1.2.2**	**1.2.3**	**1.2.4**
Systems	Walt	X			X			X
RMA	Bill							
Software	Al, Tom		X			X		X
Hardware	Carl, Fred							
Mechanical	Steve			X			X	X
Civil	Bo							
Test	John	X	X	X	X	X	X	X
ILS	Don							
Publications	Starkey	X						
Data Mgmt	Mary	X						
Config Mgmt	Barry	X						
Quality	Bob	X	X	X	X	X	X	X

---

[354] **Author Comment:** Responsibility should never be allocated to people without authority. Authority and responsibility must always be provided resources and a budget.

Links are made between the WBS and the organization chart. A responsibility assignment matrix (RAM) shows these links. The organization indentured list is placed on one axis of the table and the WBS elements are placed on the other axis (rows and columns). Typically the organization is shown in the rows and the columns have the WBS elements. The RAM shows the responsibility of each WBS element with the placement of an X at an intersecting cell. Alternatively each cell can show the planned hours or expected cost in place of each X.

When developing the very first instance of the WBS, previous work (projects, programs, etc) should be reviewed. There are always lessons to be learned from previous work and the architecture of the WBS should be treated with the same level of respect and analysis as the system architecture.

If a common WBS structure exists within the organization then metrics can be gathered and used to support cost estimating for future work. The metrics also can be used to tune and optimize the process. If the WBS is obfuscated and or unique to each project even though the projects have similar cost structures then the chaotic organization will not offer the most effective systems.

When responding to a customer they may offer a WBS as part of the request for proposal. This WBS may not fully match the WBS templates available in the organization. This should be clearly communicated and a common ground negotiated. If needed a translation table can be created so that the customer can have reporting based on their WBS and the organizations common WBS can be preserved for future cost estimates and process tuning. The organization should not sacrifice its future for the unique needs of a customer or group of customers.

The following is a list of common WBS elements that may apply to various systems. These elements start at the second level. The first level is the system to be developed by name.

Table 72 Common Level 2 and 3 WBS Elements

- Storage
    - Planning and Preparation
    - Storage
    - Transfer and Transportation
- Systems Engineering and Program Management
- System Test and Evaluation
    - Development Test and Evaluation
    - Operational Test and Evaluation
    - Mock-ups
    - Test and Evaluation Support
    - Test Facilities
- Training
    - Equipment
    - Services

Table 72 Common Level 2 and 3 WBS Elements

Level 2	Level 3
	Facilities
Data	
	Technical Publications
	Engineering Data
	Management Data
	Support Data
	Data Depository
Peculiar Support Equipment	
	Test and Measurement Equipment
	Support and Handling Equipment
Common Support Equipment	
	Test and Measurement Equipment
	Support and Handling Equipment
Operational/Site Activation	
	System Assembly, Installation and Checkout on Site
	Contractor Technical Support
	Site Construction
	Site/Ship/Vehicle Conversion
Industrial Facilities	
	Construction/Conversion/Expansion
	Equipment Acquisition or Modernization
	Maintenance (Industrial Facilities)
Initial Spares and Repair Parts	

The following is a list of unique WBS elements that may apply to a system. These elements start at the first level. The first level is the system to be developed by name.

Table 73 Unique Level 1 and 2 WBS Elements

Level 1	Level 2	Level 3
Vehicle System		
	Primary Vehicle	
		Frame
		Suspension and Steering
		Power Package and Drive Train
		Accessories
		Body
		Automatic and Remote Piloting
		Special Equipment
		Navigation
		Communications
		Integration, Assembly, Test and Checkout
	Secondary Vehicle	
		(Same as Primary Vehicle)
Ship System		

Table 73 Unique Level 1 and 2 WBS Elements

Level 1	Level 2	Level 3
	Ship	
		Hull Structure
		Propulsion Plant
		Electric Plant
		Sanitation
		Command and Surveillance
		Auxiliary Systems
		Outfit and Furnishings
		Restaurants, Shopping and Entertainment
		Integration Engineering
		Ship Assembly and Support Services
Electronic and or Automated Software System		
	Prime Mission Product (PMP)	
		Subsystem 1...n (Specify Names)
		PMP Applications Software
		PMP System Software
		Integration, Assembly, Test and Checkout
	Platform Integration	
Aircraft System		
	Air Vehicle (AV)	
		Airframe
		Propulsion
		AV Applications Software
		AV System Software
		Communications/Identification
		Navigation/Guidance
		Central Computer
		Fire Control
		Data Display and Controls
		Survivability
		Airline Operations Systems
		Automatic Flight Control
		Central Integrated Checkout
		Sanitation
		Entertainment
		Auxiliary Equipment
Space System		
	Launch Vehicle	
		Propulsion (Single Stage Only)
		Stage I
		Stage II...n (As Required)
		Strap-On Units (As Required)
		Shroud (Payload Fairing)
		Guidance and Control

Table 73 Unique Level 1 and 2 WBS Elements

Level 1	Level 2
	Integration, Assembly, Test and Checkout
Orbital Transfer Vehicle	
	Propulsion (Single Stage Only)
	Stage I
	Stage II...n (As Required)
	Strap-On Units (As Required)
	Guidance and Control
	Integration, Assembly, Test and Checkout
Space Vehicle	
	Spacecraft
	Payload I...n (As Required)
	Reentry Vehicle
	Orbit Injector/Dispenser
	Integration, Assembly, Test and Checkout
Ground Command, Control, Comm and Mission Equipment	
	Sensor I...n (As Required)
	Telemetry, Tracking and Control
	External Communications
	Data Processing Equipment
	Launch Equipment
	Auxiliary Equipment
Flight Support Operations and Services	
	Mate/Checkout/Launch
	Mission Control
	Tracking and C3
	Recovery Operations and Services
	Launch Site Maintenance/Refurbishment

40.6 Exercises

1. Identify goals for a potential project. Develop a process and methodology for this project. Include a block diagram of the process and methodology. Develop a schedule using a Gantt chart for this project. Develop a PERT for this project. Develop a WBS for this project. Identify how the goals, process, schedule, and PERT are impacted.
2. If you had to do it over again where would you start, the WBS, PERT, etc?
3. Create a graphical WBS using a tree block diagram.
4. Take the graphical tree block diagram WBS and convert it into an indentured list.
5. Create a WBS from the perspective of a system developer.
6. Create a WBS from the perspective of a customer.
7. Are there differences between the developer and customer WBS examples you created? If so why if not why not?
8. Are there differences and or changes that resulted from the conversion from

graphical to indentured list view? If so what and why, if not why not?

40.7 Additional Reading

1. Earned Value Management Implementation Guide, DOD, Defense Contract Management Agency, October 2006.
2. Earned Value Management System (EVMS), DOE G 413.3-10, U.S. Department of Energy, EVMS Gold Card May 06 2008.
3. Work Breakdown Structure Handbook, DOD, MIL-HDBK-881, 2 January 1998.
4. Work Breakdown Structures for Defense Materiel Items, DOD, MIL-STD-881 1 November 1968, MIL-STD-881A 25 April 1975, MIL-STD-881B 25 March 1993.

41 Earned Value Management System

Earned Value Management System (EVMS) is a tool that is used to track the progress of a project or a program. It was used as a financial analysis method on U.S. Government programs in the 1960's but its roots trace back to Frank Bunker Gilberth (1868-1924). In the 1970's and 80's the U.S. Government used Earned Value Management (EVM) to address cost and schedule risk on cost plus contracts. It was also used for new high technology systems during that time. In the 1990's EVM became mandatory for certain U.S. Government programs[355].

This section is a continuation of the section on Planning and Tracking Progress. The practices of developing schedule, PERT charts, and WBS information products needs to be understood before moving into EVMS. It is also a good idea to review the Basis of Estimate (BOE) section. The basic steps to EVM or an EVM System (EVMS) are:

- **Define the Work**: The approach for defining the work is to develop a WBS. A good WBS has mutually exclusive elements so that work is easily associated in only one work element. The most detailed elements of the WBS hierarchy or indentured list is called activities or tasks.
- **Assign Value**: Each activity is assigned a planned value (PV). The PV is derived from and tracks to the Basis of Estimates (BOEs). The PV can be in units of currency, labor, or both.
- **Assign Earning Rules**: Earning rules are assigned to each activity. There are different standard rules that are used in the EVMS plan and are covered in the cost account management section.
- **Execute the Project and Track Progress**: As the project is executed cost accounts are tracked and compared to the planned activities. The cost accounts are the WBS elements that have been assigned value and earning rules. The rules and status assessment let you take credit for work and fundamentally get paid.

EVM tracks the project on a cost schedule basis but does not measure the project

[355] Preparation and Submission of Budget Estimates OMB Circular: A-11. Earned Value Management, NASA Policy Directive 9501.3, August 2002. Mandatory Procedures for Major Defense Acquisition Programs and Major Automated Information System Acquisition Programs, DOD-5000.2R, April 2002. Program And Project Management For The Acquisition Of Capital Assets, U.S. Department Of Energy, DOE Order 413.3, July 2006.

quality. For example a project may be ahead of schedule under budget but the stakeholders may be unhappy and the system is considered a failure. It is never considered high quality and may never go operational.

EVM works with other tools to reduce the risk of project failure. These tools broadly fall into the category of systems practices. The systems practices coupled with domain knowledge yield other metrics that when used with EVM can lead to success. The non-EVM metric items are varied, broad, and unique to the work or domain. They are surfaced with the EVM planning effort and may be considered as part of the EVM System (EVMS).

Most organizations narrow the EVMS to only include EVM and keep the technical performance measurement (metrics) as a separate system or entirely ignore them and view the management problem as purely a financial exercise. This is a difficult situation and the approach for keeping them separate or merged is at the discretion of the organization and they can actually make effectively work.

Author Comment: Next time you build your custom dream house see if you can manage your project purely from a financial (cost schedule) point of view. Making domain experts subservient to financial managers will lead to the eventual collapse of any organization. Domain experts have vision to move into ever evolving futures and they understand what needs to be done at all times including what to do when unplanned and unforeseen events surface. Flipping the pyramid and making domain masters' report to people who do not understand their business and or technologies is like asking children to lead a family. It does not work.

41.1 EVM Plan and Budget

The EVM plan is usually captured in a tool that is able to offer an integrated schedule, PERT Charts, and WBS. The ultimate goal is to use the WBS to create work packages. The work packages include a description of the work, the products, start end dates, and a unique charge number that is used to track labor hours and or material costs. The control account level is usually one level above the work package / planing package level. So several work packages may share a common cost account charge number (code) used by staff to capture their work time. The practices used to capture the EVM plan are:

- Project schedule including dependencies using Gantt and PERT charts
- WBS and budget derived from the proposal BOEs

When developing the schedule and WBS some suggest the order of development is WBS, activity list (items with duration), sequence via PERT, schedule with dates, schedule with resources. Others suggest the WBS should be last. The reality is the process is iterative as each view impacts other views.

As part of EVM, activity sequencing is very important and is usually not addressed when EVM is not used. The sequencing establishes the activity dependencies. There are four types of schedule dependencies (activity A is before B in the time line):

- **Finish to Start (FS)**: Activity A must finish before activity B can start.
- **Start to Start (SS)**: Activity B can start as soon as activity A starts.
- **Finish to Finish (FF)**: Activity A cannot finish until activity B finishes.
- **Start to Finish (SF)**: Activity A cannot start before activity B finishes.

When developing the budget, ideally the proposal BOEs form the budget captured in the WBS. However after contract award the costs and WBS shift from the original proposal during the final negotiations. So there needs to be a traceable reconciliation between the original proposal BOEs, original WBS and the new WBS with the new work and the new assigned values.

The Contract Budget Base (CBB) is the total budget less fee for all authorized work. The performance measurement baseline (PMB) is the time-phased sum of all the allocated budgets. Management reserve is the amount of contract budget withheld by management. Undistributed budget is contractually authorized funds not yet allocated to WBS elements.

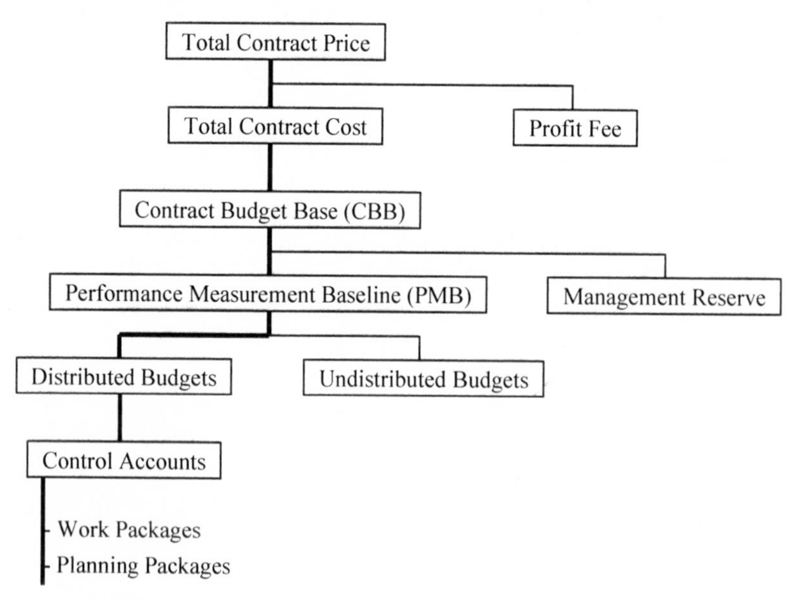

Figure 175 Major EVM Elements

Author Comment: When a project is executed cost accounts should rarely be emptied, especially in an engineering setting. The team should conclude that they have converged, there are no more issues, the quality is reasonable, and adding other perspectives have not changed these conclusions. If the cost account is emptied then there is the strong possibility that the team just stopped midstream. Surplus budget is re-distributed back into management reserve or other cost accounts. Consistent draining of cost accounts is a warning flag that insufficient funds have been provided to execute the project or program. This is a serious issue and more money must be provided to properly execute the project or the project must be reduced in size so that individual cost accounts contain sufficient funds to ethically[356] complete the tasks.

41.2 Planned Value (PV)

PV is the work scheduled in the plan. PV can be cumulative or current. Cumulative PV is the sum of the approved budget for work to be performed to date. Current PV is the approved budget for work scheduled to be performed during a given period (days, weeks, months, or years). The period is typically a month. The budget at complete is the total value of the work of the approved budget or the total PV. PV is the cost and schedule baseline also referred to as BCWS (Budgeted Cost for Work Scheduled).

	Jan	Feb	Mar	Apr	May
PV	X	X	X	X	X

Status Date = End of Mar

Cum PV = 3X **Cur PV = 1X** **BAC = 5X**

41.3 Actual Cost (AC)

AC is the money spent for executing the work. AC can be cumulative or current. Cumulative AC is the sum of the actual cost for work performed to date. Current AC is the actual cost during a given period (days, weeks, months, or years). The period is typically a month. AC is also referred to as ACWP (Actual Cost of Work Performed).

	Jan	Feb	Mar	Apr	May
PV	X	X	X	X	X
AC	Y	Y	Y		

Status Date = End of Mar

Cum PV = 3X Cur PV = 1X BAC = 5X

Cum AC = 3Y **Cur AC = 1Y**

[356] Running a sweat shop or hiding deficiencies is not only unethical it can lead to lawsuits and criminal charges of fraud and abuse.

41.4 Earned Value (EV)

EV is a claim of the value of the work done to date. EV can be cumulative or current. Cumulative EV is the sum of the claimed value of the work performed to date. Current EV is the claimed value of the work performed in a given period (days, weeks, months, or years). The period is typically a month. Cost Account Managers (CAM) using Earned Value Rules and their assessments claim the values. EV is also referred to as BCWP (Budgeted Cost for Work Performed).

	Jan	**Feb**	**Mar**	**Apr**	**May**
PV	X	X	X	X	X
AC	Y	Y	Y		
EV	Z	Z	Z		

Status Date = End of Mar

Cum PV = 3X	Cur PV = 1X	BAC = 5X
Cum AC = 3Y	Cur AC = 1Y	
Cum EV = 3Z	**Cur EV = 1Z**	

PV is the cost and schedule baseline. AC is the actual cost. EV identifies what the project has accomplished using EV rules and CAMs. Estimate at Complete (EAC) is the actual cost to date plus an objective estimate of the remaining authorized work. The EAC will be revisited as the EVM discourse continues.

EAC = Actual Cost (AC) + Estimate to Compete (ETC)

41.5 Earned Value Rules

Cost Account Management tracks the cost schedule progress of an activity in EVMS. As part of EVMS planning the criteria for determining progress is selected and used to track progress. The criteria or EVMS earning rules are:

- Fixed Formula
- Milestone Weights
- Milestone Weight with Percent Complete
- Units Complete
- Percent Complete
- Level of Effort

41.5.1 Fixed Formula

Fixed formula is used for accounting periods of less than three months. It applies a percent complete to the start and end of each activity. The percentages are typically:

- **0/100**: Nothing earned for start, 100% earned when complete
- **25/75**: 25% earned for start, 75% earned when complete
- **50/50**: 50% earned for start, 50% earned when complete

This approach is best for short-term activities and is easy and quick to status. It is not very effective for long-term high value packages.

41.5.2 Milestone Weights

Milestone Weights assigns value to each milestone. When a milestone is fully completed the budgeted value is earned. It is used for long duration activities and typically assigns a milestone for each month or accounting period. It is best if the milestones are tangibles and traceable such as site inspection, pour foundation, frame house, roof, electrical and plumbing, interior finishing, landscaping, cleanup, and final repair. The problem surfaces when intangibles define a milestone such as phase 1 design complete, phase 2 design complete, etc. Another disadvantage is the ability to take partial credit for work in progress. Its advantage is the forced surfacing of objective measurable milestones, which otherwise might not be recognized and identified.

41.5.3 Milestone Weight with Percent Complete

Milestone Weight with Percent Complete allows credit to be taken before a milestone is reached. It is used when there are long periods between milestones (no milestone falling in every monthly interval). The disadvantages of this approach are that the cost account manager (CAM) must assess the percent complete and document the assessment methodology for each milestone. Using a ratio of actual schedule date and planned milestone date is artificial and does not capture the actual status. For example the work may not have even started or the work might have hit a significant issue.

41.5.4 Units Complete

Units Complete uses a physical count to determine what is earned. This means that someone must determine a justified final count as part of the plan. This tends to be limited to a production setting. It has been used to track software development using countable elements such as lines of code, number of functions, number of function points, etc. It also has been used to track information products based on page count. Units counted include number of:

- Requirements
- Test cases
- Operational concepts
- Use Cases
- Operational Threads
- Functional flow diagrams
- Display screens
- Written pages
- Lines of code
- Diagrams
- Tables
- Figures
- Drawings
- Parts

41.5.5 Percent Complete

Percent Complete is a subjective assessment of the percent an activity has progressed or completed. The value is placed on the work activity. The logic for assessing each activities percent complete is documented. The disadvantage of this approach is the subjective nature of the assessment. The danger is that the assessment will be a ratio of the current date versus the complete date and an understanding of the actual work completed is unknown. The advantage of this approach is that detailed planning is not required.

41.5.6 Level of Effort

Level of Effort (LOE) is based on the passage of time. Value is earned and is always equal to the monthly planned event. So the present value (PV) is always equal to the earned value (EV). This method is used to track constant overhead or sustaining activities such as management, administration, secretarial support, project cost accounting, etc. The advantage of this approach is that no status is required. The disadvantage is the misuse of LOE as an earned value method. When misused EVMS is unable to accurately track progress.

41.6 Cost Account Management

So who should perform the cost accounting function? If the people doing the work, leading the work, or even strategizing the work are given the CAM status task, they are closest to the activity and are able to give the most reasonable assessment. However, this is a tedious time consuming activity taking away their valuable talents from the immediate work to be performed. Also there is the risk of a compromised assessment as the assessor assess their progress. It is not independent. Nor is there the ability to apply broad experience from many different CAM activities if it is given to the people performing and responsible for the work.

So this is a difficult problem. The best EVM program is sufficiently planned so that independent CAM staff can effectively assess progress with very limited interviews of the people performing, leading, and managing the work. They are able to access physical products and the information products, which are snapshots in time of the completed work, and assess the status of an entire project or program.

41.7 Variance Analysis

Variance analysis is used to determine what is happening on a project. The analysis uses the original plan and an assessment of where the project is by one or more CAM personnel to produce a series of variance metrics.

41.7.1 Cost Variance (CV)

CV is the difference between the earned value of the work performed and the actual cost. It identifies the earned value of the work performed for the dollars worth of work scheduled. If the result is positive the project us under run or under budget. It cost less to accomplish the work than was budgeted and planned. If the result is negative the project is over run or over budget. Cost variance percentage (CV%) identifies the percentage of cost variance from what has been earned. **CV = EV - AC, CV% = CV/EV, Positive = Under Run, Negative = Over Run**

41.7.2 Schedule Variance (SV)

SV is the difference between the earned value of work performed and the work scheduled. It is the value of work performed minus the value of the work scheduled. Schedule variance percentage (SV%) identifies the percentage change of the current schedule from the planned schedule. **SV = EV - PV, SV% = SV/PV, Positive = on schedule, ahead of schedule, Negative = behind schedule**

41.7.3 Cost Performance Index (CPI)

CPI is a measure of how cost efficiently the team is spending the money. It is a ratio of the value of the work performed versus the actual cost. If the CPI is less than 1.0 the cost is greater than budgeted or the team is less cost efficient. If the CPI is greater than 1.0, then the cost is less than budgeted or the team is more cost efficient. A CPI of 1.04 suggests getting $1.04 worth of work for each $1.00 spent. Typically the CPI variance in excess of plus or minus 7% to 10% needs a formal variance analysis report to explain what is happening on the project and why it is different from the plan. **CPI = EV/AC, < 1.0 less cost efficient, > 1.0 more cost efficient**

Table 74 CPI Drivers

Negative	Plus
• Work more complex than originally known • More design review comments and rework than originally planned • Unable to draw system boundary, poor requirements, increasing scope • Increased labor, material, and or overhead costs	• New processes and tools showing benefits • Shorter learning curves than anticipated • Team bonding and dynamics better than anticipated • Economies of scale within maturing organization surfacing

There are two approaches for understanding the CPI. The first is from a technical point of view and the second is from a fundamental point of view. A technician tracks the CPI and it associated numbers. There is no attempt to understand what may be behind the numbers. A fundamentalist attempts to describe the qualitative environment behind the numbers. The ideal approach is to view this from a technical and fundamental point of view and use root cause analysis to try to explain both the good and bad. This model is borrowed from stock investing strategies of technicians and fundamentalists[357].

41.7.4 Schedule Performance Index (SPI)

The SPI is a measure of how efficiently the team is executing the schedule. It is the ratio of the claimed value of the work performed versus the work scheduled. If the SPI is less than 1.0 the project is behind schedule, greater that 1.0 and the project is ahead of schedule. An SPI of 0.90 is stated as either 10% behind schedule or working at 90% efficiency. Typically the SPI variance in excess of plus or minus 7% to 10% needs a formal variance analysis report to explain what is happening on the project and why it is different from the plan. **SPI = EV/PV, < 1.0 behind schedule, > 1.0 ahead of schedule**

Table 75 SPI Drivers

Negative	Plus
• Same as CPI drivers • Personnel shortage • Revised execution plan • Supporting organizations behind schedule • Late vendor delivery • Late customer feedback direction	• Same as CPI drivers • More personnel able to effectively execute plan • Faster supporting organizations delivery • Faster vendor delivery • Faster customer feedback direction

41.8 Estimate at Completion (EAC)

The EAC is also referred to as the Latest Revised Estimate (LRE). It is an estimate of the cost to complete a cost account and a project (or program). So the EAC is developed for projects as well as Control Accounts and Work Packages.

The EAC is usually calculated on a monthly basis or when a significant change happens to the project. Typically the accounting department runs the tools that house the plan and gather EV status from the CAMs. The CAMs review the individual EACs, make corrections as needed, and then the program manager presents the total EAC to the appropriate stakeholders.

There are different methods for calculating the EAC. The selected method is based on the dollar value of the project, the risk, accounting system available and

[357] Fundamentals are things about a company, such as: what does it produce, where is it located, what is its plan, how are they growing, what are they investing in, how do they hire people, how big are they, etc. Technicians ignore all that information. Technical analysts only look at the stock price over a period of time. They do not care about company strategy, its management team etc. They just look at the stock price and use various mathematical formulas to guess where a stock price is headed and explain poor performance by using other numbers as supporting data.

the accuracy of the estimates. Each project needs to determine which EAC formula best fits the project's size and complexity.

- **EAC = Actual Cost (AC) + Estimate to Complete (ETC)**: This formula assumes that all remaining work is independent of the burn-rate incurred to date and will complete as planned. Use the AC for accounts that are complete and use the BAC for accounts that are still open and less than 50% along in the schedule.
- **EAC = Actual Cost (AC) + New Estimate to Complete**: This formula assumes there is new knowledge about the work to be completed. When using this formula expert opinion, project and industry trends, and other forms of objective analysis are provided. This formula needs new information not present in the plan. The information needs to be gathered by the CAM and vetted by the stakeholders.
- **EAC = [AC/EV x (work completed and in progress)]+ [cost of work yet to begin]**: This formula assumes that the new work not yet begun will be completed as planned. The work completed and in progress is the BAC for the accounts started and completed.
- **EAC = AC/EV x BAC**: This formula is the easiest to use, but it assumes that the "burn-rate" will be the same for the remainder of the project.
- **EAC = BAC/CPI**: This formula is also easy to use and assumes that the "burn-rate" remains constant for the remainder of the project. Note that 1/CPI = AC/EV.
- **EAC = AC + [1/CPI (BAC-EV)]**: This calculation uses the Actual Cost (AC) Budget at Completion (BAC), Earned Value (EV) and the Cost Performance (CPI) Index to calculate EAC.

Table 76 EAC Calculation Methods and VAC

Calculations	**EAC**	**VAC**
EAC = Actual Cost (AC) + Estimate to Complete (ETC)	$1,484,613	$4,590
EAC = Actual Cost (AC) + New Estimate to Complete	$1,490,121	-$918
EAC = [AC/EV x (work completed and in progress)]+ [cost of work yet to begin]	$1,478,628	$10,575
EAC = AC/EV x BAC	$1,455,555	$33,648
EAC = BAC/CPI	$1,455,555	$33,648
EAC = AC + [1/CPI (BAC-EV)]	$1,455,555	$33,648

Though all the EAC calculations are correct, choosing the appropriate formula for your project is important. Once selected it should be used consistently with all future reports so that the comparisons between each reporting period have the same ground rules. All these formulas use existing information in the EVMS database except for the formula that uses a new estimate to compete.

41.9 Variance at Complete (VAC)

The VAC is the budget at complete minus the estimate at complete. If the result is positive, the project is under budget or under run. If the result is negative, the project is over budget or over run. **VAC = BAC - EAC, Positive = Under Run, Negative = Over Run**

41.10 EVMS Reports and Status

There are many reports that can be provided from the EVM system. The following tables are suggested reports for viewing the data. The first report captures the EVMS plan and the status. The second reports shows the cost and schedule performance. The third report offers an estimate to complete based on the current EVMS data.

Table 77 EVMS Plan Status Example

as of 3/31	PV	EV	AC	BAC
Account 1	$138,546	$138,546	$142,650	$138,546
Account 2	$73,494	$73,494	$64,800	$73,494
Account 3	$78,732	$59,472	$56,250	$148,689
Account 4	$53,649	$26,829	$27,900	$107,298
Accounts 5-n	$0	$0	$0	$1,021,176
Project Total	$344,421	$298,341	$291,600	$1,489,203

Table 78 EVMS Variance Report

as of 3/31	CV	SV	CV%	SV%	CPI	SPI
Account 1	-$4,104	$0	-3.0%	0.0%	0.97	1.00
Account 2	$8,694	$0	11.8%	0.0%	1.13	1.00
Account 3	$3,222	-$19,260	5.4%	-24.5%	1.06	0.76
Account 4	-$1,071	-$26,820	-4.0%	-50.0%	0.96	0.50
Accounts 5-n	-	-	-	-	-	-
Project Total	$6,741	-$46,080	2.3%	-13.4%	1.02	0.87

Table 79 EVMS Estimate at Complete Report Example

as of 3/31	BAC	CPI	EAC	VAC
Account 1	$138,546	0.97	$142,650	-$4,104
Account 2	$73,494	1.13	$64,800	$8,694
Account 3	$148,689	1.06	$140,634	$8,055
Account 4	$107,298	0.96	$111,581	-$4,283
Accounts 5-n	$1,021,176	-	-	-
Project Total	$1,489,203	1.02	$1,455,555	$33,648

Table 79 EVMS Estimate at Complete Report Example

as of 3/31	BAC	CPI	EAC	VAC
Note: Used EAC = BAC/CPI				

EVMS like all models is complex and needs to be summarized for all the stakeholders. An approach is to produce a progress report[358] that traces to the EVMS report. The report includes:

- **Work Summary**: Describes the work performed during the reporting period. Includes positive and negative events.
- **Schedule**: Identifies if the program is ahead, behind, or on schedule and why. If behind schedule, include the approach for moving forward. This may mean the schedule must shift.
- **Studies**: Describes studies started, performed, and completed during reporting period and key findings.
- **Experimental Exploratory Work**: Describes experimental and exploratory work performed during reporting period and its impacts. This is work outside the original EVMS plan.
- **Design**: Descriptions and illustrations of designs, changes, and problems or issues encountered.
- **Test**: Describe dry run and or formal tests performed. Include major failures, their root causes, and planned fixes if known.
- **Issues**: describe difficulties or problems encountered which could negatively impact the effort. Offer proposed solutions to address the issues.
- **Plan Summary**: Summarize the plan steps followed to execute the tasks.
- **Completion Dates**: provide completion dates for each major task.
- **Percent Complete**: Provide percent complete for the tasks and the overall effort.
- **Additional Information**: Additional information, which may cause a change in the schedule or costs.

41.11 EVMS Parameters Summary

The following is a summary of the EVMS parameters. It is a quick reference card and includes additional EVMS parameters.

PERFORMANCE BASELINE COMPONENTS

PB = CBB +contingency + non-contract costs

[358] Program Progress Report, DOD Data Item Description, DI-MGMT-80555A, November 2006.

PB = performance baseline (TPC, total project cost)
Contingency = = held by customer & based on technical and programmatic risks
CBB = PMB + MR = contract budget base
PMB = CAs + UB + SLPPs = performance measurement baseline
MR = = management reserve (held by contractor based on contractor risks)
SLPP = = summary level planning package
AUW = = authorized unpriced work (contractually approved but not yet negotiated)
CA = WPs + PPs = control account (includes AUW, lowest WBS element assigned)
UB = = undistributed budget (activities not yet distributed to CA)
WP = = work package (near-term, detail-planned activities within a CA)
PP = = planning package (far-term activities within a CA)

EVMS BASIC COMPONENTS

AC = actual cost = ACWP = actual cost of work performed
EV = earned value = BCWP = budgeted cost for work performed
PV = planned value = BCWS = budgeted cost for work scheduled
BAC = cumulative PV = cumulative BCWS = budget at completion

VARIANCES

CV = EV - AC = BCWP - ACWP = cost variance
SV = EV - PV = BCWP - BCWS = schedule variance
CV % = (EV - AC)/EV = (BCWP - ACWP)/BCWP = cost variance, percentage
SV % = (EV - PV)/PV = (BCWP - BCWS)/BCWS = schedule variance, percentage
VAC = BAC - EAC = variance at completion

OVERALL STATUS

% scheduled = PVcum/BAC = BCWScum/BAC
% complete = EVcum/BAC = BCWPcum/BAC
% budget spent = ACcum/BAC = ACWPcum/BAC
WR = BAC - EVcum = BAC - BCWPcum = work remaining
BR = BAC - ACcum = BAC - ACWPcum = budget remaining

PERFORMANCE INDICES

CPI = EV/AC = BCWP/ACWP = cost performance index
SPI = EV/PV = BCWP/BCWS = schedule performance index
TCPIBAC = WR/BR = to complete performance index, BAC
TCPIEAC = WR/ETC = to complete performance index, EAC

COMPLETION ESTIMATES

EAC = BAC/CPIcum = estimate at completion, general
EACCPI = ACcum + WR/CPIcum = estimate at completion, CPI
EACcomposite = ACcum + WR/(CPIcum* SPIcum) = estimate at completion, composite
LRE = = latest revised estimate (contractor's, assessed monthly, annual bottoms-up)
ETC = EAC - ACcum = estimated to complete

Author Comment: EVMS is a good tool for reporting status but it cannot be used to control or manage a project. Everyone will disagree with the statement except for those who have seen massive programs and projects fail under EVMS. The reason for this unusual lack of symmetry is hysteresis. As a project starts to go off track the cost and schedule curves will remain relatively static. Once the cost and schedule curves show a problem the project is severely broken. Also when corrective actions are attempted the cost and schedule curves will continue to show a problem even though the fix actually may be working, again because of hysteresis.

If this delay relationship is recognized and EVMS is still used to manage and control a program rather that just track status, practitioners will fall into the trap of micro-management. They will laboriously identify fine tasks in hopes of getting it right. The reality is the fine tasks are always wrong as the project moves within the problem solving space. There are thousands of micro tasks that surface in the execution of any project, even small projects.

The reality is that domain experts using unique domain metrics manage and track a project path. These domain experts with their unique metrics have the least amount of hysteresis. They also have the luxury of being able to use any relevant metric item that makes sense and represents the current state of affairs.

A perfect example of this is a restaurant that decides to change the quality of the food. This transition can be a project with a cost and schedule. The deliverables can be tracked as numbers of meals produced per week or per month. When the food quality starts to change the customers immediately complain about the terrible food but the drop in revenue will take several weeks as the word gets out and the deliverables, meals in this case, drops because of rejection. EVMS will not show a problem for several weeks or even months. Different recipes may be tried with the poor quality food forcing the schedule to drag out. Once the word is out and the goodwill is lost, it may take several months to rebuild the restaurant business. This rebuild is not shown for months, but the customers will provide immediate feedback on the fantastic food and they are recognized as they return the following week. This is domain knowledge using domain metrics, practiced by people who love and understand the restaurant business.

So why should you use EVMS in the first place?

EVMS is used as a check against the domain experts just in case they are failing. This is a catastrophic failure once detected, because it should never reach the EVMS metrics. There should be sufficient checks and balances with the domain management area to ensure they are always successful. If EVMS shows a problem then the domain area needs serious corrective action in the form of changed process, assuming there are no personnel issues.

41.12 Exercises

1. List domain specific metrics that can be reported in parallel with EVMS status.
2. If you do not have an accounting department should you use EVMS?
3. If you do not have dedicated CAMS reporting to the program management office should you use EVMS?
4. Can a CAM survive long just doing CAM tasks? If not what other assignments can be given to a CAM to stimulate the desire to keep living and be excited about waking up every morning?
5. Is EVMS an example of an overly complex mechanism / model? If so how can it be simplified and yet preserved?

41.13 Further Reading

1. Earned Value Management Implementation Guide, DOD, Defense Contract Management Agency, October 2006.
2. Earned Value Management System (EVMS), DOE G 413.3-10, U.S. Department of Energy, EVMS Gold Card May 06 2008.

42 Major Milestones Reviews And Audits

A major milestone typically has several items being delivered. As part of this delivery there are formal review meetings or audits that are used to describe these items, the findings, and handoff the products. There are project and program specific milestones unique to each project or program and can include anything the project planners identify. Some examples of milestones are:

- **When building a house**: land preparation, foundation, framing, roof, plumbing, electricity, finishing, landscaping
- **When developing a new consumer product**: market research, product concepts, markets tests, initial market release, full scale production
- **When developing a senior system design project**: find project idea, gather data, identify architecture alternatives, develop model, test model, run model, select architecture

So are there generic milestones, reviews, and audits that planners might consider when planning an effort? One approach is to use a standard that tries to capture the essence of any project. The following is a list of possible generic milestones applicable to any project[359].

- System Requirements Review
- System Design Review
- Software Requirements Review if applicable
- Preliminary Design Review
- Critical Design Review
- Test Readiness Review
- Functional Configuration Audit
- Physical Configuration Audit
- Formal Qualification Review
- Installation and Checkout
- Operational Readiness Demonstration

These nouns and verbs tend to be self-explanatory, however there are further

---

[359] This section is selectively extracted from Technical Reviews And Audits Systems Equipments And Computer Software, MIL-STD-1521A and B June 1976 June 1995.

clarifications associated with these potential milestones up to the point of formal qualification review. In many ways the following review descriptions suggest the type of analysis or work to be performed in a systems effort. There is a pleural of nouns and verbs that might apply to your activity[360].

Author Comment: By reviewing this information the level of conciseness will significantly expand and that the team will detect and account for items that other teams will miss. The danger is that this knowledge will translate to cost, but the project risk will be significantly reduced. The only way to deal with this issue is to make sure the sponsor is fully educated so that when they perform their cost realism and vendor selection process, the answer is clear that the enlightened team should be selected. This education is significant and begins during the pre-proposal effort.

Critical thinking also needs to be applied to this information. Just because something is suggested does not mean it should be used for the project. This information[361] is basically a view written down by people who had just gone to the Moon. They did their best to capture their experience. The worst thing the U.S. may have done was to abandon various standards and guides that tried to capture this work as part of cost reduction initiatives. The reality is many do not even know where to even start without this information.

In general reviews and audits need to consider:

1. Meeting agenda plans
2. Conference rooms
3. System practices data, specifications, drawings, manuals, schedules, design, and test
4. Specialty study results
5. Trade study results
6. Risk analysis results
7. Prototypes, mockups, breadboards, in-process hardware
8. Test methods and data
9. Meeting minutes

42.1 System Requirements Review (SRR)

The SRR is held very early in the system development. It can happen at anytime but is normally held after the functional analysis and preliminary requirement allocation. It is used to determine initial direction and progress of the effort and its convergence on an optimum and complete system. The functional analysis and

---

[360] MIL-STD-490 also contains a large number of nouns and verbs that stimulate thinking that might apply to the activity. This is captured in the specification section of this text.

[361] MIL-STD-490, MIL-STD-499, MIL-STD-1521, etc.

allocation includes operations, maintenance, training, hardware, software, facilities, manufacturing, personnel, and human factors.

Since many efforts start with a customer that offers an A-Level specification, the A-level specification is fully vetted and accepted by both parties based on the analysis. This suggests that requirements have been added, deleted, and modified based on what has been learned to get to the initial architecture concept. A separate SRR can be held for each of the subsystems depending on the nature and complexity of the system.

42.2 System Design Review (SDR)

The essence of the SDR is to present the selected functional architecture approach. It is held to evaluate the optimization, traceability, correlation, completeness, and the risk of the allocated requirements, including the corresponding test requirements in meeting the system subsystem requirements, the functional baseline.

The review includes all the system requirements including operations, maintenance, test, training, hardware, computer software, facilities, personnel, and preliminary logistic support considerations. The review also includes the work to date including mission and requirements analysis, functional analysis, requirements allocation, manufacturing methods process selection, program risk analysis, system cost effectiveness analysis, logistics support analysis, trade studies, intra- and inter-system interface studies, integrated test planning, specialty discipline studies, and configuration management. A technical understanding is reached on the validity and the degree of completeness of the following information products:

- System Subsystem Specification(s)
- System Cost
- Preliminary Operational Concept Description(s)
- Preliminary Software Requirements Specification(s)
- Preliminary Interface Requirements Specification(s)
- If appropriate, Prime Item Development Specification(s)
- If appropriate, Critical Item Development Specification(s)

The SDR is the final review before moving to the preliminary design of the system or the detailed requirements analysis of software. The SDR reviews operational mission requirements, system subsystem specification requirements, allocated performance requirements, programming and manufacturing methods, processes planning, and ensures that the information products are necessary and sufficient. The SDR should:

1. Ensure that the System Subsystem Specification is adequate and cost

effective in satisfying validated mission requirements.

2. Ensure that the allocated requirements represent a complete and optimal synthesis of the system requirements.
3. Ensure that the technical program risks are identified, ranked, avoided, and reduced through:
   a. Tradeoffs particularly for sensitive mission requirements versus engineering realism and manufacturing feasibility to satisfy the anticipated production quantities of related performance requirements
   b. Subsystem component hardware proofing
   c. A responsive test program
   d. Implementation of comprehensive engineering disciplines such as worst case analysis, failure mode and effects analysis, maintainability analysis, producibility analysis and standardization
4. Identify how the final combination of operations, manufacturing, maintenance, logistics and test and activation requirements have affected overall program concepts, quantities and types of equipment, unit product cost, computer software, personnel, and facilities
5. Ensure that a technical understanding of requirements has been reached and technical direction is provided

42.3 SRR SDR Review Details

SRR	SDR
a. Mission and Requirements Analysis b. Functional Flow Analysis c. Preliminary Requirements Allocation d. System Cost Effectiveness Analysis e. Trade studies (addressing system functions in mission and support hardware firmware software). f. Architecture Synthesis g. Logistics Support Analysis h. Specialty Discipline Studies electrical, software, civil, mechanical, chemical, biological, environmental engineering, reliability, maintainability, safety, security, interoperability, compatibility, survivability, vulnerability, test, quality, energy management, environmental considerations, sustainability. i. System Interface Studies j. Generation of Specification k. Program Risk Analysis l. Integrated Test Planning m. Producibility Analysis Plans n. Technical Performance Measurement	(1) Systems Engineering: a. Mission and Requirements Analysis b. Functional Analysis c. Requirements Allocation d. System Cost Effectiveness e. Synthesis f. Survivability Vulnerability g. Reliability Maintainability Availability (RMA) h. Electromagnetic Compatibility i. Logistic Support Analysis, integrated logistics support including maintenance concept, support equipment concept, logistics support concept, maintenance, supply, software support facilities, etc. j. System Safety (emphasis placed on system hazard analysis and identification of safety test requirements) k. Security l. Human Factors m. Transportability including Packaging and Handling n. System Mass Properties o. Standardization

SRR	**SDR**
Planning o. Engineering Integration p. Data Management Plans q. Configuration Management Plans r. System Safety s. Human Factors Analysis t. Value Engineering Studies and Quality Impacts u. Life Cycle Cost Analysis v. Preliminary Manufacturing Plans w. Personnel Requirements Personnel Analysis x. Milestone Schedules Describe progress and problems in: (1) Risk identification and risk ranking (the interrelationship among system effectiveness analysis, technical performance measurement, intended manufacturing methods, and costs are discussed). (2) Risk avoidance reduction and control (the interrelationships with tradeoff studies, test planning, hardware proofing, and technical performance measurement are discussed). (3) Significant tradeoffs among stated system subsystem specification requirements constraints and resulting engineering design requirements constraints, manufacturing methods process constraints, and logistic cost of ownership requirements constraints and unit production cost design-to-cost objectives. (4) Identify computer resources of the system and partitioning the system into hardware and software. Tradeoff studies conducted to evaluate alternative approaches and methods for meeting operational needs and to determine the effects of constraints on the system. Evaluations of logistics, technology, cost, schedule, resource limitations, intelligence estimates, etc., made to determine their impact on the system. Specific tradeoffs related to computer resources are addressed: a. Candidate programming languages and computer architectures evaluated in light of requirements for approved higher	p. Electronic Warfare q. Value Engineering and Quality Impacts r. System Growth Capability s. Program Risk Analysis t. Technical Performance Measurement Planning u. Producibility Analysis and Manufacturing v. Life Cycle Cost Design to Cost Goals w. Quality Assurance Program x. Environmental Conditions, Shock, Vibration, Temperature, Humidity, etc. y. Training and Training Support z. Milestone Schedules aa. Software Development Procedures, Software Development Plan (SDP), Software Test Plan (STP), and other identified plans, etc. (2) Results of significant trade studies: a. Sensitivity of selected mission requirements versus realistic performance parameters and cost estimates b. Operations design versus maintenance design, including support equipment impacts c. System centralization versus decentralization d. Automated versus manual operation e. Reliability Maintainability Availability f. Commercially available items versus new developments g. Standard items versus new development h. Testability trade studies (Allocation of fault detection isolation capabilities between elements of built in test, on board on-site fault detection isolation subsystem, separate support equipment, and manual procedures) i. Size and weight j. Desired propagation characteristics versus reduction interference to other systems (optimum selection frequencies) k. Performance logistics trade studies l. Life cycle cost reduction for different computer programming languages m. Functional allocation between hardware, software, firmware and

SRR	SDR
order languages and standard instruction set architectures. b. Alternative approaches evaluated for implementing security requirements. If an approach has been selected, discuss how it is the most economical balance of elements, which meet the total system requirements. c. Alternative approaches identified for achieving the operational and support concepts, and for joint operations or opportunities for cross systems support. (5) Producibility and manufacturing considerations which could impact the program decision such as critical components, materials and processes, tooling and test equipment development, production testing methods, long lead items, and facilities personnel skills requirements. (6) Significant hazard consideration made to develop requirements and constraints to eliminate or control these system associated hazards.	personnel procedures n. Life Cycle Cost system performance trade studies to include sensitivity of performance parameters to cost o. Sensitivity of performance parameters versus cost p. Cost versus performance q. Design versus manufacturing consideration r. Make versus buy s. Software development schedule t. On-equipment versus off-equipment maintenance tasks, including support equipment impacts u. Common versus peculiar support equipment (3) Updated design requirements for operations maintenance functions and items. (4) Updated requirements for manufacturing methods and processes. (5) Updated operations maintenance requirements for facilities. (6) Updated requirements for operations maintenance personnel and training.

SRR and SDR
(7) Specific actions to be performed include evaluations of: a. System design feasibility and system cost effectiveness b. Capability of the selected configuration to meet requirements of the System Subsystem Specification c. Allocations of system requirements to subsystems configuration items d. Use of commercially available and standard parts e. Allocated inter and intra system interface requirements f. Size, weight, and configuration of hardware to permit economical and effective transportation, packaging, and handling consistent with applicable specifications and standards g. Specific design concepts which may require development toward advancing the state-of-the-art h. Specific subsystems components which may require "hardware proofing" and high-risk long-lead time items i. The ability of inventory items to meet overall system requirements and their compatibility with configuration item interfaces j. Planned system design in view of providing multi-mode functions k. Considerations given to: (1) Interference caused by the external environment to the system and the system to the external environment. (2) Allocated performance characteristics of all system transmitters and receivers to identify potential intra-system electromagnetic (EM) incompatibilities. (3) Non-design, spurious and harmonic system performance characteristics and their

SRR and SDR
effect on electromagnetic environments of operational deployments. l. Value Engineering studies, preliminary Value Engineering Change Proposals (VECPs) and impact on quality. (8) Review the Preliminary Operational Concept Document, System Subsystem Specification, Hardware Development Specifications, preliminary Software Requirements, and Interface Requirements Specifications for format, content, technical adequacy, completeness and traceability correlation to the validated mission support requirements. All entries marked "not applicable (N/A)" or "to be determined (TBD)" are identified and explained by the contractor. (9) Review test documents, including hardware subsystem and system test plans, to ensure that the proposed test program satisfies the test requirements of all applicable specifications. All entries labeled "not applicable (N/A)" or "to be determined (TBD)" in the test section of any applicable specification are identified and explained by the contractor. (10) Review the system, hardware, and software design for interaction with the natural environment. If any effect or interaction is not completely understood and further study is required, or it is known but not completely compensated for in the design, the proposed method of resolution shall also be reviewed. All proposed environmental tests are reviewed for compatibility with the specified natural environmental conditions. (11) Maintenance functions developed to determine that support concepts are valid, technically feasible, and understood. In particular, attention is given to: a. RMA considerations in the updated System Subsystem Specification b. Maintenance design characteristics of the system c. Corrective and preventive maintenance requirements d. Special equipment, tools, or material required e. Requirements or planning for automated maintenance analysis f. Item Maintenance Analysis compatibility with required maintenance program when weapon is deployed g. Specific configuration item support requirements h. Forms, procedures, and techniques for maintenance analysis i. Maintenance related tradeoff studies and findings (includes commercially available equipment, software fault diagnostic techniques) j. Logistic cost impacts k. Support procedures and tools for computer software which facilitate software modification, improvements, corrections and updates l. Hardness critical items processes m. Support equipment concept. (12) High-risk areas or design concepts requiring possible advances of the state-of-the-art. Prepared test programs and existing simulation test facilities are reviewed for sufficiency and compatibility. (13) The optimization, traceability, completeness, and risks associated with the allocation of technical requirements, and the adequacy of allocated system requirements as a basis for proceeding with the development of hardware and software configuration items. Include any available preliminary Software Requirements and Interface Requirements Specifications. (14) For manufacturing hardware only: a. The production feasibility and risk analyses addressed at the SRR are updated and expanded. This effort should review the progress made in reducing production risk and evaluate the risk remaining for consideration in the Full Scale Development Phase. Estimates of cost and schedule impacts shall be updated.

SRR and SDR
b. Review of the Production Capability Assessment includes: A review of production capability is accomplished which will constitute an assessment of the facilities, materials, methods, processes, equipment and skills necessary to perform the full scale development and production efforts. Identification of requirements to upgrade or develop manufacturing capabilities is made. Requirements for Manufacturing Technology programs are identified as an element of this production assessment. c. Present the management controls and the design manufacturing engineering approach to assure that the equipment is producible. d. Present a review of tradeoff studies for design requirements against the requirement for producibility, facilities, tooling, production test equipment, inspection, and capital equipment for intended production rates and volume. e. The analysis, assessments and tradeoff studies should recommend any additional special studies or development efforts as needed.

42.4 Software Specification Review (SSR)

The SSR is a formal review of software requirements in the Software Requirements Specification and the Interface Requirements Specification(s). Normally, it is held after System Design Review but prior to the start of software preliminary design. A collective SSR for a group of software configuration items, treating each configuration item individually, can be held. Its purpose is to establish the allocated baseline for preliminary software design by demonstrating the adequacy of the Software Requirements Specification (SRS), Interface Requirements Specification(s) (IRS), and Operational Concept Description (OCD). The following items are reviewed.

- Functional overview of the software, including inputs, processing, and outputs of each function
- Overall software performance requirements including execution time, storage requirements, and similar constraints
- Internal control and data flow between each of the software functions within the software
- External interface requirements between the internal software configuration items and external to the system
- Qualification requirements that identify applicable levels and methods of testing for the software requirements
- Any special delivery requirements
- Quality factor requirements including correctness, reliability, efficiency, integrity, usability, maintainability, testability, flexibility, portability, reusability, and interoperability
- Mission requirements of the system and its associated operational and support environments
- Functions and characteristics of the computer
- Milestone schedules

- Updates since the last review to previously delivered software related items including actions or procedures deviating from approved plans

42.5 Hardware Requirements Review (HRR)

The HRR[362] is a formal review of hardware requirements in the Hardware Requirements Specification and the Interface Control Document(s). In the past this was done as part of Preliminary Design Review. However with the introduction of high density programmable and application specific hardware devices[363] with resulting massive functionality, an HRR is suggested as part of risk reduction.

The HRR is held after System Design Review but prior to the start of hardware preliminary design. A collective HRR for a group of hardware configuration items, treating each configuration item individually can be held. Its purpose is to establish the allocated baseline for preliminary hardware design by demonstrating the adequacy of the Hardware Requirements Specification (HRS), Interface Control Document(s) (ICD), and Operational Concept Description (OCD).

42.6 Preliminary Design Review (PDR)

The PDR is a formal technical review of the basic design approach for a configuration item or for a related group of configuration items. It is held after the Hardware Development Specification(s), the Software Top Level Design Document (STLDD), the Software Test Plan (STP), the Hardware Test Plan, and preliminary versions of the Computer System Operator's Manual (CSOM), Software User's Manual (SUM), Computer System Diagnostic Manual (CSDM), and Computer Resources Integrated Support Document (CRISD) are available, but prior to the start of detailed design.

For each configuration item the review can be performed as a single event, or spread over several events, depending on the nature and the extent of the development of the configuration item and on provisions specified in the contract Statement of Work.

A collective PDR for a group of configuration items, treating each configuration item individually, can be held; such a collective PDR can be spread over several events, such as for a single configuration item.

The technical program risk associated with each configuration item is reviewed on a technical, cost, and schedule basis. For software, a technical understanding is reached on the validity and the degree of completeness of the STLDD, STP, and the preliminary versions of the CSOM, SUM, CSDM, and CRISD.

---

[362] MIL-STD-1521 is hardware oriented, a separate hardware requirements review is not identified. This is an example of starting with a base and doing what needs to be done.

[363] Field Programmable Gate Arrays (FPGA), Application-Specific Integrated Circuit (ASIC).

42.7 Critical Design Review (CDR)

The CDR is held for each configuration item prior to fabrication production coding release. It ensures that the detail design solutions, as reflected in the Draft Hardware Product Specifications, Software Detailed Design Documents (SDDD), Database Design Documents (DBDD), Interface Design Documents (IDD), and engineering drawings satisfy requirements established by the hardware Development Specification and Software Top Level Design Documents (STLDD).

CDR is held after the Computer Software Operator's Manuals (CSOM), Software User's Manuals (SUM), Computer System Diagnostic Manuals (CSDM), Software Programmer's Manuals (SPM), and Firmware Support Manuals (FSM) have been updated or newly released. For complex large configuration items the CDR may be conducted on an incremental basis where progressive reviews are conducted versus a single CDR.

The overall technical program risks associated with each configuration item are reviewed on a technical (design and manufacturing), cost and schedule basis. For software, a technical understanding is reached on the validity and the degree of completeness of the SDDD, IDD, DBDD, STD, CRISD, SPM, and FSM, and preliminary versions of the CSOM, SUM, and CSDM.

42.8 PDR CDR Review Details

42.8.1 Hardware

PDR Hardware	CDR Hardware
a. Preliminary design synthesis of the hardware Development Specification for the item being reviewed. b. Trade-studies and design studies results (see SDR for a representative listing). c. Functional flow, requirements allocation data, and schematic diagrams. d. Equipment layout drawings and preliminary drawings, including any proprietary or restricted design process components and information. e. Environment control and thermal design aspects f. Electromagnetic compatibility g. Power distribution and grounding design aspects h. Preliminary mechanical and packaging design of consoles, racks, drawers, printed circuit boards, connectors, etc. i. Safety engineering considerations	a. Adequacy of the detail design reflected in the draft hardware Product Specification in satisfying the requirements of the Hardware Configuration Items (HWCI) Development Specification for the item being reviewed. b. Detail engineering drawings for the Hardware including schematic diagrams. c. Adequacy of the detailed design in the following areas: (1) Electrical design (2) Mechanical design (3) Environmental control and thermal aspects (4) Electromagnetic compatibility (5) Power generation and grounding (6) Electrical and mechanical interface compatibility (7) Mass properties (8) Reliability Maintainability Availability (9) System Safety Engineering (10) Security Engineering (11) Survivability Vulnerability (including

PDR Hardware	CDR Hardware
j. Security engineering considerations k. Survivability Vulnerability (including nuclear) considerations l. Preliminary lists of materials, parts, and processes m. Pertinent reliability maintainability availability data n. Preliminary weight data o. Development test data p. Interface requirements contained in configuration item Development Specifications and interface control data (e.g., interface control drawings) derived from these requirements. q. Configuration item development schedule r. Mock-ups, models, breadboards, or prototype hardware when appropriate s. Producibility and Manufacturing Considerations (e.g., materials, tooling, test equipment, processes, facilities, skills, and inspection techniques). Identify single source, sole source, and diminishing source. t. Value Engineering Considerations and Value Engineering Change Proposals (VECPs) analysis including quality impacts. u. Transportability, packaging, and handling considerations v. Human Engineering and Biomedical considerations (including life support and Crew Station Requirements). w. Standardization considerations	nuclear) (12) Producibility and Manufacturing (13) Transportability, Packaging and handling (14) Human Engineering and Biomedical Requirements (including Life Support and Crew Station Requirements) (15) Standardization (16) Design versus Logistics Tradeoffs (17) Support equipment requirements d. Interface control drawings e. Mock-ups, breadboards, and or prototype hardware f. Design analysis and test data g. System Allocation Document for Hardware inclusion at each scheduled location. h. Initial Manufacturing Readiness (for example, manufacturing engineering, tooling demonstrations, development and proofing of new materials, processes, methods, tooling, test equipment, procedures, reduction of manufacturing risks to acceptable levels). i. Preliminary VECPs and or formal VECPs include quality impacts j. Life cycle costs k. Detail design information on all firmware to be provided with the system. l. Verify corrosion prevention control considerations to Ensure materials have been chosen that will be compatible with operating environment. m. Findings Status of Quality Assurance Program

PDR Hardware
x. Description and characteristics of commercially available equipment, including any optional capabilities such as special features, interface units, special instructions, controls, formats, etc., (include limitations of commercially available equipment such as failure to meet human engineering, safety, and maintainability requirements of the specification and identify deficiencies). y. Existing documentation (technical orders, commercial manuals, etc.,) for commercially available equipment and copies of contractor specifications used to procure equipment is made available for review. z. Firmware to be provided with the system: microprogram logic diagrams and reprogramming instruction translation algorithm descriptions, fabrication, packaging (integration technology gate density, device types such as CMOS, PMOS, ASIC, FLASH, FPGA), and special equipment and support software needed for developing, testing, and supporting the firmware and VHDL.

PDR Hardware
aa. Life Cycle Cost Analysis ab. Armament compatibility ac. Corrosion prevention control considerations ad. Findings Status of Quality Assurance Program ae. Support equipment requirements.

42.8.2 Software

PDR Software	CDR Software
a. Functional flow. The computer software functional flow embodying all of the requirements allocated from the Software Requirements Specifications and Interface Requirements Specifications to the individual Top-Level Software Units (TLSU) of the Software Configuration Items (CSCI). b. Storage allocation data. This information is presented for each Software item as a whole, describing the manner in which available storage is allocated to individual TL SW Units. Timing, sequencing requirements, and relevant equipment constraints used in determining the allocation are to be included. c. Control functions description. A description of the executive control and start recovery features of the Software including method of initiating system operation and features enabling recovery from system malfunction.	a. Software Detailed Design, Database Design, and Interface Design Documents. In cases where the CDR is conducted in increments, complete documents to support that increment shall be available. b. Supporting documentation describing results of analyses, testing, etc. as mutually agreed by the contracting agency and the contractor. c. System Allocation Document for Software inclusion at each scheduled location. d. Computer Resources Integrated Support Document. e. Software Programmer's Manual f. Firmware Support Manual g. Progress on activities required by Software PDR. h. Updated operation and support documents (CSOM, SUM, CSDM). i. Schedules for remaining milestones. j. Updates since the last review to all previously delivered software related documents.

PDR Software
d. Software structure. Describe the top-level structure of the Software, the reasons for choosing the components described, the development methodology to be used within the constraints of the available computer resources, and any support programs required to develop maintain the Software structure and allocation of data storage. e. Security. Identify unique security requirements and a description of the techniques to be used for implementing and maintaining security within the Software. f. Special software architecture needs. Identify any special software architecture needs such as re-entrancy requirements and a description of the techniques for implementing. g. Computer software development facilities. Identify the availability, adequacy, and planned use of computer software development facilities. h. Computer software development facility versus the operational system. Provide information relative to unique design features which may exist in a TLCSC to allow use within the computer software development facility, but which will not exist in the TLCSC installed in the operational system. Provide information on the design of support programs not explicitly required for the operational system but which will be generated

PDR Software
to assist in the development of the Software configuration items. Provide details of the Software Development Library controls. i. Development tools. Describe any special simulation, data reduction, or utility tools that are not delivered, but which are planned for use during software development. j. Test tools. Describe any special test systems, test data, data reduction tools, test computer software, or calibration and diagnostic software that are not deliverable, but which are planned for use during product development. k. Description and characteristics of commercially available computer resources, including any optional capabilities such as special features, interface units, special instructions, controls, formats, etc. Include limitations of commercially available equipment such as failure to meet human engineering, safety and maintainability requirements of the specification and identify deficiencies. l. Existing documentation (commercial manuals, etc.) for commercially available computer resources and copies of contractor specifications used to procure computer resources are made available for review. m. Support resources. Describe those resources necessary to support the software and firmware during operational deployment of the system, such as operational and support hardware and software, personnel, special skills, human factors, configuration management, test, and facilities space. n. Operation and support documents. The preliminary versions of the CSOM, SUM, CSDM, and CRISD are reviewed for technical content and compatibility with the top-level design documentation.

42.8.3 Support Equipment

PDR Support Equipment (SE)	CDR Support Equipment (SE)
a. Review considerations applicable to the Hardware and Software configuration items. b. Verify testability analysis results. For example, on repairable integrated circuit boards are test points available so that failure can be isolated to the lowest level of repair. c. Verify that externally furnished SE is planned to be used to the maximum extent possible. d. Review progress of long-lead time SE items, identified through interim release and SE Requirements Document (SERD) procedures.	a. Review requirements for SE. b. Verify maximum considerations externally provided SE c. Identify existing or potential SE provisioning problems d. Determine qualitative and quantitative adequacy of provisioning drawings and data e. Review reliability of SE f. Review logistic support requirements for SE items g. Review Calibration requirements h. Review documentation for SE.

PDR Support Equipment
e. Review progress toward determining total SE requirements for installation, checkout, and test support requirements. f. Review the reliability maintainability availability of support equipment items. g. Identify logistic support requirements for support equipment items and rationale for their selection. h. Review calibration requirements. i. Describe technical manuals and data availability for support equipment.

PDR Support Equipment
j. Verify compatibility of proposed support equipment with the system maintenance concept. k. If a Logistic Support Analysis (LSA) is not done, then review the results of SE trade-off studies for each alternative support concept. For existing SE and printed circuit boards testers, review Maintainability data resulting from the field use of the equipment. Review the cost difference between systems using single or multipurpose SE vs. proposed new SE. Examine technical feasibility in using existing, developmental, and proposed new SE. For mobile systems, review the mobility requirements of support equipment. l. Review the relationship of the computer resources in the system subsystem with those in Automatic Test Equipment (ATE). Relate this to the development of Built In Test Equipment (BITE) and try to reduce the need for complex supporting SE. m. Verify on-equipment versus off-equipment maintenance task trade study results, including support equipment impacts. n. Review updated list of required support equipment.

42.8.4 Engineering Data

PDR Engineering Data	CDR Engineering Data
Review Level 1 engineering drawings for ease of conversion to higher levels and, if available, review Level 2 and 3 drawings for compliance with requirements.	Continuing from results of Preliminary Design Review (PDR), review data.

42.8.5 Detailed Evaluation of Electrical Mechanical Designs

PDR Hardware Details	CDR Hardware Details
a. Determine that the preliminary detail design provides the capability of satisfying the performance characteristics paragraph of the Hardware Development specifications. b. Establish compatibility of the Hardware operating characteristics in each mode with overall system design requirements if the Hardware is involved in multi-mode functions. c. Establish existence and nature of physical and functional interfaces between Hardware and other items of equipment, computer software, and facilities.	1. Detailed block diagrams, schematics, and logic diagrams are compared with interface control drawings to determine system compatibility. Analytical and available test data shall be reviewed to Ensure the hardware Development Specification has been satisfied. 2. Provide information on firmware, which is included in commercially available equipment or to be included in developed equipment. Firmware in this context includes the microprocessor and associated sequence of microinstructions necessary to perform the allocated tasks.

CDR Hardware Details
As a minimum, the information presented during CDR provides descriptions and status for the following: a. Detailed logic flow diagrams b. Processing algorithms c. Circuit diagrams d. Clock and timing data (e.g., timing charts for micro- instructions) e. Memory (e.g., type (RAM, PROM), word length, size (total and spare capacity)) f. Micro-instruction list and format

CDR Hardware Details
g. Device functional instruction set obtained by implementation of firmware. h. Input output data width (i.e., number of bits for data and control.) i. Self-test (diagnostics) within firmware. j. Support software for firmware development: (1) Resident assembler, (2) Loader, (3) Debugging routines, (4) Executive monitor, (5) Non-resident diagnostics, (6) Cross assembler and higher level language on host computer, (7) Instruction simulator

42.8.6 Detailed Evaluation of Software Designs

PDR Software Details	CDR Software Details
a. Determine whether all interfaces between the CSCI and all other configuration items both internal and external to the system meet the requirements of the Software Requirements Specification and Interface Requirements Specification(s). b. Determine whether the top-level design embodies all the requirements of the Software Requirements Specification and Interface Requirements Specification(s). c. Determine whether the approved design methodology has been used for the top-level design. d. Determine whether the appropriate Human Factors Engineering (HFE) principals have been incorporated in the design. e. Determine whether timing and sizing constraints have been met throughout the top-level design. f. Determine whether logic affecting system and nuclear safety has been incorporated in the design.	Present the detailed design (including rationale) of the CSCI to include: a. The assignment of CSCI requirements to specific Lower- Level Software Units, the criteria and design rules used to accomplish this assignment, and the traceability of Unit and LLSU designs to satisfy CSCI requirements, with emphasis on the necessity and sufficiency of the Units for implementing TLSU design requirements. b. The overall information flow between software Units, the method(s) by which each Unit gains control, and the sequencing of Units relative to each other. c. The design details of the CSCI, TLSUs, LLSUs, and Units including data definitions, timing and sizing, data and storage requirements and allocations. d. The detailed design characteristics of all interfaces, including their data source, destination, interface name and interrelationships; and, if applicable, the design for direct memory access. The contractor shall also give an overview of the key design issues of the interface software design, and indicate whether data flow formats are fixed or subject to extensive dynamic changes.

CDR Software Details
e. The detailed characteristics of the database. Database structure and detailed design, including all files, records, fields, and items. Access rules, how file sharing will be controlled, procedures for database recovery regeneration from a system failure, rules for database manipulation, rules for maintaining file integrity, rules for usage reporting, and rules governing the types and depth of access shall be defined. Data management rules and algorithms for implementing them shall be described. Details of the language required by the user to access the database shall also be described.

42.8.7 Electromagnetic Compatibility

PDR Electromagnetic Compatibility	CDR Electromagnetic Compatibility
Review HWCI design for compliance with electromagnetic compatibility electromagnetic interference (EMC EMI) requirements. Use Electromagnetic Compatibility Plan as the basis for this review. Check application of MIL-STDs and MIL-Specs cited by the system equipment specification(s) to the HWCI Subsystem design. Review preliminary EMI test plans to assess adequacy to confirm that EMC requirements have been met.	a. Review EMC design of all HWCIs. Determine compliance with requirements of the Electromagnetic Compatibility Plan and HWCI specifications. b. Review system EMC including effects on the electromagnetic environment (inter-system EMC) and intra-system EMC. Determine acceptability of EMC design and progress toward meeting EMC requirements. c. Review EMC test plans. Determine adequacy to confirm EMC design characteristics of the system HWCI subsystem.

42.8.8 Design Reliability

PDR Design Reliability	CDR Design Reliability
1. Identify the quantitative reliability requirements specified in the hardware Development and Software Requirements Specification(s), including design allocations, and the complexity of the CSCIs. 2. Review failure rate sources, derating policies, and prediction methods. Review the reliability mathematical models and block diagrams as appropriate. 3. Describe planned actions when predictions are less than specified requirements. 4. Identify and review parts or components which have a critical life or require special consideration, and general plan for handling. Present planned actions to deal with these components or parts. 5. Identify applications of redundant HWCI elements. Evaluate the basis for their use and provisions for "on-line" switching of the redundant element. 6. Review critical signal paths to determine that a fail-safe fail-soft design has been provided. 7. Review margins of safety for HWCIs between functional requirements and design provisions for elements, such as: power supplies, transmitter modules, motors, and hydraulic pumps. Similarly, review structural elements; i.e., antenna	1. Review the most recent predictions of hardware and software reliability and compare against requirements specified in hardware Development Specification and Software Requirements Specification. For hardware, predictions are substantiated by review of parts application stress data. 2. Review applications of parts or configuration items with minimum life, or those which require special consideration to Ensure their effect on system performance is minimized. 3. Review completed Reliability Design Review Checklist to Ensure principles have been satisfactorily reflected in the configuration item design. 4. Review applications of redundant configuration item elements or components to establish that expectations have materialized since the PDR. 5. Review detailed HWCI reliability demonstration plan for compatibility with specified test requirements. The number of test articles, schedules, locations, test conditions, and personnel involved are reviewed to Ensure a mutual understanding of the plan and to provide overall planning information to activities concerned. 6. Review the failure data reporting procedures and methods for determination

PDR Design Reliability	CDR Design Reliability
pedestals, dishes, and radomes to determine that adequate margins of safety is provided between operational stresses and design strengths. 8. Review Reliability Design Guidelines for HWCIs to Ensure that design reliability concepts are available and used by equipment designers. Reliability Design Guidelines include, as a minimum, part application guidelines (electrical derating, thermal derating, part parameter tolerances), part selection order of preference, prohibited parts materials, reliability apportionments predictions, and management procedures to ensure compliance with the guidelines.	of failure trends. 7. Review the thermal analysis of components, printed circuit cards, modules, etc. Determine if these data are used in performing the detailed reliability stress predictions. 8. Review on-line diagnostic programs, off-line diagnostic programs, support equipment, and preliminary technical orders (and or commercial manuals) for compliance with the system maintenance concept and specification requirements. 9. Review software reliability prediction model and its updates based upon test data and refined predictions of component usage rates and complexity factors.

PDR Design Reliability
9. Review for HWCIs preliminary reliability demonstration plan: failure counting ground rules, accept-reject criteria, number of test articles, test location and environment, planned starting date, and test duration. 10. Review elements of reliability program plan to determine that each task has been initiated toward achieving specified requirements. 11. Review the reliability controls.

42.8.9 Design Maintainability

PDR Design Maintainability	CDR Design Maintainability
1. Identify the quantitative maintainability requirements specified in the hardware Development and Software Requirements Specifications; if applicable, compare preliminary predictions with specified requirements. 2. Review HWCI preventive maintenance schedules in terms of frequencies, durations, and compatibility with system schedules. 3. Review repair rate sources and prediction methods. 4. Review planned actions when predictions indicate that specified requirements will not be attained. 5. Review planned designs for accessibility, testability, and ease of maintenance characteristics (including provisions for automatic or operator-controlled recovery from failure malfunctions) to determine consistency	1. Review the most recent predictions of quantitative maintainability and compare these against requirements specified in the HWCI Development Specification and Software Requirements Specification. 2. Review preventive maintenance frequencies and durations for compatibility with overall system requirements and planning criteria. 3. Identify unique maintenance procedures required for the configuration item during operational use and evaluate their total effects on system maintenance concepts. Assure that system is optimized from a maintenance and maintainability viewpoint and conforms to the planned maintenance concept. This includes a review of provisions for automatic, semiautomatic, and manual recovery from hardware software failures and malfunctions.

PDR Design Maintainability	CDR Design Maintainability
with specified requirements. 6. Determine if planned HWCI design indicates that parts, assemblies, and components will be so placed that there is sufficient space to use test probes, soldering irons, and other tools without difficulty and that they are placed so that structural members of units do not prevent access to them or their ease of removal. 7. Review provisions for diagnosing cause(s) of failure; means for localizing source to lowest replaceable element; adequacy and locations of planned test points; and planned system diagnostics that provide a means for isolating faults to and within the configuration item. This review encompasses on-line diagnostics, off-line diagnostics, and proposed technical orders and or commercial manuals.	4. Identify design-for-maintainability criteria provided by the checklist in the design detail to Ensure that criteria have, in fact been incorporated. 5. Determine if parts, assemblies, and other items are so placed that there is sufficient space to use test probes, soldering irons, and other tools without difficulty and that they are placed so that structural members of units do not prevent access to them or their ease of removal. 6. Review detailed maintainability demonstration plan for compatibility with specified test requirements. Supplemental information is provided and reviewed to Ensure a mutual understanding of the plan and to provide overall planning information to activities concerned.

PDR Design Maintainability
8. Review for HWCIs the Design for Maintainability Checklist to Ensure that listed design principles lead to a mature maintainability design. Determine that design engineers are using the checklist. 9. Evaluate for HWCIs the preliminary maintainability demonstration plan, including number of maintenance tasks that are accomplished; accept-reject criteria; general plans for introducing faults into the HWCI and personnel involved in the demonstration. 10. Review elements of maintainability program plan to determine that each task has been initiated towards achieving specified requirements. 11. Ensure that consideration has been given to optimizing the system item from a maintainability and maintenance viewpoint and that it is supportable within the maintenance concept as developed. Also, for HWCIs ensure that a Repair Level Analysis (RLA) has been considered.

42.8.10 Human Factors

PDR Human Factors	CDR Human Factors
1. Present evidence that substantiates the functional allocation decisions. Cover all operational and maintenance functions of the configuration item, in particular, ensure that the approach to be followed emphasizes the functional integrity of the man with the machine to accomplish a system operation. 2. Review design data, design descriptions and drawings on system operations, equipments, and facilities to Ensure that human performance requirements of the hardware	1. Review detail design presented on drawings, schematics, mockups, or actual hardware to determine that they meet human performance requirements of the HWCI Development Specification and Software Requirements Specification. Interface Requirements Specification(s), and accepted human engineering practices. 2. Demonstrate by checklist or other formal means the adequacy of design for human performance. 3. Review each facet of design for man machine compatibility. Review time cost

PDR Human Factors	CDR Human Factors
Development and Software Requirements Specifications are met. Examples of the types of design information to be reviewed are: a. Operating modes for each display station, and for each mode, the functions performed, the displays and control used, etc. b. The exact format and content of each display, including data locations, spaces, abbreviations, the number of digits, all special symbols (Pictographic), alert mechanisms (e.g., flashing rates), etc. c. The control and data entry devices and formats including keyboards, special function keys, cursor control, etc.	effectiveness considerations and forced tradeoffs of human engineering design. 4. Evaluate the following human engineering biomedical design factors: a. Operator controls b. Operator displays c. Maintenance features d. Anthropometry e. Safety features and emergency equipment f. Work space layout g. Internal environmental conditions (noise, lighting, ventilation, etc.) h. Training equipment i. Personnel accommodations

PDR Human Factors
d. The format of all operator inputs, together with provisions for error detection and correction. e. All status, error, and data printouts - including formats, headings, data units, abbreviations, spacing, columns, etc. Present in sufficient detail to allow stakeholders to judge adequacy from a human usability standpoint, and design personnel to know what is required, and test personnel to prepare tests. 3. Make recommendations to update the System Subsystem, or Software Requirements Specification and Interface Requirements Specification(s) in cases where requirements for human performance need to be more detailed. 4. Review man machine functions to Ensure that man's capabilities are utilized and that his limitations are not exceeded.

42.8.11 System Safety

PDR System Safety	CDR System Safety
1. Review results of configuration item safety analyses, and quantitative hazard analyses (if applicable). 2. Review results of system and intra-system safety interfaces and trade-off studies affecting the configuration item. 3. Review safety requirements levied on subcontractors. 4. Review known special areas of safety, peculiar to the nature of the system (e.g., fuel handling, fire protection, high levels of radiated energy, high voltage protection, safety interlocks, etc.).	1. Review configuration item detail design for compliance to safety design requirements. 2. Review acceptance test requirements to ensure adequate safety requirements are reflected in the system. 3. Evaluate adequacy of detailed design for safety and protective equipment devices. 4. Review configuration item operational maintenance safety analyses and procedures.

PDR System Safety

PDR System Safety
5. Review results of preliminary safety tests (if appropriate). 6. Review adequacy and completeness of configuration item from design safety viewpoint. 7. Review compliance of commercially available configuration items or configuration item components with system safety requirements and identify modifications to such equipment.

42.8.12 Natural Environment

PDR Natural Environment	CDR Natural Environment
1. Review planned design approach toward meeting climatic conditions (operating and non-operating ranges for temperature, humidity, etc.) that are specified in the HWCI Development Specification. 2. Ensure that the stakeholders clearly understand the effect of, and the interactions between, the natural environment and HWCI design. In cases where the effect and interactions are not known or are ambiguous, ensure that studies are in progress or planned to make these determinations. 3. Current and forecast natural environment parameters may be needed for certain configuration items; e.g., display of airbase conditions in a command and control system, calculation of impact point for a missile, etc. ensure compatibility between the configuration item design and appropriate meteorological communications by comparing characteristics of the source (teletype, facsimile, or data link) with that of the configuration item. Ensure that arrangements or plans to obtain needed information have been made and that adequate display of natural environmental information is provided.	1. Review detail design to determine that it meets natural environment requirements of the hardware Development Specification. 2. Ensure that studies have been accomplished concerning effects of the natural environment on, or interactions with, the HWCI. Studies which have been in progress are complete at this time. 3. Determine whether arrangements have been made to obtain current and or forecast natural environment information, when needed for certain HWCIs. Assure compatibility of HWCI and source of information by comparing electrical characteristics and formats for the source and the HWCI.

42.8.13 Equipment and Parts Standardization

PDR Equipment and Components	CDR Equipment and Components
a. Review current and planned actions to determine that equipment or components for which standards or specifications exist are used whenever practical. (Standard items with internal stock numbers should have first preference). b. Review specific tradeoffs or modifications that maybe required of existing designs if existing items are, or will be, incorporated in the HWCI. c. Existing designs are reviewed for use or non-use based on the potential impact on the overall program in the following areas: (1) Performance, (2)	1. Determine that every reasonable action has been taken to fulfill the use of standard items (standard internal part numbers should be first preference) and to obtain approval for use of non-standard or non-preferred items. Accordingly, the following criteria are evaluated: a. Data sources that were

PDR Equipment and Components	CDR Equipment and Components
Cost, (3) Time, (4) Weight, (5) Size, (6) Reliability, (7) Maintainability, (8) Supportability, (9) Producibility, (7) Sustainability d. Review HWCI design to identify areas where a practical design change would materially increase the number of standard items. e. Ensure that Critical Item Specifications are prepared for hardware items identified as engineering or logistics critical.	reviewed. b. Factors that were considered in the decision to reject known similar or existing designs. c. Factors that were considered in decisions to accept any existing designs which were incorporated, and the tradeoffs, if any, that had to be made.

42.8.14 Parts Standardization and Interchangeability

Parts Standardization and Interchangeability PDR	Parts Standardization and Interchangeability CDR
a. Review procedures to determine if maximum practical use will be made of parts built to approved standards or specifications. The potential impact on the overall program is to be evaluated when a part built to approved standards and specifications cannot be used for any of the following reasons: (1) performance, (2) weight, (3) size, (4) reliability maintainability availability, (5) supportability, (6) survivability (including nuclear), (7) sustainability b. Identify potential design changes that will permit a greater use of standard or preferred parts and evaluate the tradeoffs. c. Ensure understanding of parts control program operations for selection and approval of parts in new design or major modifications. d. Review status of the Program Parts Selection List. e. Review status of all non-standard parts identified. f. Review pending parts control actions that may cause program slippages, such as non-availability of tested parts.	a. Determine whether there are any outstanding non-standard or non-preferred parts approval requests and action necessary for approval or disapproval. (Status of parts control program operations). b. Identify non-standard-non-preferred parts approval problems and status of actions toward resolving the problems. c. Review potential fabrication production line delays due to non-availability of standard or preferred parts. In such cases, determine whether it is planned to request use of parts, which may be replaced by standard items during subsequent support repair cycles. Assure that appropriate documentation makes note of these items and that standard replacement items are provisioned for support and used for repair. d. Require certification that maximum practical interchangeability of parts exists among components, assemblies, and HWCIs. Reservations concerning interchangeability are identified, particularly for hardness critical items. e. Sample preliminary drawings and cross check to Ensure that parts indicated on the drawings are compatible with the Program Parts Selection List.

42.8.15 Assignment of Official Nomenclature

Assignment of Official Nomenclature PDR	Assignment of Official Nomenclature CDR
a. Ensure understanding of	a. Determine whether official nomenclature and

Assignment of Official Nomenclature PDR	Assignment of Official Nomenclature CDR
procedure for obtaining assignment of nomenclature and approval of nameplates. b. Determine that a nomenclature conference has been held and agreement has been reached with the stakeholders on the level of nomenclature; i.e., system, set, central, group, component, sub-assembly, unit, etc.	approval of nameplates have been obtained to extent practical. b. Determine whether Request for Nomenclature, has been processed to the agreed level of indenture. c. Ensure that approved nomenclature has been reflected in the Development and Product Specifications. d. Identify problems associated with nomenclature requests together with status of actions towards resolving the problems. e. Ensure that a software inventory numbering system has been agreed to and implemented to the CSCI level.

42.8.16 Value Engineering

Value Engineering PDR	Value Engineering CDR
Review the in-house incentive Value Engineering Program, which may include but not be limited to the following: a. Value Engineering organization, policies and procedures including impact on quality assessments. b. Value Engineering Training Program with emphasis on maintaining or improving quality rather than irresponsible financial application. c. Potential Value Engineering projects, studies and VECPs including impact on quality assessments. d. Schedule of planned Value Engineering tasks events. e. Policies and procedures for subcontractor Value Engineering Programs.	1. Review status of all VECPs presented per the terms of the contract. 2. Identify the impact on quality for each VECP. 3. Review any new areas of potential Value Engineering considered profitable to challenge. 4. If required by contract (funded VE program), review the actual Value Engineering accomplishments against the planned VE program.

42.8.17 Sustainability[364]

Sustainability PDR and CDR
1. Review community sustainability and accountability analysis. 2. Review technology assessment findings. 3. Review key sustainability goals, requirements, and issues. 4. Review sustainability tradeoff criteria and sustainable architecture approaches. 5. Review internal and external sustainability issues and drivers. 6. Review new sustainability performance requirements. 7. Review use of recycled, recovered, or environmentally preferable materials. 8. Review list of and justification of toxic and hazardous materials used in the system. 9. Review list and justification of prohibited materials used in the system. 10. List indirect cost shifting out of the system to external systems and stakeholders. 11. Review maintenance logistics safety reliability quality, etc 12. Review the production and transport sustainability performance results for the

[364] Sustainability is not part of MIL-STD-1521B, it is being offered in this text.

Sustainability PDR and CDR
selected design: PCF, TCF, CCF, PEC, TEC, CEC, etc. Review the production and transport sustainability performance ratios for the selected design: PCCR, TCCR, PCER, TCER, etc.

42.8.18 Ethics[365]

Ethics PDR and CDR
1. Review possible technology abuses associated with the system. 2. Review list of possible stakeholders negatively impacted by the system. 3. List indirect cost shifting out of the system in development. 4. Review fundamental ethical questions: Is there a conflict of interest? Is a stakeholder getting a hidden benefit? Is a stakeholder being harmed? Is there a hidden stakeholder? Is information being suppressed? Is the team solving the obvious problems others seem to ignore? Are checks and balances being removed or made ineffective? Is a stakeholder being marginalized? Is staff being marginalized? Is there a misuse of technology? Is physical harm being permitted Is loss of life being permitted Is physical harm being permitted because it makes financial sense? Is physical loss of life being permitted because it makes financial sense? Do the needs of the one out weight the needs of the many and do the needs of the many out weight the needs of the one

42.8.19 Transportability

PDR Transportability	CDR Transportability
1. Review HWCI to determine if the design meets size and weight to permit economical handling, loading, securing, transporting, and disassembly for shipment using in house and commercial carriers. Identify potential outsized and overweight items. Identify system items defined as being hazardous. Ensure packaging afforded hazardous items complies with hazardous materials regulations. 2. Identify HWCIs requiring special temperature and humidity control or those possessing sensitive and shock susceptibility characteristics. Determine special transportation requirements and availability for use with these HWCIs.	1. Review transportability evaluations accomplished for those items identified as outsized, overweight, sensitive, and or requiring special temperature and humidity controls. 2. Review actions taken as a result of the above evaluation to ensure adequate facilities and military or commercial transporting equipment are available to support system requirements during Production and Deployment Phases. 3. Review design of special materials handling equipment, when required, and action taken to acquire equipment. 4. Ensure DOD Certificates of Essentiality for movement of equipment

[365] Ethics is not part of MIL-STD-1521B, it is being offered in this text.

PDR Transportability	CDR Transportability
3. Review Transportability Analysis to determine that transportation conditions have been evaluated and that these conditions are reflected in the design of protective, shipping, and handling devices. In addition to size and weight characteristics, determine that analysis includes provisions for temperature and humidity controls, minimization of sensitivity, susceptibility to shock, and transit damage.	have been obtained for equipment exceeding limitations of criteria established in contract requirements. 5. Ensure transportability approval has been annotated on design documents and remain as long as no design changes are made that modify significant transportability parameters. 6. Identify equipment to be test loaded for air transportability of material in Cargo Aircraft.

42.8.20 Test

PDR Test	CDR Test
1. Review all changes to the System Subsystem, HWCI Development, Software Requirements, and Interface Requirements Specifications subsequent to the established Allocated Baseline to determine whether test section of all these specifications adequately reflects these changes. 2. Review test concepts for Development Test and Evaluation (DT&E) testing (both informal and formal). Information includes: a. The organization and responsibilities of the group that will be responsible for test. b. The management of in-house development test effort provides for: (1) Test Methods (plans procedures) (2) Test Reports (3) Resolution of problems and errors (4) Retest procedure (5) Change control and configuration management (6) Identification of any special test tools not deliverable c. The methodology to be used to meet quality assurance requirements qualification requirements, including the test repeatability characteristics and approach to regression testing. d. The progress status of the test effort since the previous reporting milestone. 3. Review status of all negative or provisional entries such as "not applicable (N A)" or "to be determined (TBD)" in the	1. Review updating changes to all specifications subsequent to the PDR, to determine whether the test section of the specifications adequately reflects these changes. 2. Review all available test documentation for currency, technical adequacy, and compatibility with the test section of all Specification requirements. 3. For any development model, prototype, etc., on which testing may have been performed, examine test results for design compliance with hardware Development, Software Requirements, and Interface Requirements Specification requirements. 4. Review quality assurance provisions qualification requirements in HWCI Product, Software Requirements, or Interface Requirements Specifications for completeness and technical adequacy. The test sections of these specifications include the minimum requirements that the item, materiel, or process must meet to be acceptable. 5. Review all test documentation that supports test requirements of the HWCI Product Specifications for compatibility, technical adequacy, and completeness. 6. Inspect any breadboards, mockups, or prototype hardware available for test program implications. 7. Review Software Test Descriptions to ensure they are consistent with the Software Test Plan and they thoroughly

PDR Test	CDR Test
test section of the System Subsystem, hardware Development, Software Requirements or Interface Requirements Specifications. Review all positive entries for technical adequacy. Ensure that associated test documentation includes these changes. 4. Review interface test requirements specified in the test section of the hardware Development, Software Requirements, and Interface Requirements Specifications for compatibility, currency, technical adequacy, elimination of redundant test. Ensure that all associated test documents reflect these interface requirements.	identify necessary parameters and prerequisites to enable execution of each planned software test and monitoring of test results. As a minimum, test descriptions identify the following for each test: a. Required preset hardware and software conditions and the necessary input data, including the source for all data. b. Criteria for evaluating test results. c. Prerequisite conditions to be established or set prior to test execution. d. Expected or predicted test results.

PDR Test
5. Ensure that all test planning documentation has been updated to include new test support requirements and provisions for long-lead time support requirements. 6. Review contractor test data from prior testing to determine if such data negates the need for additional testing. 7. Examine all available breadboards, mock-ups, or devices which will be used in implementing the test program or which affect the test program, for program impact. 8. Review plans for software Unit testing to ensure that they: a. Address Unit level sizing, timing, and accuracy requirements. b. Present general and specific requirements that will be demonstrated by Unit testing. c. Describe the required test-unique support software, hardware, and facilities and the interrelationship of these items. d. Describe how, when, and from where the test-unique support items will be obtained. e. Provide test schedules consistent with higher level plans. 9. Review plans for CSC integration testing to ensure that they: a. Define the type of testing required for each level of the software structure above the unit level. b. Present general and specific requirements that will be demonstrated by CSC integration testing. c. Describe the required test-unique support software, hardware, and facilities and the interrelationship of these items. d. Describe how, when, and from where the test-unique support items will be obtained. e. Describe CSC integration test management, to include: (1) Organization and responsibilities of the test team (2) Control procedures to be applied during test (3) Test reporting (4) Review of CSC integration test results (5) Generation of data to be used in CSC integration testing. f. Provide test schedules consistent with higher level plans. 10. Review plans for formal CSCI testing to ensure that they: a. Define the objective of each CSCI test, and relate the test to the software requirements being tested. b. Relate formal CSCI tests to other test phases. c. Describe support software, hardware, and facilities required for CSCI testing; and

PDR Test
how, when, and from where they will be obtained. d. Describe CSCI test roles and responsibilities. e. Describe requirements for Government-provided software, hardware, facilities, data, and documentation. f. Provide CSCI test schedules consistent with higher- level plans. g. Identify software requirements that will be verified by each formal CSCI test.

42.8.21 Maintenance and Data

PDR Maintenance and Data	CDR Maintenance and Data
1. Describe System Maintenance concept for impact on design and SE. Review adequacy of maintenance plans. Coverage provided for On Equipment (Organizational), Off Equipment - On Site (Intermediate), Off Equipment - Off Site (Depot) level maintenance of Externally Furnished Equipment, and Internally Furnished Equipment.	1. Review adequacy of maintenance plans. 2. Review status of unresolved maintenance and maintenance data problems since the PDR. 3. Review status of compliance with the Reliability, Maintainability Data Reporting and Feedback Failure Summary Reports.

PDR Maintenance and Maintenance Data
2. Determine degree of understanding of the background, purpose, requirements, and usage of Maintenance (failure) Data Collection and Historical Status Records. 3. Describe method of providing Maintenance, Failure, Reliability, Maintainability Data. 4. Describe how requirements are submitted for Equipment Classification (EQ CL) Codes when this requirement exists. 5. Review plans for and status of Work Unit Coding of the equipment. Work Unit codes are available for documenting Maintenance Data commencing with configuration item Subsystem Testing.

42.8.22 Spare Parts and Externally Furnished Property

PDR Spare Parts and Externally Furnished Property	CDR Spare Parts and Externally Furnished Property
1. Review logistics and provisioning planning to Ensure full understanding of scope of requirements in these areas and that a reasonable time-phased plan has been developed for accomplishment. Of specific concern are the areas of: provisioning requirements, externally provided equipment usage, and spare parts, and support during installation, checkout, and test. 2. Review provisioning actions and identify existing or potential provisioning problems - logistic critical and long lead time items are identified and evaluated against use of the interim release	1. Review provisioning planning through normal logistics channels and Administrative Contracting Officer (ACO) representative (Industrial Specialist) to ensure its compatibility (content and time phasing) with contractual requirements. The end objective is to provision by a method which ensures system supportability at operational date of the first site. Also accomplish the following: a. Ensure understanding of contractual requirements, including time phasing, instructions from logistics support agencies, interim release authority and procedure, and responsibility to deliver

PDR Spare Parts and Externally Furnished Property	CDR Spare Parts and Externally Furnished Property
requirements. 3. Review plans for maximum screening and usage of externally provided equipment, and extent plans have been implemented. 4. Review progress toward determining and acquiring total installation, checkout, and test support requirements.	spare repair parts by need date. b. Determine that scheduled provisioning actions, such as, guidance meetings, interim release and screening are being accomplished adequately and on time. c. Identify existing or potential provisioning problems.

CDR Spare Parts and Externally Furnished Property
2. Determine quantitative and qualitative adequacy of provisioning drawings and data. Verify that Logistics Critical items are listed for consideration and that adequate procedures exist for reflecting design change information in provisioning documentation and Technical Orders. 3. Ensure support requirements have been determined for installation, checkout, and test for approval by contracting agency. Ensure screening has been accomplished and results are included into support requirements lists. 4. Determine that adequate storage space requirements have been programmed for on-site handling of Installation and Checkout (I&C), test support material, and a scheme has been developed for down streaming and joint use of insurance (high cost) or catastrophic failure support items.

42.8.23 Packaging Special Design Protective Equipment

PDR Packaging SDPE	CDR Packaging SDPE
1. Analyze all available specifications (System Subsystem, HWCI Development, Software Requirements, Interface Requirements, and Critical Items) for packaging requirements for each product fabrication and material specification. 2. Evaluate user operational support requirements and maintenance concepts for effect and influence on package design. 3. Establish that time phased plan for package design development is in consonance with the development of the equipment design. 4. Review planned and or preliminary equipment designs for ease of packaging and simplicity of package design, and identify areas where a practical design change would materially decrease cost, weight, or volume of packaging required. 5. Review requirements for SDPE necessary to effectively support configuration item during transportation, handling and storage processes. Ensure SDPE is categorized as a configuration	1. Review proposed package design to ensure that adequate protection to the HWCI, and the media on which the CSCI is recorded, is provided against natural and induced environments hazards to which the equipment will be subjected throughout its life cycle, and to ensure compliance with contractual requirements. Such analysis includes, but not be limited to, the following: a. Methods of preservation b. Physical mechanical shock protection including cushioning media, shock mounting and isolation features, load factors, support pads, cushioning devices, blocking and bracing, etc. c. Mounting facilities and securing hold-down provisions d. Interior and exterior container designs. e. Handling provisions and compatibility with aircraft materials handling system f. Container marking g. Consideration and identification of

PDR Packaging SDPE	CDR Packaging SDPE
item using specifications conforming to the types and forms as prescribed. Review SDPE development product specifications for adequacy of performance interface requirements. 6. Determine initial package design baselines, concepts, parameters, constraints, etc., to the extent possible at this phase of the configuration item development process. 7. Ensure previously developed and approved package design data for like or similar configuration items are being used. 8. Establish plans for trade studies to determine the most economical and desirable packaging design approach needed to satisfy the functional performance and logistic requirements. 9. Verify the adequacy of the prototype package design. 10. Review the packaging section of the Specification to ensure full understanding of the requirements. Identify package specification used for hazardous materials.	dangerous hazardous commodities 2. Review design of SDPE HWCI to determine if a category I container is required. The analysis of the proposed container or handling, shipping equivalent encompasses as a minimum: a. Location and type of internal mounting or attaching provisions b. Vibration - shock isolation features, based on the pre-determined fragility rating (or other constraint of the item to be shipped.) c. Service items (indicators, relief valves, etc.) d. Environmental control features e. External handling, stacking and tie-down provisions with stress ratings. f. Dimensional and weight data (gross and net) g. Bill-of-material h. Marking provisions including the center-of-gravity location i. For wheeled SDPE (self-powered or tractor trailer) the overall length, width, and height with mounted item, turning radius, mobility, number of axles, unit contact load, number of tires, etc.

CDR Packaging SDPE
j. Position and travel of adjustable wheels, titling, or other adjustments to facilitate loading. 3. Review the results of trade studies, engineering analyses, etc., to substantiate selected package SDPE design approach, choice of materials, handling provisions, environmental features, etc. 4. Ensure that package SDPE design provides reasonable balance between cost and desired performance. 5. Review all preproduction test results of the prototype package design to Ensure that the HWCI is afforded the proper degree of protection. 6. Review Packaging Section of the HWCI Product Specification for correct format, accuracy and technical adequacy. 7. Review contractor procedures to assure that the requirements for Preparation for Delivery of the approved HWCI Product Specification, will be incorporated into the package design data for provisioned spares.

42.8.24 Technical Manuals

PDR and CDR Technical Manuals
1. Review status of the "Technical Manual Publications Plan" to Ensure that all aspects of the plan have been considered to the extent that all concerned agencies are apprised of the technical manual coverage to be obtained under this procurement. The suitability of

PDR and CDR Technical Manuals
available commercial manuals and or modifications thereto shall also be determined. 2. Review the availability of technical manuals for validation verification during the latter phases of DT&E testing. 3. If a Guidance Conference was not accomplished or if open items resulted from it, then review as applicable provisions for accomplishing TO in-process reviews, validation, verification, prepublication, and post publication reviews.

42.8.25 System Allocation Document

PDR and CDR System Allocation Document
1. Review the Draft System Allocation Document for completeness and technical adequacy to extent completed. 2. The format provides the following minimum information: a. Drawing Number b. Issue c. Number of Sheets d. Location e. Configuration Item Number f. Title g. Part Number h. Serial Number i. Specification Number j. Equipment Nomenclature k. Configuration Item Quantity l. Assembly Drawing

42.8.26 Design Producibility and Manufacturing

PDR Design Producibility and Manufacturing	CDR Design Producibility and Manufacturing
1. Demonstrate and present evidence that manufacturing engineering will be integrated into the design process. a. Provide evidence of performing producibility analyses on development hardware trading off design requirements against manufacturing risk, cost, production, volume, and existing capability availability. Evidence of such analyses must conclusively demonstrate that in-depth analyses were performed by qualified organizations individuals and the results of those analyses will be incorporated in the design. b. Preliminary manufacturing engineering and production planning demonstrations address: material and component selection, preliminary production sequencing, methods and flow concepts, new processes, manufacturing risk, equipment and facility	1. Review the status of all producibility (and productivity) efforts for cost and schedule considerations. 2. Review the status of efforts to resolve manufacturing concerns identified in previous technical reviews and their cost and schedule impact to the production program. 3. Review the status of Manufacturing Technology programs and other previously recommended actions to reduce cost, manufacturing risk and industrial base concerns. 4. Identify open manufacturing concerns that require additional direction effort to minimize risk to the production program. 5. Review the status of manufacturing engineering efforts, tooling and test equipment demonstrations, proofing of

PDR Design Producibility and Manufacturing	CDR Design Producibility and Manufacturing
utilization for intended rates and volume, production in-process and acceptance test and inspection concepts. (Efforts to maximize productivity in the above areas should be demonstrated.)	new materials, processes, methods, and special tooling test equipment. 6. Review the intended manufacturing management system and organization for the production program in order to show how their efforts will effect a smooth transition into production.

PDR Design Producibility and Manufacturing
c. Management systems to be used will ensure that producibility and manufacturing considerations are integrated throughout the development effort. 2. The producibility and manufacturing concerns identified in the SRR and the SDR are updated and expanded to: a. Provide evidence that concerns identified in the Manufacturing Feasibility Assessment and the Production Capability Estimate are addressed and that resolutions are planned or have been performed. b. Make recommendations including manufacturing technology efforts and provide a schedule of necessary actions to resolve open manufacturing concerns and reduce manufacturing risk.

42.9 Test Readiness Review

The TRR is a formal review of the readiness to begin formal testing. It is conducted after test procedures are available and integration testing is complete. The purpose of TRR is to determine if testing can start. A technical understanding is reached on the informal test results, and on the validity and the degree of completeness of the System Manuals (operator, user, maintenance, etc). The following items are reviewed.

- Changes to specifications, design documents, test plans and descriptions
- Dry run test results
- Test procedures used in conducting testing, including retest procedures for test anomalies and corrections
- Test resources, facilities, tools, personnel, supporting materials
- Review traceability between requirements and their associated tests
- Test limitations, test problems, rules for retest in when anomalies surface
- Summary of known problems including all known discrepancies
- Rules for capturing discrepancies if they surface
- Schedules
- Review of updates to all evolving and previously delivered items

At the conclusion of the TRR a decision is made if the team can proceed to test. If the TRR findings suggest the test can proceed, testing is started and results are captured for the test result submission.

42.10 Functional Configuration Audit (FCA)

The objective of the FCA is to verify that the configuration item's actual performance complies with its specifications. Test data is reviewed to verify that the configuration item performs as required by its functional allocated configuration identification.

For software, a technical understanding is reached on the validity and the degree of completeness of the Software Test Reports, and as appropriate, Computer System Operator's Manual (CSOM), Software User's Manual (SUM), and the Computer System Diagnostic Manual (CSDM).

The FCA for a complex configuration item may be conducted on a progressive basis throughout the item's development and culminates at the completion of the qualification testing of the configuration item with a review of all discrepancies at the final FCA. The FCA is held on that configuration of the configuration item which is representative (prototype or pre-production) of the configuration to be released for production. When a prototype or pre-production article is not produced, The FCA is conducted on a first production article. For cases where configuration item qualification can be only determined through integrated system testing, FCA's for such configuration items are not complete until all the integration testing is complete.

42.11 Physical Configuration Audit (PCA)

The PCA is a formal examination of the as-built version of a configuration item against its design documentation to establish the product baseline. After successful completion of the audit, all subsequent changes are processed by engineering change action. The PCA also determines that the acceptance testing requirements in the documentation is adequate for acceptance of production units of a configuration item by quality assurance activities. The PCA includes a detailed audit of engineering drawings, specifications, technical data and tests used in production of hardware configuration items and a detailed audit of design documentation, listings, and manuals for software configuration items. The review includes an audit of the released engineering documentation and quality control records to make sure the as built or as coded configuration is reflected by this documentation. For software, the Software Product Specification and Software Version Description are part of the PCA review.

The PCA is held on the first article of configuration items and those that are a re-procurement of a configuration item already in the inventory as identified and selected jointly by the buyer and seller. A PCA is conducted on the first configuration item to be delivered by a new seller even though PCA was previously accomplished on the first article delivered by a different seller.

The product baseline is established when there is a formal approval of the configuration item Product Specification and the satisfactory completion of a PCA.

A final review is made of all operation and support documents (i.e., Computer System Operator's Manual (CSOM), Software User's Manual (SUM), Computer System Diagnostic Manual (CSDM), Computer Programmer's Manual (SPM), Firmware Support Manual (FSM). The review checks format, completeness, and conformance with data item descriptions[366].

42.12 Formal Qualification Review (FQR)

The FQR is held to verify the actual performance of the configuration items of the system through test compliance with the specifications, and to identify the test reports and data, which document results of qualification tests of the configuration items. The point of certification is determined by the buyer and depends on the nature of the program, risk aspects of the particular hardware and software, and progress in successfully verifying the requirements of the configuration items.

When feasible, the FQR is combined with the FCA at the end of configuration item subsystem testing, prior to PCA. If sufficient test results are not available at the FCA to ensure the configuration items will perform in their system environment, the FQR is held (post PCA) during System testing whenever the necessary tests have been successfully completed. For non-combined FCA FQRs, traceability, correlation, and completeness of the FQR are maintained with the FCA and duplication of effort avoided.

In cases where the FQR and the FCA can be accomplished in a single combined Audit Review, certification of the configuration items is accomplished after completion of the FCA and the certification is considered as accomplishment of the FQR.

When the system is not ready for FQR at the time of FCA, the FQR is delayed until it is determined that sufficient information on the system's qualification is available. The FQR may be delayed up to the end of System testing if necessary. When a separate FQR is necessary the FQR team is assembled in the same manner, as that required for the FCA team. No duplication of FCA effort occurs at the FQR; however, a review of the FCA minutes is performed and the FQR is considered as an extension of FCA.

42.13 FCA PCA Review Details

FCA	PCA
The schedules for the FCA are recorded on the configuration item development record. A configuration item cannot be audited without authentication of the functional and allocated baseline. In	a. Test manager in attendance b. Identification of items to be accepted: (1) Nomenclature (2) Specification Identification

---

[366] Data Item Descriptions (DID) identify the format and content of data deliverables such as documents. These are information products. The goal is to have a consist look and feel to all information products and ensure expected content is provided.

FCA	PCA
addition, the final draft Product Specification for the configuration item is submitted for review prior to FCA. **<u>Seller Responsibility</u>** a. Test manager in attendance b. Identification of items to be audited: (1) Nomenclature (2) Specification identification number (3) Configuration item number (4) Current listing of all deviations waivers against the configuration item, either requested of, or approved by the buyer. (5) Status of Test Program to test configured items with automatic test equipment. **<u>Procedures and Requirements</u>** The test procedures and results are reviewed for compliance with specifications. The following testing information is available for the FCA team: a. Test plans, specifications, descriptions, procedures, reports for configuration item. b. Complete list of successfully accomplished functional tests during which pre-acceptance data was recorded. c. Complete list of successful functional tests if detailed test data are not recorded. d. Complete list of functional tests required by the specification but not yet performed. (To be performed as a system or subsystem test). e. Preproduction and production test results. Testing accomplished with the approved test procedures and validated data (witnessed) is sufficient to ensure configuration item performance as set forth in the specification and meet the quality assurance provisions qualification requirements contained in the specification test section. For those performance parameters that cannot be completely verified during testing, adequate analysis or simulation is completed. The results of the analysis or simulations are sufficient to ensure configuration item performance as outlined in the specification. Test reports, procedures, and data used by the FCA team are made a matter of record in the FCA minutes.	Number (3) Configuration item Identifiers (4) Serial Numbers (5) Drawing and Part Numbers (6) Identification Numbers (7) Code Identification Numbers (8) Software inventory numbering system c. A list delineating all deviations waivers against the configuration item either requested or contracting agency approved. The PCA cannot be performed unless data pertinent to the configuration item being audited is provided to the PCA team at time of the audit. Required information includes: a. Configuration item product specification. b. A list delineating both approved and outstanding changes against the configuration item. c. Complete shortage list. d. Acceptance test procedures and associated test data. e. Engineering drawing index including revision letters. f. Operating, maintenance, and illustrated parts breakdown manuals. g. Proposed Material Inspection and Receiving Report. h. Approved nomenclature and nameplates. i. Software Programmer's Manuals (SPMs), Software User's Manuals (SUMs) Computer System Operator's Manual (CSOM), Computer System Diagnostic Manual (CSDM), and Firmware Support Manual (FSM). j. Software Version Description Document. k. FCA minutes for each configuration item. l. Findings Status of Quality Assurance Programs. Data describing item configuration: a. Current approved issue of hardware development specification, Software Requirements Specification, and Interface Requirements Specification(s)

FCA	PCA
A list of the contractor's internal documentation (drawings) of the configuration item is reviewed to ensure that the contractor has documented the physical configuration of the configuration item for which the test data are verified. Drawings of HWCI parts, which are to be provisioned, should be selectively sampled to assure that test data essential to manufacturing are included on, or furnished with, the drawings. Configuration Items (CIs) which fail to pass quality assurance test provisions are to be analyzed as to the cause of failure to pass. Appropriate corrections are made to both the CI and associated engineering data before a CI is subjected to requalification. A checklist is developed which identifies documentation and hardware and computer software to be available and tasks to be accomplished at the FCA for the configuration item. Acknowledge accomplishment of partial completion of the FCA for those configuration items whose qualification is contingent upon completion of integrated systems testing. For CSCIs the following additional requirements apply: a. The contractor provides the FCA team with a briefing for each CSCI being audited and delineates the test results and findings for each CSCI. As a minimum, the discussion includes CSCI requirements that were not met, including a proposed solution to each item, an account of the engineering change proposals (ECP) incorporated and tested as well as proposed, and a general presentation of the entire CSCI test effort delineating problem areas as well as accomplishments. b. An audit of the formal test plans descriptions procedures are made and compared against the official test data. The results are checked for completeness and accuracy. Deficiencies are documented and made a part of the FCA minutes. Completion dates for all discrepancies are clearly established and documented. c. Audits of the Software Test Reports are performed to validate that the reports are	to include approved specification change notices and approved deviations waivers. b. Identification of all changes actually made during test. c. Identification of all required changes not completed. d. All approved drawings and documents by the top drawing number as identified in the configuration item product specification. All drawings are of the category and form specified in the contract. e. Manufacturing instruction sheets for HWCIs. Identify difference between the physical configurations of the selected production unit and the Development Units used for the FCA and to certify or demonstrates differences do not degrade the functional characteristics of the selected units. PCA Procedures and Requirements. Drawing and Manufacturing Instruction Sheet Review Instructions: a. A representative number of drawings and associated manufacturing instruction sheets for each item of hardware, reviewed to determine their accuracy and ensure that they include the authorized changes reflected in the engineering drawings and the hardware. Inspection of drawings and associated manufacturing instruction sheets may be accomplished on a valid sampling basis. The purpose of this review is to ensure the manufacturing instruction sheets accurately reflect all design details contained in the drawings. Since the hardware is built in accordance with the manufacturing instruction sheets any discrepancies between the instruction sheets and the design details and changes in the drawings are also reflected in the hardware. b. The following minimum information is recorded for each drawing reviewed: (1) Drawing number title (include revision letter) (2) Date of drawing approval (3) List of manufacturing instruction

FCA	PCA
accurate and completely describe the CSCI tests. d. All ECPs that have been approved are reviewed to ensure that they have been technically incorporated and verified. e. All updated to previously delivered documents are reviewed to ensure accuracy and consistency throughout the documentation set. f. Preliminary and Critical Design Review minutes are examined to ensure that all findings have been incorporated and completed. g. The interface requirements and the testing of these requirements are reviewed for CSCIs. h. Review database characteristics, storage allocation data and timing, and sequencing characteristics for compliance with specified requirements.	sheets (numbers with change letter titles and date of approval) associated with this drawing. (4) Discrepancies comments (5) Select a sample of part numbers reflected on the drawing. Check to insure compatibility with the Program Parts Selection List, and examine the HWCI to insure that the proper parts are actually installed. c. As a minimum, the following inspections are accomplished for each drawing and associated manufacturing instruction sheets: (1) Drawing number identified on manufacturing instruction sheet should match latest released drawing. (2) List of materials on manufacturing instruction sheets should match materials identified on the drawing. (3) All special instructions called on the drawing should be on the manufacturing instruction sheets. (4) All dimensions, tolerances, finishes, etc., called out on the drawing should be identified on the manufacturing instruction sheets.

PCA
(5) All special processes called out on the drawing should be identified on the manufacturing instruction sheets. (6) Nomenclature descriptions, part numbers and serial number markings called out on the drawing should be identified on the manufacturing instruction sheets. (7) Review drawings and associated manufacturing instruction sheets to ascertain that all approved changes have been incorporated into the configuration item. (8) Check release record to ensure all drawings reviewed are identified. (9) Record the number of any drawings containing more than five outstanding changes attached to the drawing. (10) Check the drawings of a major assembly black box of the hardware configuration item for continuity from top drawing down to piece-part drawing. Review of all records of baseline configuration for the HWCI by direct comparison with engineering release system and change control procedures to establish that the configuration being produced does accurately reflect released engineering data. This includes interim releases of spares provisioned prior to PCA to ensure delivery of currently configured spares. Review engineering release and change control system, to ascertain that they are adequate to properly control the processing and formal release of engineering changes. The minimum needs and capabilities set forth below are required of his engineering

PCA
release records system. The formats, systems, and procedures are to be used. Information in addition to the basic requirements is to be considered part of the contractor's internal system. Contract Administration Office (CAO) Quality Assurance Representative (QAR) records can be reviewed for purpose of determining the contractor's present and most recent past performance. As a minimum, the following information is contained on one release record for each drawing number, if applicable: a. Serial numbers, top drawing number, specification number; b. Drawing number, title, code number, number of sheets, date of release, change letter, date of change letter release, engineering change order (ECO) number. The release function and documentation is capable of determining: a. The composition of any part at any level in terms of subordinate part numbers (disregard standard parts); b. The next higher assembly using the part number, except for assembly into standard parts; c. The composition of the configuration item or part number with respect to other configuration items or part numbers; d. The configuration item and associated serial number on which subordinate parts are used. (This does not apply to contractors below prime level who are not producing configuration items); e. The accountability of changes which have been partially or completely released against the configuration item; f. The configuration item and serial number effectively of any change. g. The standard specification number or standard part numbers used within any non-standard part number; h. The specification document and specification control numbers associated with any subcontractor, vendor, or supplier part number. The engineering release system and associated documentation is capable of: a. Identifying changes and retaining records of superseded configurations formally accepted; b. Identifying all engineering changes released for production incorporation. These changes are completely released and incorporated prior to formal acceptance of the configuration item; c. Determining the configuration released for each configuration item at the time of formal acceptance. Engineering data is released or processed through a central authority to ensure coordinated action and preclude unilateral release of data. Engineering change control numbers are unique. Difference between the configuration of the configuration item qualified and the configuration item being audited is a matter of record in the minutes of the PCA. For HWCI acceptance tests data and procedures comply with its product specification. The PCA team determines any acceptance tests to be re-accomplished, and reserves the prerogative to have representatives witness all or any portion of the required audits, inspections, or tests. HWCIs which fail to pass acceptance test requirements are repaired if necessary and retested in the manner specified by the PCA team leader in accordance with the product specification. Data confirming the inspection and test of subcontractor equipment end items at point of manufacture. Such data should have been witnessed by a customer representative. The PCA team reviews the prepared back-up data (all initial documentation which accompanies the configuration item) for correct types and quantities to ensure adequate

PCA
coverage at the time of shipment to the user. Configuration items which have demonstrated compliance with the product specification are approved for acceptance as follows: The PCA team certifies by signature that the configuration item has been built in accordance with the drawings and specifications. As a minimum, the following actions are performed by the PCA team on each CSCI being audited: a. Review all documents which will comprise the Software Product Specification for format and completeness b. Review FCA minutes for recorded discrepancies and actions taken c. Review the design descriptions for proper entries, symbols, labels, tags, references, and data descriptions. d. Compare Top-Level Software Units design descriptions with Lower-Level Software Unit descriptions for consistency e. Compare all lower-level design descriptions with all software listings for accuracy and completeness f. Check Software User's Manual(s), Software Programmer's Manual, Computer System Operator's Manual, Firmware Support Manual, and Computer System Diagnostic Manual for format completeness and conformance with applicable data item descriptions. (Formal verification acceptance of these manuals should be withheld until system testing to ensure that the procedural contents are correct) g. Examine actual CSCI delivery media (card decks, tapes, disks. etc.,) to insure conformance with the Software Requirements Specification. h. Review the annotated listings for compliance with approved coding standards

42.14 Exercises

1. Identify reviews and audits for a research and development program.
2. Identify reviews and audits for a development program.
3. Identify reviews and audits for an industrial product.
4. Identify reviews and audits for a consumer product.
5. Identify reviews and audits for building your custom house.
6. Are there differences between reviews and audits you just identified?
7. Identify sustainability elements in the previous reviews and audits.
8. Identify ethics elements in the previous reviews and audits.

42.15 Further Reading

1. Specification Practices, Military Standard, MIL-STD-490 30 October 1968, MIL-STD-490A 4 June 1985.
2. System Engineering Management, Military Standard, MIL-STD-499 17 July 1969, MIL-STD-499A 1 May 1974.
3. Technical Reviews And Audits Systems Equipments And Computer Software, MIL-STD-1521A June 1976, MIL-STD-1521B June 1995.

43 Statement of Work

At some point a Statement of Work (SOW) needs to be developed either because the capability to build a system, subsystem, or component may not exist in house or more effective solutions are possible from other sources. The SOW identifies what work products need to be developed and when they are to be delivered. It also may contain the contract terms and conditions.

The SOW must be consistent with the system requirements and both information products must be offered at the same time. Since it contains the description of the work to be performed, its management, and delivery dates, at no time should any technical requirements appear in the SOW. The SOW should not dictate the process, dictate architecture, or assume architecture. Each SOW is unique to the activity it captures. There are general categories of information that many SOW documents should contain.

The SOW is contractually binding. As part of the contract award process there is a Best and Final Offer (BAFO) which includes negotiating the final work. It is at this time that the SOW may be modified but in a structured way so that all the submitting bids are treated equally. Changing the SOW after the start of a project or program is not suggested.

43.1 Characteristics of a Good SOW

The SOW is read and interpreted by engineers, scientists, accountants, lawyers, contract specialists, managers, and other stakeholders. Everyone needs to understand the SOW. It is used during solicitation, contract award and administration. It is important that the words be clear and consistent to avoid confusion. The authoring of a good SOW is not a trivial exercise. The following is considered to be a reasonable list of goals for a good SOW.

- Allow contractors and customers to clearly understand the work requirements and needs
- Allow bidders to accurately cost their proposals and submit high quality technical proposals
- Provide a baseline for the development of other parts of the contract, such as the proposal evaluation criteria, technical proposal instructions and independent cost estimates
- Minimize the need for changes which can increase cost and delay completion

- Allow the customer and contractor to assess performance
- Avoid claims and disputes under the contract

The SOW is part of the contract and is subject to contract law. Ambiguous work statements can result in protests, unsatisfactory performance, delays, claims, disputes, and increased contract costs. It is the basis for successful performance by the contractor and effective administration of the contract by the customer. It lets all potential bidders compete equally for a contract and is used to determine if a contractor meets the work requirements. The SOW should:

- State in clear understandable terms the work to be done in developing or producing the goods to be delivered or services to be performed
- Use explicit terms so stakeholders clearly understand needs
- Identify goods or services needed to satisfy a requirement and define what is required in specific performance-based quantitative terms
- Facilitate preparation of responsive proposals and delivery of required goods or services
- Aid customer in vendor selection and contract administration after award

43.2 SOW Style Guide

The need for a good SOW is obvious. An approach is to develop a style guide that offers guidance to staff writing the SOW. The style guide should include a statement on imperatives, example sentence syntax, suggested words and phrases, and word and phrases to avoid.

43.2.1 SOW Imperatives

The imperative is signaled with a key word like shall or must. Imperatives are words and phrases that signal the work that must be provided. This is not unlike the imperatives used in a technical specification.

Use "shall" whenever a provision is mandatory; "Will" expresses a declaration of purpose or intent. Use "shall" when describing a provision binding on the contractor. Use "will" to indicate actions by the customer. Do not use "should" or "may" because they leave the decision for action up to the contractor.

See the specification section of this text for other imperatives. Basically the same rules apply to the SOW as were described for specifications.

43.2.2 SOW Sentence Structure Suggestions

Simple words, phrases, and sentences should be used. Well-understood words and phrases improve the SOW by minimizing confusion and misunderstanding. The SOW should state the outputs in clear, concise, commonly used, easily understood, measurable terms. Be brief, precise, and consistent. Obviously there are no absolutes. The following are general rules for writing requirement statements.

- Write understandable requirements for all stakeholders
- Avoid technical language
- Use simple wording in concise sentences
- Use active rather than passive voice, "The contractor shall...."
- Use verbs that identify work and performance that answer: "What are the work requirements?"
- When selecting appropriate work words, explicitly define the total nature of the work requirement
- Use consistent terminology throughout the SOW

43.2.3 SOW Words and Phrases to Avoid

There are words and phrases that should be avoided. The SOW should be stated explicitly in structured order avoiding words that allow for multiple interpretations.

- Do not use "Any," "Either," "And", "Or"; these words suggest the contractor can make a choice which may not support the intent of the customer
- Do not use "etc." it has no meaning
- Do not use pronouns repeat noun to avoid misinterpretation

This is a list of phrases that have multiple meanings. They should be avoided because they add confusion and add not value.

- To the satisfaction of the contracting officer
- As determined by the contracting officer
- In accordance with instructions of the contracting officer
- As directed by the contracting officer
- In the opinion of the contracting officer
- In the judgment of the contracting officer
- Unless otherwise directed by the contracting officer
- To furnish if requested by the contracting officer
- All reasonable requests of the contracting officer shall be compiled with
- Photographs shall be taken when and where directed by the contracting officer
- In strict accordance with
- In accordance with best commercial practice
- In accordance with best modern standard practice
- In accordance with the best engineering practice

- Workmanship shall be of the highest quality
- Workmanship shall be of the highest grade
- Accurate workmanship
- Securely mounted
- Installed in a neat and workmanlike manner
- Skillfully fitted
- Properly connected
- Properly assembled
- Good working order
- Good materials
- In accordance with applicable published specifications
- Products of a recognized reputable manufacturer
- Tests will be made unless waived
- Materials shall be of the highest grade, free from defects or imperfections, and of grades
- Approved by the contracting officer
- Kinks and bends may be cause for rejection
- Carefully performed
- Neatly finished
- Metal parts shall be cleaned before painting
- Suitably housed
- Smooth surfaces
- Pleasing lines
- Of an approved type
- Of standard type
- Any phrases referring to "The inspector"

See the specification section of this text for other words and phrases to avoid. Basically the same rules apply to the SOW as were described for specifications.

43.2.4 Work Words

Use verbs that identify work and performance task requirements and answer the explicit question: "What are the work requirements?" Use Active verbs. Examples include analyze, audit, calculate, create, design, develop, erect, evaluate, explore, interpret, investigate, observe, organize, perform, and produce. Avoid Passive verbs that can lead to vague statements. For example, the phrase "the contractor shall perform," is preferred in lieu of "it shall be performed" because the latter does not definitively state which party shall perform. The following sample list contains words, which have the inherent value of work.

Table 80 SOW Work Words

analyze	solve by analysis	interpret	explain the meaning of
annotate	provide with comments	inquire	ask, make a search of
ascertain	find out with certainty	integrate	to add parts to make whole
attend	be present at	investigate	search into; examine closely
audit	officially examine	judge	decide; form an estimate of
build	make by putting together	make	cause to come into being
calculate	find out by computation	maintain	to keep in existing state, to continue in, carry on
consider	think about, to decide	manufacture	fabricate from raw materials
construct	put together; build	modify	to change, alter
control	direct; regulate	monitor	to watch or observe
contribute	give along with others	notice	comment upon, review
compare	find out likeness or differences	observe	inspect, watch
create	cause to be; make	originate	initiate, to give rise to
determine	resolve; settle; decide	organize	integrate, arrange in a coherent unit
differentiate	make a distinction between	perform	do, carry out, accomplish
develop	bring into being or activity	plan	devise a scheme for doing, making, arranging activities to achieve objectives
define	make clear; settle the limits	probe	investigate thoroughly
design	perform an original act	produce	give birth or rise to
evolve	develop gradually, work out	pursue	seek, obtain or accomplish
examine	look at closely; test quality of	reason	think, influence another's actions
explore	examine for discovery	resolve	reduce by analysis, clear up
extract	take out; deduce, select	record	set down in writing or act of electronic reproduction of communications
erect	put together; set upright	recommend	advise, attract favor of
establish	set up; settle; prove beyond dispute	review	inspection, examination or evaluation
estimate	approximate an opinion of	revise	to correct, improve
evaluate	find or fix the value of	study	careful examination or analysis
fabricate	build; manufacture, invent	seek	try to discover; make an attempt
form	give shape to; establish	search	examine to find something
formulate	to put together add express	scan	look through hastily, examine intently
generate	produce, cause to be	screen	to separate, present, or shield
identify	to show or to find	solve	find an answer
implement	to carry out, put into practice	test	evaluate, examine

Table 80 SOW Work Words

install	place; put into position	trace	to copy or find by searching
inspect	examine carefully or officially	track	observe or plot the path of
institute	set up; establish, begin	update	modernize, make current

43.2.5 Product Word List

Many contracts include information products. The following is a list of information products that may be delivered as part of a contract.

Table 81 SOW Product Word List

Agenda	CD / DVD	Manuals	Records
Audio	USB Files	Manuscript	Reports
Books	Findings	Materials	Reproducible
Cards	Forms	Minutes Outlines	Requests
Certificates	Graphics	Pamphlets	Sheets
Charts	Guides	Plans	Specifications
Decks	Handbooks	Procedures	Standards
Disc-Magnetic	Illustrations	Proposals	Systems
Documentation	Ledgers	Publications	Tapes
Drafts	Lists	Recommendations	Transparencies
Drawings	Logs	Recordings	Visual Aids

43.3 General SOW

1.0 Introduction and Overview
1.1 Background
1.2 Scope of Work
1.3 Objectives
2.0 References
3.0 Task Descriptions
4.0 Deliverables
5.0 Schedules, PERT Charts
6.0 Travel
7.0 Progress and Compliance
8.0 Transmittal, Delivery, and Accessibility
9.0 Notes

43.4 Request for Proposal SOW

The following SOW is for a request for proposal and includes evaluating a section on proposed solutions. This SOW is usually referred to as Instructions for Proposal (IFPP).

1.0 Introduction and Overview
1.1 Background
1.2 Scope of Work
1.3 Objectives
2.0 References
3.0 Contract Task Descriptions
4.0 Contract Deliverables
5.0 Contract Performance Period, Schedules, PERT Charts
6.0 Proposal Evaluation Criteria
7.0 Proposal Instructions

43.5 System Intensive SOW

The following SOW outline is for a system intensive activity:

1.0 Scope
1.1 Locations
1.2 General Tasks
2.0 References
3.0 Specific Systems Tasks
3.1 Concept Analysis
3.2 Architecture Analysis
3.3 Human Factors
3.3 Requirements Management
3.4 Functional Analysis
3.5 Synthesis
3.6 Trade Studies
3.7 Interface Management
3.8 Specialty Engineering
3.9 Integrity of Analysis
3.10 System Development/Integration
3.11 System Production
3.12 System Performance
3.13 Sustainment
3.14 Provisioning
3.15 Investment Analysis
3.16 Quality Assurance
3.17 Technology Assessment, Infusion, Transfer
3.18 Lifecycle Management
3.19 Risk Management
3.20 System Security
3.21 Organizational Metrics
3.22 Configuration Management
3.23 Verification and Validation
3.24 Lifecycle Engineering
3.25 Maintain the System Engineering Process

3.26 Training
3.27 Operational Support

4.0 Program Management
4.1 Organization
4.2 Risk Management
4.3 Contract Work Breakdown Structure (CWBS)
4.4 Earned Value Management
4.5 Network Logic Schedule

5.0 Meeting Agendas and Minutes
5.1 Program Management Reviews
5.2 Technical Exchange Meetings (TEM)
5.3 System Engineering Reviews
5.4 Document Library

6.0 Schedule
7.0 Deliverables
8.0 Travel
9.0 Contract Administration
10.0 Other Direct Costs
11.0 Furnished Property

43.6 Developmental SOW A

3.0 Requirements
3.1 Program Management
3.1.1 Program Management Plan

3.1.2 Performance Analysis and Reporting
3.1.2.1 Program Schedule
3.1.2.2 Program Management Status Reports

3.1.3 Technical Meetings, Program Reviews and Conferences
3.1.3.1 Post Award Conference
3.1.3.2 Program Management Reviews
3.1.3.3 System Baseline Reviews
3.1.3.4 Technical Interchange Meetings

3.1.4 Contract Work Breakdown Structure
3.1.5 Data Management
3.1.6 Quality Assurance

3.1.7 Configuration Management
3.1.7.1 Configuration Identification
3.1.7.2 Configuration Control
3.1.7.3 Configuration Audits

3.2 Systems Engineering Management
3.2.1 System/Subsystem Description
3.2.1.1 Human Engineering Design
3.2.1.1.1 Approach
3.2.1.1.2 Planning and Execution
3.2.1.1.3 Objective
3.2.1.1.4 Scope
3.2.1.1.5 Human Factors Engineering Program Emphasis Areas

3.2.2 Product Engineering

3.2.2.1 Hardware
3.2.2.1.1 Parts Control
3.2.2.1.2 Manufacturing
3.2.2.1.3 Reliability
3.2.2.1.3.1 Reliability Prediction
3.2.2.1.3.2 Total Logistic Prediction

3.2.2.2 Software
3.2.2.2.1 Software Development Planning
3.2.2.2.2 Software Documentation
3.2.3 Interface Requirements
3.2.3.1 Technical Data Report Document

3.3 End Items

3.3.1 Prime Mission Equipment
3.3.1.1 First Article Systems
3.3.1.1.1 Configuration 1: Insert Name of Configuration
3.3.1.1.2 Configuration 2: Insert Name of Configuration
3.3.1.2 Follow-On Production Systems

3.3.2 Support and Test Equipment
3.3.2.1 Program Support Facility
3.3.2.2 Special Tools
3.3.2.3 Other Support and Test Equipment

3.3.3 Spares and Piece Parts
3.3.3.1 Site Spares
3.3.3.2 Depot Spares
3.3.3.3 Piece Parts

3.4 System Test and Evaluation
3.4.1 Test and Evaluation Program

3.4.2 Test and Evaluation Management

3.4.2.1 Test Conduct
3.4.2.1.1 Test Program Requirement Verification
3.4.2.1.2 Test Support
3.4.2.1.3 Test Readiness Reviews
3.4.2.1.4 Problem Reporting System
3.4.2.1.5 Regression Testing

3.4.3 System Test Program
3.4.3.1 Development, Test & Evaluation
3.4.3.1.1 Quality Control Inspections
3.4.3.1.2 Contractor's Preliminary Test
3.4.3.1.3 First Article Inspections
3.4.3.1.4 First Article Design Verification
3.4.3.1.5 First Article System Test
3.4.3.2 Production Acceptance Test & Evaluation
3.4.3.2.1 Production Test
3.4.3.2.2 Site Acceptance Test
3.4.3.2.3 Test Equipment

3.5 Integrated Logistics Support (ILS)
3.5.1 ILS Program
3.5.1.1 ILS Management Team
3.5.1.2 ILS Program Planning
3.5.1.3 Logistics Guidance Conference
3.5.1.4 Site Maintenance Staffing Values File Update
3.5.1.5 Logistics Support Analysis Records (LSAR)
3.5.1.5.1 LSAR Data Candidates
3.5.1.5.1.1 Data Candidates for LSAR Database
3.5.1.5.1.2 Maintenance Information
3.5.1.5.2 LSAR Data Selection Criteria for Developed Items

3.5.1.6 Provisioning
3.5.1.6.1 Provisioning Guidance Conference

3.5.1.7 Tools, Test and Support Equipment
3.5.1.7.1 Selection Criteria Priorities
3.5.1.7.2 Common Support Equipment
3.5.1.7.3 Special Tools, Support and Test Equipment

3.5.1.8 Technical Manuals
3.5.1.8.1 Technical Manual Development Plan
3.5.1.8.2 Non-Developmental Item Technical Manuals
3.5.1.8.3 Contractor Developed Technical Manuals
3.5.1.8.3.1 Manuals
3.5.1.8.3.2 Manual Verification and Validation
3.5.1.8.3.2.1 Verification

3.5.1.8.3.2.2 Validation

3.5.2 Training and Training Support
3.5.2.1 Training Planning and Management
3.5.2.1.1 Training Guidance Conference
3.5.2.1.2 Contracting Officer's Training Representative
3.5.2.2 Training Program Requirements
3.5.2.3 Access to Contractor's Facilities and Data
3.5.2.4 System Orientation Manual
3.5.2.5 Performance Examinations
3.5.2.6 Training Subject Areas
3.5.2.6.1 System Software Maintenance Training
3.5.2.6.1.1 Student Prerequisites
3.5.2.6.1.2 Training Description
3.5.2.6.1.3 Training Objectives
3.5.2.6.1.4 Training Materials
3.5.2.6.1.5 Special Software Maintenance Training Requirements
3.5.2.6.2 Second Level Engineering Training
3.5.2.6.2.1 Student Prerequisites
3.5.2.6.2.2 Training Description
3.5.2.6.2.3 Training Objectives
3.5.2.6.2.4 Training Materials
3.5.2.6.3 Hardware Maintenance Training
3.5.2.6.3.1 Student Prerequisites
3.5.2.6.3.2 Training Description
3.5.2.6.3.3 Training Objectives
3.5.2.6.3.4 Special Hardware Maintenance Training Requirements
3.5.2.6.3.4.1 Training Materials
3.5.2.6.3.4.2 Resident Training Option
3.5.2.6.3.4.3 Computer Based Instruction Option
3.5.2.6.3.4.4 On-the-Job Training Option

3.5.2.6.4 Depot Repair/Support Training
3.5.2.6.4.1 Training Content
3.5.2.6.4.2 Training Materials

3.5.2.6.5 System Operations Course
3.5.2.6.5.1 Student Prerequisites
3.5.2.6.5.2 Training Description
3.5.2.6.5.3 Training Objectives
3.5.2.6.5.4 Training Materials

3.5.2.6.6 Test Team Training
3.5.2.6.6.1 Student Prerequisites
3.5.2.6.6.2 Training Description
3.5.2.6.6.3 Training Objectives
3.5.2.6.6.4 Training Materials

3.5.2.6.7 On-Site Representative and Support Services Training
3.5.2.6.7.1 Student Prerequisites
3.5.2.6.7.2 Training Description
3.5.2.6.7.3 Training Objectives
3.5.2.6.7.4 Training Materials

3.5.2.7 Training Equipment and Materials

3.5.2.8 Developed Training

3.5.3 Maintenance Support
3.5.3.1 Hardware Maintenance
3.5.3.2 Software Maintenance
3.5.3.3 Support Transition
3.5.3.3.1 Software Support Transition
3.5.3.3.2 Hardware Depot Level Transition

3.6 Site Implementation
3.6.1 Site Preparation
3.6.2 Implementation Task Plan
3.6.4 Site Installation and Checkout
3.6.4.1 Test Configuration Review
3.6.4.2 Site Acceptance Testing
3.6.4.3 Retrofit Plan/Retrofit Kit Installation

3.7 Technical and Engineering Services
3.7.1 Test Rights
3.7.2 Preparation of Work Scopes for Minor Work
3.7.2 Labor Descriptions

4.0 Acronyms
5.0 Definitions

43.7 Developmental SOW B

3.0 Limited Production and Deployment Requirements
3.1 Program Management

3.1.1 Project Control
3.1.1.1 Program Metrics
3.1.1.2 Contract Work Breakdown Structure

3.1.2 Meetings, Reviews, and Conferences
3.1.2.1 Post Award Conference

3.1.2.2 System Engineering Reviews and Working Groups

3.1.2.2.1 Critical Design Review
3.1.2.2.2 Test Readiness Reviews
3.1.2.2.3 Post-Test Reviews

3.1.2.3 Other Reviews
3.1.2.3.1 Program Management Reviews
3.1.2.3.2 Technical Interchange Meetings
3.1.2.3.3 Logistics Guidance Conference
3.1.2.3.4 Provisioning Conference

3.1.3 Cost Management

3.1.4 Configuration Management Program
3.1.4.1 Contractor's Configuration Management Plan
3.1.4.2 Configuration Identification
3.1.4.2.1 Developmental Configuration
3.1.4.2.2 Configuration Baselines
3.1.4.3 Configuration Identifiers
3.1.4.3.1 Part/Item Identification Numbers
3.1.4.3.2 Software Identification Numbers
3.1.4.3.3 Type Designations
3.1.4.4 Configuration Control
3.1.4.4.1 Engineering Change Proposal
3.1.4.4.2 Requests for Deviation
3.1.4.4.3 Requests for Waiver
3.1.4.4.4 Parts Substitution
3.1.4.4.5 Change Description
3.1.4.4.6 Notice of Revision
3.1.4.5 Configuration Status Accounting
3.1.4.6 Discrepancy Reporting System
3.1.4.7 Configuration Audits
3.1.4.7.1 Functional Configuration Audit
3.1.4.7.2 Physical Configuration Audit
3.1.4.8 Bar Code Marking

3.1.5 Data Management Program
3.1.5.1 Integration of Contractor's Data Management Effort
3.1.5.2 Data Control and Status Accounting
3.1.5.2.1 Controls over Subcontractor Data

3.1.6 Quality Assurance

3.2 Systems Engineering and Design Development
3.2.1 Systems Engineering Documentation
3.2.1.1 System Design Description
3.2.1.2 Interface Control Documentation
3.2.1.3 Human Engineering Design

3.2.1.3.1 Approach
3.2.1.3.2 Planning and Execution
3.2.1.3.3 Objective
3.2.1.3.4 Scope
3.2.1.3.5 Human Factors Engineering Program Emphasis Areas

3.2.2 Software Development
3.2.2.1 System Software Design
3.2.2.2 Software Documentation
3.2.2.2.1 Software Development Plan
3.2.2.2.2 Software Requirements Description
3.2.2.2.3 Software Design Description
3.2.2.2.4 Software Test Plan
3.2.2.2.5 Software Test Description
3.2.2.2.6 Software Test Report
3.2.2.2.7 Software Version Description
3.2.2.2.8 Software User Manual
3.2.2.2.9 Software Product Specifications
3.2.2.2.10 Interface Design Document
3.2.2.2.11 Technical Data Report Document
3.2.2.3 Development of Non-Operational Software

3.2.3 Hardware Identification and Selection

3.2.4 Access to System Development Environment
3.2.5 Reliability Program
3.2.6 Maintainability Program
3.2.7 Electromagnetic Interference and Susceptibility
3.2.8 Facilities
3.2.9 Security

3.3 System Test and Evaluation

3.3.1 Development Test and Evaluation
3.3.1.1 Design Qualification Tests
3.3.1.2 Formal Qualification Tests
3.3.1.3 First Article Tests
3.3.1.3.1 Factory Acceptance Tests
3.3.1.3.2 Site Acceptance Tests
3.3.1.3.3 Delta DT&E Tests
3.3.1.3.3.1 System Maintainability
3.3.1.3.3.2 System Reliability
3.3.1.3.3.3 System Alignment
3.3.1.3.3.4 Algorithm Implementation
3.3.1.3.3.5 Interface Tests
3.3.1.3.3.6 Continuous Operation

3.3.2 Production Acceptance Test and Evaluation
3.3.3 Operational Test and Evaluation
3.3.4 Test Equipment
3.3.5 Re-testing
3.3.6 Test and Evaluation Training Program Development
3.3.6.1 Developmental Test and Evaluation Briefing
3.3.6.2 Operational Testing and Evaluation Training

3.4 Limited Production Installations
3.4.1 Limited Production Installation Management
3.4.2 Site Survey
3.4.3 Contractor Site Engineering Reports
3.4.3.1 Red-lined Installation and Facility Drawings
3.4.3.2 Final As Built Drawings
3.4.4 Facility Site Preparation and Physical Facility Modification
3.4.4.1 Test Equipment
3.4.4.2 Priority and Coordination of Activities
3.4.4.3 Site Security
3.4.5 Site Acceptance Test
3.4.5.1 Clean Up

4.0 Integrated Logistics Support
4.1 ILS Program Management
4.1.1 ILS Management Team
4.1.2 ILS Program Planning
4.1.3 Maintenance Concept

4.2 Logistics Management Information (LMI)
4.2.1 LMI Requirements
4.2.2 LMI Data Reviews
4.2.3 LMI Acceptance Criteria
4.2.4 Level of Repair Analysis

4.3 Supply Support
4.3.1 Site Spares
4.3.1.1 Tools and Test Equipment

4.4 Technical Data
4.4.1 Technical Drawings and Associated Lists
4.4.1.1 Product Drawings
4.4.1.2 COTS Manuals and Documentation
4.4.1.3 COTS Hardware Description Document
4.4.2 In-Process Review
4.4.3 Operator and Maintenance Technical Instruction Books
4.4.3.1 Technical Instruction Books
4.4.3.2 Formal Draft Operator's Manual Manuscript

4.4.3.3 Formal Draft Operator's Manual Manuscript
4.4.3.4 Formal Draft Site-Level Maintenance Manual Manuscript
4.4.3.5 Second-Level Engineering Maintenance Manual Manuscript
4.4.3.6 Final Manuscript for Review
4.4.3.7 Final Reproducible Manuscript for Technical Instruction Book
4.4.3.8 Support/Test Equipment Manuals
4.4.3.9 Technical Instruction Books Validation

4.5 Status Reporting

5.0 Maintenance
5.1 Contractor Depot Logistics Support Requirements
5.2 E&R LRU Repair Procedure
5.3 Lay-In Spares
5.4 Post-Commissioning Technical Site Support
5.5 Second Level Engineering Support
5.6 Contractor Transition Plan

6.0 System Support Equipment/Facilities Services
6.1 Special Test Equipment
6.2 Computer Resources Support
6.3 Contractor Support Services

7.0 Full Production and Deployment Requirements
7.1 Program Management
7.1.1 Meetings, Resources, and Conferences
7.1.2 Configuration Management Program
7.1.3 Data Management Program

7.2 Systems Engineering
7.3 Full Production Installations
7.4 Training
7.5 Status Reporting
7.6 Testing

8.0 Acronyms

43.8 Other SOW Samples

There are SOWs that are not associated with a system specification. These SOWs tend to request services. They include general requirements associated with the performance of the work to be accomplished. The requirements are offered very early, next to the tasks in the SOW.

1.0 Introduction and Overview
2.0 Requirements

3.0 Tasks and Services

The follow is a list of other SOW samples that may or may not be paired with a system specification:

- **Non development Item:** Quality Control, Configuration Management, Configuration Management, Configuration Identification (CI), Configuration Baselines, Configuration Control, Configuration Audits, Functional Configuration Audit (FCA), Physical Configuration Audit (PCA), Configuration Status Accounting (CSA) Information, Engineering Change Proposals (ECP), System Refresh and Upgrade, System Technology Infusion
- **Software Development:** Software Development Plan, Size, Cost, Schedule, and Critical Computer Resource Estimations, Work Breakdown Structure, Risk Management, Project Tracking, Metrics, Software Development Folders/Files (SDF), Information Product or Contract Data Requirements List Tracking, Requirements Analysis, Software Design, Implementation, Coding, Testing, Documentation, Enhancements and Corrections, Reviews, Modeling and Simulation, Transition Support, Software Quality Assurance, Configuration and Data Management, Configuration Management and Control, Data Management, Independent Verification and Validation, Status and Project Support, Post Project Review, Furnished Information, Facilities, and Equipment, Other Special Requirements
- **Commercial Hardware:** Quality Assurance, Meetings, Conferences, and Reviews, Configuration Control, Reliability, Maintainability, and Availability (RMA), Human Factors Engineering, Procurement of Hardware and Software, Inkjet Printer, Other Materials, Documentation, Pre-Installation Activities, Site Surveys, Site Installation and Integration Planning, Factory Acceptance Test Planning, Site Acceptance Test Planning, Pre-Installation Preparation, Installation, Integration and Acceptance Testing, Installation and Integration, Site Acceptance Testing, Post-Installation Activities, Site Acceptance Test Report, Software and Documentation, Decommissioning of Equipment, Training Support, Integrated Logistics Support, Support Equipment, Packaging, Handling, Storage, and Transportation (PHS&T), Interim Contractor Maintenance and Logistics Support, Operational Hardware Maintenance Services, Replacements of Consumable Items, Software Maintenance, Maintenance Log, Support Facilities, Corrective Action Plan, Lowest Replaceable Unit Identification and Marking, Engineering Support
- **Program Implementation:** System Integration, Facility Engineering, System Deployment, Environmental and Site Surveys, Investment Analysis
- **Technical Management and Admin Support:** Technical Support Tasks,

Managerial Support Tasks, Administrative Support, Other Direct Costs, Facilities, Travel, Electronic Communications, Miscellaneous, Program Management, Documentation, Reporting Requirements

- **Financial management Support:** Business Management Support, Investment Analysis, Sustainment Cost Analysis, Cost Estimating
- **Instructional Services:** Instructor Qualifications, Instructor Recruitment, Developmental Instructor Program, Special Techniques Training, Instructor Certification, Shadowing Assignments, Additional Instructor Qualifications, Curriculum Development Training, Curriculum Approval, Instructional Approach, Academic Accreditation, Student Progress, Seminars and Workshops, Training Programs, Correspondence Study Programs, Observance of Legal Holidays and Administrative Leave, Facilities, Equipment, Materials and Services, Cardio-Pulmonary Resuscitation (CPR) Training, Curriculum A, Curriculum B
- **Depot Logistics Support:** Reporting Requirements, Performance Metrics, Meeting Support, Exchange and Repair (E&R), Contractor Repair, Testing, Inspection, and Acceptance, Piece Parts, Expendables and Consumables, Warranty, Requisitioning Procedures, Logistics Priorities and Response Time, Quality Assurance Program, Contractor Repaired Items Quality Assurance, Audit of the Inventory of Property, Common Spares Management, Parts Obsolescence, OEM Second Sources, Site Licenses and Data Rights, Replacements When Identical Parts Become Unavailable, Bar-coding, Definitions, Contractor Depot Logistics Support, Exchange and Repair Items, Customer Care Center Representative, Floor Replaceable Unit, Furnished Property Material, Logistics and Inventory System, Lowest Replaceable Unit, Repairable, Serviceable, Expendable, Site Spare Parts, Personnel at Sites, Logistics Center, Packaging, Handling, Storage, and Transportation
- **IT Support:** Labor Categories and Skill Sets, Software Applications Development And Maintenance Support, Software Applications Support Services, Asset Management and Procurement Support, Security Clearances, Transitional Implementation Plan, Weekly/Monthly Status Reports, Oral Reports, Automated Reports
- **Copier Fax Printer Maintenance:** Response Time, Repair, Maintenance, Monitoring, and Work Documentation
- **Furniture:** Workstations, Electrical Systems, Power Requirements, Telecommunications and Data Requirements, Free Standing/Conventional Case Goods, Reception Area, Conference Room, File and Storage Systems, Installation
- **Investment Assessment Support Services:** Program Management, Scheduling, Baseline Management, Earned Value Management, Program

and Technical Risk Management, Investment Alternative Analysis, Organizational and Performance Metrics, Configuration Management, Knowledge Management, Strategic Planning, Business Process Modeling, Modeling and Performance Analyses, Operational Assessment Models, Maintenance and Operations Support, Contract Management, Contractor Responsibilities, Interrelationships of Contractors

- **Cost and Financial Support Services:** Program Level Financial Management Support, Financial Baseline Maintenance, Budget Formulation, Budget Presentation, Budget Execution, Budget Forecasting, Business Management and Financial Support, Financial Reporting Support, Cost and Benefit Analysis, Cost and Schedule Risk Assessments
- **Business Process Re-engineering:** Organization Assessment, Alternatives, Recommendations, Pilot Implementation, Full Scale Rollout
- **ISO Registration:** Lead Auditor Training Program, Ad hoc Instructional Sessions, Training and Session Materials, Advisory Services, Document Review, Audit Activities, Technical and Activity Management, Location of Work Performance, Equipment, Materials and Facilities, Training Support
- **Moving Services**: Excess Property Pickup, Furniture Installations/Reconfigurations, Moving Services Activities, Event Activities, Special Furniture Cleaning, Special Furniture Repair, Special Moving Activities, Special Handling Moving Activities, Contract Work Hours, Points of Contact
- **Local Travel Services:** Vehicle Marking/Identification, Vehicle Operators, Qualification of Employees, Trip Record, Schedule Delays, Large Capacity Vehicles, Equipment Deficiencies
- **Health and Fitness Services:** Health and Fitness Specialist, Cleaning Fitness Center Equipment, Facility, Furnished Property and Services
- **Guard Services:** Guard Supervisor Requirements and Qualifications, Guard Staffing Requirement, Furnished Equipment, Firearms Requirements, Training, Guard Operations, Contractor Guard Manual, Incident Reporting
- **Food Services:** Facilities, Hours, Responsibilities, Utilities, Catering, Exclusivity
- **Custodian:** Work Schedule, Interference with Business, Protection of Property, Environmental Management, Products and Materials, Basic Services, Space Cleaning, Floor Care, Restroom Services, Other Services, Service Call Work, Service Calls for Cleaning, Service Calls for Lighting, Stripping and Rewaxing Floors, Shampooing Carpets and Rugs, Cleaning Light Fixtures, High Dusting/Cleaning, Cleaning Exterior Glass, Cleaning Interior Glass, Cleaning Venetian Blinds, Government Furnished Property and Services, Contractor Furnished Items, Definitions
- **Window Cleaning:** Holidays, Security, Lost and Found Property, Key

Control, Materials, Equipment and Tools, Utilities, Storage Space, Janitor's Closets, and Locker Rooms, Coordination, Quality Control, Unforeseen Facility Closures, Safety and Fire Prevention, Personnel Training

- **Grounds Maintenance:** Purpose, Work Outside Regular Hours, Work Control, Herbicides and Pesticides, Contractor's Superintendent, Uncertified Personnel, Pesticide Use Records, Continuity of Services., Maintenance Level I (Administrative Areas), Maintenance Level II, Maintenance Level III, Maintenance Level IV (Railroad and Power line Right of Way), Grass Cutting, Edging, Plant and Shrub Pruning, Cultivation and Mulching of Shrubs, Hedges, and Flower Beds, Fertilization, Trash and Litter Collection and Disposal., Raking, Tree Removal, Stump and Above Ground Root Removal, Seeding, Erosion Control, Under Brushing, Tree Pruning, Irrigation, Ditch Cleaning, Tree and Shrub Establishment, Trees, Shrubbery, Severe Shrub Pruning, Vegetation Removal, Sod, Sprigging, Furnished Facilities, Furnished Equipment, Furnished Material, Availability of Utilities
- **Weather Observation Services:** Location, Surface Observations, Aviation Observations, Records, Facility, Telephone Service, Forms, Records, Supplies, Regulatory Handbooks and Manuals
- **Building Maintenance:** Work Schedule Control Records and Reports, Building Monitors, Staffing, Standards, Major Repair, Replacement Modernization Renovation, Manufacturer and Installer Warranties, As Built Drawings, Weather or Vandalism Damages, Work Outside Regular Hours, Backlogged Calls and Work, Emergency Urgent Routine Calls, Beyond Scope of Call, Materials and Equipment, Historical Data, Preventive Maintenance, Lighting, Start-up Shut-down of HVAC Systems, Labor Requirements, Material Requirements, Construction and Weight Handling Equipment Requirements, Review of Proposed Work Scopes, Labor Material Equipment Costs, Urgent Routine Work, Handbooks, Travel Zone Maps, Floors and Floor Coverings, Interior Walls Ceilings and Trim, Doors, Stairs and Stairwells, Traverse and Curtain Rods, Venetian Blinds and Shades, Cabinets and Countertops, Exterior Trim and Walls, Roofing, Gutters and Downspouts, Exterior Concrete and Masonry, Exterior Accessories, Stairs, Doors, Windows, and Screens, Swimming Pools, Signs, Certificates, Painting, Plumbing, Electrical, Lock Smith, Heating Ventilation Air Conditioning Refrigeration, Security Fences and Wire Cages, Metal Working, Welding, Machinist Tasks, Food Service Equipment
- **Elevator Maintenance:** Work Outside Regular Hours, Preventive Maintenance Inspection And Service, Service Work, Inspections, Testing, And Certification
- **Electrical Panel Replacement:** Construction Phasing, Coordination, Field

Engineering, Cutting And Patching, Submittal Procedures, Shop Drawings, Conferences, Samples, Control Of Installation, Temporary Facilities, Barriers And Fencing, Water Control, Protection Of Existing Equipment And Existing Building, Security, Work Hours, Starting Systems Or Equipment, Final Cleaning, Project Record Documents, Codes And Permits, Products, Inspection

- **Lighting:** Temporary Utilities, Protection Of Existing Utilities And Cables, Coordination, Layout, Liability Insurance, Site Clearing And Grubbing, Excavation And Backfill, Access Roads, Walkways, Site Surfacing, Base Course For Slab On Grade, Concrete Forms, Steel Reinforcement, Concrete, Finishes, Epoxy Floor Coating, Rubber Matting, Painting, Heating, Ventilating And Air Conditioning, Engine Generator Installation, Above Ground Storage Tank, Electrical Installations, Lightning Protection System, Interior Electrical Work, Generator
- **Roofing:** Licenses/Certification, Insurance Coverage, Hours of Operation, Motor Vehicle Operations, Environmental Management, Records, Monthly Reports, Warranty, Roofing System Inspection, Roofing System Maintenance, Roofing System Repair, Installed Equipment, Hazardous Materials Inspection, Job Site Inventory, Waste Material and Debris, Safety, Material Selection Samples, Site Areas Inspection
- **Lightning protection Services:** Barriers And Fencing, Codes And Permits, Conduct Of Work, Connections, Construction Progress Schedule, Cutting And Patching, Disconnect, Down Conductors, Field Engineering, Final Cleaning, Fire Stopping, Grounding And Bonding Products, Inspection And Testing Laboratory Services, Inspection Of Site By Contractor, Installation, Installing Fire-Resistive Joint Sealant, Installing Through-Penetration Fire stops, Lightning Protection, Miscellaneous Conductors, Project Record Documents, Protection Of Existing Equipment And Existing Building, Protection Of New Work, Quality Assurance, Control Of Installation, Roof And Down Conductors, Security, Shop Drawings, Temporary Facilities, Transient Protection, Wire And Cable Conductors, Work Hours
- **Architectural Engineering:** Drawings, Specification, Cost Estimate, Design Analysis, Electronic Deliverables, Engineering Services, Meetings, Site Investigation, Existing Conditions Survey and Report, Geo technical Investigation and Report, Hazardous Materials Investigation and Report, Environmental Protection Requirements Investigation, Report, Permit Applications and Supporting Documents, Building Finishes Package, Quality Assurance, Facility Management and Contract Administration Support, Contract Deliverables, Design Submissions, Schematic Concept Submission, Design Development Submission, Final Submission, Bidding Document Submission, Labor Category Descriptions, Key Personnel, Minimum Qualifications

- **Paving Roads:** Protection of Existing Utilities and Cables, Coordination, Layout, Liability Insurance, Permits, Site Clearing and Grubbing, Excavation and Backfill, Access Roads, Site Surfacing, Base Course, Concrete Forms, Steel Reinforcement, Concrete, Materials, Execution
- **OSHA Services:** Health and Safety Plan, Project Coordination, Laboratory Analytical Data
- **Tiered environmental impact study:** Identify Impacted Stakeholders, Conduct Stakeholder Meetings, Prepare Distribute Coordination Letters, Receive, Analyze, Summarize and Document Stakeholder Comments, Determine Purpose and Need, Develop Alternatives, Determine the Affected Environment, Conduct Environmental Analysis of Alternatives to Compare Environmental Consequences, Prepare a Draft Environmental Impact Statement, Circulate Draft Environmental Impact Statement for Public Comment, Collect, Categorize and Respond to Comments, Conduct Public Hearings, Prepare Final Environmental Impact Statement, Circulate the Final Environmental Impact Statement, Prepare Draft Record of Decision,, Second Tier EIS Tasks, Conduct Stakeholder Meetings, Determine Purpose and Need, Develop Alternatives, Determine The Affected Environment, Conduct Environmental Analysis of Alternatives to Compare Environmental Consequences, Prepare Draft Environmental Impact Statement, Circulate Draft Environmental Impact Statement for Public Comment, Collect, Categorize and Respond to Comments, Conduct Pubic Hearings, Prepare Final Environmental Impact Statement, Circulate Final Environmental Impact Statement, Prepare Draft Record of Decision, Community Involvement Tasks, Measurement Tasks, Support Tasks
- **Lead Based Paint Abatement:** Hours Of Operation, Motor Vehicle Operations, Mission Interference, Security Requirements, Pre-Construction Conference, Projector Inspector, Warranty Of Work, Government Data And Services, Management Plans, Services, Contractor Information And Project Data, Liability Insurance, Licenses And Permits, Plans, Plan Guidance, Plans And Report/Data Requirements, Contractor Quality Control (CQC) Plan (When Applicable), Protective Safety Equipment, Auxiliary Generator, Construction Equipment, Daily Log, Activity Hazard Analysis, Replacement Of Doors, Windows, And Etc., Potential Hazards Monitoring, Lead-Based Paint Removal Requirements, Heat Blower Guns, Vacuum Blasting Techniques, Certified Industrial Hygienist/Industrial Hygienist, Sampling Results, Wipe Concentration Levels, Lead-Based Paint Control Area, Work Site Area, Official Visual Inspection, Final Cleaning Activities, References And Standards

Why the emphasis on these other SOW examples?

This is an extremely important issue related to system practices. Many managers come from non-system engineering or practices arenas. They may find themselves in settings needing systems practices but they instead use practices normally associated with extremely stable services. They need to know that applying these practices will severely damage the project or program. For example it is not uncommon for organizations entering the systems engineering arena to place technical requirements in the SOW, as might be found in a custodian service contract.

Similarly a system will include subsystems that can be implemented using extremely stable services. For example if a system includes facilities then simple and stable but extremely important services are needed (e.g. Custodial, Building Maintenance, etc). Creating complex SOWs based on separate requirements is unnecessary.

43.9 Internal SOW

In extremely large vertically integrated companies the subsystems and components can be purchased from internal organizations. At what point should the SOW be developed in this setting?

If the organization responsible for developing the system decides to solicit external companies, then the SOW is a given and also should be offered to the internal company organization. If the organization responsible for developing the system decides to not solicit external companies, then the SOW should be developed for the internal organization if it is loosely coupled to the organization responsible for the system. Loosely coupled suggests:

- Different geographic location outside of 30 minute range
- Different profit center
- Different business specialty area
- If it is a product division with established products
- If there is a history of problems between the organizations

43.10 After Contract Award

After contract award, the contract SOW becomes the standard for measuring contractor performance. Stakeholders must be aware of the contractual and legal implications of its language and statements. It defines the contract and is subject to the interpretations of contract law. This suggests final legal review prior to contract award and acceptance.

The customer and the contractor will use it to determine their rights and obligations after contract award. In a dispute concerning performance, rights, or obligations, clearly defined requirements help enhance the legal enforceability of a

SOW. The SOW has a high level of precedence in the solicitation document and contract.

43.11 Exercises

1. Develop a SOW outline for developing the architecture of an ocean wind energy system, an urban wind energy system, and rural wind energy system.
2. Develop a SOW outline for building and operating an ocean wind energy system, an urban wind energy system, and rural wind energy system.

43.12 Additional Reading

1. Handbook for Preparation of Statement of Work, MIL-HDBK-245C 10 September 1991, MIL-HDBK-245D 3 April 1996.
2. Standard Practice Data Item Descriptions (DIDs), DOD, DOD-STD-963A 15 August 1986, MIL-STD-963B 31 August 1997.
3. Work Breakdown Structure, DOD, MIL-HDBK-881, 2 January 1998.
4. Work Breakdown Structures For Defense Materiel Items, DOD, MIL-STD-881 1 November 1968, MIL-STD-881A 25 April 1975, MIL-STD-881B 25 March 1993.

44 Procuring Goods Services Systems

There are three broad approaches to procuring goods and services. The first approach is to buy an existing product either from the manufacturer, distributor, or retailer. A product can be either a hard good, soft good, and or service. This is the simplest form of procurement with the least risk. If a product does not exit in a complete form then development needs to happen using a contract. There are two additional procurement approaches and they are based on contracts. The second approach to procurement is to enter into a contract as part of a competitive bid process. This is a complex process, which has risk. The third approach is to enter into a sole source contract with a vendor. This may simplify the procurement process but it has enormous risk because all transparency and visibility is removed.

Author Comment: Everyone has an obligation to be an educated smart consumer. Otherwise inefficient predatory monopolistic vendors will rise and yield an unsustainable industrial base. The vendors have an obligation to support the consumer by offering clear non-obfuscated descriptions and specifications of their solutions. So this is a serous systems practice not to be taken lightly or relegated to a closed room. "The mind is not a vessel to be filled but a fire to be ignited"[367].

44.1 Buyers

Commodity based goods and services available through multiple distribution sources can be handled using a buyer or team of buyers. The products and services are well-known and stable with industry accepted specifications. The costs may vary depending on quantity and need time. A buyer assigned to a class of such commodities can negotiate and make the purchases with little complication. In some cases the buy includes certain quality factors such as full parts inspection or parts screening either by the vendor, the buying organization, or a third part.

44.2 Request for Informatio n

A Request for Information (RFI) is a formal request from industry for information about their products, systems, components, subsystems, services, etc. It is used to gather data on the current state of the industrial base so that an effective Request for Proposal (RFP) can be developed.

It involves offering a broad statement of the possibility of a project or program.

---

[367] On Listening to Lectures, Mestrius Plutarchos (ca. 46 - 127) a Greek historian writer.

This statement triggers interest, and information is offered by all parties that may be followed up with formal presentations.

The vendors use the RFI to prepare for a future RFP. This preparation may include soliciting various stakeholder needs, developing internal plans, and even investing in research and development in anticipation of the RFP. It is also an opportunity for the vendors to influence the RFP structure and requirements via an educational interchange that may include white papers, presentations, samples, demonstrations, facility tours, etc.

44.3 Request for Proposal

A Request for Proposal (RFP) is a formal vehicle that is used to solicit bids from vendors offering systems, subsystems, services, components, etc. It is the culmination of the RFI and the start of a project or program. The RFP includes an A-Level Specification that houses the technical requirements, a Statement of Work (SOW) that describes the work to be done, deliverables to be provided, and a contract terms and conditions, which may be part of the SOW.

- **Statement of Work**: The SOW is described elsewhere in this text.
- **A-Level Specification**: The A-Level Specification is described elsewhere in this text.
- **Contract Terms and Conditions**: The terms and conditions of the contract are unique to each industry. They include can include incentives, penalties, and expected vendor behavior.

44.4 Cost Realism

Whenever proposals are received they need to be evaluated for cost realism[368]. At the simplest level the proposals should offer the same goods and services so the costs compared. Bids that are significantly lower or higher are suspect proposals and need to be evaluated to determine if they are offering more functionality, better performance, higher quality, or a long history of previous work. It is possible that many proposal submissions may be coming from organizations with little or no history and only one proposal with a significantly different cost but long history of similar work is offered. So cost realism is not just looking for a statistically different point in a graph. It needs to consider the fundamentals of the offering organizations.

Usually the extreme lowest bidder is eliminated out of the competition because of cost realism. However if all vendors are low and one is high, the highest bidder

[368] Contract Pricing Reference Guides, Volume 1 - Price Analysis, Volume 2 - Quantitative Techniques for Contract Pricing, Volume 3 - Cost Analysis, Volume 4 - Advanced Issues in Contract Pricing, Volume 5 - Federal Contract Negotiation Techniques, Federal Acquisition Institute (FAI) and the Air Force Institute of Technology (AFIT), 2011.

may be selected because internal cost analysis suggests its cost is reasonable and the lower cost vender bids are not reasonable. They are trying to buy into the contract and not stating that they are willing to invest their own funds to make up the difference.

44.5 Vendor Product Evaluation and Selection

"We agree to support the Apollo 11 mission" is a quote from the movie "The Dish"[369]. It is about how the Parkes Observatory in Australia was used to relay the live television broadcast of man's first steps on the moon, during the Apollo 11 mission in 1969[370].

Vendor product evaluation and selection is the receiving end of someone's system practices. It is the successful application of those practices which results in a product, system, component, service, etc that is now being evaluated for inclusion in another system.

The vendor product evaluation and selection is a phased approach where a course tradeoff analysis is performed to narrow the list as time progresses and more is learned. The process begins by reviewing vendor data. This may be followed by a vendor invitation to gather your requirements while they offer a presentation and perhaps a demonstration. There may or may not be a formal request for information (RFI). After the presentation they may leave samples for further evaluation. This may be followed up by visits to the vendor location. At this point a Request for Proposal (RFP) is produced and the vendors respond with a formal proposal.

Prior to final selection there always should be a site visit to validate the information gathered and confirm perceptions by the team. When selecting a vendor there are three broad categories of tradeoff criteria: (1) Technical, (2) Non Technical, and (3) Sustainability.

The non-technical criteria include:

- **Company Stability**: As represented by years in business, gross revenue or size of company, profits, level of research and development, layoff history (brain drain). Layoff history is an extremely serious issue and must be properly explained. If an organization does not value its employees it will

---

[369] The Dish, Movie, Director: Rob Sitch, Writing credits: Santo Cilauro, Tom Gleisner, Jane Kennedy, Rob Sitch, Warner Brothers, 27 April 2001.

[370] Six hundred million people witnessed Neil Armstrong's first steps on the Moon through television pictures transmitted to Earth from the lunar module, Eagle. Parkes Radio Telescope pictures were so good that NASA stayed with them for the rest of the 2½-hour moonwalk after 10 minutes of transmission from Honeysuckle Creek tracking station near Canberra, and NASA's Goldstone station in California. A violent squall struck Parkes and the telescope was buffeted by strong winds that swayed the support tower and threatened the integrity of the telescope structure.

not value its customers.

- **Location**: Are there multiple locations, any export import issues, transport issues, access to universities, major metropolitan areas, transportation (roads, airports, railroads, and or ports).
- **Market Share**: Is the market dominated by the vendor, is the vendor a number two, or number three. If not dominant, why.
- **Product Stability**: Years in product business area, maturity of the product, degree of obsolescence, degree of state of the art, proven operations.

The technical criteria include:

- **Features**: Including features not found in other offerings, features with possible future applications, features that push the state of the art.
- **Technology Readiness Level**: What is the TRL rating using one of the standard scales described in this text, TRL 1 to 10.
- **Technology Maturity**: What is the technology maturity level as described in this text, bleeding edge, leading edge, state of the art, dated, or obsolete.
- **Performance**: Areas that exceed other offerings, performance that is or exceeds the state of the art.

The sustainability criteria include:

- **Total Energy**: This is a measure of the total energy used in production, shipping, operating, and disposal. Many items consume more energy in production, shipping, and disposal than operation. Examples include personal computers, which consume less energy in operation than production, shipping, and disposal because of its short life. Other examples are items that do not consume energy such as screws. A refrigerator will consume more energy in operation than production because of its long life.
- **Total Carbon Release**: Just like energy, this is the carbon footprint of a product and its allocation to production, transport, operation, and disposal. The carbon footprint will track total energy.
- **Internal Sustainability**: This is a measure of the product, system, services, component, subsystem ability to survive through the years. It is a measure of its own sustainability.
- **External Sustainability**: This is a measure of the impact on other systems. This traditionally is negative environmental impact such as pollution but it also can be impacts on quality of life. Examples include stressing existing educational, health care, transportation, power, water, sanitation, communications and other systems.
- **Cost Shifting**: This is a measure of the degree of cost shifting to other

stakeholders. It is most egregious if the stakeholders are not aware of the cost shifting. Examples include fracture gas drilling where gas and or oil may be extracted but surrounding water wells may be destroyed rendering other properties worthless.

In addition the tradeoff criteria identified in the architecture selection should be considered. Just as in the case of architecture selection the approach to vendor product selection should be the most effective solution not the lowest cost. So the MOE should be used as described in the architecture selection section.

MOE = Sum of Vendor Criteria / Vendor Cost

Other criteria that may surface include:

- **Product**: performance suitability, transparency (documentation, specifications, etc), functional match (function analysis, % requirements delivered, etc), security, safety, maintainability, update cycle, maturity, upward compatibility, quality, reliability, architectural compatibility, portability, resource utilization efficiency, maintenance costs fees, interoperability (proprietary elements force commitment to vendor)
- **Vendor**: reputation (credibility, stability, longevity, credentials, management), technical support (scope and responsiveness), willingness to negotiate changes, training support, competitive standing, periodic vendor release dates accommodate target systems delivery dates, references
- **Stakeholder Acceptability**: familiarity with candidate, open attitude to new technology offered, training to develop candidate expertise

44.6 Sole Source

Sometimes there is a need to offer a contract to a vendor using a sole source approach. In this setting there is no competition. Instead a strong case is internally made to justify the sole source approach. The following comes from the US Federal Acquisition Rules (FAR)[371] and is offered as a guide to justify a sole source contract:

- Only one source and no other supplies or services can satisfy requirements
- Unusual and compelling urgency
- Industrial mobilization; engineering, developmental, or research capability; or

[371] FAR 6.302. Federal Acquisition Regulation, General Services Administration, Department Of Defense, National Aeronautics And Space Administration, (This edition includes the consolidation of all Federal Acquisition Circulars through 2001-27), March 2005.

expert services

- International agreement
- Authorized or required by statute
- National security
- Public interest

44.7 Exercises

1. What approach did you use to buy your last article of clothing?
2. What approach did you use to buy your last computer, furniture item, automobile or other expensive item?
3. How important is company stability when evaluating proposals?
4. Prepare a simple RFI package for a regional alternative energy system, not to exceed 5 pages. What are the most important elements in the RFI?
5. Prepare a response to the RFI, not to exceed 2 pages. What are the most important elements in the RFI response?
6. Prepare an RFP, not to exceed 5 pages. What are the most important elements in the RFP?
7. Form groups of 3 or more students and pick an RFP from the group. Prepare separate proposal responses to the RFP, not to exceed 5 pages each. What are the most important elements in the proposals?
8. Gather proposals responding to the same RFP, perform an evaluation, and select the best proposal.

44.8 Additional Reading

1. Contract Pricing Reference Guides, Volume 1 - Price Analysis, Volume 2 - Quantitative Techniques for Contract Pricing, Volume 3 - Cost Analysis, Volume 4 - Advanced Issues in Contract Pricing, Volume 5 - Federal Contract Negotiation Techniques, Federal Acquisition Institute (FAI) and the Air Force Institute of Technology (AFIT), 2011.
2. Federal Acquisition Regulation, General Services Administration, Department Of Defense, National Aeronautics And Space Administration, (This edition includes the consolidation of all Federal Acquisition Circulars through 2001-27), March 2005.

45 Proposals and Business Plans

Proposals are in response to a sponsor with money. It is usually part of an open competition in a competitive bid process. Occasionally there are unsolicited proposals, however there is still a sponsor with some mechanism to access funds if the proposal makes sense for the sponsor. This may result in a sole source contract.

Business plans are based on someone deciding there is a market need that can be filled. This new insight surfaces because a new technology or new product idea surfaces. There is no sponsor with money waiting to be given to the organization. So initial development happens using funds that may be completely lost.

Many treat proposals separately from business plans. I decided to merge proposals and business plans to stimulate thinking about the ramifications of each approach in a high technology society, especially as related to sustainability.

Today many traditional high technology U.S. companies rely exclusively on proposals and funding from U.S. government sponsors rather than market analysis and a business plan. In the past, shortly after World War II that was not the case. Many high technology companies were commercial companies that had divisions supporting the U.S. government. Today we have companies that work exclusively for the U.S. government using the proposal mechanism or sole source contracts. Meanwhile in the 1980's the U.S. decided to use commercial off the shelf technology (COTS) rather than develop new technologies and products. It also decided to privatize many government functions. So now there is a dilemma - who will develop the new technologies and products, especially sustainability technologies and products in this century?

45.1 Bid no Bid Market Analysis

There are several questions that need to be answered before deciding to prepare a proposal or enter a market. Answering these questions involves finding the stakeholders, identifying their needs, understanding the competitors, and understanding your organization. The questions are:

- What is the size of the market
- Who are the competitors
- Who are the top two competitors
- How mature is the market: emerging, new, established, or diminishing
- Is the market saturated
- How mature is the technology: bleeding edge, leading edge, state of the art,

dated, or obsolete
- How will you address the needs
- What organizational capabilities are needed to address the need and do you have those capabilities
- Do you need to build organizational capabilities, products, and or technologies to address the need
- What separates you from the competition

There is a win probability associated with each opportunity. If there are no incumbents[372] then the probability is equally split across all the potential bidders or market participants. If there is an incumbent then 51% of the win probability is assigned to the incumbent and the remaining equally split across the other participants. If there are no other participants then the incumbent should be assigned a 75% win probability[373].

Each opportunity represents some benefit to an organization. The benefit of why the organization must pursue the opportunity should be clearly stated. The possible reasons to pursue an opportunity are:

- Entry into new strategic business with extensive future
- Core business well within the bounds of the organization, it is ours to lose
- The organization is the incumbent, it is ours to lose
- Opportunity will increase organizations capability
- Opportunity will enhance one or more products
- Opportunity will allow product expansion into new lines
- Opportunity will grow one or more technologies

There are reasons why an organization should not pursue an opportunity. Each pursuit takes away resources from the other opportunities. The goal is to balance the opportunities. For example if only core business or incumbent status work is pursued with no new strategic business mix, the organization risks catastrophic failure if the core market should suddenly shift, shrink, or disappear. The same arguments exist for organizational capabilities and technologies.

The approach for gathering data in a bid no bid decision overlaps market analysis but is not the same. The bid no bid data gathering activity includes:

- Trade journal and literature searches

---

[372] An incumbent is a vendor that has provided a previous solution to the same soliciting organization or market.

[373] The Probability Based Decision (PBD) as discussed in this text can be used to determine a more traceable and reasoned win probability.

- Technical and trade conferences
- Competitor public information analysis, marketing brochures, websites, papers, etc
- Customer and user site visits
- Request for information (RFI) reviews and responses
- Bidder conferences

The market analysis data gathering activities include:

- Market surveys
- Analysis of similar products
- Competitor public information analysis, marketing brochures, websites, papers, etc
- Test market runs using prototypes

Once a decision is made to pursue an opportunity, then either a proposal or a business plan is developed. The proposal is submitted to the soliciting organization and the business plan is submitted to either the external funding source or internal funding authority.

45.2 Proposal

Typically the capture or pre-proposal manager used during the pre-proposal effort is supplemented with a proposal manager. The pre-proposal manager is the visionary that has led the team to the point where a proposal can be developed and seriously considered by a sponsor. The proposal manager works with the pre-proposal manager and the team to develop the story that clearly shows why the sponsor should fund the organization[374].

A proposal must address every item in the Request for Proposal (RFP). A compliance matrix is developed from the Statement of Work, Instructions for Proposal Preparation (IFPP), and the System Specification. Every item is satisfied with compliance. Any non-compliance or deviation causes the proposal to be rejected without analysis.

This may seem like a drastic action but customers invest significant resources in Requests for Information (RFI) and dialog long before an RFP hits the street. They need to be able to compare offerings that are consistent. Any non-compliance introduces comparison issues that are not easy to resolve short of losing control of the contract before it even starts.

---

[374] Sequential Thematic Organization of Publications (STOP): How to Achieve Coherence in Proposals and Reports, Hughes Aircraft Company Ground Systems Group, Fullerton, Calif., J. R. Tracey, D. E. Rugh, W. S. Starkey, Information Media Department, ID 65-10-10 52092, January 1965.

Author Comment: Some organizations become dysfunctional and arrogant management thinks they can mistreat customers like they mistreat staff and suppliers. This usually leads to proposals that are very poor and the organization starts to suffer. It is cultural in nature and very difficult to correct short of letting the organization disappear.

45.3 Basis of Estimate

The Basis of Estimate (BOE) is used to determine the cost of a good or service. It is also used as evidence to show how the cost was determined. If a Work Break Down Structure (WBS) exists, a BOE is created for each WBS element. Unlike functional decomposition which is mostly top down the BOE is mostly bottom up. Part of this is because a top down view exists and has provided insight into the work to be performed. Usually during pre-proposal phase the functional decomposition starts to surface the organizational structure to perform the work and the WBS. By proposal time there should be a reasonable view of how the work will be performed and the BOE paper stack forms the foundation of the proposal cost volume.

The BOE is a description of how a cost is determined. There is typically one BOE for each WBS element. The WBS element represents a cost account that is tracked during the execution of an activity. So there may be history of previous BOE, WBS, and actual cost data that can be used to prepare the current BOE. In a mature organization there is always some previous history than can be accessed and considered in the preparation of a current BOE. If there is no history, then that is noted and a logical method is offered for determining the cost in the BOE.

45.3.1 BOE Elements

A BOE is a form with several fields that capture the cost information. The number of form fields varies and can be as simple as a task name, date, author, task description, cost rational, and the final estimate. For example servicing an automobile at the local garage uses a simple BOE form. The following is a list of fields that can be used for an organizations standard BOE form or forms:

- **Source**: Project name, date, and originator.
- **Task**: WBS number, task name (good or service), and task description.
- **Type**: The content of the BOE is noted as hours, dollars, hours and dollars.
- **Purchase**: If there is a purchase its source is noted catalog, Internet, subcontractor.
- **Travel**: Destination transportation (air, rail, car, sea), logging, meals, local transportation (car rental)
- **Cost Rational**: The logic used to determine the cost in hours and or dollars. It can point to quotes, invoices, and other supporting data. If automated cost tools are used, the tools and parameters are identified. Include revisions,

reviews, and management oversight. Include costs for machines, tools, software and other support elements. Identify assumptions such as using existing capital resources and indirect costs borne by another organization.

- **Labor**: Administrative, manager, technician, engineering, engineering support, executive, secretary, unskilled labor, skilled labor, etc.
- **Skill**: Entry, journeyman, mentor, staff, senior staff, level 7, level 6, level 5, level 4, level 3, level 2, etc.
- **Experience**: 0-5 years, 5-10 years, 10-20 years, 20+ years.
- **Suggested Earned Value Method**: level of effort (LOE), 50/50, 0/100.
- **Totals**: By labor categories such as Levels 7 to 2 and item costs such as for tools, parts list, etc.
- **Management Overhead**: Typically a percent and is in addition to the management oversight.
- **Distribution**: anticipated ramp up and down is noted and can be a step function, linear increase, triangle, gaussian, etc.
- **Attachments**: Quotes, invoices, prices lists, etc.

More complex work needs more complex BOE forms. Whatever fields are used in the BOE they need to be consumed by either a manual process or automated tool. For example if staffing distributions are offered then staffing distributions need to be consumed by the cost and pricing tools.

When developing a BOE it is an ideal time to capture the worst, expected, and best cases. The expected BOE is usually based on the following kinds of assumptions:

- Zero learning curve because staff is available
- Zero risk, if there is risk then the risk range needs to be determined and the middle range used for the expected BOE

There is a range of optimistic estimates. Extreme care needs to be taken when offering the optimistic estimate. Optimistic estimates are based on the following assumptions:

- Zero rework after reviews
- Zero rework after elements are integrated
- Zero schedule delays of dependent elements
- Zero defects during test
- Best risk case
- New processes and tools that have been proven to show productivity improvements but have not been rolled out in the enterprise or industry

The worst case estimate is obviously the opposite of the expected and optimistic assumptions. In addition other assumptions may surface:

- Equipment failure
- Insufficient staffing
- Loss of capability
- Economic hardships

45.3.2 BOE Risk Assignments

Each BOE can be associated with risk elements. This can be captured with a supplementary BOE risk assessment form that is associated with the BOE. The following is a list of possible risk areas and assessments:

Table 82 BOE Cost Risk Assignments

WBS Element #: **Element Name:**

Risk Item	Risk	Weight
Design Risk		
1. Concept only	15%	1
2. Conceptual Design Phase: some drawings; many sketches	8%	1
3. Preliminary Design > 50 % complete; some analysis complete	4%	1
4. Detailed Design > 50% Done	0%	1
Technical Risk		
1. New design; well beyond current state-of-the art	15%	2 or 4
2. New design of new technology; advances state-of-the art	10%	2 or 4
3. New design some R&D does not advance the state-of-the-art	8%	2 or 4
4. New design different from existing designs or technology	6%	2 or 4
5. New design nothing exotic	4%	2 or 4
6. Extensive modifications to an existing design	3%	2 or 4
7. Minor modifications to an existing design	2%	2 or 4
8. Existing design and off-the-shelf hardware	1%	2 or 4
If Design or Manufacturing concerns use weight = 2 If Design and Manufacturing concerns use weight = 4		
Cost Risk		
1. Engineering judgment	15%	1 or 2
2. Top-down estimate from analogous programs	10%	1 or 2
3. In-house estimate for item with minimal experience and minimal in-house capability	8%	1 or 2
4. In-house estimate for item with minimal experience but related to existing capabilities	6%	1 or 2
5. In-house estimate based on previous similar experience	4%	1 or 2
6. Vendor quote (*or industrial study*) with some design sketches	3%	1 or 2
7. Vendor quote (*or industrial study*) with established drawings	2%	1 or 2

Risk Item	Risk	Weight
8. Off-the-shelf or catalog item	1%	1 or 2
If Cost or Labor concerns use weight = 1 If Cost and Labor concerns use weight = 2		
Schedule Risk		
1. Delays completion of critical path subsystem item	8%	1
2. Delays completion of non-critical path subsystem item	4%	1
3. No schedule impact on any other item	2%	1

Prepared by: **Date:**

Comments:

The actual risk level can be hidden from the BOE authors. The authors select the risk item in each category and assign the weights. The risk level can be applied during pricing.

45.3.3 Cost Estimating

The labor portion of BOE can be prepared in terms of hours or in terms of labor costs. For simple applications such as an automobile service estimate, the costs are shown simultaneously with the labor hours. For more complex efforts the costs need to be derived from the initial BOE data. The cost estimating includes the following:

- Apply the labor rate for each labor category estimate in the BOE and calculate the total labor costs
- Apply a common management reserve to each BOE
- Verify management overheads in each BOE are consistent and appropriate
- Apply overhead rate and calculate the overhead costs
- Apply general and administrative (G&A) rate and calculate the G&A costs
- Apply cost of money (inflation plus true cost of money) and calculate the cost of money costs
- Apply fee (profit) rate and calculate the fee
- Apply the risk level (may be proprietary) for each BOE risk rating and calculate the risk costs
- Total all the costs

Example 64 Translating Initial BOE Data to BOE Costs

At some point a BOE based on labor hours needs to be reduced to cost. This can happen coincident with the BOE labor hour development or as part of a separate process. To reduce the BOE to costs the labor categories and rates need to be disclosed so that the calculation can be made. Also the overhead, G&A, cost of money, and fee rates need to be disclosed. To simplify the BOE and its traceability to cost, a BOE for the same WBS element can be developed for each labor category.

So there may be several BOE forms for a single WBS each representing different personnel, labor categories, and collections of equipment lists. Some may argue the WBS should be further decomposed. However in the simple case of developing a hardware specification for an Air Traffic Control Console, many people with different skills and skill levels are involved.

Table 83 Translating Initial BOE Data to BOE Costs

WBS: 1.1 **Name**: Common Console Requirements Specification	
Cost Elements	**Calculations**
Level 7 Engineer Hours	2100
Level 7 Engineer Rate/Hour (from rate card)	$50
Level 7 Engineering Dollars	$105,000
Level 5 Engineer Hours	210
Level 5 Engineer Rate/Hour (from rate card)	$70
Level 5 Engineering Dollars	$14,700
Total Engineering Dollars	**$119,700**
Overhead Rate (from rate card)	100%
Overhead Dollars	$119,700
Total Burdened Dollars	$ 239,400
G&A Rate (from rate card)	10%
G&A Dollars	$ 23,940
Subtotal	**$263,340**
Cost of Money Rate (from rate card)	5%
Cost of Money Dollars	$13,167
Total Cost	$276,507
Fee Rate (from rate card)	10%
Fee Dollars	$ 27,651
Total Price	**$304,158**
Risk from risk assessment card	
Design Risk	15%
Technical Risk	15%
Cost Risk	15%
Schedule Risk	2%
Total Risk Rate	47%
Total Risk Dollars	$142,955
Total Price with Risk	**$447,112**

45.3.4 Pricing

Just because a product or service is calculated at a certain cost it does not mean an organization needs to offer the product or service at that price. The price can be higher or lower. If the price is significantly lower the sponsor may need to be notified so that when they perform cost reasonableness analysis the proposal is not disqualified because of cost unreasonableness.

Some organizations are willing to buy into an effort for long term strategic goals. They will apply money from profits, research and development, and or new business. Obviously in a diminishing market or an unhealthy organization, research and development, new business, and profits are sacrificed in that order. Usually the problem is not detected until the layoffs start and then the profits disappear and losses are the norm.

So pricing is a very critical activity. Everyone in the organization should know the impact of buying into any effort and how it translates into lost opportunity. In a monopoly, oligopoly, or true industry leader situation the price can be arbitrarily set beyond normal fee levels[375].

45.4 New Business or New Product Plan

A new business or product plan is used to attract investment capital, secure loans, and assist in attracting strategic business partners. It shows whether or not a business area or product is viable. It is a realistic look at every aspect of the new business or product and shows that all the issues have been addressed before actually launching the new business or product.

As a management tool, it helps track, monitor, and evaluate progress. It is a living document that is modified with knowledge and experience. By using the business plan to establish timelines and milestones, progress can be tracked and compared with projections to actual accomplishments.

As a planning tool, it guides the team through the various phases of the new business or product. A thoughtful plan can help identify roadblocks and obstacles so that they are avoided.

The core elements of the plan are the business description, financial data and supporting documents. It includes an executive summary, market analysis, company description, organization and management, marketing and sales management, service or product line, funding request, financials, and an appendix with supporting information. The following is a sample outline.

- **Executive Summary**: market, products, strategy, management team, (owners, board members, external advisors), objectives, mission, keys to

[375] In 1985 Boeing was awarded the Peace Shield command, control, communications, and intelligence (C3I) system. In 1991 Hughes Aircraft Corporation assumed the project, which had been subject to delays. This was an extremely painful turn of events since Hughes was the obvious first choice but various conditions led to the Boeing award. No one knows how much Hughes eventually charged for the effort but rumors were that it was significant. The published number was stated in excess of $1 billion. Source: this author, Walt.

success, competitive edge

- **Company Summary**: ownership, history, locations and facilities
- **Products**: product descriptions (1, 2, n), industry impacts, competitive comparison, sales literature, sourcing, technology, future products
- **Market Analysis Summary**: market segmentation, target market segment strategy, (market needs, market trends, market growth), industry analysis, (industry participants, distribution patterns, competition and buying patterns, main competitors)
- **Strategy and Implementation Summary**: strategy pyramids, value propositions, competitive edge, marketing strategy, (positioning statement, pricing strategy, promotion strategy, distribution strategy, marketing programs), sales strategy, (accomplishments, sales forecast), strategic alliances, milestones
- **Management Summary**: organizational structure, management team, (owners, board members, external advisors), management team gaps, personnel plan
- **Financial Plan**: key tasks, important assumptions, key financial indicators, break even analysis, projected profit and loss, projected cash flow, projected balance sheet, business ratios
- **Summary**: appendix product brochures

45.5 Product Life Cycle

What is the difference between a product life cycle and a system life cycle? Is there a difference between a system and a product? What are the differences between a new product, established products, and a system? These are interesting questions and form the basis for a discussion of product life cycle.

A product cannot exist until there are at least two distinct and separate buyers. However taking something and converting it into a product, or productizing it requires resources. The resources are used to identify either the super set of required capabilities so that each potential customer is satisfied or a core set of required capabilities and optional capabilities that satisfy each potential customer. Options are usually developed to allow for different price points, which allow more market participants with less money to use some aspect of the product.

Typically a product starts out small with a limited number of features. As the product matures more features are added. At some point a stakeholder might suggest another use for the product. For example a fixed installation product may be suggested for a mobile environment. At this point a product line is born.

The products and product lines can be represented using marketing brochures or fishbone charts. Typically fishbone charts are used internally. They can be standalone fishbone or interconnected fishbone charts showing the heritage of various product lines. The fishbone also can be simplified as nodes in a chart similar to a PERT[376] chart. Each product line shows its heritage and its date of release.

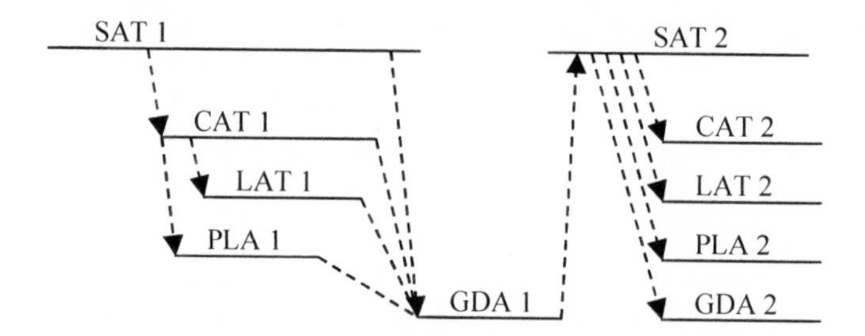

Figure 176 Heritage of Different Product Lines

A product line has a life cycle just like a system. Additionally, the product line life cycle is the same as for a system. Just like in a system, it is multidimensional and different words are used to convey different aspects of life cycle. For example research prototype, pre-production prototype, production, decommissioning, and disposal applies. So do concept development, requirements, design, implementation, test, operation and maintenance, and disposal.

A system can become a product. For example Hughes Aircraft built air defense and air traffic control systems that were productized and sold to different nation states around the world. Within these systems there were subsystems such as displays, RADARS, etc which were also productized and sold to other vendors for their system solutions.

Moving to a product mind set for a system solution actually introduces new challenges that make the system better. The challenges force certain system characteristics such as growth, adaptability, maintainability, methodical precise practices, etc because everything needs to be duplicated and repeated. Small twists in the individual solutions for each new customer are then rolled back into the product lines.

So a new product is like a new system rolled out for the first time. An established product is like a system that has been productized. In this environment capability maturity just emerges. It is not a chore or a certificate to achieve it is a way of business. Otherwise productization of a system and subsystems will fail.

Author Comment: It is impossible to achieve capability maturity levels above level 2 without a stable large market that allows for the emergence of multiple products and product lines from a organization.

---

[376] PERT Program or Project Evaluation and Review Technique is used in project planing to show task sequencing and dependencies. The PERT graphic is easily adapted to represent other sequences and dependencies like a product line heritage.

45.6 Exercises

1. Will you use STOP for your proposals? If so why, if not why not?
2. Should you use R&D funds for short-term pursuits, mid-term pursuits, long-term pursuits, or to offset the price of a bid?
3. What is your new business development process? How does it compare with other students? How does it compare with systems engineering or development processes? How do these answers compare with your answers in the early part of this text?
4. What is your research and development process? How does it compare with other students? How does it compare with systems engineering or development processes? How do these answers compare with your answers in the early part of this text?
5. If revenue is derived 100% from government proposals, can a commercial product or product-line get established? If yes, why, and if no, why not?
6. Is it appropriate for a technology or product to be in search of a system or is it always a system in search of a technology or product? Can you give examples of each? If not why not?
7. Is it appropriate for a system to be in search of a need or is it always a need in search of a system? Can you give examples of each? If not why not?
8. What can you do if your sole source sponsor disappears?
9. Prepare a company capability presentation for a possible new business area or market. Include company history, products, and capabilities. Prepare a companion presentation of how you plan to penetrate the new market and the first customer. Can defense plants be converted to support industrial work? If so why, if not why not?

45.7 Additional Reading

1. Contract Pricing Reference Guides, Volume 1 - Price Analysis, Volume 2 - Quantitative Techniques for Contract Pricing, Volume 3 - Cost Analysis, Volume 4 - Advanced Issues in Contract Pricing, Volume 5 - Federal Contract Negotiation Techniques, Federal Acquisition Institute (FAI) and the Air Force Institute of Technology (AFIT), 2011.
2. R&D Productivity Second Edition; Hughes Aircraft, June 1978, AD Number A075387, Ranftl, R.M., "R&D Productivity", Carver City, CA: Hughes Aircraft Co., Second Edition, 1978, OCLC Number: 4224641 or 16945892. ASIN: B000716B96.
3. Sequential Thematic Organization of Publications (STOP): How to Achieve Coherence in Proposals and Reports, Hughes Aircraft Company Ground Systems Group, Fullerton, Calif., J. R. Tracey, D. E. Rugh, W. S. Starkey, Information Media Department, ID 65-10-10 52092, January 1965.

46 Systems Management

There a several systems management practices that can be used to plan, control, and track systems development and operations. Some of practices include organization drivers, transition and systems management planning, tools, and ethics.

46.1 Organization Settings

You can talk about systems, measure systems, buy and sell systems, or build and actively participate with systems. For example, in an educational system you can test students to gather metrics, talk about the educational system, buy and sell it as in privatize it, or you can educate the students. If more money and time is spent on anything other than educating the students then the system is broken. All activities other than educating the students is overhead that should be carefully managed.

When people don't know what to do or are unwilling to do the real work then measuring, buying and selling, and talking about the system become dominant. The trick is to redirect the energy into the true elements of the system and return to a reasonable balance of talking, measuring, buying and selling, but mostly building and actively participating in the system.

When attempting to apply systems practices in an organization it is important to understand the drivers or ideology behind the organization. This suggests practices and the level of sophistication used for each practice. Organizations can be grouped and characterized as follows:

- **Systems Driven**: organizations place the highest priority on systems practices. It has a separate systems group that tends to drive and lead all other groups. A systems engineering Management Plan (SEMP) drives all other plans. It responds to stakeholder needs for a project, program, or market.
- **Technology Driven**: organizations push technology to its limits. It is less concerned with responding to large stakeholder groups and instead responds to stakeholders requesting performance or function outside the envelope of existing solutions. Technology management practices dominate. There is evidence of technology maturity curves and technology readiness levels throughout the organization.
- **Market Driven**: organizations respond to a broad set of stakeholders in a market area. Unless there is a monopoly the main theme is getting to market before the competition. In a high technology setting there are elements of

being technology driven. The systems practices tend to be limited towards stakeholder market assessments, some technology assessment and architecture selection.

- **Finance Driven**: organizations focus exclusively on financial metrics. Tools like EVMS are used to track progress and cost. Value engineering becomes paramount, as cost cutting is the primary objective. It may be appropriate in a saturated market that will soon disappear. It also may be appropriate after the first system is delivered and successfully operational. The EVMS and production process for pushing out systems does need a learning curve but by three systems of the exact same type, the EVMS should be reasonably established and value engineering offering lower cost elements. This approach has its roots in large-scale manufacturing.
- **Management Drive**: organizations adopt elements of finance driven organizations. There is a tool like EVMS to track cost and schedule. The Program or Project Management Plan (PMP) drives all other plans. This plan is in channel conflict with the SEMP. Typically the SEMP, if it exists, is subservient to the PMP. This essentially means that management can and does override systems findings if it is not in the best interest of management (one stakeholder out of many). It becomes ugly when costs are shifted to other stakeholders especially if they are unaware of the shifting. In this setting management damage control can dominate[377] and rule if things are not going well because of wrong decisions not based on systems practices (holistic, transparent, all stakeholder inclusion).
- **Administrative Driven**: organizations are found in established successful systems. It is basically a healthy management driven organization with none of the negative aspects of unrealistic cost schedule goals or damage control. EVMS may be used but not at a micro-management level with emphasis on just meeting cost and schedule regardless of damage. Budgets and schedules are realigned to match goals and real world operations.
- **Marketing Driven**: organizations have marketing personnel involved in day to day activities, even though they are unqualified to be in those roles. This is usually because of a market that is saturated or being displaced by new technology and a severe imbalance in the organization. Panic sets in and the main organization is blamed for the problems rather than acknowledging that the market is gone; time for something new.
- **Domain Expert Driven**: organizations find previous operator and or maintainer experts and place them in charge of the organization. IBM in the 1960's and 1970's advocated this approach and justified it by suggesting it

[377] On Bullshit, from Princeton professor Harry G. Frankfurt, Winner of the 2005 Bestseller Awards, Philosophy Category, On Bullshit, Harry G. Frankfurt, Princeton University Press, January 2005, ISBN 9780691122946.

was easy to train people in software but it would take years to train an operator or maintainer because of the need for real world On-the-Job-Experience (OJE). The problem is the domain experts cannot visualize beyond their own experience and usually lack exposure to system practices. This leads to ossification.

Some of these organizations are positive and some are negative. The goal should be to blend the most positive aspects of all these organizational settings into the target organization. For example a Systems, Technology, and Market driven organization with excellent Administration may be an ideal goal for many.

46.2 Transition

Aside from architecture, transition is probably the single most important aspect of system development. Most assume that one day someone decides a need should be filled and a system is developed to fill the need. However, most systems are rarely developed as a single event project or program. That only happens after the first system is developed and then rolled out to other communities or countries. However developing the first system has massive transition that is both consciously planned and managed or stumbled into as everyone tries to focus on the final goal.

For example we transitioned into a system that took people to the moon, there were multiple projects within multiple programs. Each was a piece of a larger project or program. In many cases one project or program was built on the experience and knowledge of previous projects and programs.

- Project Mercury ran from 1959 through 1963. The goal was to put a human in orbit around the Earth.
- Project Gemini ran from 1965 and 1966. The objective was to develop techniques for space travel to the moon, specifically Apollo. It included missions long enough for a trip to the Moon and back, space walks, and new orbital maneuvers including rendezvous and docking.
- Within the Apollo program, Apollo 11 landed humans on the Moon on July 20, 1969.

So who plans the transition? Is this a management exercise? Is this a technical exercise? Transition planning requires both technical and management participation.

To support transition someone needs to define the first set of transition points. This is not unlike identifying architecture alternatives. Some transition approaches are small steps. Other transition approaches are large chunks. There are advantages and disadvantages to each transition approach. Some may require development of throw away system elements. Others may use the transition elements for backup needs. Just like in the case of architecture the suggested practice is to:

- Identify the alternatives, in this case it is a series of pictures that show the transition story for each approach. Keep each transition strategy on a single sheet of paper or you may get lost in irrelevant details.
- Prepare the advantages and disadvantages list. Again keep it on a single piece of paper. After it is developed let it sit for a while, then revisit the list with a new perspective.
- Identify the transition tradeoff criteria. The transition tradeoff criteria are not that different from the architecture tradeoff criteria. However there may be some unique transition criteria that are included.
- Identify Life Cycle Cost (LCC) for each transition approach.
- Perform the tradeoff using the Measure of Effectiveness (MOE) to select the best transition approach
- Perform a sensitivity analysis. This includes arbitrarily taking some tradeoff cells to their extremes.

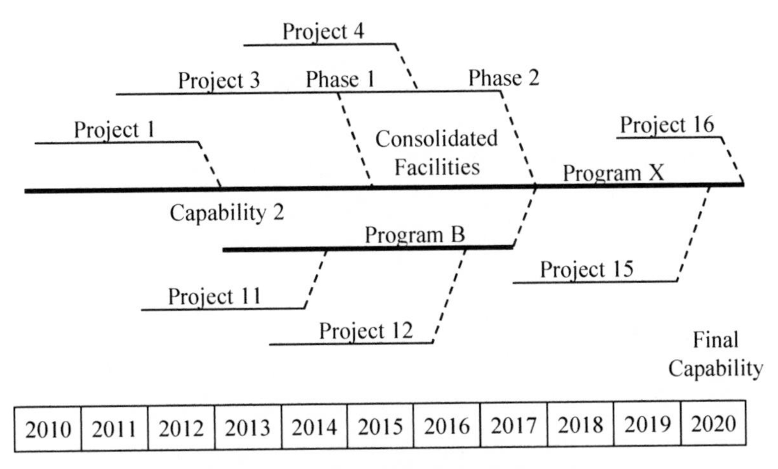

Figure 177 Strategic Plan PERT Chart

When the transition strategy is developed it is important to consider the human elements in the system. Each transition step will require training not only in new operations, but also maintenance and systems management. There should be a period of stability between each transition leap. Without this period of stability, when the next transition step is made no one will be able to determine if problems are associated with the new transition step or the pervious transition step. What should that period of stability be? I think it ideally should be one year of stable operations before the next transition step is rolled out into the operational community. Once a system makes the first successful transition, the time frame between the next transition step can be shortened, however this will have significant impact on the development effort which has plans based on the original transition plan of one year of stable operations.

In the case of a product or product line the transition strategy is based on product features and the roll out strategy is based on anticipated market saturation. This transition strategy can be shown with a Fishbone Diagram, PERT chart, or series of pictures that show each transition state.

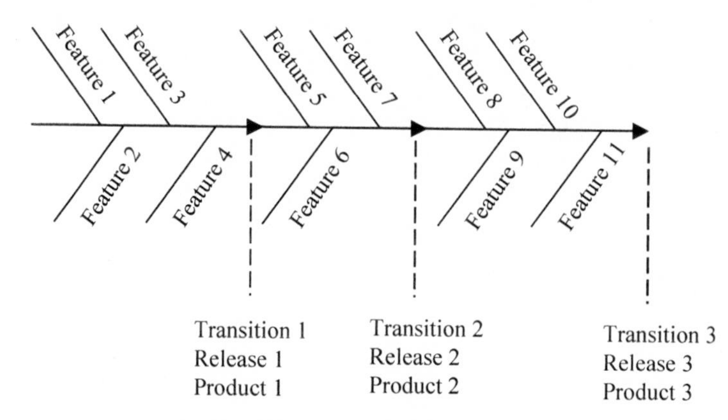

Figure 178 Transition Fishbone Diagram

46.3 Systems Engineering Management Plan

The Systems Engineering Management Plan (SEMP)[378] describes the efforts for planning, controlling, and conducting a full integrated engineering effort. It is used to understand and evaluate the engineering work. The SEMP contains three parts.

Part I System Engineering: describes the system engineering process used for the definition of system design and test requirements. This includes system engineering to define the system performance parameters and the preferred system configuration to satisfy the requirements; Planning and controls of the technical program task; Management of a totally integrated effort of design engineering, (all disciplines), test engineering, logistics engineering and production engineering. A narrative supplemented by graphical presentations describes the plans and procedures for the following elements of the system engineering process:

- Functional allocation, trade studies, design optimization and effective analysis, synthesis, interface compatibility, logistic support analysis, producibility analysis, training for users operators and maintainers.
- Requirements allocation including methods for documenting allocated requirements for designers, integrators, and test personnel as well as for review. Includes how requirement allocation is developed, maintained and

[378] Systems Engineering Management Plan (SEMP), DOD Data Item Description, DI-MGMT-81024 August 1990.

used throughout the life of the program.

- Specification generation describing baseline control, including procedures used during requirement allocation, design, configuration management and test with attention on hardware, software, and firmware integration.
- Other system engineering tasks describing the plan and procedures for other system engineering tasks.

Part II Technical Program Planning and Control: identifies organizational responsibilities and authority for system engineering management. This includes control of subcontracted engineering; levels of control established for performance and design requirements and control methods to be used; plans and schedules for the design, development, assembly, integration, test and evaluation functions, and control of documentation. These areas are applicable to both hardware and software engineering activities.

- **Program risk analysis**: Includes an analysis of any risk, which may be associated with the design, development, test and evaluation requirements. The analysis identifies critical areas and identifies the need for prototyping, testing or backup development to minimize technical risk. The risk analysis identifies test requirements, technical performance measurement parameters and critical milestones.
- **Engineering program integration**: Describes the planning and control functions for conducting a totally integrated engineering effort.
- **Contract work breakdown structure and specification tree**: Describes the manner in which the system engineering management develops the technical elements of the contract work breakdown structure (CWBS) and other contractual tasks required to form a complete CWBS. Develop a complete specification tree and relate it to the CWBS.
- **Assignment of responsibility and authority**: Identify the key personnel for each of the technical work breakdown structure (WBS) elements, clear definition of their responsibilities, the vehicles or documents used to state these assignments, and their standards or measures of accomplishments. Existing and proposed procedures establishing the authority, lines of communication and specific functions of these and other organizations associated with engineering policies and their implementation.
- **Program reviews**: Describe the reviews to assess, re-optimize, and redirect the effort.
- **Design reviews**: Plan and schedule for all design reviews under terms of the contract. Describe documentation provided prior to and at the various design reviews and how it relates to other contractually required data. This includes technical manuals, drawings, specifications, software firmware

documentation, user's manuals, and how all these final products relate to each other. The total documentation release system is discussed including signoff.

- **Interface control**: Describe the procedures for interface control of the contracted segment with other system segments performed by other system participants including organizations that furnish equipment, facilities, software and personnel.
- **Documentation control**: Describe methods for controlling change to internal technical data not subject to control by the formal configuration management system. Provide detail to establish its consistency with the configuration management and change control requirements of the contract.
- **Engineering testing**: Identify what engineering efforts lead to the system test documentation. Include a discussion of test engineering effort not included in the other documentation.
- **Tradeoff studies**: Identify the major tradeoff studies to be performed and the general plan for their accomplishment. Include the method for identifying, performing, and documenting the results of tradeoff studies.
- **Technical performance measurement**: Describes the plan for technical performance tracking and reporting including:

  - ❑ Identification of technical performance characteristics and technical program achievement parameters for each of the identified work elements of the CWBS. Note routine reporting proposed parameters.
  - ❑ The methods, equations or models for transforming parameter values of lower-level elements to higher-level elements and their sensitivities.
  - ❑ A planned profile for each of the parameters, the profile being an anticipated and time-phased variation, if any, for these parameters during the design, development, fabrication assembly and testing period. Note significant technical performance measurement events.
  - ❑ A description of how technical performance measurement is related to cost and schedule performance measurement.
  - ❑ A sample technical performance report for an out-of-tolerance technical parameter. This includes a planned value, demonstrated value, specification requirement, and current estimate and variance analysis.
  - ❑ A proposed list of technical performance reports used for reporting. It identifies if one report is used for all reportable parameters or if separate reports are used on a subsystem or individual parameter basis or combinations.
  - ❑ Identification of technical performance achievements by developing parameters, which address: subsystem hardware delivery and operation, computer equipment delivery and operation, subsystem software development through each phase of activity. Subsystem hardware

software integration, specification and statement of work (SOW) requirements, computer programs documentation plan, identification and acquisition of all design critical data.

- Describe the plan and procedures for other technical program planning and control tasks to be accomplished.

Part III Engineering Integration: Describe the methods to integrate the engineering efforts. Include a summary of each specialty program and cross-reference the individual plans covering the specialty programs. Discuss engineering specialty integration and relationships of engineering with overall logistic efforts, including fault isolation methods (automatic, semiautomatic, and manual) and the documentation, and how support equipment is identified.

46.4 Design & Implementation

The original front-end systems team should include representatives from the entire project or program. As the effort unfolds different disciplines increase in staffing size while others shrink. At no time should a particular group go to zero staffing. That is essentially removing a critical stakeholder from the solution and leads to serious problems.

Normally when people think of design and implementation they do not think of systems, systems practices, or systems staff. Systems personnel and systems practices become critical during the design and implementation phases of a project or program. Rolling systems staff off the labor plan for the design and implementation phase removes a critical stakeholder form the effort.

The systems staff are not only keepers of the system vision and intent of the requirements, but they are also keepers of the process. That is their stakeholder role. In the pressure to design and implement the solution the team lacking the systems perspective will deviate and cut corners not realizing that their ignorance of a broader picture can lead to complete failure. In addition to the systems stakeholder role, the design and implementation phase of a project or program uses the following system practices:

- Requirement changes, tracking, analysis, management and incorporation into the baseline
- System Requirements Management Database (SRDB) controlled updates and reissue of new baseline requirements and specifications
- Updates to studies impacted by design and implementation details that surface with this next levels of understanding
- Reviewing design and implementation information products for architecture and requirement deviations or violations

- Developing test plans and procedures for sell off and certifications
- Developing manuals and other information products for maintenance, training, installation, parts, support
- Developing training course material
- Reviewing prototypes, models, and simulations

46.5 Technical Interchange Meetings

Technical Interchange Meetings (TIMs) are periodic forums on large programs or projects that promote the interchange of views and information regarding different aspects of a system in development. The TIMs are usually organized by different areas and have a periodic schedule, such as monthly. The TIM is a forum to gather diverse stakeholders and educate so that everyone has a similar view at the end of the TIM. Examples of TIMS are:

- Operational sites meetings discuss current and new system capabilities
- Lab meetings of models or prototypes to discuss impacts and approaches
- Developers facility meetings to discuss the system and the progress
- Subcontractors meetings discussing their subsystems and their progress
- Topics include software, hardware, maintenance, human factors, logistics, training, transition, test, etc

46.6 Working Groups

Working groups are formed for the express purpose of producing a product, usually a document. They are not meant to educate or include all stakeholders. The working group product may be part of a project or program that also influences or drives a broader industry standard. The working group usually consists of experts in the field.

46.7 Action Items

Action items are a list of things to be done. They are maintained in a table like structure with several attributes such as: number, very descriptive title, origination date, source, close date, and comment.

Action items are usually generated as part of a meeting. They translate to work that may not have been anticipated in the original plan. As a result action items tend to be ignored unless they come directly from a customer.

The problem is if a group of people thinks they need to gather action items and management does not acknowledge the action items. If management continues to follow the EVMS plan and what they think is needed to close the project, important work will be missed. An alternative is to acknowledge the action items and assess if an EVMS re-plan is needed. If a re-plan is not needed then the action items should be systematically addressed and closed within the existing planned structure.

46.8 To Do Lists

A to-do-list is a technique used to bring a project in crisis into stability. It is usually used to get to a milestone that everyone believes will be missed. Many times managers have disappeared from the fray and only dedicated project staff is still engaged and has a strong desire to succeed.

A natural leader that has risen in the crisis usually develops a to-do-list. The list identifies the things that need to be done between the current date and the milestone. It is prioritized and has names assigned to each item on the list. Many times when this happens management reappears and helps to facilitate working off the items on the to do list. Unfortunately the original conditions, which put the project in crisis, remain and the to-do-lists keep reappearing if the natural leaders are still available.

46.9 Goals Advanced Concepts

Using goals to manage an effort is similar to EVMS except the detail is significantly reduced and emphasis is shifted away from cost and schedule to producing something of use or value within a pre-defined financial box. A financial box is a defined set of money and schedule. The box rarely is permitted to expand. So the drive is to accomplish the most effective things before the money is exhausted.

Within the box are resources that know the final goal. They establish a prioritized list of goals that lead to the final goal. The prioritized list of goals is usually small and ranges from four to ten goals leading to the final goal.

Goals need to be clearly stated and measurable. They are listed on a single piece of paper or chart with the anticipated completion date. New projects, programs, and products begin with a set of goals. If an individual goal is met within the partitioned financial box, then the next goal is addressed. If however an activity gets stuck at a particular goal, then an assessment is made to increase the size of the financial box, suspend or cancel the project, program, or product.

An activity may be suspended if it is dependent on something that is not available or mature enough to use. This is typically a case where a technology needs to mature and its anticipated maturity level may be reached in a relatively short period of time such as one to two years.

Author Comment: If Earned Value Management System (EVMS) is used outside a stable known production setting, such as in small research and big development (little R big D), the lower levels eventually trace to goals established by the people that lead to an EVMS traceable activity. They prioritize the goals and the emphasis is to complete the goals. If the financial box for the EVMS task is too small, this usually leads to compromises where the characteristics of the goals shift and result in what most perceive as low quality. This is one of the traps and failings of EVMS when applied outside of stable production settings. Success is declared at the interim steps but once the system is subjected to external scrutiny it is a failure.

46.10 Problem Solving Innovation and Team Mix

In the end the goal of an individual or a team is to engage in effective problem solving and innovation. I mix the two together because problem solving can become toxic where the problem is de-scoped to such a level that it goes away. This typically is associated with conservative low risk teams. The context or system boundary keeps shrinking until everyone declares success. The problem with this approach is that all progress stops.

Many times to solve a problem, innovation is needed. Innovation involves increasing the system boundary or scope. It also includes pulling elements from outside the current accepted domain. This typically is associated with progressive high-risk teams.

The trick is to blend individuals from both perspectives to form a well-balanced team that can close on a problem without de-scoping it and yet introduce new and innovative elements.

Table 84 Team Mix Characteristics

Conservative – Less Innovative	Progressive – More Innovative
• Very tightly related focused ideas	• Very loosely related unfocused ideas
• Remove connections to shrink scope	• New connections including external
• Refine the system	• Replace the system
• Incremental change	• Radical change
• Build on assumptions	• Challenge and create new assumptions
• Encourage continuity and stability	• Encourage radical shifts
• Risk averse wants to close	• Thrive on risk wants the challenge

Other individual and team characteristics include consensus building or ignore consensus, detail or non-detail oriented, technical or non-technical. The ideal team is progressive and innovative with emphasis on consensus, detail and non-detail elements, with technical and non-technical perspectives.

46.11 Maslow's Hierarchy of Needs

Maslow's Hierarchy of Needs suggests different levels of needs and that the lower needs must be satisfied before people will seek the next higher level need[379]. For example a person lacking food, safety, love, and esteem would probably first try to satisfy the need for food than safety. The needs are typically represented in a pyramid called Maslow's Pyramid where the lowest level needs are at the bottom and the highest level needs are at the top. The needs from lowest to highest are:

---

[379] Abraham Maslow proposed this in a paper in 1943. A.H. Maslow, A Theory of Human Motivation, Psychological Review 50(4), 1943.

1. **Physical**: This is the most basic need for air, water, food, clothing, and shelter. This also includes the instinctive need for sex.
2. **Safety**: If there is physical stability then safety surfaces, as the next need. This includes physical safety and security of the body such as personal security, health and well being, avoiding accidents, illness, and their adverse impacts. There is also an element of stability that suggests justice, fairness, consistency, reliability, predictability manifested in elements such as employment, personal finances / savings / investments, resources, family, health, property, morality or laws.
3. **Love**: If there is a safe environment then love and belonging surfaces as the next need. This is a social need and includes feelings of belonging. There is a strong desire for friendship, intimacy, and family leading to friends, boy or girl friend, wife, children. This also includes social groups such as clubs, office clicks, religious affiliations, professional organizations, sports teams, mentors, close colleagues, and confidants.
4. **Esteem**: If there is an environment of love then esteem surfaces as the next need. This is an interesting gate because the musical rock group the Beatles[380] had a song, which clearly stated that "all you need is love". This need includes self esteem, confidence, achievement, respect of others, self respect, reputation, prestige, self confidence, worth, strength, usefulness, to be needed in the world. There were many themes in movies and music of the 1950's and 1960's where the pursuit of esteem led to disastrous consequences as egotistical characters would cross moral boundaries and lose their souls. Terms such as saving face, damage control and, what will the neighbors think are examples of severely misplaced esteem.
5. **Self-Actualization**: Even if all the other needs are satisfied, there may still be discontent and restlessness. Musicians need to make music, artists need to paint, and poets need to write for ultimate happiness. What a person can be, he must be, this need is called self-actualization[381]. It refers to the desire for self-fulfillment.

Maslow's view of self actualization is further refined and based on his approach of studying extremely healthy accomplished people which he viewed as self actualized. This was contrary to previous approaches of studying people with impairments. From these studies using his qualitative biographical analysis approach key characteristics started to surface:

---

[380] "All You Need Is Love" by John Lennon credited to Lennon and McCartney. Performed by The Beatles on the first live global television broadcast via satellite on June 25 1967, Our World, seen by 400 million in 26 countries.

[381] Kurt Goldstein, originated idea of self-actualization in his book: The Organism 1934.

- Morality, lack of prejudice, ends don't justify the means
- Creativity, spontaneity, acceptance of facts
- Differentiate fake and dishonest from real and genuine
- Problem solving rather than surrender or avoidance
- Nonconformity, humility, respect

He identified a way of thinking and called it Being-cognition or B-cognition. The B-values are:

- **Truth**, rather than dishonesty
- **Goodness**, rather than evil
- **Beauty**, not ugliness or vulgarity
- **Unity**, wholeness, transcendence of opposites, rejection of arbitrary or forced choices
- **Aliveness**, not deadness or mechanization of life
- **Uniqueness**, not bland uniformity
- **Perfection and necessity**, not sloppiness, inconsistency, or accident
- **Completion**, rather than incompleteness
- **Justice and order**, not injustice and lawlessness
- **Simplicity**, not unnecessary complexity
- **Richness**, not environmental impoverishment
- **Effortlessness**, not strain
- **Playfulness**, not grim, humorless, drudgery
- **Self-sufficiency**, not dependency
- **Meaningfulness**, rather than senselessness

The final take away from Maslow's pyramid is the idea of maturity and progressing to a mature state. It is a common sense idea and it is easily seen in nature where babies progress to children, teenagers, young adults, adults, middle age, and old age. The same holds true for companies and organizations. The progression might be as follows from least mature to most mature:

- **Level 1**: Collection of bodies, temporaries or consultants, working a project
- **Level 2**: Full time employees, rented facilities and equipment, a few projects
- **Level 3**: Employees, owned facilities and equipment, some programs
- **Level 5**: See level 3 plus a product
- **Level 6**: See level 3 plus multiple product lines
- **Level 7**: See level 3 plus multiple product lines and a technology
- **Level 8**: See level 3 plus multiple product lines and multiple technologies

46.12 Ethics

Ethical considerations while developing and managing systems should be part of all the information products. This is especially true for the information products at the higher levels such as system architecture documents, operational concepts, and sustainability analysis. The following are guidelines that can be used to stimulate ethical discussions in the system and management of the activity:

- Is there a conflict of interest?
- Is a stakeholder getting a hidden benefit?
- Is a stakeholder being harmed?
- Is there a hidden stakeholder?
- Is information being suppressed?
- Is there evidence of abuse of power?
- Is the environment totalitarian dictatorial intimidating harassing?
- Is the environment secretive where there is no reason to be secretive?
- Is the team not solving the obvious problems that others seem to ignore?
- Is the team ignoring obvious problems others seem to solve?
- Are checks and balances being removed or made ineffective?
- Is a stakeholder being marginalized?
- Is staff being marginalized?
- Is there a misuse of technology?
- Do the needs of the one out weight the needs of the many and do the needs of the many out weight the needs of the one[382]
- Is physical harm being permitted
- Is loss of life being permitted
- Is physical harm being permitted because it makes financial sense?
- Is physical loss of life being permitted because it makes financial sense?

46.13 Exercises

1. Provide examples of systems, technology, and market driven organizations.
2. Draw a Strategic Plan PERT Chart of your educational and professional background to date or of a program.
3. Draw a Fishbone diagram of a product.
4. Describe the channel conflict between a SEMP and a project management plan. How do you propose to resolve the channel conflict?

---

[382] An interesting take away from a popular movie: Star Trek II: The Wrath of Khan, Writers: Gene Roddenberry, Harve Bennett, Jack B. Sowards, Samuel A. Peeples (uncredited), Jack B. Sowards (screenplay) and Nicholas Meyer (screenplay uncredited), 1982.

5. If systems engineers are present during design and implementation are designers and implementers present during the concept stage of a project or program? If so why and if not why not?
6. Layout a program and identify the timing of TIMS.
7. Identify how working groups will interact with the program.
8. Develop an action item table.
9. What can you do if action items are not closing?
10. Develop a to do list. How is it different from the action items?
11. Layout a project or program using goals. Is this easier than EVMS? When should EVMS be started and used?
12. Can effective systems be developed if people are worried about food, clothing, or shelter? How about safety? How about love and belonging? How about esteem? How about if they are fully satisfied?
13. Is it ethical to have a layoff and show a profit?
14. Is it ethical to have multiple layoffs for 2 years, show a profit but shrink the organization by 75%?
15. Is profit a reasonable metric that should be communicated to stakeholders to show the health of an organization?
16. If the profit metric is being manipulated to hide trouble in an organization, such as a shifted market or forced obsolescence, is this unethical?
17. Is it ethical to have companies derive almost 100% of their revenues from government customers?
18. Is it ethical to move an organization from being a multiple technology and product line company to a state where there are only temporary employees?
19. Is it ethical to focus on saving face, damage control and, what others think?

46.14 Additional Reading

1. A Theory of Human Motivation, Abraham H. Maslow, Psychological Review 50(4), 1943.
2. On Bullshit, Harry G. Frankfurt, Princeton University Press, January 2005, ISBN 9780691122946.
3. Systems Engineering Management Plan (SEMP), DOD Data Item Description, DI-MGMT-81024, March 1987, August 1990.
4. Systems Engineering Plan (SEP) Preparation Guide, DOD Data Item Description, August 2005.
5. Systems Engineering Plan (SEP), DOD Data Item Description, DI-SESS-81785 October 2009.
6. The Organism, Kurt Goldstein, 1934.

47 Sustainability and Progress

Sustainable development has become a significant topic of discussion in recent years. Some have argued that it is the next great challenge and that we should focus our energies towards sustainability in this new century. So what is sustainable development?

The Brundtland Commission[383] defined sustainable development as development that "meets the needs of the present without compromising the ability of future generations to meet their own needs." Sustainable development is usually divided into social, economic, environmental and institutional. The social, economic, environmental areas address key principles of sustainability, while the institutional area addresses key policy and capacity issues.

It is obvious that a people should try to live well now and in the future. The issue is the mechanisms they have at their disposal to accommodate that life. One of those mechanisms is systems engineering to develop the most effective solution.

Although it seems obvious to systems engineers that systems practices should be used to address the sustainability challenges of this new century, others not versed in systems practices have other views. For example product-oriented people feel that their products are a perfect fit in a particular niche of sustainability. Technology people think they have a magic technology that will solve a challenging sustainability issue. Financial people think that tax incentives or virtual markets will spur innovation and let business rise to the occasion of sustainability needs. We as systems practitioners know that none of these will work in isolation. Instead we realize that systems engineering is the only way to start to chip away at this problem. So what are the systems practices that we need to modify and perhaps introduce so that we can start to address the sustainability challenges?

This section is different from the other sections. There is significantly more opinion and reflection in this section. That is not to say the other sections were purely academic.

In many ways the other sections of this book are dedicated to this section. Sustainability and progress is probably the most important task for humanity. The question is how can we maintain our sustainability and progress in an ever more complex and fragile world built on our systems.

---

[383] Brundtland Commission, or the World Commission on Environment and Development, known by its Chair Gro Harlem Brundtland, convened by the United Nations in 1983 published Our Common Future, also known as Brundtland Report, in 1987.

47.1 Progress

Is progress and sustainability part of the same goal? That is an interesting question. Progress was used extensively in the last century. It was progress that led to lights in the dark, safe water, sewage, safe abundant food[384], refrigeration, telephones, radio, television, washers, dryers, garbage disposals, automobiles, airplanes, etc. This was extremely difficult and people made enormous sacrifices to put these systems in place. They significantly benefited the future generation. The question is what progress is this generation making for the future generations.

"Progress is movement in a forward direction. It is a combination of achievement and forward vision, which serves its creator without fear or unforeseen consequence. There can be no progress if these rules are not respected. Technology is not progress unless it is combined in the direction of humanity. Pseudo progress can destroy the natural foundations of human life. It can lead to a society of chaotic beings that can result in global suicide"[385].

Misuse of technology as previously discussed in this text is a serious consideration when addressing progress and sustainability. The Earth has a limited set of resources and one of the most important resources is people. Misusing their abilities and energies is not sustainable. It is critical that we not waste time with useless adventures that may lead to financial fortunes for some in a time of plenty and lose our opportunity to do what must be done before the times get challenging. It was only a slightly over 100 years ago that humanity lived in the dark. We should not take our modern world for granted.

47.2 Growth

Many equate and discuss growth in terms of economics and finance such as gross domestic product (GDP). However, just like EVMS, GDP as a measure of growth is very misleading.

Money is a method of storing wealth so that a simple common element can be used to exchange goods and services. It does not relate to or show the quality or usefulness of the goods and services. The argument is that the goods and services will always follow the need and quality based on the consumer and their choices. This assumes a rational consumer. However what happens if consumers become

---

[384] Many people are still hungry but most have concluded this is a political problem. This may soon change as the population continues to grow and potentially outstrips our ability to provide safe food even in a perfect political situation.

[385] Paraphrased from an interview with Burkhard Heim (February 9, 1925 - January 14, 2001) a German theoretical physicist. He devoted a large portion of his life to the pursuit of a unified field theory. One of his childhood ambitions was to develop a method of space travel, which contributed to his motivation to find such a theory.

irrational either because of their own failings or because of external forces? What happens if the money becomes a means to an end rather than a mechanism to fund useful projects[386]?

If the focus is on the quality of life as a market need that can be targeted and filled with the introduction of washers, dryers, toasters, personal transportation, entertainment[387], etc then the introduction of useful goods and services is easily explained. However with massive technology and the ability to produce goods and services anywhere, technology abuse can overcome our sustainable systems, and flood markets with useless products. At this point GDP becomes a useless metric.

47.3 Space Program

There never would have been an environmental movement that started in the 1970's without the U.S. space program and its unbelievable mission of going to the moon. The picture of the fragile Earth in a sea of blackness while seen from a dead space body inspired visionaries to take a whole new view of our world and our place in the system. This is a classic example of emergence and justification of technological growth for the sake of growth.

Figure 179 Earth Rising Seen from Moon and Neil Armstrong on Moon

The sad part is the space program was an outgrowth of defense. It only grew to its level of performance because of the concept of Dual use Technology and a presidential initiative. It was not viewed as something that made sense and the next step for humanity to follow. So even though we depend on satellites for every aspect of our life, few are willing to fund the fundamentals of trying to survive in

---

[386] In my humble option this is the cause of all depressions. The money gets locked up in the hands of a few and the velocity of money drops to such a low level that all useful work stops. It's not that the few are evil it is just that they cannot duplicate the creativity and innovation of millions of people at the working levels.

[387] Many educational and soul-searching works can be captured on film and in music.

the ultimate hostile environment even though that knowledge is easy to equate with us being able to effectively manage our resources on Earth.

47.4 Internal versus External System Sustainability

Internal system sustainability is the ability of a system to sustain itself. Bad systems should fail and disappear, however what should a people do if good systems fail and disappear.

Table 85 Internal Sustainability and Maturity Levels

Food 1. None 2. Hunter Gatherers 3. Family Farmers 4. Regional Farms 6. Large Farms 7. Large green houses 8. Vertical Farms	Water 1. Stream 2. River 3. Well 4. Town Square 5. Piped 6. Treated Piped	Waste Management 1. None 2. Raw into River 3. Processed into River except during rain 4. Processed into River 5. Effluent watering system
Health Care 1. None 2. Doctor 3. Clinic 4. Hospital 5. Research Hospital	Construction 1. None 2. Mud Hut 3. Durable Materials 4. Office Buildings 5. Sky Scrapers	Employment 1. None 2. Investor 3. Worker 4. Owner
Education 1. None 2. Traditions 3. Verbal Apprentice 4. Elementary 5. High School 6. College	Transportation 1. Foot 2. Animal 3. Train 4. Public Transit 5. Automobile 6. Plane 7. Space Ship	Energy 1. None 2. Individual Intermittent 3. Individual Continuos 4. Group Intermittent 5. Group Continuous
Communications 1. Verbal 2. Smoke 3. Print 4. Telegraph 5. Radio 6. Telephone 7. Satellite 8. Cell Phone 9. Internet	Security 1. None 2. Random 3. Random + Police 4. Police 5. Police + Military 6. None Needed	Wealth Exchange Finance Investments 1. Barter 2. Precious items 3. Money 4. Equities - Stocks, Mutual Funds 5. Interest Bearing - Bonds, Banks 6. Trusts - Real-estate Investment Trusts 7. Taxes Bonds - For infrastructure

Who decides the difference between a good and a bad system? For example what if the clean safe water delivery systems in the U.S. were permitted to fail and go away to be replaced with bottled water? All communities have a maturity level and it should never be assumed that a community can sustain its' current level of existence without serious effort.

External sustainability is a measure of a systems impact on its surrounding community. This is the traditional ecological view of sustainability. Some will point to a scale where one balances the other, however it is not an either or proposition. The easy way out is totally unacceptable. Somehow internal and external sustainability must be addressed.

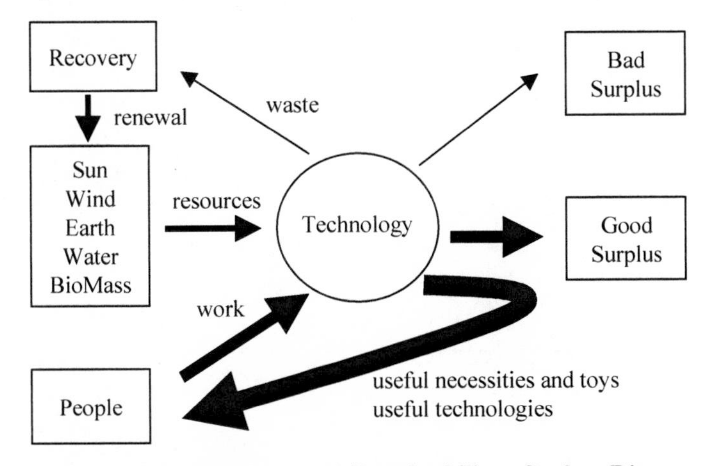

Figure 180 Technology and Sustainability - Sankey Diagram

It is easy to slip into simple economic views of the problem. However stepping back and doing things like drawing simple context diagrams and then attempting to go into the context diagrams may yield important insights.

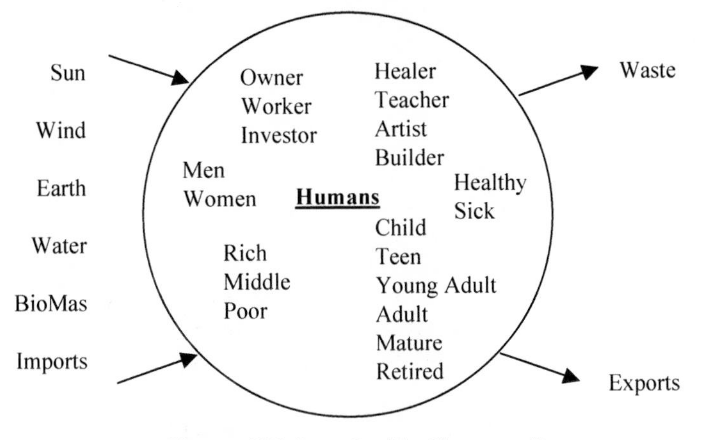

Figure 181 Sustainable Community

47.5 Sustainability V-Diagram

The System V-Diagram has been used to communicate some aspect of systems engineering to systems practitioners and non-system practitioners. It shows traceability through the various specification levels and the links to the associated, verification, validation and integration levels. One of the primary messages is that decomposition is from the top down while test and integration is from the bottom up. It also shows when system handoff or delivery occurs so that system validation can proceed. It tends to be modified to show some particular aspect of systems engineering. So what might we say about the V-Model if we start to introduce sustainability into the picture?

At one level it is obvious that sustainability would find itself moving into each of the individual blocks. However once the system becomes operational, eventually shuts down, and is disposed of, the traditional System V-Diagram does not fully address sustainability. A suggestion is to modify it to include sustainability for the full life cycle of a system and not limit it to just development, verification, and validation.

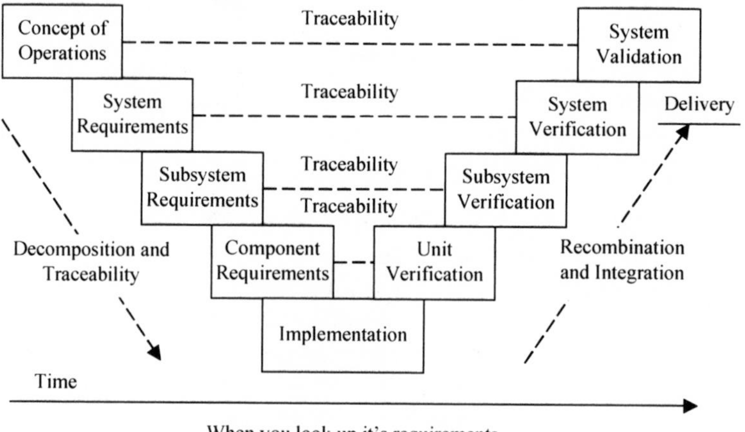

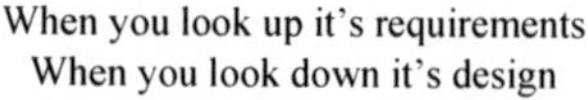

When you look up it's requirements
When you look down it's design

Figure 182 Traditional System V-Diagram

In this extended view of the life cycle of the system the operations, shutdown and disposal are added resulting in the Sustainable System V-Diagram. This extended view suggests a need for one or more information products associated with "community sustainability". This is not unlike what happened with the rise of the environmental movement and the introduction of environmental impact studies when new or modified structures are proposed in a community[388].

---

[388] Community is the collection of stakeholders that are impacted by the system.

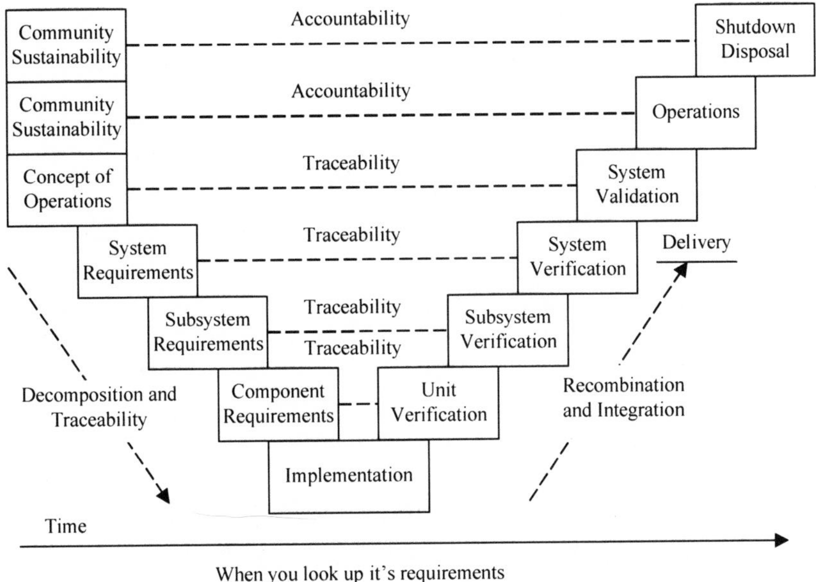

Figure 183 Sustainable System V-Diagram

47.6 Community Sustainability and Accountability

In the past many systems development efforts started with the Operational Concept. The Sustainable Systems V-Diagram suggests that systems developments start with Community Sustainability and be concurrently developed with the operational concepts. Obviously each system is unique but there may be generic items of interest that should be in each Community Sustainability Concept and Accountability document. The following is a possible list for these common sustainable items of interest:

- Who are the stakeholders
- What are the stakeholder needs
- What are the key sustainability issues in this system
- What are the sustainability goals in this system
- What are the key sustainability and stakeholder requirements
- What is the accountability path for poor sustainability

The community sustainability concept can suggest where sustainability issues are addressed in various systems practices during the development stages of the system. Some of the common systems practices that should address sustainability are:

- Sustainable Requirements
- Technology Assessment, Innovation, and Stability
- Modeling and Prototyping
- Maintainability, Logistics, Safety, Reliability, Quality
- Life Cycle Cost
- Architecture Identification, Tradeoff, and Selection

Although this list is logically sequenced, the greatest impact that systems practices can have are in the areas of architecture and technology. This is the role of the systems integrator, to pick the most effective architecture and technology for the system and then integrate the system. The question is can current systems integrators start to stimulate technology growth in these areas.

47.7 The Search for Sustainable Requirements

Systems practitioners are good at surfacing new kinds of metrics and ways of characterising performance. Granted this is usually after a long process of surfacing functional requirements and characteristics. But at some point performance surfaces.

In this particular case our challenge is to perform literature searches, vendor searches, and vendor assessments of current offerings in alternative energy sources, transportation vehicles, fuels, etc. Once the vendors and their offerings are identified the next step is to start to find discriminators between these various offerings. These discriminators start to surface new performance specification categories. For example in the simple case of wind turbines the following performance may be identified and compared:

- Minimum Useful Wind Speed or Cut In Speed
- Maximum Useful Wind Speed or Cut Out Speed
- Peak Theoretical Power Per Day, Week, Month, Season, Year
- Average Power Per Day, Week, Month, Season, Year
- Power Distribution Per Day, Week, Month, Season, Year
- Power Per Footprint Area, Per Linear Height, Per Unit Of Weight
- Set Up Time In Transportable Settings

At one level the broad choices of wind turbines include Horizontal-axis wind turbines (HAWT), Vertical-axis wind turbines (VAWT). They come in different sizes and in the case of VAWT the wind capture approaches vary. At another level there is the choice of Micro Turbines and a massively distributed system

The same process needs to be applied to solar photo voltaic, solar thermal, alternative transportation, alternative fuels, and other traditional system elements now subjected to sustainability needs. What are the performance characteristics of existing products, are these measures capturing what is really needed or should new performance metrics or requirements be created to effectively represent what is actually needed?

The simpler case of sustainability and requirements is probably associated with functionality and characteristics. The following is a small set of possible requirements to consider:

- Use recycled, recovered, or environmentally preferable materials
- Toxic and Hazardous materials are not used
- Avoid prohibited materials: lead, radioactive material, glass fibers, ozone depleting chemicals or substances, toxic materials that give off toxic fumes when burned or exposed to high temperatures. Registry of Toxic Effects of Chemical Substances (RTECS)[389], 29 CFR 1910 OSHA standards [OSHA].
- Avoid materials on EPA 17[390] list [EPA 1999]:

1. Benzene
2. Carbon tetrachloride
3. Chloroform
4. Dichloromethane
5. Methyl ethyl ketone
6. Methyl isobutyl ketone
7. Tetrachloroethylene
8. Toluene
9. 1,1,1-Trichloroethane
10. Trichloroethylene
11. Xylenes
12. Cadmium and cadmium compounds
13. Chromium and chromium compounds
14. Cyanide compounds
15. Lead and lead compounds
16. Mercury and mercury compounds
17. Nickel and nickel compounds

Once various vendor solutions are identified and characterised, the next steps might be to start to identify the performance requirements that are a candidate for improvement. At this point the systems integrator moves from commercial product consumer to technology developer while attempting to foster or push technology development and growth.

---

[389] Registry of Toxic Effects of Chemical Substances (RTECS) toxicity database compiled without reference to validity or usefulness of studies. Center for Disease Control, The National Institute for Occupational Safety and Health (NIOSH) maintained it as a freely available publication until 2001.

[390] The 33/50 Program targeted 17 priority chemicals and set as its goal a 33% reduction in releases and transfers by 1992 and a 50% reduction by 1995 measured against a 1988 baseline.

47.8 New Sustainability Performance Requirements

In keeping with the system practice of identifying new measures of performance for new systems or new views of systems, new performance requirements should be surfaced when addressing sustainability. The following is a suggested list of new sustainability performance requirements. They are based on key ratios associated with any product or system. The ratios use the expected load on the system. So first the loads are calculated, then the key ratios are determined. The sustainability performance loads are as follows:

- **Packaging Carbon Footprint (PCF):** The package carbon load calculation includes production, recycling, and disposal.
- **Transport Carbon Footprint (TCF):** The transport carbon load includes the fuel during transport, the building and maintenance of the transport vehicles, and the infrastructure to support the vehicles.
- **Content Carbon Footprint (CCF):** The content carbon load includes production, recycling, and disposal.
- **Packaging Energy Consumption (PEC):** The package energy load calculation includes production, recycling, and disposal.
- **Transport Energy Consumption (TEC):** The transport energy load includes the fuel during transport, the building and maintenance of the transport vehicles, and the infrastructure to support the vehicles.
- **Content Energy Consumption (CEC):** The content energy load includes production, recycling, and disposal.

These loads should be minimized in the system. In addition the system should be in balance. To maintain this balance key ratios also should be minimized. The sustainability performance ratios are:

- **Packaging / Content Carbon Ratio (PCCR):** This is the ratio of Packaging Carbon Footprint (PCF) versus Content Carbon Footprint (CCF). This can be expressed as carbon ratio of package versus content or PCCR = PCF/CCF.
- **Transport / Content Carbon Ratio (TCCR):** This is the ratio of Transport Carbon Footprint (TCF) versus Content Carbon Footprint (CCF). This can be expressed as carbon ratio of package versus content or TCCR = TCF/CCF.
- **Packaging / Content Energy Ratio (PCER):** This is the ratio of Packaging Energy Content (PEC) versus Content Energy Content (CEC). This can be expressed as carbon ratio of package versus content or PCCR = PEC/CEC.
- **Transport / Content Energy Ratio (TCER):** This is the ratio of Transport Energy Content (TEC) versus Content Energy Content (CEC). This can be

expressed as carbon ratio of package versus content or TCER = PEC/CEC.

Example 65 Unsustainable Packaging

The goal is to minimize the total sustainability performance loads and the sustainability performance ratios in the system. For example, if the total load of PEC + TEC + CEC is minimized and yet the packaging is so small that the content is "less" than the package, then the question must be asked - is system violating basic common sense related to sustainability? The following is a list of packaging practices and their characteristics:

- Does it make sense to place 3 machine screws in a package? Wouldn't a bin of screws and nuts be more effective? Does every product item need a bar code at the final point of sale? Would a bar-coded bag of 100 nuts sold to the retailer preserve the same metrics for both the wholesaler and retailer?
- It is not unusual to shrink packages and keep prices the same during difficult economic times. Is this reasonable, sustainable, or even ethical in a time of stress?
- Cross packaging of dependent products leads to similar inefficiencies. Does it make sense to place 3 machine screws in one package and 4 complementary nuts in another package?

Once multiplied by millions or hundreds of millions this translates into enormous waste of resources and can be measured as high PCER and PCCR ratios.

Example 66 Unsustainable Transport

Does it make sense to manufacture an item and transport it across a continent or around the world? The high TCER and TCCR ratios clearly show the sustainable approach. So how did we get into a situation where products are constantly moving across the planet? This is a difficult question and fundamental to the question of being systems driven or finance driven.

In the U.S. during the early 1970's there was a significant argument against the new Environmental Protection Agency (EPA) because the new environmental regulations the EPA needed to enforce were forcing rust belt industries to relocate to countries with little or no environmental regulations. This was the start of offshore production and one of the major reasons for new trends against government bureaucracy and its role in U.S. society[391].

In a perfect world the financial analysis results would match the system analysis results. However in an imperfect world with artificial markets and indirect cost shifting we see our current state of affairs. The system analysts can build the models, perform the analysis and show the inefficiency in the markets and its

[391] Author perception of the current events of the time.

impact on sustainability. It is then up to policy makers to determine how to proceed.

Example 67 Think Local Production and Distribution

Reducing energy and carbon ratios translates into local production and distribution of products. The issue is how to factor the true costs of producing a product into a mechanism so that location and distribution is not driven by the most clever scheme of offsetting indirect costs but the lowest impact on the whole system. This is an example of making sure the context diagram is large enough to represent the true system and not some artificial boundary driven by a limited set of stakeholders.

One approach is to clearly attempt to measure the performance of a product and its impact on the whole system. The TCER, TCCR, PCER, and PCCR are relatively easy to calculate. The issue is what to do once the performance numbers are produced. Policy makers may force a tax or consumers might choose to purchase based on the ratios. Then there is the question of how to validate the claimed ratios.

The simple act of just thinking local might be all that is needed, but that does not necessarily mean a local production facility has the lowest ratios. It may help a people with regulations such as the U.S. but it is of little use to a people with little or no regulations.

47.9 Technology Assessment, Innovation, Stability

An interesting question is can systems practices spur innovation. Perhaps some simple thought experiments not uncommon in systems practices might answer that question. Thought Experiment 1: Why do wind turbines primarily use propellers?

- What does it mean if a 200-foot propeller is used on a 300-foot wind turbine tower
- What about using a ducted fan or a centrifugal (squirrel-cage) fan or other methods of capturing the wind
- Should it be just one fan structure on the 400-foot tower or multiple structures
- Are there wind gradients across the diameter of the fan and what do those wind gradients do to efficiency
- In a multiple wind turbine farm, should only one type of wind turbine be used
- What is the ideal physical placement of each turbine in a large wind farm

Clearly someone should be modeling the different wind turbine architectures and characterizing them for different environments. Perhaps there even should be simulation test facilities using large wind tunnels to try these different wind turbine architectures in semi live settings but under controlled conditions. Thought Experiment 2: How should a wind turbine interface to the grid?

- Use fixed rotation wind turbines but then what are the lost opportunities
- Use solid state electronics but what about extremely large wind turbines
- Use motor generators, rotary converters, or double fed induction generators to interface to the power grid but what about losses

These issues are associated with technology assessment. Technology assessment includes identifying the Technology Maturity Levels and Technology Readiness Levels[392] of various system elements. These assessments tie in closely with modeling and prototyping, the primary goals being to mature technologies that have promise of positively impacting a system with one or more sustainability challenges.

47.10 Maintenance Logistics Safety Reliability Quality

Current systems are extremely maintainable, supportable, safe, reliable, and of high quality. Decades of evolution bought them to this state. New systems based on displacement technology that do not have the luxury of decades of evolution must also be highly maintainable, supportable, safe, reliable, and of high quality.

Although many systems practitioners are very familiar with Maintainability, Logistics, Safety, Reliability, and Quality many sustainability practitioners lack these basic understandings. Many solutions offered by the sustainability community are non-starters because they are not maintainable. For example solar blankets have been proposed to slow down glacier melt. However it is obvious the approach did not consider maintainability or production. The lesson is that sustainable solutions must not compromise standards of Maintainability, Logistics, Safety, Reliability, and Quality.

47.11 Life Cycle Cost

Cost shifting is a prctice where the true system costs are shifted to other stakeholers who may be unaware of their new found responsibility. The Life Cycle Cost (LCC) equation is modified to remove any possibility of cost shifting. A suggestion is to consider the followng cost elements to repersent total cost.

LCC = R+D+P+O+M+W+S+T Where:

R = Research
D = Development
P = Production
O = Operation
M = Maintenance
W = Waste
S = Shut Down and Decommissioning
T = Disposal

[392] As previously discussed in this text.

An alternative view is: LCC = R+D+P+O+M+S+I+E or PROMISED Where:

R = Research
D = Development
P = Production
O = Operation
M = Maintenance
S = Shut Down & Disposal
I = Infrastructure
E = Environment & Waste

Author Comment: Infrastructure (I) represents the cost of establishing the infrastructure so that others can offer system solutions. Examples are government funded research, development, and operations of air traffic control systems and research and development of air planes for defense that are then transferred to civilian use. In this setting it is absurd to think a commercial airline company is a real business that funds all its costs. It becomes even more absurd when managers of these entities think they should derive revenue by charging for toilet services while in flight. It would be impossible for these companies to exist if the true cost of flying were factored into the business. They could not exist. The same is true of all the businesses that benefit from the rural electrification program, interstate highways, roads, space satellite systems and a plethora of other infrastructure projects that no business could justify in a reasonable business mode. But that does not mean the Infrastructure costs should not be identified for various system solutions. To the contrary they should be clearly identified so that a system does not become a burden on the people[393].

47.12 Architecture Identification Tradeoff Selection

At some point architecture alternatives, tradeoffs, and selection need to be addressed. Are there any unique tradeoff criteria to consider when selecting an architecture? How can the architectures be depicted? What process can be reasonably followed to select the architecture especially when sustainability surfaces? The following is a list of possible new sustainable related criteria to consider in architecture development:

- Sustainability
- Short Term and Long Environmental Impact: Air, land, water, sea, space, and outer atmosphere
- Air, water, and land pollution
- Noise[394] and visual pollution
- Cost Shifting

---

393 What is the Infrastructure cost of a catastrophic Nuclear Power Plant disaster?

394 Noise pollution has been a real issue addressed by air traffic control systems for decades.

- Technology Maturity, Stability, Growth Potential
- Maintainability, Producibility, Supportability
- Aesthetics, form, user acceptability

To start the process, identify the architecture alternatives. The identification includes the name of the architecture, a simple picture and a few words that capture the essence of the architecture approach. This is limited to a single page for each architecture approach.

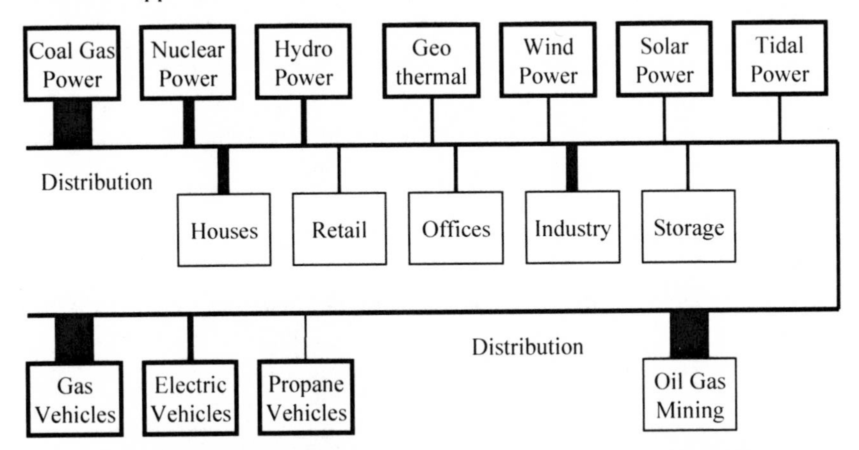

Figure 184 Sustainable Architecture Depiction - Sankey Diagram

What should be considered in the architecture of a power generation system for a community[395]? If the focus is on a new wind farm that will produce power does it make sense to use one type of wind turbine in a homogeneous system or should the system solution use multiple sizes and types of wind turbines in a heterogeneous architecture arrangement? Should the architecture use different energy generation technologies? What are the maintenance and support impacts of heterogeneous architectures and are they mitigated by the advantages?

Identifying the architecture alternatives, surfacing the tradeoff criteria and selecting the most effective architecture is one of the most important practices in systems engineering. It is especially important when questions of sustainability surface.

47.13 Value Systems

When there are different architecture alternatives what methods can be used to pick the best approach? Within this question is the concept of value systems. There are different value systems that can be used to make a selection; most of them based

[395] The Rural Electrification Act of 1936, supplied the infrastructure and funding to electrify isolated U.S. farms. May 20, 1936, S. 3483, Public, No. 605.

on financial metrics. However there is an alternative to using strictly financial metrics to make a selection. Except for the MOE, the following is a list of financial value systems.

Lowest Cost
Highest Profit
ROI - Greatest Return on Investment
TCO - Total Cost of Ownership
IRR - Internal Rate of Return
EVA - Economic Value Added
ROV - Real Option Value
ROA - Return on Assets
ROIE - Return on Infrastructure Employed
MOE - Measure of Effectiveness

Although we are familiar with various value systems, many outside the systems engineering community are not familiar with the Measure of Effectiveness (MOE) that can be calculated with each architecture approach. The Architecture MOE is the sum of the tradeoff ratings divided by the total cost or LCC.

MOE = Sum of Tradeoff Criteria / Total Cost or LCC

The Architecture MOE moves the architecture selection discussion to a different level. One solution might be lowest cost, another highest cost, and yet a third with a mid-level cost. The same applies to the tradeoff rating where one solution might have the highest rating, another the lowest, and yet a third with a middle rating. The most effective solution is the one with the best Architecture MOE.

47.14 Advantages Disadvantages List

In the beginning of the architecture trade study there is nothing, just a blank sheet of paper. The first step is to identify architecture alternatives, no matter how bizarre. One of the alternatives is the current approach. Another alternative is the dream approach. They form two extremes. The other approaches are everything in between. The alternatives are listed on a single sheet of paper and two columns are added. They are labeled advantages and disadvantages.

This is kept on a single sheet of paper so it can be easily visualized allowing the analysts to quickly get to the key issues. The objective is to minimize irrelevant information and quickly cut down on the alternatives. If there are only two alternatives, something is wrong. If there are ten alternatives, something is wrong. The answer is in between, and should include the impossible alternatives.

Table 86 List of Advantages Disadvantages

Architecture	Advantages	Disadvantages
Approach 1	Advantage 1 Advantage 2 Advantage 3	Disadvantage 1 Disadvantage 2
Approach 2	Advantage 1	Disadvantage 1

Table 86 List of Advantages Disadvantages

Architecture	Advantages	Disadvantages
	Advantage 2	Disadvantage 2
Approach n	Advantage 1	Disadvantage 1

The advantages-disadvantages or plusses-minuses tables are very important. It is at this time that the key tradeoff criteria start to surface. These tradeoff criteria are used in the next phase of the architecture trade study.

47.15 Sustainability Tradeoff Criteria

Using the advantages and disadvantages table, start to surface tradeoff criteria that naturally flow into the tradeoff matrix. Although there are tradeoff criteria unique to sustainability, the architecture tradeoff should still consider system relevant criteria. These criteria can be general and applicable to any architecture.

Resilience
Robustness
Ruggedness
Survivability
Brittleness
Flexibility
Reconfigureability
Scalability
Stability
Controllability
Elegance
Symmetry
Beauty
Simplicity
Reasonableness

Transition
Interoperability
Usability
Availability
Fault tolerance
Graceful degradation
Performance
Capacity
Growth
Technology insertion
Ability to meet requirements
Performance
User acceptability
Produce-ability
Testability

Reliability
Maintainability
Training
Supportability
Survivability
Comfort
Sustainability
Effectiveness
Safety
Security
Vulnerability
Deployment
Shutdown
Disposal

They also can include unique items associated with a domain like sustainability:

Carbon Pollution
Water Pollution
Land Pollution
Air Pollution
Noise Pollution

Visual Pollution
Fuel Sustainability
Regeneration
Progress
Model Results

Internal Sustainability
External Sustainability
Survivability
Population Growth
Standard Of Living

Social Mobility
Freedom and Liberty
Quality Of Life
Happiness

47.16 Sustainability Tradeoff Matrix

List the tradeoff criteria in rows and place the alternatives in columns. Rate each criterion for each alternative. The values can be 1-3, 1-4, 1-10, or a ranking of each alternative relative to the other alternatives. If using high, medium, or low, they can be translated to numbers. In the beginning of the trade study, fill in each cell of the matrix with a rating even if there is no supporting data. It is an educated guess based on current knowledge and perceptions. Play with each approach. Do this in one day.

Now comes the hard part of moving the content of the tradeoff matrix to fully vetted content approved by all the stakeholders. Examine the criterion and architecture alternatives. Examine each intersection or cell. Identify studies, techniques, methods and approaches to convert the initial gut-based ratings into ratings backed by sound scientific and engineering principles. This is performed as a group using hard science, soft science, and everything else in that order to back up the numbers. Document the logic even if it is just bullets on a chart.

Table 87 Architecture Tradeoff Matrix

Criteria	**A1**	**A2**	**AN**	**Wt**	**A1**	**A2**	**AN**
Criterion 1	5	7	9	1	5	7	9
Criterion 2	5	5	8	5	25	25	40
Criterion 3	5	5	6	1	5	5	6
Criterion 4	5	6	6	2	10	12	12
Criterion 5	5	9	8	3	15	27	24
Criterion Y	5	9	8	1	5	9	8
Total	30	41	45	-	65	85	99

Take snapshots and change the cells of the tradeoff matrix. Add weights to the criteria based on continued refined analysis of the problem. Some criteria may disappear and others may surface. Don't be afraid to call the team in and have everyone enter their view of the rating for each cell, no matter how detached they may be from the detailed studies. At some point some criteria will become a wash and go away while others become very different or new criterion are added.

Sum the total for each approach. This is the rating for each architecture alternative. Keep it simple at first and do not use weights until there are some initial results. As more insight is gained apply a different weight to each criterion and change the total.

Develop initial costs and total life cycle costs (LCC) for each approach. Take each rating and divide it by the initial cost. That is the initial Architecture MOE. It is a measure of goodness of each approach for each dollar spent. Do the same thing for the life cycle cost and see if they are different. Pick the architecture that has the highest MOE when all costs are considered (the LCC). This yields the biggest advantage for each unit of cost.

Table 88 Architecture MOE

Criteria	**A1**	**A2**	**AN**	**Wt**	**A1**	**A2**	**AN**
Total	30	**41**	**45**		65	**85**	**99**
LCC	1	1.2	1.4		1	1.2	1.4
MOE	30	**34**	**32**		65	**71**	**71**

Table 88 Architecture MOE

Criteria	A1	A2	AN	Wt	A1	A2	AN
The LCC is normalized. Adding cost shows that Arch 2 is more effective than Arch N. Weights shows Arch 2 becomes less effective and matches Arch N.							

What should the tradeoff criteria include? That is really something the team decides. It is part of the discovery process. However, a word of caution: the tradeoff criteria should not include cost or requirements. It is a given that all solutions will satisfy the known requirements at some cost. Cost should be used at the bottom of the tradeoff where each approach's total rating is divided by cost. This essentially identifies the goodness of each approach per unit of cost: this is called the Architecture Measure of Effectiveness or Architecture MOE.

The tradeoff study is not about the numbers in the matrix. It is about the journey to populate the matrix. Anyone can go into a closed room, fill in a matrix, and emerge to dictate that this is the answer. That is a failed effort.

A team populates the tradeoff matrix where each architecture camp makes their position. As they find weakness in their architecture alternative, they then proceed to modify it until the weakness is mitigated or completely removed. As long as the architecture concept remains in place it matures and moves from a straw man approach that is easy to knock down to an iron man then stone man architecture that is difficult to knock down. This holds for all the architecture approaches considered in the tradeoff matrix. So the tradeoff matrix is a framework to capture the quantitative study results and the qualitative arguments. It summarizes the arguments, positions, and journey that are captured in the architecture study.

47.17 Sustainability Push Versus Pull

Things either can be pushed into the mainstream or pulled in by the mainstream. This applies to new ideas, technologies, architectures, processes, methods, products, etc. The problem arises when a new technology that is good surfaces and the mainstream rejects it in favor of the status quo especially if the status quo is less effective and everyone knows it, this displacement technology[396]. There are also cases where new technologies arrive and there is no status quo but the stakeholders still reject the technology. The following are examples of challenges faced by previous generations to establish new technologies that are now the mainstream:

- Electricity versus gas
- AC versus DC electricity
- Automobile versus horse
- Television versus movie house
- Automobile versus public transit
- Airplane cargo carrier versus truck

[396] The early Internet was an example of a massive displacement technology.

- Telephone versus letter
- Radio versus telegraph
- Airplane versus railroad
- Airplane oceanic versus ship
- Cell phone versus land line phone
- Internet versus print media versus retail versus sales staff, etc

It is interesting to list home appliances and entertainment:

- Indoor water
- Indoor sewage
- Electricity
- Telephone
- Radio
- Automobile
- Washer
- Television
- Dryer
- Garbage disposal
- Dish washer
- Stereo record player
- Stereo FM radio and record player
- Reel to reel tape recorder
- Cassette tape recorder
- Video tape recorder
- Laser video player
- Cell phone
- Compact disk (CD) player
- Digital video disk (DVD) player
- Personal computer
- Internet
- MP3 player
- Global Positioning System (GPS)
- Internet phone

So what tools, techniques, strategies can be used to convert from push to pull and if the conversion is not happening what can be done to enable that conversion. The following are some possibilities:

- Great marketing campaign, but what does that mean? Use the Veblen effect[397]; appeal to the ego and exclusivity of the less enlightened.
- Broadcast the Measure of Effectiveness (MOE) findings far and wide. This was a technique used for new systems such as satellites.
- Establish small example and point to the exceptional advantages. Niagara Falls provided AC electric power for the first time on a large scale[398]. This was used to show the benefits of AC electricity.
- Educate the stakeholders. This was a technique used to successfully introduce computers in the last century.
- Use the Socratic method of discourse where questions and answers stimulate critical thinking and illuminate ideas. The goal is to lead to the obvious conclusions.

---

[397] The Veblen effect is described in this text.

[398] The Electrical World, A weekly Review of Current Progress in Electricity and Its Practical Applications, Volume 29, WJJC (The W.J. Johnston Company), Library of Princeton University, January 2 to June 26 1897.

The alternative to converting from a push to a pull scenario is to force the push. Some of the techniques are:

- Convert a waste element to a revenue producer. This is a form of recycling that happens at the production level.
- Legislation which forces the new technology to be adopted or makes the old technology illegal or tightly controlled.
- Tax incentives in the form of reduced taxes or tax credits.
- Create a virtual market. Sulfur dioxide is used as an example to establish other commodity markets for harmful by products such as carbon dioxide.

Author Comment: With the introduction of mass media there is information overload so demonstrations of new technology may not move the technology from a push to a pull. A classic example is the manned space program which resulted in humans stepping foot on the moon. Even though that was a spectacular demonstration of technologies, it was soon abandoned. Six Apollo missions landed astronauts on the Moon, the last in December 1972. In these six Apollo space flights, 12 men walked on the Moon. Apollo began after President John F. Kennedy's 1961 address to Congress declaring a national goal of landing a man on the Moon by the end of the decade.

47.18 Moving Forward

After World War II the USA fell into the cold war. This led to massive defense spending and the space race. To justify some of this expenditure the people convinced themselves of the concept of dual use[399]. Although inefficient, dual use did lead to practical systems needed for the emerging modern society. Interstate freeways became the backbone of the country as new cities emerged along its corridors. Defense RADAR was applied to civilian air traffic control and weather services. Computers and massive communications from the SAGE system gave birth to our highly automated and interconnected way of life. And certainly the space race yielded satellites and other technologies that transformed our world.

The problem is dual use has become a dated concept. There is little flowing into practical applications of everyday life as defense has taken on the concept of using commercial elements and the space program has turned into low level activities[400] that are not pushing any technologies to their limits. So the dilemma is who will pay

---

[399] Dual Use Technology, Dual Use Processes, Dual Use Products all refer to Dual Use where peaceful and military use is possible. It became part of the national dialog after World War II. Early references are not easily found. A Survey of Dual-Use Issues, IDA Paper P-3176 Prepared for Defense Advanced Research Projects Agency (DARPA).

[400] The Space Shuttle has been flying since 1981, some suggest NASA has turned into a taxi service for satellites rather than a technology driver and developer.

for the new technologies and systems needed for our sustainable future?

The argument starting in the 1980's has been that the markets will provide for new technologies and systems, as they are needed. The idea is that the financial sector, which stores capital, would make capital investments available for new ventures. However, capital was not released for investment in new technology, development, and systems. Instead capital was being used to chase other capital via various financial investment products[401].

Infrastructure	**Manufacturing**	**Services**
Housing		
Water	Durable Goods	Value Added
Sewage	Automobiles	Education
Lighting	Computers	Health Care
Power	Televisions	Restaurants
Roads		Intellectual Prop
Trains	Non-Durable Goods	
Automobiles	Toilet paper	Overhead
Planes	Detergent	Insurance
Satellites	Food	Investments
Space Ships		

Figure 185 Infrastructure Manufacturing Services

So where is the intelligent application of capital? Who decides who gets funded and under what conditions?

Outside the free market another approach is to use government funds to support projects people think are needed for their society. An example is government provided healthcare in some countries. As the details are uncovered it becomes more complex and many people immediately point to socialism as the approach with a negative connotation.

So the argument is socialism versus capitalism when deciding how to apply capital to a problem. Whether it is taxes and a politician deciding where to apply the money or investments and savings with a financial executive deciding where to apply the money the result is the same. Projects that must be funded for sustainability are not funded. The only way out of this intellectual box is to accept the following:

- Dual use technology no longer exits
- It's not about socialism or capitalism

---

[401] My Life as a Quant: Reflections on Physics and Finance, Emanuel Derman, Wiley, 2007, SBN 0470192739. How I Became a Quant: Insights from 25 of Wall Street's Elite, Richard R. Lindsey, Wiley, 2009 ISBN 0470452579. The Quants: How a New Breed of Math Whizzes Conquered Wall Street and Nearly Destroyed It, Scott Patterson, Crown Business, 2010, ISBN 0307453375

- Someone needs to pay for indirect costs that no business can ever justify

A bold approach is to get government back into the research business and directly fund sustainable projects in much the same way government funded defense and space in the dual use technology days. There even may be resurgence in dual use as defense realizes some of the sustainable projects may directly impact their needs. Also, perhaps NASA should return back to manned space missions. After all, not only is the extremely hostile space environment the ultimate teacher for sustainability but humanities future also may lie in the stars.

47.19 Going Backward

The fundamental element of the Earth is dirt. There is nothing wrong with dirt. There are billions of organisms living on dirt, in dirt, within dirt, between dirt, at one with the dirt. Humans at one time lived in the dirt and they certainly can return back to living in the dirt. The problem is billions of existing people cannot survive on a dirt-based system.

So what uplifted humanity away from the dirt and allowed 6 billion people to survive? The answer is technology applied in our systems.

The problem is our current technologies in our systems have reached their limits and our technologies and systems are causing long-term problems for our Planet. Some think that we should consciously reduce our population or the Earth will reduce our population involuntarily. However who gets to decide whom goes and who stays and under what conditions should all this devolution happen?

It's time to grow up. Asking for a conscious reduction in population is wrong. At what point do we stop our devolution and say everything is now in balance? Perhaps there is a reason for 6 billion people that may not become apparent until a small asteroid hit makes 5.5 billion people go away.

Necessities			**Luxuries**
	Transportation	Entertainment	
Food	Cars	Movies	Jewelry
Clothing	Trains	Concerts	
Shelter	Planes	Theatre	
Infrastructure			
	Information	Services	
	Computers	Health Care	
	Televisions	Education	
	Radios		

Figure 186 From Necessities to Luxuries

It is obvious that dual use practiced after World War II yielded technologies that allowed 6 billion people to emerge. However dual use was never very efficient and has been abandoned. So the issue is who funds the projects needed for our sustainability - survival?

Business does not care about large-scale technology or the long term survival and growth of humanity. They only care about profits, which tend to be tied to short term needs - and that is perfectly appropriate. There is no relationship between profit and large-scale sustainability projects.

So the question is who will make the investments needed for a future that not only gracefully supports 6 billion people today but also allows for reasonable growth and a happy healthy future?

Table 89 Our Progress to Date

Tech Level	Waste	Surplus	People
Hunter Gatherers	None	None	0.1 million
Agricultural	None	Food, Clothing	1 million
Industrial	1 Earth	Food, Clothing, Infrastructure	200 million
Technological	4 Earth's	Food, Clothing, Infrastructure, Information	6000 million

47.20 Sustainability Practitioners

Although sustainable development is not new and it is obvious that we practiced some form of sustainable development to get us to this point in our history, it appears that something new has surfaced to cause such interest. Perhaps it is our increasing sensitivity to our technologies and systems and their impact on our world. Perhaps we are detecting that our current systems and technologies have reached their limits and it is time for the next step.

This text has suggested the next step in this quest for sustainability must include systems practices. It offers some suggestions to systems practitioners as they attempt to address sustainability in their system solutions. Although system practices are known in the systems community and the community gets very exited over small details of some aspect of a systems practice, the reality is that many outside the community need basic introduction to the field, especially those attempting to offer sustainable solutions.

47.21 Sustainable Air Traffic Control Example

In the early 1980's the FAA had concluded that the current Air Traffic Control (ATC) System would start to saturate; reach capacity limits in certain parts of the country that would then affect all flights. The National Air Space (NAS) Plan or Brown Book was developed in response to this future projection of capacity limits. The Office of Technology Assessment (OTA)[402] performed a review of the NAS plan in 1982 and concluded that it was primarily technology driven and emphasized

---

[402] The OTA was established with The Technology Assessment Act of 1972. It was de-funded in 1995 as part of the 104th Congress "Contract with America".

Enroute ATC[403]. According to the OTA: "The national air space system is a three-legged stool made up of airports, the ATC system, and procedures for using the airspace".

The original FAA intent was to significantly raise the automation level of the Enroute systems while the planes were within Enroute airspace. Effectively the planes would be metered and spaced so that the airport peak loads could be spread out in time[404]. It was a continuation of the flow control concepts at the national level that were found to be very effective.

It did not matter that the OTA may not have fully understood the impact of the planned automation. It was reasonable to add the other two legs of the ATC system (Airports and Procedures) to the NAS plan. A revised plan was issued the following year in 1983. Many years ago I concluded there are only a few ways to increase capacity of an ATC system.

- **Cities:** The first is to build new cities away from existing cities with new airports some distances away from existing cities. The FAA has no control over how and where cities should be built. This is a bigger issue and may actually be the source of the ATC capacity problem and other problems. So the only answer might be to start to think broader and bigger and realize new cities are needed.
- **Bigger Planes:** The second is to build bigger airplanes and make sure they are filled to capacity. The FAA could encourage the airlines to buy bigger airplanes and increase seating capacity. However, some airports may be unable to handle the bigger airplanes and huge numbers of people arriving and departing. History suggests that is what may have actually happened as the casual traveler compares air travel of the 1970's and early 1980's with air travel today.
- **Procedures:** The third is to change the procedures. The procedures could be modified. Approach departure routes could abandon sound abatement rules and subject ground neighborhoods to the stresses of loud noise and possible fear of airplane accidents. Some might even suggest changes to reduce minimum separation standards. Others might suggest removing airplanes from standard air routes[405] to significantly improve capacity. However

---

[403] Review of the FAA 1982 National Airspace System Plan, August 1982, NTIS order #PB83-102772. Library of Congress Catalog Card Number 82-600595, U.S. Government Printing Office, Washington, D.C.

[404] In the early 1970's the FAA established a Central Flow Control Facility to prevent clusters of congestion from disrupting the nationwide air traffic flow.

[405] Air routes are like railroad tracks in the sky. It makes sense given that railroads existed before airplanes. It is easy to see how people from the previous century would adopt this approach for order, efficiency, and safety. Flying from visual point to visual point or

removing airplanes from air routes would require new navigational aids like Global Positioning System (GPS)[406] and a new control system. Basic things like Flight Plan Aided Tracking, Fix Posting Determination, etc all embedded within the existing automated system would need changes.

- **Automation:** The fourth is to increase automation to allow more airplanes to be pumped through the system. The reality is the system was very automated circa 1980. Many of the manual activities had been addressed with the introduction of the computer. Also automation is most effective where the airplanes have time to respond and maneuver.

The job at the tower is to basically watch the airplane on the taxiway and get it safely and quickly to and from the runway via the taxiways. Some airports can be expanded with additional runways, more taxiways, and larger physical terminals. However many airports that mattered reached their capacity limits.

The Terminal RADAR Approach Control Facility (TRACON) also has little options to increase capacity. The airplanes fly fast and 30 miles away from an airport does not leave much time for metering or other techniques to increase capacity.

If the airport peak loads could be spread out where there was time to respond, such as when the airplane is Enroute, then the system capacity could increase. So the automation was most appropriate for the Enroute centers.

The problem is the automation of the Enroute centers was captured within the Advanced Automation System Program (AAS) and the Automated En Route ATC (AERA) system concept. From the beginning, the GAO report found the AERA concept was flawed. So the ambitious goal of automation failed as AAS was restructured to just replace the existing hardware and software with new technology while essentially staying within the same levels of automation. Portions of the GAO report findings are included in this text and summarized with author comments as follows[407].

From Executive Summary

- *"Growth.* —FAA's traffic forecasts have been too high in the past and there are

---

navigation beacon to navigation beacon made perfect sense.

[406] First experimental Block-I GPS satellite was launched in 1978. In 1996, U.S. President Bill Clinton issued a policy directive declaring GPS to be a dual-use system. The policy presents a vision for management and use of GPS for military, civil, commercial, and scientific interests, both national and international where the Department of Defense, Department of Transportation, and Department of State manage it as a national asset.

[407] Review of the FAA 1982 National Airspace System Plan, August 1982, NTIS order #PB83-102772. Library of Congress Catalog Card Number 82-600595, U.S. Government Printing Office, Washington, D.C.

questions about the methodologies and assumptions underlying the projections on which the NAS Plan is based. Overestimation may have led FAA to foreclose technological options and accelerate the implementation schedule unnecessarily. It may also have led FAA to overestimate the user-fee revenues that will be available to pay for the proposed improvements."

Author Comment: Growth is an interesting concept. It is unclear how aviation may have changed if the original AAS program were permitted to evolve. Everyone in the community at the time knew that the introduction of the computer into air traffic control was sold as a labor saving device. However what really happened is the system was transformed to a new level allowing more people to fly while at the same time increasing the labor force as new kinds of activities surfaced. This not only happened in air traffic control but also in every segment of the world that adopted computer technology. What would have happened if AAS as originally envisioned emerged? We may never know.

- "*En Route Computer Replacement.* —FAA's option analysis issued in January 1982 supports upgrading the 10 en route computers that face capacity problems.[408] The NAS Plan, released at about the same time, calls instead for replacing the computer hardware (called rehosting the software) in all 20 centers as a part of a long-term plan to increase productivity and reliability as well as capacity. OTA does not find persuasive the reasons advanced by FAA for rejecting the previously preferred option of upgrading only selected en route centers. In addition, the choice of a host computer now may limit the options available to the contractor for the sector suite and software. OTA conferees were sharply divided in their views on this question. Some felt that the choice of a host computer now might limit future ability to benefit from a distributed computer architecture, local area networking, and new techniques in software development. Others believed that, if the host is chosen judiciously, the transition to a new system embodying these advanced and desirable features could be made without difficulty."

Author Comment: The proposed Hughes architecture was highly distributed in anticipation of the coming automation. Hughes was concerned that the cost of the rehost program would eventually stop AAS and the full-scale automation envisioned with AERA. In hindsight they were probably correct.

- "*Automation.* —While the NAS Plan envisions substantial cost savings due to extensive automation, supporting analysis is not provided in the plan. This analysis is probably still in progress and may take some time to complete, but it would be useful for the interim results to be made available to assist in congressional review of the automation portions of the overall plan. In addition, there is concern on the

---

[408] Federal Aviation Administration, "Response to Congressional Recommendations Regarding the FM's En Route Air Traffic Control Computer System, " report to the Senate and House Appropriations Committees pursuant to Senate report 96-932, DOT/FAA/ AAP-823, January 1982.

part of some experts about the ability of human operators to participate effectively in such a highly automated system and to intervene in the event of system error or failure."

Author Comment: Hughes was very aware of the role of automation in air defense systems. It was common knowledge that moving from a manual system to an automated system significantly reduced fighter intercept time and could mean the difference between winning and losing a battle. Everyone was also aware of the manual air defense and air traffic control system prior to RADAR and its affect on aircraft separation. So the practitioners in the business did not need to be sold on the concept of AERA. It was intuitively obvious that using machines to find ideal trajectories was significantly safer and faster than using humans. This was all lost in the politics of the day, as air traffic controllers feared losing their jobs and existing companies feared losing their contracts.

- "*Satellites.* —Satellite technology has significant potential applications for communication, and eventually for surveillance and navigation. FAA does not see a role for satellites in the period covered by the NAS Plan. FAA's decision against satellites appears to have been driven by timing and present cost effectiveness, rather than technology readiness or long-term system advantages."

Author Comment: Satellites were interesting and exciting. However no one at the time could figure out how satellites could allow more aircraft to be pumped through the system. The USA fortunately spent the money to provide RADAR coverage down to the ground level using gap filler RADARs where they were needed. Hughes had more satellite experience than anyone, but they were unable to see an application to address the ATC capacity problem. Hughes was extremely technology driven, with no real interest in products, so rather than trying to sell unneeded satellites, which they easily could have, the team continued to work the AAS program. While some were concerned about machines failing when offering advanced automation no one seemed concerned about satellite failure.

- "*User Effects.* —A great many of the proposed ATC system improvements are directed to the needs of traffic operating under the instrument flight rules (IFR), particularly while en route at cruise altitude. These improvements will benefit FAA itself by automating functions and reducing labor costs. The principal beneficiaries among users will be air carriers and larger business aircraft. Personal general aviation (GA) users could receive improved weather information, an important benefit; but in order to obtain this benefit and other operational advantages of the new system, more avionics will be required, and there would be restrictions on access to airspace by aircraft not so equipped. The Department of Defense (DOD) too, is concerned about the cost of new ATC avionics and feels that the new plan must be carefully coordinated with the military services to ensure that their mission needs and responsibilities for administration of the airspace are integrated with those of FAA."

Author Comment: The problem was system capacity. Some radicals were even suggesting taking GA out of the equation entirely so that the system could be sustainable in the future. The cost of avionics for DOD was a red hearing. The DOD budget at the time was at least 10 times more than the FAA budget. This cost would have been trivial compared to the cost of a typical DOD aircraft.

- *"Cost and Funding.* —Implementing the improvements proposed in the 1982 NAS Plan would more than double FAA's facilities and equipment budget through 1987, compared to the last 10 years. FAA has not yet released cost estimates for completing the proposed programs, but it seems likely that expenditures of like magnitude will be needed in the years beyond 1987. FAA proposes to recover 85 percent of its total budget through user fee revenues and a drawdown of the uncommitted Trust Fund balance. The user fee schedule would perpetuate the existing crosssubsidy from airline passengers and shippers of air cargo to GA. Business aviation would benefit particularly because of the extensive use these aircraft make of the IFR system. In addition, higher user fees may dampen the growth of aviation, thereby reducing the revenues expected to pay for the proposed improvements."

Author Comment: Many organizations were after the FAA trust fund coming from the gate fees in the 1980's and 1990's. For years there were movements to privatize the FAA while shifting all the indirect costs to the taxpayers via various privatization schemes. The reality is air travel is very high technology with huge indirect costs paid for by the taxpayer. From research for new aircraft via DOD to RADAR and other such systems that have no commercial equivalents that business can justify as investment.

<u>From Technologies</u>

- "Rand's principal conclusion is that the goal of full automation sought under AERA is a questionable research and development strategy that may present serious problems with regard to safety, efficiency, and increased productivity. An ATC system in which computers make most of the time-critical decisions in controlling aircraft, while the human operator serves in a managerial and back-up role, implies a needlessly complete and irrevocable commitment to automation. Rand argues for an alternative approach, called "shared control, " that would construct the future ATC system as a series of independently operable, serially deployable modules that would aid—not replace— the human controller and keep him routinely involved in the minute-to-minute operation of the system."

Author Comment: This killed the automation program. Placing the human in the minute-to-minute operation of the system would have been like placing people on the ground with binoculars to radio in positions (like they did at one time) rather than use RADAR. It would have been like not digitizing RADAR and using a computer to generate tracks, instead just staying with the proven broad band system and letting everyone use grease pencils and shrimp boats (like they did at one time).

AERA is represented as a phased implementation of progressively more elaborate automation and is structured as AERA I, II, and III.

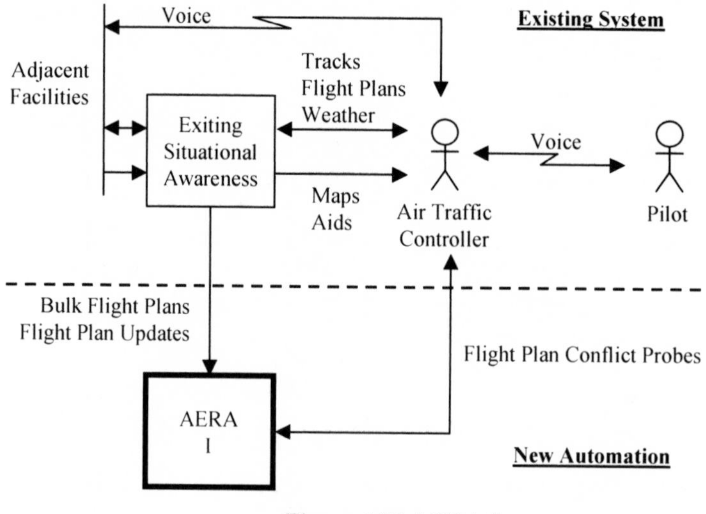

Figure 187 AERA I

The AERA I goal is to access the flight plan database and use it to perform flight plan conflict probes prior to providing a clearance that would need an update to a flight plan. This automation is of a strategic nature because the database is slowly updated, on the order of minutes. The automation includes graphical projections of the trajectories of the aircraft.

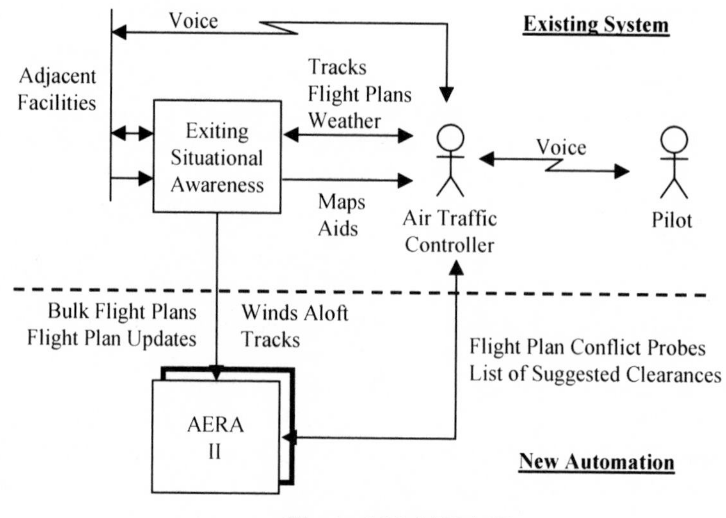

Figure 188 AERA II

In AERA II the goal is to access the tactical database populated by tracks (from

RADAR) and winds aloft data to work with the flight plan database and offer a list of suggested clearances. The graphical projections of the trajectories of the aircraft now include the tactical information associated with the tracks. With this real time information a list of possible clearances is offered to a controller.

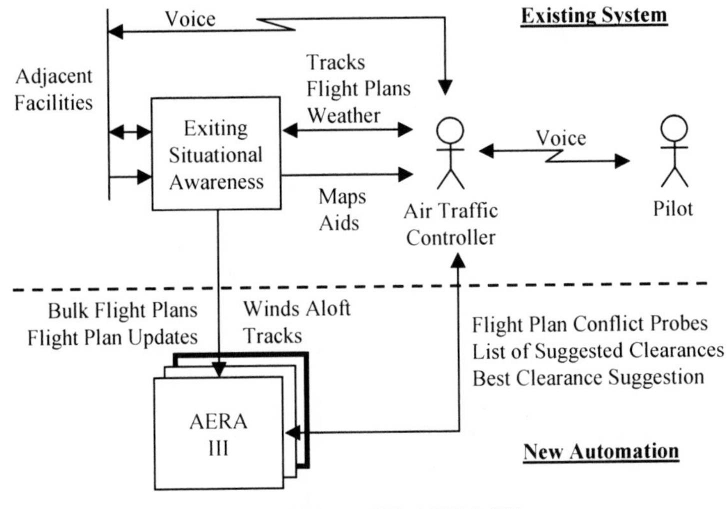

Figure 189 AERA III

In AERA III additional automation is added to offer a best clearance suggestion from the list of clearances. It is the choice of the controller to use or not use the information provided by the automation system. If the system could actually achieve this level of automation then the clearances could be data linked directly to the pilot with the air traffic controller becoming more of a system monitor rather than an active decision maker in the system.

Author Comment: It is obviously very easy to draw the AERA pictures. However turning these pictures into reality is much more difficult. The AAS program was supposed to provide the foundation for AERA. But AAS failed, in part because of the complexities of not only replacing an existing system that evolved over decades, but also because of the technical challenges of AERA, and the political challenges of its suggested levels of automation.

Today air travel is not a pleasant experience. All flights are booked to maximum capacity. There are no open seats as existed in the 1980's. Even the galleys have been removed to add more seating. Delays were common in the past but today we see delays of hours on the Tarmac. This was unheard of in the past. The bottom line is the system has reached its limits. The question now is will there be restrictions to reduce the number of passengers (airplanes in the sky) or will there be a new system[409]? The following is a summary of the SE2020 solicitation:

"December 4, 2009. The FAA is soliciting bids from companies interested in competing for NextGen support contracts with an approximate combined value of $7 billion, the largest award in the agency's history. Under the umbrella awards, called System Engineering 2020 (SE2020), the FAA will award as many as five separate contracts for research and development and systems engineering work that will help the agency deliver NextGen.

The SE2020 contracts will be awarded to teams of companies, up to three of which will perform research and development work and two of which will perform systems engineering work. This work will complement and enhance major NextGen initiatives already under way, such as Automatic Dependent Surveillance - Broadcast, System Wide Information Management and Data Communications. Contract teams will focus on a series of operational capabilities, including Trajectory Based Operations, Collaborative Air Traffic Management and Reduced Weather Impact. The goal is to achieve early NextGen successes to improve safety and bring greater efficiencies to the nation's airspace system."

The following is a summary description of the NextGen program:

"NextGen is an umbrella term for the ongoing, wide-ranging transformation of the National Airspace System (NAS). At its most basic level, NextGen represents an evolution from a ground-based system of air traffic control to a satellite-based system of air traffic management. This evolution is vital to meeting future demand, and to avoiding gridlock in the sky and at our nation's airports.

NextGen will open America's skies to continued growth and increased safety while reducing aviation's environmental impact.

We will realize these goals through the development of aviation-specific applications for existing, widely-used technologies, such as the Global Positioning System (GPS) and technological innovation in areas such as weather forecasting, data networking and digital communications. Hand in hand with state-of-the-art technology will be new airport infrastructure and new procedures, including the shift of certain decision-making responsibility from the ground to the cockpit.

When fully implemented, NextGen will allow more aircraft to safely fly closer together on more direct routes, reducing delays and providing unprecedented benefits for the environment and the economy through reductions in carbon emissions, fuel consumption and noise."

Hindsight is always 20/20. That was a common saying in the ATC community circa 1980. However the GAO report, which rejected the AERA concept because it was too risky, was the wrong answer. It is interesting they did not give a history of air traffic control and reference the SAGE system. Without that context no one would realize that air traffic control advances were always driven with new and innovative technologies. Certainly satellites were new and innovative and GAO

---

[409] SE2020 and NextGen are the FAA's new initiative to build a new air traffic control system.

pointed out that the FAA ignored satellites in the plans. However GAO and others failed to identify how satellites could translate to increased capacity[410].

Perhaps the most significant failure of the GAO report was to realize that air traffic control is a classic control loop with the human in the feedback loop. There were many automated control loops in everyday practice by 1982. For example radios used automatic frequency and gain control circuits so that they would not drift off frequency and the sound not vary in loudness. Television used these same closed loop control mechanisms to make sure the picture maintained correct brightness, contrast, and color. A simple example of removing any of these automated control systems and placing the human in the loop to control and balance the sound or video by turning knobs would have clearly demonstrated how the current air traffic control system was limited by the human. The poor air traffic controller continuously turns knobs to provide safe and efficient separation services.

The real tragedy is the option offered was an all or nothing option for AERA. Rather than allowing the AERA technology to continue to evolve with limited operational tests, it was abandoned. Hughes with their air defense intercept algorithms could have substantially moved the AERA concept forward. A simple approach of awarding production to IBM for the status quo and a new research and development activity associated with AERA concepts would have been interesting.

Few are aware that at that time there were proponents of moving the entire air traffic control function into the airplane cockpit. The idea was to provide full situational awareness in the cockpit and let the pilots negotiate the airspace.

The AERA technology was the next logical step in automation. It did not propose moving the function into the cockpit. It is unclear how AERA compared to the SAGE system, which introduced the computer into the system with incredible levels of automation and communications networking. It is unclear how it compared with the manned space program and the mission to the moon.

What is clear is that with the removal of AERA from the system goals many challenges disappeared. It is possible that with AERA many of the computer companies that went out of business in the late 1980's could have remained and perhaps new forms of computing could have surfaced. Perhaps the Internet could have arrived 10 years earlier. So the impact of this decision had far ranging consequences that no one can really identify except to say that progress certainly stopped and the current system internal sustainability is now in question.

47.22 Sustainable Air Traffic Control Lesson

AAS is a very important lesson for all sustainability practitioners. In this case the institutions were in place, they were very effective, the money was there, and

[410] Satellites were great for countries with no infrastructure, hostile environments, or spread out over vast bodies of water. It is like cellular phone communications, the infrastructure can be quickly and easily established.

everyone had the best of intentions. Yet the big goal was never achieved and the system stagnated.

What happened was the age-old battle of the status quo and the new. Eventually the status quo won and the new was abandoned. There were actually camps that formed. IBM and the air traffic controllers represented the status quo. IBM was awarded the AAS production contract. Hughes and many engineers represented the new. Hughes lost the AAS production contract and moved on to field air traffic control systems in Canada and the Pacific Rim. The Pacific Rim shortly followed by Canada were the first to embrace satellites both for data link and position keeping.

So the status quo is extremely powerful and at times dangerous. In the 1980's it was not uncommon for older people to always refer to status quo and the bureaucracy. The bureaucracy was a thing to be avoided. However that is easier said than done.

In the 1980's there was a huge gap between technologists and everyone else in society. Today that gap is significantly smaller partly because the engineers have abandoned much of their high tech views with downsizing[411] and partly because everyone else has become more technically literate with the Personal Computer, Cell Phones, and the Internet. Most are now familiar with their version of SAGE. The sad part is the system that everyone thought was high risk, developed by extremely high tech people, was dwarfed by what was to come and is now common in this century. The following points summarize the whole sad situation:

- In the early 80's they were asked to create a new architecture that would not fail and support growth and technology insertion into the next century.
- The solution was a highly distributed system with over 300 computers per facility with 30 facilities across the USA.
- Never did they visualize hundreds of millions of computers connected together into a single machine, the Internet.

47.23 Beyond Sustainability Regenerative Systems

Regenerative systems go beyond sustainability. As part of their normal operations they improve the environment while operating and leave the environment in a better state after their operation. The following are examples of regenerative systems:

- Systems that are absolutely waste free and consume waste
- Systems that use water and return it back to the environment cleaner than when it arrived into the system
- Systems that add back the original bio-diversity to land and then add more

---

[411] Sustainable Development Possible with Creative System Engineering, Walter Sobkiw, 2008, ISBN 0615216307.

bio-diversity to land than what was possible before the system introduction

The concept of regenerative systems was started in the late 1970's by professor John T. Lyle[412] who challenged graduate students to envision a community based on living within the limits of available renewable resources without environmental degradation.

47.24 Final Comments

The Space Act of 1958 included the forces of defense, control with ownership, and the elusive "for all mankind" idea. I remember the space race like yesterday. Many people including myself only believed in the "for all mankind" part of the equation. People were very excited about the early space walks and the first moon landing. But as time passed people started to believe that this was easy. Meanwhile we had enormous problems of war and hunger on Earth. So many people started asking the fundamental question of why we were spending so much energy and resources on space when there were needs here on Earth.

Going into space required enormous intellect, will power, resources, etc. Few realized that shutting it down would be a huge act of hubris. Today few realize that if you shutdown something you never may be able to recreate it regardless of our desires. This is emergence at its most significant level.

There is no question that the military used space funding to further their needs. There is also no question that those who think they have a right to own everything furthered their needs. The problem is where does the "for all mankind" fit into the equation and what does it really mean.

Today when we point to hunger, want, and other ills we immediately shift to the idea that this is all rooted in our politics. We think our technologies and systems are mature enough to feed and provide for everyone. However what if that is not the case. What if we use politics as an excuse and in reality we are engaged in an act of hubris. What if we really are lacking the technologies, know how, and systems to beat the elements and live well on Earth.

This book was started in December of 2009. By spring of 2011 it was complete and we had witnessed the earthquake and tsunami in Japan with the resulting nuclear disaster. Not only were we not able to gracefully deal with the earthquake and tsunami[413] but the nuclear disaster was also beyond our control. This is a classic case of being humbled by Mother Nature and our best technologies failing.

Were placed on this Earth to be the stewards of all its' life including human life? Were we placed here on Earth to bring life to other bodies in our solar system?

---

[412] Regenerative Design for Sustainable Development, John Tillman Lyle, Wiley Professional, December 1 2008, ISBN 0471178438

[413] Even though the Japanese earthquake preparedness was state-of-the-art and an example to the rest of the world.

Should we even consider such a challenge? Will that work help us be better stewards on Earth? These ideas are a tall order and I am not sure we can ever achieve these goals. But I do know we will need all our systems practices moving forward.

47.25 Exercises

1. Draw a functional block diagram of a house and a city. Draw a floor plan of a house and a layout of a city. Make an artist rendering of a house and a city. Identify the subsystems of a house and a city. How are the house and city similar and different in each system view?
2. What are the conditions of the roads, water, sewage, power, healthcare, hospitals, clinics, food, farming, fishing, housing, transportation, road, airport, railroad, educational, university, radio, television, communications, telephone, cell phone, Internet, financial, farming, entertainment, retail, employment, industrial, and satellite systems in your community? Make a table and address every system and subsystem. What other systems and subsystems should we be concerned about addressing?
3. Is it possible to develop a process that will force unethical people to behave ethically? How does this relate to sustainability?
4. Is it possible to create a process for a high technology organization so that people who have no education, experience, or background can successfully execute the process? What about a low-tech organization? If no why not, if yes then why? How does this relate to sustainability?
5. How can compensation incentives be changed to force people to execute an accepted process? What can you do if the compensation incentives encourage departure or bypassing of standard process elements? Can you provide examples? How does this relate to sustainability?
6. How can you duplicate or replace in 5 years an existing but stressed infrastructure system that took 50 years to evolve or develop?
7. Identify performance requirements for a television, stereo, car, computer, cell phone, house, kitchen, city, hospital, school, university, bridge, regional water, national electric power, solar power, and wind power systems. Place each on a standalone sheet of paper and then consolidate as needed.
8. What other performance requirements and for which systems do you think we need performance requirements for our survivability, internal sustainability, and external sustainability and why?
9. Does Earth have a set of performance requirements and if so what are they?
10. Do you think a performance requirement of having the Earth support 100 billion people is a reasonable performance requirement? If so why, if not why not? In an ideal Earth system how long would it take for the population to reach 100 billion people?
11. Can you provide examples of simple answers to what are impossible

problems such as using a ball of twine to find your way through a labyrinth? Are these examples of reason? How does this relate to systems practices? How does this relate to sustainability?

12. What is the most sustainable LRU level? Which community is more sustainable, one with 5 hardware stores in a 5-mile radius or one with 1 hardware store in a 20-mile radius?
13. Do your parking lots have sufficient spaces for the associated buildings? If not how could someone make this timing and sizing mistake after 60+ years of experience? Is this sustainable?
14. Do your parking lots have sufficient entry and exit point for fast safe traffic movement? If not how could someone make this timing and sizing mistake after 60+ years of experience? Were they minimized for security reasons? Is this secure? Is this sustainable?
15. Draw a concept diagram for a parking lot servicing 100+ retail stores. Compare it with your classmates. What are the differences and issues? Which parking lot is ideal and why?
16. Draw a context diagram for a house, city, town, shopping center, healthcare system, educational system, and Earth. Identify 3+ functional and performance requirements (total 6+) for each. How do they compare?
17. Draw a context diagram for the following power plants: coal fired, methane, hydro, nuclear, solar, wind, and geothermal. Identify 3+ functional and performance requirements (total 6+) for each. How do they compare?
18. What are the key functional and performance characteristics of your city and Earth? What are the key functional and performance requirements of an ideal Earth and city? How do they compare?
19. If small is relative, is building a housing community of 1000+ houses small once compared to the rebuilding of Europe after World War II or sending a man to the moon? If yes why, if not why not?
20. Do you need to formally run systems practices after years of experience on successful large systems for small commercial products? If yes why, if not why not?
21. Can you successfully develop small commercial products without first being exposed to successfully applied systems practices? If yes why, if not why not?

47.26 Additional Reading

1. A Survey of Dual-Use Issues, IDA Paper P-3176 Prepared for Defense Advanced Research Projects Agency (DARPA), March 1996.
2. Brundtland Commission, or the World Commission on Environment and Development, known by its Chair Gro Harlem Brundtland, convened by the United Nations in 1983 published Our Common Future, also known as

Brundtland Report, in 1987.

3. How I Became a Quant: Insights from 25 of Wall Street's Elite, Richard R. Lindsey, Wiley, 2009 ISBN 0470452579.
4. My Life as a Quant: Reflections on Physics and Finance, Emanuel Derman, Wiley, 2007, SBN 0470192739.
5. Regenerative Design for Sustainable Development, John Tillman Lyle, Wiley Professional, December 1 2008, ISBN 0471178438.
6. Response to Congressional Recommendations Regarding the FM's En Route Air Traffic Control Computer System, report to the Senate and House Appropriations Committees pursuant to Senate report 96-932, Federal Aviation Administration, DOT/ FAA/ AAP-823, January 1982.
7. Review of the FAA 1982 National Airspace System Plan, August 1982, NTIS order #PB83-102772. Library of Congress Catalog Card Number 82-600595, U.S. Government Printing Office, Washington, D.C.
8. Sustainable Development Possible with Creative System Engineering, Walter Sobkiw, 2008, ISBN 0615216307.
9. The Electrical World, A weekly Review of Current Progress in Electricity and Its Practical Applications, Volume 29, WJJC (The W.J. Johnston Company), Library of Princeton University, January 2 to June 26 1897.
10. The Quants: How a New Breed of Math Whizzes Conquered Wall Street and Nearly Destroyed It, Scott Patterson, Crown Business, 2010, ISBN 0307453375.

48 Appendix A Public Laws and Dual Use

After World War II a number of US Government organizations were established by public law. Today we take these organizations for granted and they appear to provide services that offset significant costs from industry to government.

They were started with the military mission in mind and allowed to venture into non-military missions using the concept of Dual Use. Even today, it is obvious that no business or industry could ever justify entry into any of the services offered in the early years of these government organizations.

In the past three decades there have been several proposals to sell these assets to private organizations, however the true costs that should be returned to the original investors, generations of tax payers, does not enter the dialog. It is obviously trillions of dollars with good will that is incalculable. These organizations and their results led to our modern sustainable way of life. Sustainability is a very complex subject, but avoiding war, revolution, and famine is a first step in sustainability.

48.1 Federal-Aid Highway Act of 1956

The Federal-Aid Highway Act of 1956 (Public Law 84-627) has a long history dating back to the early 1900's. On the one hand there were forces that saw the need for roads across the country on the other side there were forces that did not want to use taxes to pay for these roads. In 1956 the merging of commercial forces with the concept of pay roads and public good government forces with the concept of taxes to pay for the roads merged when the forces of defense justified the project as an attempt to survive nuclear war. Forget about the absurdity of some thinking that roads would be needed prior to or after a nuclear war, the concept of Dual Use started to surface and allowed for the USA to move into a new modern world.

48.1.1 Federal-Aid Highway Act of 1956 Abstract

National Interstate and Defense Highways Act (1956)

AN ACT

To amend and supplement the Federal-Aid Road Act approved July 11, 1916, to authorize appropriations for continuing the construction of highways; to amend the Internal Revenue Code of 1954 to provide additional revenue from the taxes on motor fuel, tires, and trucks and buses; and for other purposes.

Be it enacted by the Senate and House of Representatives of the United States of America in Congress assembled,

TITLE I—FEDERAL-AID HIGHWAY ACT OF 1956

SEC. 101. SHORT TITLE FOR TITLE I.

This title may be cited as the "Federal-Aid Highway Act of 1956".

SEC. 102. FEDERAL-AID HIGHWAYS.

(a) (1) AUTHORIZATION OF APPROPRIATIONS.—For the purpose of carrying out the provisions of the Federal-Aid Road Act approved July 11, 1916 (39 Stat. 355), and all Acts amendatory thereof and supplementary thereto, there is hereby authorized to be appropriated for the fiscal year ending June 30, 1957, $125,000,000 in addition to any sums heretofore authorized for such fiscal year; the sum of $850,000,000 for the fiscal year ending June 30, 1958; and the sum of $875,000,000 for the fiscal year ending June 30, 1959. The sums herein authorized for each fiscal year shall be available for expenditure as follows:

(A) 45 per centum for projects on the Federal-aid primary high- way system.

(B) 30 per centum for projects on the Federal-aid secondary high- way system.

(C) 25 per centum for projects on extensions of these systems within urban areas.

(2) APPORTIONMENTS.—The sums authorized by this section shall be apportioned among the several States in the manner now provided by law and in accordance with the formulas set forth in section 4 of the Federal-Aid Highway Act of 1944; approved December 20, 1944 (58 Stat. 838) : Provided, That the additional amount herein authorized for the fiscal year ending June 30, 1957, shall be apportioned immediately upon enactment of this Act.

(b) AVAILABILITY FOR EXPENDITURE.—Any sums apportioned to any State under this section shall be available for expenditure in that State for two years after the close of the fiscal year for which such sums are authorized, and any amounts so apportioned remaining unexpended at the end of such period shall lapse: Provided, That such funds shall be deemed to have been expended if a sum equal to the total of the sums herein and heretofore apportioned to the State is covered by formal agreements with the Secretary of Commerce for construction, reconstruction, or improvement of specific projects as provided in this title and prior Acts: Provided further, That in the case of those sums heretofore, herein, or hereafter apportioned to any State for projects on the Federal-aid secondary highway system, the Secretary of Commerce may, upon the request of any State, discharge his responsibility relative to the plans, specifications, estimates, surveys, contract awards, design, inspection, and construction of such secondary road projects by his receiving and approving a certified statement by the State highway department setting forth that the plans, design, and construction for such projects are in accord with the standards and procedures of such State applicable...

SEC. 108. NATIONAL SYSTEM OF INTERSTATE AND DEFENSE HIGHWAYS.

(a) INTERSTATE SYSTEM.—It is hereby declared to be essential to the national interest to provide for the early completion of the "National System of Interstate Highways", as authorized and designated in accordance with section 7 of the Federal-Aid Highway Act of 1944 (58 Stat. 838). It is the intent of the Congress that the Interstate System be completed as nearly as practicable over a thirteen-year period and that the entire System in all the States be brought to simultaneous completion. Because of its primary importance to the national defense, the name of such system is hereby changed to the "National System of Interstate and Defense Highways". Such National System of Interstate and Defense Highways is hereinafter in this Act referred to as the "Interstate System".

(b) AUTHORIZATION OF APPROPRIATIONS.—For the purpose of expediting the construction, reconstruction, or improvement, inclusive of necessary bridges and tunnels, of the interstate System, including extensions thereof through urban areas, designated in accordance with the provisions of section 7 of the Federal-Aid Highway Act of 1944 (58 Stat. 838), there is hereby authorized to be appropriated the additional sum of $1,000,000,000 for, the fiscal year ending June 30, 1957 , which sum shall be in addition to the authorization heretofore made for that year, the additional sum of $1,700,000,000 for the fiscal year ending June 30, 1958, the additional sum of $2,000,000,000 for the fiscal year ending June 30, 1959, the additional sum of $2,200,000,000 for the fiscal year ending June 30, 1960, the additional sum of $2,200,000,000 for the fiscal year ending June 30, 1961, the additional sum of $2,200,000,000 for the fiscal year ending June 30, 1962, the additional sum of $2,200,000,000 for the fiscal year ending June 30, 1963, the additional sum of $2,200,000,000 for the fiscal year ending June 30, 1964, the additional sum of $2,200,000,000 for the fiscal year ending June 30, 1965, the additional sum of $2,200,000,000 for the fiscal year ending June 30, 1966, the additional sum of $2,200,000,000 for the fiscal year ending June 30, 1967, the additional sum of $1,500,000,000 for the fiscal year ending June 30, 1968, and the additional sum of $1,025,000,000 for the fiscal year ending .June 30, 1969...

48.2 National Aeronautics and Space Act of 1958

National Aeronautics and Space Act of 1958 (Public Law 85-568) established the National Aeronautics and Space Administration (NASA). It is interesting to see the three forces of public good via government, defense, and business embodied in the act. Embodied in the effort is also the concept of Dual Use. NASA relied heavily on Dual Use arguments to continue funding after the Apollo Program and the moon landings.

- The public good is captured by the statement: "The Congress hereby declares that it is the policy of the United States that activities in space should be devoted to peaceful purposes for the benefit of all mankind."
- The defense aspect captured by the statement: "The Congress declares that the general welfare and security of the United States require that adequate provision be made for aeronautical and space activities."
- The business element captured in the unusual assignment of the intellectual property to the administrator rather than the newly formed entity: "Upon any application as to which any such statement has been transmitted to the Administrator, the Commissioner may, if the invention is patentable, issue a patent to the applicant unless the Administrator, within ninety days after receipt of such application and statement, requests that such patent be issued to him on behalf of the United States. If, within such time, the Administrator files such a request with the Commissioner, the Commissioner shall transmit notice thereof to the applicant, and shall issue such patent to the Administrator unless the applicant within thirty days after receipt of such notice requests a hearing before a Board of Patent Interferences on the question whether the Administrator is entitled under this section to receive such patent."

In 2007 the space act was modified. The defense component was de-emphasized. No longer was there a need to meet with the president and department of defense representatives. NASA was given the task of using commercial space services and creating indemnification mechanisms. It was also given the task of developing Hybrid Vehicles.

48.2.1 Space Act of 1958

National Aeronautics and Space Act of 1958, Public Law #85-568, 72 Stat., 426. Signed by the President on July 29, 1958, Record Group 255, National Archives and Records Administration, Washington, D.C; available in NASA Historical Reference Collection, History Office, NASA Headquarters, Washington, D.C.

An Act

To provide for research into problems of flight within and outside the earth's atmosphere, and for other purposes.

Be it enacted by the Senate and House of Representatives of the United States of America in Congress assembled,

TITLE I -- SHORT TITLE, DECLARATION OF POLICY, AND DEFINITIONS

SHORT TITLE

Sec. 101. This act may be cited as the "National Aeronautics and Space Act of 1958".

DECLARATION OF POLICY AND PURPOSE

Sec. 102. (a) The Congress hereby declares that it is the policy of the United States that activities in space should be devoted to peaceful purposes for the benefit of all mankind.

(b) The Congress declares that the general welfare and security of the United States require that adequate provision be made for aeronautical and space activities. The Congress further declares that such activities shall be the responsibility of, and shall be directed by, a civilian agency exercising control over aeronautical and space activities sponsored by the United States, except that activities peculiar to or primarily associated with the development of weapons systems, military operations, or the defense of the United States (including the research and development necessary to make effective provision for the defense of the United States) shall be the responsibility of, and shall be directed by, the Department of Defense; and that determination as to which such agency has responsibility for and direction of any such activity shall be made by the President in conformity with section 201 (e).

(c) The aeronautical and space activities of the United States shall be conducted so as to contribute materially to one or more of the following objectives:

(1) The expansion of human knowledge of phenomena in the atmosphere and space;

(2) The improvement of the usefulness, performance, speed, safety, and efficiency of aeronautical and space vehicles;

(3) The development and operation of vehicles capable of carrying instruments, equipment, supplies and living organisms through space;

(4) The establishment of long-range studies of the potential benefits to be gained from, the opportunities for, and the problems involved in the utilization of aeronautical and space activities for peaceful and scientific purposes.

(5) The preservation of the role of the United States as a leader in aeronautical and space science and technology and in the application thereof to the conduct of peaceful activities within and outside the atmosphere.

(6) The making available to agencies directly concerned with national defenses of discoveries that have military value or significance, and the furnishing by such agencies, to the civilian agency established to direct and control nonmilitary aeronautical and space

activities, of information as to discoveries which have value or significance to that agency;

(7) Cooperation by the United States with other nations and groups of nations in work done pursuant to this Act and in the peaceful application of the results, thereof; and

(8) The most effective utilization of the scientific and engineering resources of the United States, with close cooperation among all interested agencies of the United States in order to avoid unnecessary duplication of effort, facilities, and equipment.

(d) It is the purpose of this Act to carry out and effectuate the policies declared in subsections (a), (b), and (c).

DEFINITIONS

Sec. 103. As used in this Act--

(1) the term "aeronautical and space activities" means(A) research into, and the solution of, problems of flight within and outside the earth's atmosphere, (B) the development, construction, testing, and operation for research purposes of aeronautical and space vehicles, and (C) such other activities as may be required for the exploration of space; and

(2) the term "aeronautical and space vehicles" means aircraft, missiles, satellites, and other space vehicles, manned and unmanned, together with related equipment, devices, components, and parts.

TITLE II--COORDINATION OF AERONAUTICAL AND SPACE ACTIVITIES

NATIONAL AERONAUTICS AND SPACE COUNCIL

Sec. 201. (a) There is hereby established the National Aeronautics and Space Council (hereinafter called the "Council") which shall be composed of--

(1) the President (who shall preside over meetings of the Council);

(2) the Secretary of State;

(3) the Secretary of Defense

(4) the Administrator of the National Aeronautics and Space Administration;

(5) the Chairman of the Atomic Energy Commission;

(6) not more than one additional member appointed by the President from the departments and agencies of the Federal Government; and

(7) not more than three other members appointed by the President, solely on the basis of established records of distinguished achievement from among individuals in private life who are eminent in science, engineering, technology, education, administration, or public affairs.

(b) Each member of the Council from a department or agency of the Federal Government may designate another officer of his department or agency to serve on the Council as his alternate in his unavoidable absence.

(c) Each member of the Council appointed or designated under paragraphs (6) and (7) of subsection (a), and each alternate member designated under subsection (b), shall be appointed or designated to serve as such by and with the advice and consent of the Senate, unless at the time of such appointment or designation he holds an office in the Federal Government to which he was appointed by and with the advice and consent of the Senate.

(d) It shall be the function of the Council to advise the President with respect to the performance of the duties prescribed in subsection (e) of this section.

(e) In conformity with the provisions of section 102 of this Act, it shall be the duty of the President to--

(1) survey all significant aeronautical and space activities, including the policies, plans, programs, and accomplishments of all agencies of the United States engaged in such activities;

(2) develop a comprehensive program of aeronautical and space activities to be conducted by agencies of the United States;

(3) designate and fix responsibility for the direction of major aeronautical and space activities;

(4) provide for effective cooperation between the National Aeronautics and Space Administration and the Department of Defense in all such activities, and specify which of such activities may be carried on concurrently by both such agencies notwithstanding the assignment of primary responsibility therefor to one or the other of such agencies; and

(5) resolve differences arising among departments and agencies of the United States with respect to aeronautical and space activities under this Act, including differences as to whether a particular project is an aeronautical and space activity.

(f) The Council may employ a staff to be headed by a civilian executive secretary who shall be appointed by the President by and with the advice and the consent of the Senate and shall receive compensation at the rate of $20,000 a year. The executive secretary, subject to the direction of the Council, is authorized to appoint and fix the compensation of such personnel, including not more than three persons who may be appointed without regard to the civil service laws or the Classification Act of 1949 and compensated at the rate of not more that $19,000 a year, as may be necessary to perform such duties as may be prescribed by the Council in connection with the performance of its functions. Each appointment under this subsection shall be subject to the same security requirements as those established for personnel of the National Aeronautics and Space Administration appointed under section 203 (b) (2) of this Act.

(g) Members of the Council appointed from private life under subsection (a) (7) may be compensated at a rate not to exceed $100 per diem, and may be paid travel expenses and per diem in lieu of subsistence in accordance with the provisions of section 5 of the Administrative Expenses Act of 1946 (5 U.S.C. 73b-2) relating to persons serving without compensation.

NATIONAL AERONAUTICS AND SPACE ADMINISTRATION

Sec. 202. (a) There is hereby established the National Aeronautics and Space Administration (hereinafter called the "Administration"). The Administration shall be headed by an Administrator, who shall be appointed from civilian life by the President by and with the advice and consent of the Senate, and shall receive compensation at the rate of $22,500 per annum. Under the supervision and direction of the President, the Administrator shall be responsible for the exercise of all powers and the discharge of all duties of the Administration, and shall have authority and control over all personnel and activities, thereof.

(b) There shall be in the Administration a Deputy Administrator, who shall be appointed from civilian life by the President by and with the advice and consent of the Senate, shall receive compensation of $21,500 per annum, and shall perform such duties and exercise such powers as the Administrator may prescribe. The Deputy Administrator shall act for, and exercise the powers of, the Administrator during his absence or disability.

(c) The Administrator and the Deputy Administrator shall not engage in any other business, vocation, or employment while serving as such.

FUNCTIONS OF THE ADMINISTRATION

Sec. 203. (a) The Administration, in order to carry out the purpose of this Act, shall--

(1) plan, direct, and conduct aeronautical and space activities;

(2) arrange for participation by the scientific community in planning scientific measurements and observations to be made through use of aeronautical and space vehicles,

and conduct or arrange for the conduct of such measurements and observations; and

(3) provide for the widest practicable and appropriate dissemination of information concerning its activities and the results thereof.

(b) In the performance of its functions the Administration is authorized--

(1) to make, promulgate, issue, rescind, and amend rules and regulations governing the manner of its operations and the exercise of the powers vested in it by law;

(2) to appoint and fix the compensation of such officers and employees as may be necessary to carry out such functions. Such officers and employees shall be appointed in accordance with the civil-service laws and their compensation fixed in accordance with the Classification Act of 1949, except that (A) to the extent the Administrator deems such action necessary to the discharge of his responsibilities, he may appoint and fix the compensation (up to a limit of $19,000 a year, or up to a limit of $21,000 a year for a maximum of ten positions) of not more than two hundred and sixty of the scientific, engineering and administrative personnel of the Administration without regard to such laws, and (B) to the extent the Administrator deems such action necessary to recruit specially qualified scientific and engineering talent, he may establish the entrance grade for scientific and engineering personnel without previous service in the Federal Government at a level up to two grades higher that the grade provided for such personnel under the General Schedule established by the Classification Act of 1949, and fix their compensation accordingly;

(3) to acquire (by purchase, lease, condemnation, or otherwise), construct, improve, repair, operate, and maintain laboratories, research and testing sites and facilities, aeronautical and space vehicles, quarters and related accommodations for employees and dependents of employees of the Administration, and such other real and personal property (including patents), or any interest therein, as the Administration deems necessary within and outside the continental United States; to lease to others such real and personal property; to sell and otherwise dispose of real and personal property (including patents and rights thereunder) in accordance with the provisions of the Federal Property and Administrative Service Act of 1949, as amended (40 U.S.C. 471 et seq.); and to provide by contract or otherwise for cafeterias and other necessary facilities for the welfare of employees of the Administration at its installations and purchase and maintain equipment therefor;

(4) to accept unconditional gifts or donations of services, money, or property, real, personal, or mixed, tangible or intangible;

(5) without regard to section 3648 of the Revised Statutes, as amended (31 U.S.C. 529), to enter into and perform such contracts, leases, cooperative agreements, or other transactions as may be necessary in the conduct of its work and on such terms as it may deem appropriate, with any agency or instrumentality of the United States, or with any State, Territory, or possession, or with any political subdivision thereof, or with any person, firm, association, corporation, educational institution. To the maximum extent practicable and consistent with the accomplishment of the purpose of this Act, such contracts, leases, agreements, and other transactions shall be allocated by the Administrator in a manner which will enable small-business concerns to participate equitably and proportionately in the conduct of the work of the Administration;

(6) to use, with their consent, the services, equipment, personnel, and facilities of Federal and other agencies with or without reimbursement, and on a similar basis to cooperate with other public and private agencies and instrumentalities in the use of services, equipment and facilities. Each department and agency of the Federal Government shall cooperate fully with

the Administration in making its services, equipment, personnel, and facilities available to the Administration, and any such department or agency is authorized, notwithstanding any other provision of law, to transfer or to receive from the Administration, without reimbursement, aeronautical and space vehicles, and supplies and equipment other than administrative supplies and equipment;

(7) to appoint such advisory committees as may be appropriate for purposes of consultation and advice to the Administration in the performance of its functions;

(8) to establish within the Administration such offices and procedures as may be appropriate to provide for the greatest possible coordination of its activities under this Act with related scientific and other activities being carried on by other public and private agencies and organizations;

(9) to obtain services as authorized by section 15 of the Act of August 2, 1946 (5 U.S.C. 55a), at rates not to exceed $100 per diem for individuals;

(10) when determined by the Administrator to be necessary, and subject to such security investigations as he may determine to be appropriate, to employ aliens without regard to statutory provisions prohibiting payment of compensation to aliens;

(11) to employ retired commissioned officers of the armed forces of the United States and compensate them at the rate established for the positions occupied by them within the Administration, subject only to the limitations in pay set forth in section 212 of the Act of June 30, 1932 as amended (5 U.S.C. 59a);

(12) with the approval of the President, to enter into cooperative agreements under which members or the Army, Navy, Air Force, and Marine Corps may be detailed by the appropriate Secretary for services in the performance of functions under this Act to the same extent as that to which they might by lawfully assigned in the Department of Defense; and

(13) (A) to consider, ascertain, adjust, determine, settle, and pay, on behalf of the United States, in full satisfaction thereof, any claim for $5,000 or less against the United States for bodily injury, death, or damage to or loss of real or personal property resulting from the conduct of the Administration's functions as specified in subsection (a) of this section, where such claim is presented to the Administration in writing within two years after the accident or incident out of which the claim arises; and

(B) if the Administration considers that a claim in excess of $5,000 is meritorious and would otherwise be covered by this paragraph, to report the facts and circumstances thereof to the Congress for its consideration.

CIVILIAN-MILITARY LIAISON COMMITTEE

Sec. 204 (a) There shall be a Civilian-Military Liaison Committee consisting of--

(1) a Chairman, who shall be the head thereof and who shall be appointed by the President, shall serve at the pleasure of the President, and shall receive compensation (in the manner provided in subsection (d)) at the rate of $20,000 per annum;

(2) one or more representatives from the Department of Defense, and one or more representatives from each of the Departments of the Army, Navy, and Air Force, to be assigned by the Secretary of Defense to serve on the Committee without additional compensation; and

(3) representatives from the Administration, to be assigned by the Administrator to serve on the Committee without additional compensation, equal in number to the number of representatives assigned to serve on the Committee under paragraph (2).

(b) The Administration and the Department of Defense, through the Liaison Committee, shall advise and consult with each other on all matters within their respective jurisdictions relating to aeronautical and space activities and shall keep each other fully and currently informed with respect to such activities.

(c) If the Secretary of Defense concludes that any request, action, proposed action, or failure to act on the part of the Administrator is adverse to the responsibilities of the Department of Defense, or the Administrator concludes that any request, action, or proposed action, or failure to act on the part of the Department of Defense is adverse to the responsibilities of the Administration, and the Administrator and the Secretary of Defense are unable to reach an agreement with respect thereto, either the Administrator or the Secretary of Defense may refer the matter to the President for his decision (which shall be final) as provided in section 201 (e).

(d) Notwithstanding the provisions of any other law, any active or retired officer of the Army, Navy, or Air Force may serve as Chairman of the Liaison Committee without prejudice to his active or retired status as such officer. The compensation received by any such officer for his service as Chairman of the Liaison Committee shall be equal to the amount (if any) by which the compensation fixed by subsection (a) (1) for such Chairman exceeds his pay and allowances (including special and incentive pays) as an active officer, or his retired pay.

INTERNATIONAL COOPERATION

Sec. 205. The Administration, under the foreign policy guidance of the President, may engage in a program of international cooperation in work done pursuant to the Act, and in the peaceful application of the results thereof, pursuant to agreements made by the President with the advice and consent of the Senate.

REPORTS TO THE CONGRESS

Sec. 206. (a) The Administration shall submit to the President for transmittal to the Congress, semiannually and at such other times as it deems desirable, a report of its activities and accomplishments.

(b) The President shall transmit to the Congress in January of each year a report, which shall include (1) a comprehensive description of the programmed activities and the accomplishments of all agencies of the United States in the field of aeronautics and space activities during the preceding calendar year, and (2) an evaluation of such activities and accomplishments in terms of the attainment of, or the failure to attain, the objectives described in section 102 (c) of this Act.

(c) Any report made under this section shall contain such recommendations for additional legislation as the Administrator of the President may consider necessary or desirable for the attainment of the objectives described in section 102 (c) of this Act.

(d) No information which has been classified for reasons of national security shall be included in any report made under this section, unless such information has been declassified by, or pursuant to authorization given by, the President.

TITLE III -- MISCELLANEOUS

NATIONAL ADVISORY COMMITTEE FOR AERONAUTICS

Sec. 301. (a) The National Advisory Committee for Aeronautics, on the effective date of this section, shall cease to exist. On such date all functions, powers, duties, and obligations, and all real and personal property, personnel (other than members of the Committee), funds, and records of that organization, shall be transferred to the Administration.

(b) Section 2302 of title 10 of the United States Code is amended by striking out "or the Executive Secretary of the National Advisory Committee for Aeronautics." and inserting in lieu thereof "or the Administrator of the National Aeronautics and Space Administration."; and section 2303 of such title 10 is amended by striking out "National Advisory Committee for Aeronautics" and inserting in lieu thereof "The National Aeronautics and Space Administration."

(c) The first section of the Act of August 26, 1950 (5 U.S.C. 22-1), is amended by striking out "the Director, National Advisory Committee for Aeronautics" and inserting in lieu thereof "the Administrator of the National Aeronautics and Space Administration", and by striking out "or National Advisory Committee for Aeronautics" and inserting in lieu thereof "or National Aeronautics and Space Administration".

(d) The Unitary Wind Tunnel Plan Act of 1949 (50 U.S.C. 511- 515) is amended (1) by striking out "The National Advisory Committee for Aeronautics (hereinafter referred to as the `Committee')" and inserting in lieu thereof "The Administrator of the National Aeronautics and Space Administration (hereinafter referred to as the `Administrator')"; (2) by striking out "Committee" or "Committee's" wherever they appear and inserting in lieu thereof "Administrator" and "Administrator's", respectively; and (3) by striking out "its" wherever it appears and inserting in lieu thereof "his".

(e) This section shall take effect ninety days after the date of the enactment of this Act, or on any earlier date on which the Administrator shall determine, and announce by proclamation published in the Federal Register, that the Administration has been organized and is prepared to discharge the duties and exercise the powers conferred upon it by this Act.

TRANSFER OF RELATED FUNCTIONS

Sec. 302. (a) Subject to the provisions of this section, the President, for a period of four years after the date of enactment of this Act, may transfer to the Administration any functions (including powers, duties, activities, facilities, and parts of functions) of any other department or agency of the United States, or of any officer or organizational entity thereof, which relate primarily to the functions, powers, and duties of the Administration as prescribed by section 203 of this Act. In connection with any such transfer, the President may, under this section or other applicable authority, provide for appropriate transfers of records, property, civilian personnel, and funds.

(b) Whenever any such transfer is made before January 1, 1959, the President shall transmit to the Speaker of the House of Representatives and the President pro tempore of the Senate a full and complete report concerning the nature and effect of such transfer.

(c) After December 31, 1958, no transfer shall be made under this section until (1) a full and complete report concerning the nature and effect of such proposed transfer has been transmitted by the President to the Congress, and (2) the first period of sixty calendar days of regular session of the Congress following the date of receipt of such report by the Congress has expired without the adoption by the Congress of a concurrent resolution stating that the Congress does not favor such transfer.

ACCESS TO INFORMATION

Sec. 303. Information obtained or developed by the Administrator in the performance of his functions under the Act shall be made available for public inspection, except (A) information authorized or required by Federal statute to be withheld, and (B) information classified to protect the national security: Provided, That nothing in this Act shall authorize the withholding of information by the Administrator from the duly authorized committees of the Congress.

SECURITY

Sec. 304. (a) The Administrator shall establish such security requirements, restrictions, and safeguards as he deems necessary in the interest of the national security. The Administrator may arrange with the Civil Service Commission for the conduct of such security or other personnel investigations of the Administration's officers, employees, and consultants, and its contractors and subcontractors and their officers and employees, actual or prospective, as he deems appropriate; and if any such investigation develops any data reflecting that the individual who is the subject thereof is of questionable loyalty the matter shall be referred to the Federal Bureau of Investigation for the conduct of a full field investigation, the results of which shall be furnished to the Administrator.

(b) The Atomic Energy Commission may authorize any of its employees, or employees of any contractor, prospective contractor, licensee, or prospective licensee of the Atomic Energy Commission or any other person authorized to have access to Restricted Data by the Atomic Energy Commission under subsection 145 b. of the Atomic Energy Act of 1954 (42 U.S.C. 2165 (b)), to permit any member, officer, or employee of the Council, or the Administrator, or any officer, employee, member of an advisory committee, contractor, subcontractor, or officer or employee of a contractor or subcontractor of the Administration, to have access to Restricted Data relating to aeronautical and space activities which is required in the performance of his duties and so certified by the Council or the Administrator, as the case may be, but only if (1) the Council or Administrator or designee thereof has determined, in accordance with the established personnel security procedures and standards of the Council or Administration, that permitting such individual to have access to such Restricted Data will not endanger the common defense and security, and (2) the Council or Administrator or designee thereof finds that the established personnel and other security procedures and standards of the Council or Administration are adequate and in reasonable conformity to the standards established by the Atomic Energy Commission under section 145 of the Atomic Energy Act of 1954 (42 U.S.C. 2165). Any individual granted access to such Restricted Data pursuant to this subsection may exchange such Data with any individual who (A) is an officer or employee of the Department of Defense, or any department or agency thereof, or a member of the armed forces, or a contractor or subcontractor, and (B) has been authorized to have access to Restricted Data under the provisions of section 143 of the Atomic Energy Act of 1954 (42 U.S.C. 2163).

(c) Chapter 37 of title 18 of the United States Code (entitled Espionage and Censorship) is amended by--

(1) adding at the end thereof the following new section:

"799. Violation of regulations of National Aeronautics and Space Administration.

"Whoever willfully shall violate, attempt to violate, or conspire to violate any regulation or order promulgated by the Administrator of the National Aeronautics and Space Administration for the protection or security of any laboratory, station, base or other facility, or part thereof, or any aircraft, missile, spacecraft, or similar vehicle, or part thereof, or other property or equipment in the custody of the Administration, or any real or personal property or equipment in the custody of any contractor under any contract with the Administration or any subcontractor of any such contractor, shall be fined not more than $5,000, or imprisoned not more that one year, or both."

(2) adding at the end of the sectional analysis thereof the following new item:

"799. Violation of regulations of National Aeronautics and Space Administration."

(d) Section 1114 of title 18 of the United States Code is amended by inserting immediately before "while engaged in the performance of his official duties" the following:

"or any officer or employee of the National Aeronautics and Space Administration directed to guard and protect property of the United States under the administration and control of the National Aeronautics and Space Administration,".

(e) The Administrator may direct such of the officers and employees of the Administration as he deems necessary in the public interest to carry firearms while in the conduct of their official duties. The Administrator may also authorize such of those employees of the contractors and subcontractors of the Administration engaged in the protection of property owned by the United States and located at facilities owned by or contracted to the United States as he deems necessary in the public interest, to carry firearms while in the conduct of their official duties.

PROPERTY RIGHTS IN INVENTIONS

Sec. 305. (a) Whenever any invention is made in the performance of any work under any contract of the Administration, and the Administrator determines that--

(1) the person who made the invention was employed or assigned to perform research, development, or exploration work and the invention is related to the work he was employed or assigned to perform, or that it was within the scope of his employment duties, whether or not it was made during working hours, or with a contribution by the Government of the use of Government facilities, equipment, materials, allocated funds, information proprietary to the Government, or services of Government employees during working hours; or

(2) the person who made the invention was not employed or assigned to perform research, development, or exploration work, but the invention is nevertheless related to the contract, or to the work or duties he was employed or assigned to perform, and was made during working hours, or with a contribution from the Government of the sort referred to in clause (1), such invention shall be the exclusive property of the United States, and if such invention is patentable a patent therefor shall be issued to the United States upon application made by the Administrator, unless the Administrator waives all or any part of the rights of the United States to such invention in conformity with the provisions of subsection (f) of this section.

(b) Each contract entered into by the Administrator with any party for the performance of any work shall contain effective provisions under which such party shall furnish promptly to the Administrator a written report containing full and complete technical information concerning any invention, discovery, improvement, or innovation which may be made in the performance of any such work.

(c) No patent may be issued to any applicant other than the Administrator for any invention which appears to the Commissioner of Patents to have significant utility in the conduct of aeronautical and space activities unless the applicant files with the Commissioner, with the application or within thirty days after request therefor by the Commissioner, a written statement executed under oath setting forth the full facts concerning the circumstances under which such invention was made and stating the relationship (if any) of such invention to the performance of any work under any contract of the Administration. Copies of each such statement and the application to which it relates shall be transmitted forthwith by the Commissioner to the Administrator.

(d) Upon any application as to which any such statement has been transmitted to the Administrator, the Commissioner may, if the invention is patentable, issue a patent to the applicant unless the Administrator, within ninety days after receipt of such application and statement, requests that such patent be issued to him on behalf of the United States. If, within such time, the Administrator files such a request with the Commissioner, the Commissioner shall transmit notice thereof to the applicant, and shall issue such patent to the Administrator unless the applicant within thirty days after receipt of such notice requests a hearing before a Board of Patent Interferences on the question whether the Administrator is entitled under this section to receive such patent. The Board may hear and determine, in accordance with rules and procedures established for interference cases, the question so presented, and its determination shall be subject to appeal by the applicant or by the Administrator to the Court of Customs and Patent Appeals in accordance with procedures governing appeals from decisions of the Board of Patent Interferences in other proceedings.

(e) Whenever any patent has been issued to any applicant in conformity with subsection (d), and the Administrator thereafter has reason to believe that the statement filed by the applicant in connection therewith contained any false representation of any material fact, the Administrator within five years after the date of issuance of such patent may file with the Commissioner a request for the transfer to the administrator of title to such patent on the records of the Commissioner. Notice of any such request shall be transmitted by the Commissioner to the owner of record of such patent, and title to such patent shall be so transferred to the Administrator unless within thirty days after receipt of such notice such owner of record requests a hearing before a Board of Patent Interferences on the question whether any such false representation was contained in such statement. Such question shall be heard and determined, and determination thereof shall be subject to review, in the manner prescribed by subsection (d) for questions arising thereunder. No request made by the Administrator under this subsection for the transfer of title to any patent, and no prosecution for the violation of any criminal statute, shall be barred by any failure of the Administrator to make a request under subsection (d) for the issuance of such patent to him, or by any notice previously given by the Administrator stating that he had no objection to the issuance of such patent to the applicant therefor.

(f) Under such regulations in conformity with this subsection as the Administrator shall prescribe, he may waive all or any part of the rights of the United States under this section with respect to any invention or class of inventions made or which may be made by any person or class of persons in the performance of any work required by any contract of the Administration if the Administrator determines that the interests of the United States will be served thereby. Any such waiver may be made upon such terms and under such conditions as the Administrator shall determine to be required for the protection of the interests of the United States. Each such waiver made with respect to any invention shall be subject to the reservation by the Administrator of an irrevocable, nonexclusive, nontransferable, royalty-free license for the practice of such invention throughout the world by or on behalf of the United States or any foreign government pursuant to any treaty or agreement with the United States. Each proposal for any waiver under this subsection shall be referred to an Inventions and Contributions Board which shall be established by the Administrator within the Administration. Such Board shall accord to each interested party an opportunity for hearing, and shall transmit to the Administrator its findings of fact with respect to such proposal and its recommendations for action to be taken with respect thereto.

(g) The Administrator shall determine, and promulgate regulations specifying, the terms and conditions upon which licenses will be granted by the Administrator for the practice by any person (other than an agency of the United States) of any invention for which the Administrator holds a patent on behalf of the United States.

(h) The Administrator is authorized to take all suitable and necessary steps to protect any invention or discovery to which he has title, and to require that contractors or persons who retain title to inventions or discoveries under this section protect the inventions or discoveries to which the Administration has or may acquire a license or use.

(i) The Administration shall be considered a defense agency of the United States for the purpose of chapter 17 of title 35 of the United States Code.

(j) As used in this section--

(1) term "person" means any individual, partnership, corporation, association, institution, or other entity;

(2) the term "contract" means any actual or proposed contract, agreement, understanding, or other arrangement, and includes any assignment, substitution of parties, or subcontract executed or entered into thereunder; and

(3) the term "made", when used in relation to any invention, means the conception or first actual reduction to practice of such invention.

CONTRIBUTIONS AWARDS

Sec. 306. (a) Subject to the provisions of this section, the Administrator is authorized, upon his own initiative or upon application of any person, to make a monetary award, in such amount and upon such terms as he shall determine to be warranted, to any person (as defined by section 305) for any scientific or technical contribution to the Administration which is determined by the Administrator to have significant value in the conduct of aeronautical and space activities. Each application made for any such award shall be referred to the Inventions and Contributions Board established under section 305 of this Act. Such Board shall accord to each such applicant an opportunity for hearing upon such application, and shall transmit to the Administrator its recommendation as to the terms of the award, if any, to be made to such applicant for such contribution. In determining the terms and conditions of any award the Administrator shall take into account- (1) the value of the contribution to the United States;

(2) the aggregate amount of any such sums which have been expended by the applicant for the development of such contribution;

(3) the amount of any compensation (other than salary received for services rendered as an officer of employee of the Government) previously received by the applicant for or on account of the use of such contribution by the United States; and

(4) such other factors as the Administrator shall determine to be material.

(b) If more than one applicant under subsection (a) claims and interest in the same contribution, the Administrator shall ascertain and determine the respective interests of such applicants, and shall apportion any award to be made with respect to such contribution among such applicants in such proportions as he shall determine to be equitable. No award may be made under subsection (a) with respect to any contribution--

(1) unless the applicant surrenders, by such means as the Administrator shall determine to be effective, all claims which such applicant may have to receive any compensation (other than the award made under this section) for the use of such contribution or any element thereof at any time by or on behalf of the United States, or by or on behalf of any foreign government pursuant to any treaty or agreement with the United States, within the United States or at any other place;

(2) in any amount exceeding $100,000, unless the Administrator has transmitted to the appropriate committees of the Congress a full and complete report concerning the amount and terms of, and the basis for, such proposed award, and thirty calendar days of regular session of the Congress have expired after receipt of such report by such committees.

APPROPRIATIONS

Sec. 307. (a) There are hereby authorized to be appropriated such sums as may be necessary to carry out this Act except that nothing in this Act shall authorize the appropriation of any amount for (1) the acquisition or condemnation of any real property, or (2) any other item of a capital nature (such as plant or facility acquisition, construction, or expansion) which exceeds $250,000. Sums appropriated pursuant to this subsection for the construction of facilities, or for research and development activities, shall remain available until expended.

(b) Any funds appropriated for the construction of facilities may be used for emergency repairs of existing facilities when such existing facilities are made inoperative by major breakdown, accident, or other circumstances and such repairs are deemed by the Administrator to be of greater urgency than the construction of new facilities.

(Sam Rayburn)
Speaker of the House of Representatives
(Richard Nixon)
Vice President of the United States and President of the Senate.

48.3 Federal Aviation Act of 1958

Federal Aviation Act of 1958 (Public Law 85-726) established the Federal Aviation Administration (FAA). The following is an excerpt from the Federal Aviation Act of 1958.

48.3.1 Federal Aviation Act of 1958 Excerpt

The Administrator shall develop, modify, test, and evaluate systems, procedures, facilities, and devices, as well as define the performance characteristics thereof, to meet the needs for safe and efficient navigation and traffic control of all civil and military aviation except for those needs of military agencies which are peculiar to air warfare and primarily of military concern, and select such systems, procedures, facilities, and devices as will best serve such need and will promote maximum coordination of air traffic control and air defense systems. The Administrator shall undertake or supervise research to develop a better understanding of the relationship between human factors and aviation accidents and between human factors and air safety, to enhance air traffic controller and mechanic and flight crew performance, to develop a human-factor analysis of the hazards associated with new technologies to be used by air traffic controllers, mechanics, and flight crews, and to identify innovative and effective corrective measures for human errors which adversely affect air safety. The Administrator shall undertake or supervise a research program to develop dynamic simulation models of the air traffic control system and airport design and operating procedures which will provide analytical technology for predicting airport and air traffic control safety and capacity problems, for evaluating planned research projects, and for testing proposed revisions in airport and air traffic control operations programs. The Administrator shall undertake or supervise research programs concerning airspace and airport planning and design, airport capacity enhancement techniques, human performance in the air transportation environment, aviation safety and security, the supply of trained air transportation personnel including pilots and mechanics, and other aviation issues pertinent to developing and maintaining a safe and efficient air transportation system.

48.4 Technology Assessment Act of 1972

The Technology Assessment Act of 1972 (Public Law 92-484) established the Office of Technology Assessment (OTA) in 1972. It was a twelve-member board, consisting of six members of Congress from each party, half from the Senate and half from the House of Representatives.

48.4.1 Technology Assessment Political Controversy

The OTA came under criticism in the early 1980's with the new directions being established by the Reagan administration. Fat City: How Washington Wastes Your Taxes, a 1980 book by Donald Lambro called OTA an "unnecessary agency" that duplicated government work done elsewhere. OTA was essentially abolished by the by 104th Congress in 1995 by de-funding it in the "Contract with America" period of Newt Gingrich's Republican ascendancy in Congress.

Prior to its de-funding it had a full-time staff of 143 people and an annual budget of $21.9 million. It closed on September 29, 1995. The move was criticized at the time, including by Republican representative Amo Houghton who commented that "we are cutting off one of the most important arms of Congress when we cut off unbiased knowledge about science and technology."

Princeton University maintains an OTA Legacy website that includes OTA publications. Federation of American Scientists launched a similar archive in 2008.

48.4.2 Technology Assessment Act

The Technology Assessment Act of 1972, Public Law 92-484, 92d Congress, H.R. 10243, October 13, 1972

AN ACT

To establish an Office of Technology Assessment for the Congress as an aid in the identification and consideration of existing and probable impacts of technological application; to amend the National Science Foundation Act of 1950; and for other purposes. Be it enacted by the Senate and House of Representatives of the United States of America in Congress assembled, That this Act may be cited as the Technology Assessment Act of 1972.

FINDINGS AND DECLARATION OF PURPOSE

SEC. 2. The Congress hereby finds and declares that:

(a) As technology continues to change and expand rapidly, its applications are

1. large and growing in scale; and

2. increasingly extensive, pervasive, and critical in their impact, beneficial and adverse, on the natural and social environment.

(b) Therefore, it is essential that, to the fullest extent possible, the consequences of technological applications be anticipated, understood, and considered in determination of public policy on existing and emerging national problems.

(c) The Congress further finds that:

1. the Federal agencies presently responsible directly to the Congress are not designed to

provide the legislative branch with adequate and timely information, independently developed, relating to the potential impact of technological applications, and

2. the present mechanisms of the Congress do not and are not designed to provide the legislative branch with such information.

(d) Accordingly, it is necessary for the Congress to—

1. equip itself with new and effective means for securing competent, unbiased information concerning the physical, biological, economic, social, and political effects of such applications; and

2. utilize this information, whenever appropriate, as one factor in the legislative assessment of matters pending before the Congress, particularly in those instances where the Federal Government may be called upon to consider support for, or management or regulation of, technological applications.

Reference Public Law 92-484 for the remaining text.

48.5 OTA Review FAA 1982 NAS Plan

48.5.1 OTA Review FAA 1982 NAS Plan

The following are excerpts for the Review of the FAA 1982 National Airspace, System Plan, August 1982, NTIS order #PB83-102772. Please see the original report for full content and context.

48.5.2 Letter

Review of the FAA 1982 National Airspace System Plan, August 1982, NTIS order #PB83-102772. Library of Congress Catalog Card Number 82-600595.

In January 1982, shortly after OTA had concluded an assessment of the airport and air traffic control system, the Federal Aviation Administration released the 1982 National Airspace System (NAS) Plan. The Transportation Subcommittee of the House Committee on Appropriations asked that OTA undertake a review of the NAS Plan, building on the results of the assessment that had been carried out at their request. OTA'S approach to conducting this review was to examine the NAS Plan at two levels—the adequacy of the plan as a whole and the appropriateness of the specific technologies selected by FAA for implementation. Our aim was to make a balanced assessment— pointing out those parts that are commendable and supported by the aviation community while also identifying alternatives that merit consideration and indicating aspects of the plan that could be improved. In so doing, it was our intent to assist the congressional review process and to make a constructive contribution to the generally shared goal of modernizing and improving the air traffic control system in the years to come.

In conducting this review, OTA held extensive consultation with representatives of the aviation community and with technical experts in the fields of computer and communication technology. Workshops on aviation growth forecasts and air traffic control technology were held, and a z-day conference of aviation experts was convened to evaluate FAA's planned modernization of the National Airspace System. The results of this consultative effort combined with analysis performed by OTA staff and the work carried out in the previous assessment form the basis for this OTA report.

In all, some 60 persons from outside OTA took part in the review of the NAS Plan. Their contributions were remarkable both for their depth and richness of insight and for the diversity of opinion on the strengths and weaknesses of the plan. We accept full responsibility for the analysis presented here, but acknowledge our debt to those who contributed so freely of their time and effort on our behalf. We are particularly grateful to the Congressional Budget Office for their assistance in analyzing traffic forecasts and funding issues.

JOHN H. GIBBONS, Director

48.5.3 Executive Summary

The National Airspace System Plan (NAS Plan) released by the Federal Aviation Administration (FAA) in January 1982 outlines the agency's most recent proposals for modernizing the facilities and equipment that make up the air traffic control (ATC) system. The plan attempts to integrate the various improvements into a single long-range program that addresses major shortcomings and reduces costs of the current system. Viewed on this high level—as a statement of policies, goals, and directions—the NAS Plan is to be commended as a significant and even bold step compared with past FAA efforts to chart the future evolution of the system.

The national airspace system is a "three-legged stool" made up of airports, the ATC system, and procedures for using the airspace. While all three need to be improved in an integrated fashion, the NAS Plan deals with only one leg—the ATC system. OTA'S assessment of the airport and ATC system found that lack of airport capacity—not ATC technology—will be the principal limit on the growth of aviation. The NAS Plan acknowledges that "capacity limitations at busy airports will be the constraining element" in the system, but it concentrates on ATC technology, and most of the proposed improvements are directed at modernization of the en route, not the terminal area, portion of the system.

FAA does intend to address the problems of airports and airspace procedures. A revised plan for airport development is to be issued later this year. FAA has also just begun a National Airspace Review (NAR), a 42-month effort that will reexamine the rules and procedures governing the use of the airspace. Still, by issuing first a plan for modernizing ATC technology, without waiting until the other efforts have more thoroughly defined needs in the area of airports and airspace procedures, FAA may be placing too much emphasis on technological solutions. This perception is reinforced by the NAS Plan itself, which gives first priority to improved technology for the en route system by the late 1980's. There is little apparent advantage in seeking to move en route traffic more expeditiously only to have it encounter delays in terminal areas, where capacity improvements are not scheduled to be made until the early 1990's.

With these reservations, the FAA plan for ATC system improvements is comprehensive. The proposed changes are technologically feasible, and they are consistent with the goals of increasing safety and productivity and accommodating future growth. Providing capacity to accommodate anticipated growth was a principal factor in developing the NAS Plan, although other factors were also involved—increased reliability, safety, productivity, and fuel savings. Still, the technological strategy and implementation schedule appear to have been driven by forecasts of aviation growth and near-term capacity problems at en route centers. FAA traffic and workload forecasts have tended to be too high in the past, however, and in some cases technological alternatives that might be equally effective or less costly than those selected by FAA appear to have been rejected because of the anticipated rate of growth in demand for ATC services. OTA'S review of the NAS Plan suggests that FAA forecasts may not be a useful guide to long-term planning and investment, and that some of these technological options may therefore warrant reexamination.

In the area of en route computer replacement, for example, some believe it would be prudent to adopt a strategy for interim steps to be taken in the 1980's that imposes no constraints on the design of the new system that will serve for the 1990's and beyond. FAA's proposed approach is to "rehost" the existing software on new computers, and then to develop new software to run on the host computers for use with the advanced sector suites to be installed by 1990. Several experts have told OTA that this approach might limit the options available in designing the new system. In their view, any interim host would have to be replaced when the new system comes on line. FAA admits this possibility, but maintains that the intent is for the host computer to serve as the basic processor for the ATC system until well into the 1990's. An alternative short-term approach would be to make selective enhancements to the present technology —i.e., upgrade the current computers in the centers where capacity problems are expected—in combination with economic or regulatory approaches to demand management, while proceeding without delay on a parallel effort to develop by 1990a totally new ATC system design that makes best use of technologies then available and that will serve until beyond the turn of the century.

As a blueprint for the modernization of the ATC system, the 1982 NAS Plan does not provide a clear sense of the priorities or dependencies among its various program elements. Nor does the plan deal explicitly with contingencies or delays caused by engineering problems or by the possible deletion of some elements due to budgetary constraints. Given the complexity and magnitude of this undertaking, FAA may have set itself an overly ambitious schedule for implementing the proposed improvements.

OTA'S review of the 1982 NAS Plan has also identified the following specific findings and issues:

- *Growth.* —FAA's traffic forecasts have been too high in the past and there are questions about the methodologies and assumptions underlying the projections on which the NAS Plan is based. Overestimation may have led FAA to foreclose technological options and accelerate the implementation schedule unnecessarily. It may also have led FAA to overestimate the user-fee revenues that will be available to pay for the proposed improvements.
- *En Route Computer Replacement.* —FAA's option analysis issued in January 1982 supports upgrading the 10 en route computers that face capacity problems.[414] The NAS Plan, released at about the same time, calls instead for replacing the computer hardware (called rehosting the software) in all 20 centers as a part of a long-term plan to increase productivity and reliability as well as capacity. OTA does not find

[414] Federal Aviation Administration, "Response to Congressional Recommendations Regarding the FM's En Route Air Traffic Control Computer System, " report to the Senate and House Appropriations Committees pursuant to Senate report 96-932, DOT/FAA/ AAP-823, January 1982.

persuasive the reasons advanced by FAA for rejecting the previously preferred option of upgrading only selected en route centers. In addition, the choice of a host computer now may limit the options available to the contractor for the sector suite and software. OTA conferees were sharply divided in their views on this question. Some felt that the choice of a host computer now might limit future ability to benefit from a distributed computer architecture, local area networking, and new techniques in software development. Others believed that, if the host is chosen judiciously, the transition to a new system embodying these advanced and desirable features could be made without difficulty.

- *Automation.* —While the NAS Plan envisions substantial cost savings due to extensive automation, supporting analysis is not provided in the plan. This analysis is probably still in progress and may take some time to complete, but it would be useful for the interim results to be made available to assist in congressional review of the automation portions of the overall plan. In addition, there is concern on the part of some experts about the ability of human operators to participate effectively in such a highly automated system and to intervene in the event of system error or failure.
- *Satellites.* —Satellite technology has significant potential applications for communication, and eventually for surveillance and navigation. FAA does not see a role for satellites in the period covered by the NAS Plan. FAA's decision against satellites appears to have been driven by timing and present cost effectiveness, rather than technology readiness or long-term system advantages.
- *User Effects.* —A great many of the proposed ATC system improvements are directed to the needs of traffic operating under the instrument flight rules (IFR), particularly while en route at cruise altitude. These improvements will benefit FAA itself by automating functions and reducing labor costs. The principal beneficiaries among users will be air carriers and larger business aircraft. Personal general aviation (GA) users could receive improved weather information, an important benefit; but in order to obtain this benefit and other operational advantages of the new system, more avionics will be required, and there would be restrictions on access to airspace by aircraft not so equipped. The Department of Defense (DOD) too, is concerned about the cost of new ATC avionics and feels that the new plan must be carefully coordinated with the military services to ensure that their mission needs and responsibilities for administration of the airspace are integrated with those of FAA.
- *Cost and Funding.* —Implementing the improvements proposed in the 1982 NAS Plan would more than double FAA's facilities and equipment budget through 1987, compared to the last 10 years. FAA has not yet released cost estimates for completing the proposed programs, but it seems likely that expenditures of like magnitude will be needed in the years beyond 1987. FAA proposes to recover 85 percent of its total budget through user fee revenues and a drawdown of the uncommitted Trust Fund balance. The user fee schedule would perpetuate the existing crosssubsidy from airline passengers and shippers of air cargo to GA. Business aviation would benefit particularly because of the extensive use these aircraft make of the IFR system. In addition, higher user fees may dampen the growth of aviation, thereby reducing the revenues expected to pay for the proposed improvements.

48.5.4 AUTOMATION AND HU MAN FACTORS

The present ATC system is very labor-intensive the operation of a highly automated system and, without significant increases in controller productivity, the cost of operating the ATC system could rise precipitously as traffic grows. The number of aircraft that a controller team can handle with the present system is limited, and the conventional solution to handling a larger volume of traffic—decreasing sector size—has practical limits. FAA looks to increased automation as the principal means of achieving higher levels of controller productivity.

AERA, which is scheduled to be implemented in the early 1990's, would change the role of the controller from that of an active participant in the control process to that of a manager who oversees Many- of the routine decision making functions now performed by humans would be automated, with the result that fewer controllers will be required for a given level of traffic. Elements of AERA are now undergoing testing, and some features will be added to the existing en route software after it has been rehosted. Other functions— those that will have the greatest impact on the role of the controller and the character of the ATC system—will not be implemented until the early part of the next decade when the redesigned software has been installed. It is this latter group of functions that may require either enhancement or replacement of the proposed host computer in the 1990's.

As envisioned by FAA, AERA is designed to increase the efficiency of airspace utilization as well as the productivity of controllers. AERA will also enable users to follow more fuel-efficient flight paths and make better use of the equipment they are now installing on their aircraft. Flight management and navigation computers, linked to AERA by a new communication link (Mode S), will eventually receive and respond to flight instructions without increasing aircrew workload. Similarly, delays in the system will be minimized by the flow control procedures, and safety will be enhanced because the system will provide for the separation of IFR from VFR traffic outside terminal areas, rather than providing separation only between IFR aircraft as is now the case.

Human factors and safety are important concerns in AERA. In a highly automated system it might be impossible to revert to manual control in the event of a system failure. Therefore, the AERA concept assumes that the functions of the future ATC system will be distributed among various elements. In the event that the main computer at an ATC facility fails, the sector suite (acquired during the second phase of system modernization) will contain enough processing power to provide at least some backup functions; other functions will be transferred in real time to neighboring centers that remain operational.

FAA has yet to refine the AERA concept completely. The distribution of functions among the various computer resources has not yet been determined, nor have the respective roles of human controllers and automated systems been defined. This task will be carried out by FAA and the contractor responsible for the design of the new system.

This point is stressed by the critics of the rehosting approach to computer replacement and those who suggest that FAA use a "clean sheet" approach to the system design. They argue that premature acquisition of host hardware for the short term could limit the options of the system design contractor in the long term. This could result in a requirement for extensive and expensive modifications of the host computers, a second wholesale computer replacement, or (since that seems unlikely) the implementation of a system that cannot take full advantage of the available technologies and design options. None of the critics suggest that replacement be deferred, and all of them recognize that at some point FAA must commit to a specific design even though there always will be a better technology available at some point in the future. Rather, their concern is that premature commitment to "rehosting" hardware could limit FAA's ability to take advantage of the best technology that is now available.

Studies of the AERA concept commissioned by FAA have generally agreed that the proposed approach is feasible. However, one study, recently completed by the Rand Corp., suggests that the AERA concept may not be sound.[415] The Rand study indicates that total commitment to automation, with the controller no longer an active part of the system, is unwarranted and could present safety problems. It suggests that the controller will not be sufficiently involved in the traffic situation to detect errors in the system and analyze them in time to take effective action. As an alternative to the AERA concept, Rand suggests a "shared control" concept in which the controller has a more active part in the control process. In the end, the level of automation proposed by Rand would be very close to that proposed under AERA, although the route to achieve that level would be different and it might not achieve the increases in productivity that would result from the implementation of the FAA plan.

FAA, on the other hand, argues that it would not be possible to achieve the incremental improvements required for the shared-control approach, and that the automated system is expected to be more reliable than a system in which human controllers are active participants. FAA maintains it would be basically unsound, beyond a point, to back up an automated system with a human one that is less reliable.

48.5.5 ATTACHMENT B-I

EXCERPT FROM "SCENARIOS FOR EVOLUTION OF AIR TRAFFIC CONTROL"[416]

VI. CONCLUSIONS

We have considered several alternative ATC futures, beginning with a Baseline case in which nothing beyond the most conservative R&D projects paid off. We have concluded that the approach of simply adding more and more controllers is ultimately counterproductive from a performance standpoint. We have examined the FAA's plan to use advanced computer science technology to construct a fully automated ATC system for application near the year 2000. The expected aircraft safety levels, fuel-use efficiency, and controller productivity have led us to question that plan and to suggest that there maybe a middle ground consisting of a highly, but not totally, automated system. We believe that pursuing the goal of full-automation AERA—with little regard for interim systems or evolutionary development-is a very questionable R&D strategy for ATC. It seems unlikely that a large-scale multi-level AERA system that can effectively handle nonroutine events, show stable behavior under dynamically changing conditions, and be virtually immune to reliability problems can be implemented in the foreseeable future. Human controllers may be required to assume control in at least some of these situations, although at present there is no conclusive evidence that they would be able to do so; indeed, some evidence and opinions from the human-factors community suggest that they would not be able to.

---

[415] Robert Wesson, et al., "Scenarios for Evolution of Air Traffic Control," The Rand Corp., R-2698-FAA, November 1981.

[416] Rand Corp., R-2698-FAA, November 1981

The AERA scenario presents serious problems for each of the three major goals of ATC—safety, efficiency, and increased productivity. By depending on an autonomous, complex, fail-safe system to compensate for keeping the human controller out of the routine decision making loop, the AERA scenario jeopardizes the goal of safety. Ironically, the better AERA works, the more complacent its human managers may become, the less often they may question its actions, and the more likely the system is to fail without their knowledge. We have argued that not only is AERA's complex, costly, fail-safe system questionable from a technical perspective, it is also unnecessary in other, more moderate ATC system designs.

Some AERA advocates assert that it is necessary to keep the human out of the time-critical loop to achieve productivity and fuel-use gains. We question that belief as well. AERA may well achieve 100 percent productivity increases in the en route high and transition sectors, and it may indeed facilitate more fuel-efficient air operations. But if the controller work force almost doubles, as expected, by the time AERA comes on-line, and AERA's domain of applicability is limited to the simplest of sector types, its ultimate effect may hardly be felt, since the actual ATC bottlenecks occur elsewhere. Further, greater fuel efficiency comes from many sources-some as simple as present-day relaxation of procedural restrictions, some as complex as the planning modules of AERA and Shared Control. AERA may meet the goals of ATC by 2000, but the costs incurred along the way will be very great-in dollars, in fundamental research that must be completed, and in restrictions on the controller's role.

Ultimately, the AERA scenario troubles us because it allows for few errors or missteps, The right choices have to be made at the right times, or a failed AERA scenario would degrade to a more costly and delayed version of the Baseline scenario. In the attempt to construct a totally automated ATC control system, unacceptably high possibilities and costs of failure overshadow the potential rewards of success. Our main conclusion is that such an overwhelming dependence on technology is simply unnecessary. If the planned AERA scenario were altered only slightly, it would be essentially equivalent to the Shared Control scenario. All of its technical building blocks are present in Shared Control:

- Air/ground datalink communication.
- Strategic planning (profile generation and alteration) and operator displays.
- Tactical execution.
- Track monitoring and alert.

Missing, however, is the right principle for piecing these building blocks together. Under AERA, they would be fully integrated into a single problem-solving system which extends its capabilities by infrequently requesting human action; under Shared Control, the building blocks would themselves be extensions of human capabilities. Operationally, this shift in perspective requires two modifications of AERA plans:

- The role of man under AERA would be expanded so that he is routinely involved in the minute-to-minute operation of the system.
- The system would be constructed as a series of independently operable, serially deployable aiding modules. The state of the art in ATC problem-solving techniques does not validate the minimal AERA human role; neither does established knowledge about human limitations or capabilities in this domain, Insisting that man be essentially automated out of such a critical control system is an unnecessarily high-risk approach.

If the system is designed to support him, we would expect the future ATC specialist to take a very active and creative role in manipulating his aiding modules. Safety could be assured by assigning the machine primary responsibility for routine separation assurance tasks at the lowest levels. The specialist should be responsible for comprehending situations at high levels of abstraction and activating modules to meet the ever-changing demands of those situations. He should be able to adjust a module's parameters and its relationships to other modules so that instead of simply monitoring the machine's preprogrammed sequence of instructions, he actually controls the outcome. He should be given the authority to determine which operation the machine performs and which he performs. He should be given the opportunity to learn all of this gradually and to influence the system's design before it is finalized.

This shift in perspective captures the spirit of this report. Specifications of module capabilities and their sequence of implementation are best left to designers who are intimately familiar with the engineering details. We have presented just one of many alternatives in which man has a significant ATC role; the details of the system design need refinement and may indeed undergo great change in the process. For example, our Shared Control scenario suggests implementing digital communications before providing any planning aids at all. Perhaps events will dictate otherwise-a late DABS introduction and an early development of automated planning techniques could reverse this sequence. Fielding a planning aid first as a stand-alone module would not compromise the Shared Control scenario in any way. The essence of the Shared Control scenario is reflected in its name-man and machine must work together and share in the overall control function of ATC.

Our key concern is that the human specialist's unique capabilities be acknowledged and the technical uncertainties of an AERA-like system be recognized and dealt with before too much of the Baseline scenario comes to pass. If this is not done, we risk relying solely on an unproven, costly technology to meet the nation's demands for ATC service. We have shown not only that there is a feasible alternative, but also that this alternative may result in lower costs, a higher level of performance, and a more satisfying role for the personnel who will be responsible for moving air traffic safely and smoothly.

49 Bibliography

1. A Logical Approach to Requirements Analysis, Dr. Peter Crosby Scott, A Dissertation in Systems, Presented to the Faculties of the University of Pennsylvania in Partial Fulfillment of the Requirements for the Degree of Doctor of Philosophy, 1993.
2. A Survey of Dual-Use Issues, IDA Paper P-3176 Prepared for Defense Advanced Research Projects Agency (DARPA), March 1996.
3. A Theory of Human Motivation, Abraham H. Maslow, Psychological Review 50(4), 1943.
4. Anthropometry of US Military Personnel, DOD-HDBK-743A, February 1991.
5. Applied Imagination: Principles and Procedures of Creative Problem Solving, A.F. Osborn, New York, NY: Charles Scribner's Son, Third Revised Edition 1963.
6. Baseline Description Document, DOD Date Item Description, DI-CMAN-81121, February 1991.
7. Brundtland Commission, or the World Commission on Environment and Development, known by its Chair Gro Harlem Brundtland, convened by the United Nations in 1983 published Our Common Future, also known as Brundtland Report, in 1987.
8. Calculating Instruments and Machines, Douglas Hartree, University of Illinois Press, 1949.
9. CMMI for Acquisition; Version 1.2; November 2007; Technical Report CMU/SEI-2007-TR-017, ESC-TR-2007-017; Capability Maturity Model(s), Carnegie Mellon University (software engineering institute).
10. CMMI for Development, August 2005; Version 1.2, CMU/SEI-2006-TR-008, ESC-TR-2006-008; Capability Maturity Model(s) - source Carnegie Mellon University (software engineering institute).
11. CMS Requirements Writer's Guide Version 4.11, Department of Health and Human Services, Centers for Medicare & Medicaid Services, August 31, 2009.
12. CMS Testing Framework Overview, Department of Health and Human Services, Centers for Medicare & Medicaid Services, Office of Information Services, Version: 1.0, January 2009.
13. Configuration Management Guidance, Military Handbook, MIL-HDBK-61 30 September 1997, MIL-HDBK-61A (SE) 7 February 2001.
14. Contract Pricing Reference Guides, Vol 1 - Price Analysis, Vol 2 - Quantitative Techniques for Contract Pricing, Vol 3 - Cost Analysis, Vol 4 - Advanced Issues in Contract Pricing, Vol 5 - Federal Contract Negotiation Techniques, Federal Acquisition Institute (FAI) and the Air Force Institute of Technology (AFIT), 2011.
15. Decision Making with the Analytic Hierarchy Process, Thomas L. Saaty, Int. J. Services Sciences, Vol. 1, No. 1, 2008.
16. Defense Acquisition Guidebook, Defense Acquisition University, August 2010.
17. Definitions of Terms For Reliability And Maintainability, DOD Standard, MIL-STD-721 25 August 1966, MIL-STD-721C 12 June 1981.
18. Designing And Developing Maintainable Products And Systems, DOD Handbook, MIL-HDBK-470A August 1997, MIL-HDBK-470 June 1995, MIL-HDBK-471 June 1995.

19. Diffusion of Innovations, Everett Rogers, 1962.
20. Discrete Time Systems, James A. Cadzow, Prentice Hall, 1973, ISBN 0132159961.
21. Earned Value Management Implementation Guide, DOD, Defense Contract Management Agency, October 2006.
22. Earned Value Management System (EVMS), DOE G 413.3-10, U.S. Department of Energy, EVMS Gold Card May 06 2008.
23. Electronic Reliability Design Handbook, Military Handbook, MIL-HDBK-338B, October 1998.
24. Electronically / Optically Generated Airborne Displays, DOD Handbook, MIL-HDBK-87213A February 2005, MIL-HDBK-87213 December 1996.
25. Engineering Change Proposal (ECP), DOD Date Item Description, DI-CMAN-80639C, September 2000.
26. Ergonomics Program Management Guidelines For Meatpacking Plants, U.S. Department of Labor Occupational Safety and Health Administration, OSHA 3 123, 1993.
27. FAA Air Traffic Control Directions, Report of the Workshop on Artificial Intelligence held at the National Academy of Sciences, Washington, D.C., October 23, 1985. Neumann, P J, Transportation Research Board, ISSN 0097-8515.
28. Federal Acquisition Regulation, General Services Administration, Department Of Defense, National Aeronautics And Space Administration, (This edition includes the consolidation of all Federal Acquisition Circulars through 2001-27), March 2005.
29. Flight Assurance Procedure Performing, A Failure Mode and Effects Analysis, NASA, GSFC-431-REF-000370, Number: P-302-720.
30. Fundamentals of Physics, David Halliday and Robert Resnick, John Wiley & Sons Inc; Revised edition (January 1, 1974), ISBN 0471344311.
31. General Principles of Software Validation; Final Guidance for Industry and FDA Staff Document issued on: January 11, 2002, U.S. Department Of Health and Human Services, Food and Drug Administration, Center for Devices and Radiological Health, Center for Biologics Evaluation and Research.
32. Guide For Achieving Reliability Availability And Maintainability, DOD, August 2005.
33. Guidelines for Nursing Homes Ergonomics for the Prevention of Musculoskeletal Disorders, U.S. Department of Labor, Occupational Safety and Health Administration, OSHA 3182-3R, 2009.
34. Guidelines for Poultry Processing Ergonomics for the Prevention of Musculoskeletal Disorders, U.S. Department of Labor Occupational Safety and Health Administration, OSHA 3213-09N, 2004.
35. Guidelines for Retail Grocery Stores Ergonomics for the Prevention of Musculoskeletal Disorders, U.S. Department of Labor, Occupational Safety and Health Administration, OSHA 3192-06N, 2004.
36. Guidelines for Shipyards Ergonomics for the Prevention of Musculoskeletal Disorders, United States Department of Labor, Occupational Safety and Health Administration, OSHA 3341-03N, 2008.
37. Handbook For Human Engineering Design Guidelines, DOD, MIL-HDBK-759C, 31 July 1995, MIL-HDBK-759B 30 October 1991.
38. Handbook for Preparation of Statement of Work, MIL-HDBK-245C 10 September 1991, MIL-HDBK-245D 3 April 1996.

39. How I Became a Quant: Insights from 25 of Wall Street's Elite, Richard R. Lindsey, Wiley, 2009 ISBN 0470452579.
40. Hughes Aircraft's Widespread Deployment of a Continuously Improving Software Process, R.R. Willis, R.M. Rova, M.D. Scott, M.I. Johnson, J.F. Ryskowski, J.A. Moon, K.C. Shumate, T.O. Winfield, Technical Report CMU/SEI-98-TR-006, ESC-TR-98-006, May 1998.
41. Human Engineering Design Criteria, DOD, MIL-STD-1472C May 1981, MIL-STD-1472D March 1989, MIL-STD-1472E October 1996, MIL-STD-1472F August 1999.
42. Human Factors for Evolving Environments: Human Factors Attributes and Technology Readiness Levels, FAA/NASA Human Factors Research and Engineering Division, FAA, DOT/FAA/AR-03/43, April 2003.
43. Human Factors Methods: A Practical Guide for Engineering And Design by Neville A. Stanton, Paul M. Salmon, Guy H. Walker, and Chris Baber, 2005.
44. Human Information Processing, Peter H. Lindsay, Donald A. Norman, Academic Press, 1977.
45. IBM's Early Computers, Charles Bashe, MIT Press, 1986.
46. Improving R&D Productivity: A Study Program and Its Applications, Robert M Ranftl, National Conference on Productivity and Effectiveness in Educational Research and Development, December 1977.
47. Installation and Acceptance Test Plan (IATP), DOD Data Item Description, DI-QCIC-80154A, January 1994.
48. Installation Test Procedures, DOD Data Item Description, DI-QCIC-80511, January 1988.
49. Integrated Logistics Support Guide, Defense Systems Management College, May 1986.
50. Integration Definition for Function Modeling (IDEF0), Federal Information Processing Standards Publication FIPS 183, 21 December 1993.
51. Integration Definition for Information Modeling (IDEF1X), Federal Information Processing Standards Publication FIPS 184, 21 December 1993.
52. John von Neumann Collected Works, Abraham Taub, Macmillan, 1963.
53. Logistic Support Analysis, DOD, MIL-STD-1388-1 October 1973, MIL-STD-1388-1A April 1983, MIL-STD-1388-2A July 1984, MIL-STD-1388-12B March 1991.
54. Man Systems Integration Standards, NASA, NASA-STD-3000, July 1995.
55. Management of the Hanford Engineer Works in World War II, How the Corps, DuPont and the Metallurgical Laboratory fast tracked the original plutonium works, Harry Thayer, ASCE Press, pp. 66-67, ISBN 0784401608.
56. Managing a Technology Development Program, James W. Bilbro & Robert L. Sackheim, Office of the Director George C. Marshall Space Flight Center, augments NASA Procedures and Guidelines, NPG 7120.5, NASA Program and Project Management Processes and Requirements.
57. Managing the Software Process, Watts S. Humphrey, Addison-Wesley Professional, 1989, ISBN 0201180952.
58. Medical Facts for Pilots, Federal Aviation Administration, Publication AM-400-98/2, August 2002.
59. Modern Control Systems, Richard C. Dorf, Addison-Wesley Publishing Company,

1967, 1974, Library of Congress CCN 67-15660, ISBN 0201016060.

60. My Life as a Quant: Reflections on Physics and Finance, Emanuel Derman, Wiley, 2007, SBN 0470192739.
61. NASA Procedural Requirements NASA Systems Engineering Processes and Requirements w/Change 1 (11/04/09), NPR 7123.1A, March 26, 2007.
62. NASA Research and Technology Program and Project Management Requirements, NPR 7120.8, NASA Procedural Requirements, February 05, 2008.
63. NASA Systems Engineering Handbook, NASA/SP-2007-6105 Rev1, December 2007.
64. National Airspace System Engineering Manual, Federal Aviation Administration, V 3.1, 2006.
65. Noise Limits, DOD Design Criteria Standard, MIL-STD-1474B12 June 1979, MIL-STD-1474C March 1991, MIL-STD-1474D February 1997.
66. Noise-Induced Hearing Loss, U.S. Department of Health and Human Services - National Institutes of Health - National Institute on Deafness and Other Communication Disorders, Publication No. 08-4233, December 2008.
67. Notice of Revision (NOR), DOD Date Item Description, DI-CMAN-80642C, September 2000.
68. On Bullshit, Harry G. Frankfurt, Princeton University Press, January 2005, ISBN 9780691122946.
69. Operational Concept Description (OCD), DOD Data Item Description, DI-IPSC-81430 1994, DI-IPSC-81430A 2000.
70. Operational Sequence Diagrams in System Design, Kurke, M. I., 1961.
71. Peopleware, Tom DeMarco, Yourdon Press, 1987. Second edition, Dorset House Publishing Company, Inc, ISBN 0932633439.
72. Physical and Quantum Electronics Series, Demetrius T. Paris, F. Kenneth Hard, McGraw-Hill Book Company, 1969, Library of Congress CCN 68-8775, ISBN 070484708.
73. Practice For System Safety, DOD Standard, MIL-STD-882D, 10 February 2000.
74. Procedures For Performing A Failure Mode Effects And Criticality Analysis, Military Standard, MIL-STD-1629A, 24 November 1980.
75. Program Progress Report, DOD Data Item Description, DI-MGMT-80555A, November 2006.
76. Quality Assurance Terms And Definitions, DOD, MIL-STD–109C 2 September 1994, MIL-STD-109B 4 April 1969.
77. Quality Program Requirements, DOD, MIL–Q-9858A 16 December 1963, MIL–Q-9858 April 1959.
78. R&D Productivity Second Edition; Hughes Aircraft, June 1978, AD Number A075387, Ranftl, R.M., "R&D Productivity", Carver City, CA: Hughes Aircraft Co., Second Edition, 1978, OCLC Number: 4224641 or 16945892. ASIN: B000716B96.
79. Regenerative Design for Sustainable Development, John Tillman Lyle, Wiley Professional, December 1 2008, ISBN 0471178438.
80. Reliability Analyses Handbook, Prepared By Project Reliability Group, JET Propulsion Laboratory, JPL-D-5703_JUL1990, July 1990.
81. Reliability Prediction of Electronic Equipment, Military Handbook, MIL-HDBK-217E 27 October 1986, MIL-HDBK-217F 2 December 1991.
82. Reliability Program For System and Equipment, Military Standard, MIL-STD-785B,

15 September 1980.

83. Request for Deviation (RFD), DOD Date Item Description, DI-CMAN-80640C, September 2000.
84. Response to Congressional Recommendations Regarding the FM's En Route Air Traffic Control Computer System, report to the Senate and House Appropriations Committees pursuant to Senate report 96-932, Federal Aviation Administration, DOT/ FAA/ AAP-823, January 1982.
85. Review of the FAA 1982 National Airspace System Plan, August 1982, NTIS order #PB83-102772. Library of Congress Catalog Card Number 82-600595, U.S. Government Printing Office, Washington, D.C.
86. Sandia Software Guidelines Volume 5; Tools, Techniques, and Methodologies; Sandia Report, SAND85–2348 l UC–32, Reprinted September 1992.
87. Sequential Thematic Organization of Publications (STOP): How to Achieve Coherence in Proposals and Reports, Hughes Aircraft Company Ground Systems Group, Fullerton, Calif., J. R. Tracey, D. E. Rugh, W. S. Starkey, Information Media Department, ID 65-10-10 52092, January 1965.
88. Site Preparation Requirements and Installation Plan, FAA Data Item Description, FAA-SI-004.
89. Software Development And Documentation, MIL-STD-498, Military Standard, 5 December 1994.
90. Software Development, Military Standard, DOD-STD-1679 01 December 1978, DOD-STD-1679A 22 October 1983.
91. Software Safety Standard, NASA-STD-8719.13B, July 2004.
92. Software System Safety Handbook, Joint Services Computer Resources Management Group, US Navy, US Army, And US Air Force, December 1999.
93. Special Issue on STOP Methodology, Journal of Computer Documentation, edited by R. J. Waite, Volume 23/3, August 1999.
94. Specification Change Notice (SCN), DOD Date Item Description, DI-CMAN-80643C, September 2000.
95. Specification Practices, Military Standard, MIL-STD-490 30 October 1968, MIL-STD-490A 4 June 1985.
96. Standard Practice Data Item Descriptions (DIDs), DOD, DOD-STD-963A 15 August 1986, MIL-STD-963B 31 August 1997.
97. Strategic Management: A stakeholder approach, Freeman, R. Edward, Pitman Publishing, 1984, ISBN 0273019139.
98. Strategies for Real-Time System Specification, Derek J. Hatley, and Pirbhai A. Imtiaz, New York: Dorset House, 1987.
99. Structured Analysis and System Specification, Tom DeMarco, Englewood Cliffs, NJ: Yourdon Press, 1978, ISBN 0917072073.
100. Structured Design, Edward Yourdon and Larry L. Constantine, New York: Yourdon Press, 1978.
101. Sustainable Development Possible with Creative System Engineering, Walter Sobkiw, 2008, ISBN 0615216307.
102. System / Segment Specification, DOD Data Item Description DI-CMAN-80008A, June 1986.
103. System Engineering Glossary - INCOSE SE Terms Glossary Document; October

1998; File: Glossary Definitions of Terms 1998-10 TWG INCOSE.doc; Prepared by: INCOSE Concepts and Terms WG, International Council on Systems Engineering (INCOSE).

104. System Engineering Management Plan (SEMP), DOD Data Item Description, DI-MGMT-81024, March 1987, August 1990.
105. System Engineering Management, Military Standard, MIL-STD-499 17 July 1969, MIL-STD-499A 1 May 1974.
106. System Software Development, Military Standard Defense, DOD-STD-2167 4 June 1985, DOD-STD-2167A 29 February 1988.
107. System Software Quality Program, Military Standard Defense, DOD-STD-2168, 29 April 1988.
108. Systems Analysis, Design, and Development: Concepts, Principles, and Practices, Charles S. Wasson, John Wiley & Sons, 2006, ISBN 0471393339.
109. Systems Engineering for Intelligent Transportation Systems, Department of Transportation, Federal Highway Administration, Federal Transit Administration, January 2007.
110. Systems Engineering Fundamentals, Supplementary Text, Defense Acquisition University Press, January 2001.
111. Systems Engineering Handbook A Guide For System Life Cycle Processes And Activities; INCOSE-TP-2003-002-03.1; version 3.1; August 2007.
112. Systems Engineering Handbook, International Council on Systems Engineering (INCOSE) INCOSE-TP-2003-016-02, Version 2a, 1 June 2004.
113. Systems Engineering Management Guide, Defense Systems Management College, January 1990.
114. Systems Engineering Management Plan (SEMP), DOD Data Item Description, DI-MGMT-81024, March 1987, August 1990.
115. Systems Engineering Plan (SEP) Preparation Guide, DOD Data Item Description, August 2005.
116. Systems Engineering Plan (SEP), DOD Data Item Description, DI-SESS-81785 October 2009.
117. Technical Papers of Western Electronic Show and Convention (WesCon), Winston W. Royce, Los Angeles, USA, 1970.
118. Technical Reviews And Audits Systems Equipments And Computer Software, MIL-STD-1521A June 1976, MIL-STD-1521B June 1995.
119. Technology Readiness Assessment (TRA) Deskbook, DOD, May 2005, July 2009.
120. Test & Evaluation of System Reliability Availability And Maintainability, DOD 3235.1-H, March 1982.
121. Test And Evaluation Handbook, Federal Aviation Administration, Version 1.0, August 21, 2008.
122. Test Inspection Reports, DI-NDTI-90909A March 1991, DI-NDTI-90909B January 1997.
123. Test Plan, Data Item Description, DI-NDTI-80566, April 1988.
124. Test Procedure, DI-NDTI-80603, June 1988.
125. The Analytic Hierarchy Process: Planning, Priority Setting, Resource Allocation (Decision-Making Series), Thomas L. Saaty, Mcgraw-Hill, January 1980, ISBN 0070543712.
126. The Army Strategy for the Environment, October 2004.

127. The Beginnings of STOP Storyboarding and the Modular Proposal, Walter Starkey, APMP Fall 2000.
128. The Computer from Pascal to Von Neumann, Herman Goldstine, Princeton University Press, 1972, ISBN 0-691-08104-2.
129. The Electrical World, A weekly Review of Current Progress in Electricity and Its Practical Applications, Volume 29, WJJC (The W.J. Johnston Company), Library of Princeton University, January 2 to June 26 1897.
130. The Fires: How a Computer Formula, Big Ideas, and the Best of Intentions Burned Down New York City-and Determined the Future of Cities, Joe Flood, May 27 2010, ISBN 1594488983.
131. The Magical Number Seven, Plus or Minus Two: Some Limits on Our Capacity for Processing Information, G.A Miller, The Psychological Review, 63, 2 (March): 81-97, 1956.
132. The Organism, Kurt Goldstein, 1934.
133. The Quants: How a New Breed of Math Whizzes Conquered Wall Street and Nearly Destroyed It, Scott Patterson, Crown Business, 2010, ISBN 0307453375.
134. The Theory of the Leisure Class, An Economic Study of Institutions, Thorstein Bunde Veblen, London: Macmillan Publishers, 1899.
135. Trusted Computer System Evaluation Criteria, DOD 5200.28-STD, December 26, 1985.
136. Two Approaches to Modularity: Comparing the STOP Approach with Structured Writing; Robert E. Horn; Visiting Scholar; Stanford University; This appeared in Journal of Computer Documentation, 1999.
137. Vitruvius The Ten Books on Architecture, Herbert Langford Warren (Illustrator), Morris Hickey Morgan (Translator) Courier Dover Publications, 1960, ISBN 0486206459.
138. Vitruvius, Fra Giocondo, Venice, 1511.
139. Vitruvius, The Ten Books on Architecture. Translated By, Morris Hicky Morgan, Illustrations And Original Designs, Prepared Under The Direction Of, Herbert Langford Warren, A.M., Nelson Robinson Jr, Cambridge, Harvard University Press, London: Humphrey Milford, Oxford University Press, 1914.
140. Vitruvius: Ten Books on Architecture, Cambridge University Press, Cambridge 1999, Editors D. Rowland, T.N. Howe, 2001 ISBN 0521002923.
141. Work Breakdown Structure Handbook, DOD, MIL-HDBK-881, 2 January 1998.
142. Work Breakdown Structures for Defense Materiel Items, DOD, MIL-STD-881 1 November 1968, MIL-STD-881A 25 April 1975, MIL-STD-881B 25 March 1993.

50 Index

A

B

D

E

F

I

J

K

L

M

N

O

P

Q

R

S

T

U

V

W

CPSIA information can be obtained at www.ICGtesting.com
Printed in the USA
BVOW02*1113100615

403894BV00004B/46/P